Which Way Out?

WHICH WAY OUT?

And Other Essays
by Arthur M. Young

ROBERT BRIGGS ASSOCIATES
Berkeley and San Francisco

To Frank Barr, Arthur Bloch, and Jack Saloma

A number of these essays were originally published in somewhat different form in *The Journal for the Study of Consciousness* edited by Dr. Charles Musès—"Time to Go Home: Some Reactions to Arthur Koestler's *The Ghost in the Machine*" in vol. 2, no. 1; "Constraint and Freedom: An Ontology Based on the Study of Dimension" in vol. 2, no. 2 and vol. 3, no. 1; "Toward a Life Science" in vol. 4, no. 1; "Which Way Out?" and "The Brain Scale of Dr. Brunler" in vol. 5, no. 2. Copyright 1969, 1971, 1972–73 by Arthur M. Young.

ROBERT BRIGGS ASSOCIATES
Berkeley and San Francisco

First Edition © 1980 by
Arthur M. Young
All rights reserved.
Printed in the United States of America

Young, Arthur M   1905-
  Which way out?

  CONTENTS: Which way out?—Modern idolatry.—Toward a life science. [etc.]
  1. Science—Philosophy—Addresses, essays, lectures. 2. Philosophy—Addresses, essays, lectures. I. Title.
B67.Y68   191   80-17525

About the Author

Arthur Middleton Young, the son of the Philadelphia landscape painter Charles Morris Young and Eliza Coxe, was born in Paris in 1905, and educated at Princeton. After having received a special course in relativity in which he was the only student, Arthur Young decided to devote himself to philosophy, and set about developing a theory of process which would give proper recognition to time in contrast to relativity as propounded by Einstein. This attempt proved premature, however, and in 1929 in search of practical experience Young embarked instead on the design and construction of a helicopter, a project which had fascinated inventors since at least the time of Leonardo da Vinci. For twelve years, Young worked alone, using small models to test his ideas. Then, in 1941, the Bell Aircraft Company contracted to build a full-scale version of his machine, and Young transferred operations to the Bell plant at Buffalo, New York. Finally, on 8 March 1946, his brainchild (now designated the Bell Model 47) received certification from the Civil Aeronautics Board in the form of the first commercial helicopter license ever awarded.

Deeply disturbed by the explosion of the atomic bomb over Japan in 1945, which seemed a sinister portent of the shape of things to come, and conscious of the ravages inflicted on the human psyche by reductionist thinking, Arthur Young now turned back to his first love, philosophy. A new paradigm of reality which would restore meaning to human aspirations was clearly an urgent necessity. Thus began a quest for understanding that has been called "one of the great spiritual journeys of our time."

In 1952, Arthur Young set up the Foundation for the Study of Consciousness; twenty years later, the Institute for the Study of Consciousness was established in Berkeley, California. Young's thinking bore fruit in his seminal books, *The Reflexive Universe* and *The Geometry of Meaning,* both published in 1976. In 1972, Young and Dr. Charles Musès edited a collection of essays titled *Consciousness and Reality.* Extracts from Arthur Young's diaries in the years 1945–48 were published in 1979 under the title of *The Bell Notes.*

Contents

Which Way Out?	1
Modern Idolatry	11
Toward a Life Science	15
Crossing the Psychic Sea	25
Where Is God?	35
Time to Go Home: Some Reactions to Arthur Koestler's *The Ghost in the Machine*	41
The Queen and Mr. Russell	58
Rotation, the Neglected Invariant	82
A Formalism for Philosophy	91
The Mind-Body Problem	127
Constraint and Freedom An Ontology Based on the Study of Dimension	142
The Brain Scale of Dr. Brunler	183
A Letter to Dean Robert J. Jahn	195

Which Way Out?

"*The universe is vast....*"
    FORTUNE MAGAZINE

"*The Kingdom of Heaven
is like a grain of mustard seed.*"
    MATTHEW 13:31

    One of the most significant discoveries of twentieth-century physics is the *finiteness* of the universe—a finding which is in conflict with a basic predisposition of the human condition: the feeling of respect accorded to magnitude, so that that which is larger than ourselves inspires fear or awe or wonder. It comes as rather a shock—and is incredible as well—to be told that the universe is not infinitely large. While this limit as a dimension of our finite existence is far beyond any practical application, the mere thought of it is disconcerting; it runs against our deepest instincts. We are not satisfied by the assurance that, though finite, the universe is still unbounded (in the same way that in moving over the surface of the earth we would never come to the edge).
    On the other hand we do not experience any reluctance to accept a limit to smallness. Even before modern physics had proved and measured the size of the atom, philosophers—notably Democritus—had postulated indivisible units. Or, if we insist that the atom has proved to be divisible, we can instance the more fundamental proton and electron as units which cannot be further subdivided. (Those who have tried have yet to succeed; but even if they do, they are still seeking a limit to smallness.) But, alas, our credulity is again given a

shock when we learn that the amount of energy associated with light corpuscles increases *as the size is reduced*. And to this reduction there is no theoretical limit! The energy necessary to create a proton is contained in a light pulse only about 10^{-13} centimeter in diameter. And the energy of a million protons would be contained in a light pulse a *million times smaller*.

The understandable objection that this is contrary to sense cannot be supported. We can easily show that in all sorts of ways the limits of sense are everywhere being transcended. Even if we were to refuse to believe that we, ourselves, are each an organization of 1,000,000,000,000,000,000,000,000,000 atoms, and that our blood cells alone are sufficient in number to give every one of the four billion inhabitants of the earth a thousand cells apiece, we still ride in automobiles powered by 300 horsepower engines and in jet airplanes whose huge engines rotate 500 times per second. And if this were not enough, there are the depressing statistics of the atom bomb, each pound of which is equivalent to a 1,000 tons of dynamite.

So we owe ourselves a confrontation with this bedazzlement with magnitude, lest we be misled instead of enlightened by science and statistics. I do not mean we should develop immunity to magnitude; I mean rather that we should try to recognize what all this business of measuring and counting and comparing *means,* so that we can put it in its proper context—especially as we already know that we cannot help carrying over from our animal heritage the instinct to fear size, and hence unconsciously make it the basis for wonder and respect, and for judgement.

Let us return to the question of the finite size of the universe. How does this extraordinary conclusion come about? The answer unlocks a number of paradoxes. Even without philosophical interpretation, we can discern that the finite dimensions of the universe stem from the finite speed of light. This follows directly from the experimental evidence of what is called "the red shift"—a displacement in the lines of the spectrum of the light from distant stars. This finding indicates that distant galaxies are moving away from us at velocities proportional to their distance. Thus, at a sufficient distance, their velocity with respect to us would approach that of light, which means that the light by which we see them could not reach us. This is verified by the equations of relativity themselves, since volume involves measurement, and the measure itself contracts with rapid movement.

In other words, the curvature of space, which limits the universe, comes about because the measure and that which is measured are not completely independent. If the velocity of light were infinite, space would be Euclidean and extend forever. Since the velocity of light is finite, space curves back on itself. This curvature not only violates the parallel postulate by making parallel lines meet, but causes two perpendicular lines ultimately to cross for a second time (a person going north on the earth's surface would ultimately meet a person going west).

The something of which both measure and that which is measured are parts is wholeness, the physico-mental universe, which, like a rubber sheet with two-way stretch, can be a lot of one thing or a lot of another, but the two factors are not independent. In this phrase "physico-mental universe" I am not creating a new concept. I am simply renaming the universe of physics, which with the advent of the concept of the dependence of measure on geometry, automatically acquires a mental aspect, because geometry is measurement, and measurement is mental.

The diagram below shows a curve in which the product of the x and y dimensions is constant. Such a curve is called a hyperbola, and it illustrates in two dimensions the property we have been discussing: an interrelationship between measure and that which is measured such that the indefinite extension of one is obtained at the expense of the extension of the other.

The diagram also illustrates the Heisenberg Uncertainty Principle, which is fundamental to quantum physics. Thus, the position of an electron, for example, can only be determined at the expense of disturbing its momentum, or velocity; since the accuracy of one measure is proportional to the disturbance

4 WHICH WAY OUT?

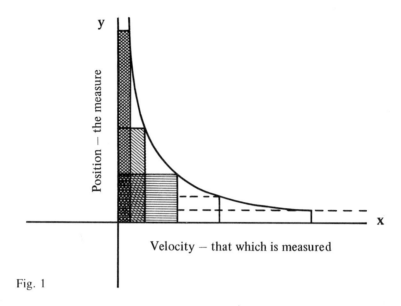

Fig. 1

of the other, it follows that the lower limit of the product of the two uncertainties is a constant. At that limit, $\Delta x \cdot \Delta y = h$ (Planck's constant). This formula states what the diagram shows: that every rectangle that can be fitted under the curve has the same area, or is constant.

We can now talk about what this means. For when we see two things vary in such a way that their product is a constant, we may be sure that that which the constant represents is something very important: it is the unchanging absolute from which the changing derives. Relativity theory evolved from the search for such absolutes, which are called *invariants*. An example is the speed of light, or again, what is called *interval*, which is the amount of space and time between two events (relativity makes the invariance of the interval a criterion for the method of measuring—all observers must use methods of measurement which will give them the same result when they measure the same thing). Quantum theory is based on such an invariant—the principle of action (the constant h in the formula above): $E \times T = h$, E representing energy and T, time.

What this tells us, in short, is that size, being an attribute, does not have any absolute significance. It tells us also that we cannot in principle isolate properties. There can be no movement without some mass, no mass without some movement. There can be no space without time, no time without space. Moreover, and this is the crux of the matter, *all existence partakes of totality* (totality being the product of measures).

From here we can go on to certain deductive conclusions which, while they may have no practical effect on scientific technique, have very important implications for the inquiry we are making. They apply, for example, to the possibility of ESP. Whatever may be thought of the subject, there are a great variety of phenomena going under the heading of ESP—such as clairvoyance, telepathy, the seeing of auras, "spiritual" healing, and the like—which have no scientific status, but for which there is considerable direct evidence. We may as well face it: no amount of testimony will establish the existence of these phenomena as long as their validity is made to depend on explanation in terms of the known laws of science and recognized scientific realities. This is not because these phenomena operate on the basis of *unknown* laws of science. It is because, if they operate, they do so only insofar as they spring from the fundamental unity of the knower and the known—*before* measure or property or laws have been sorted and allocated and assigned. Cut them into components and they cease to exist.

This does not mean that ESP faculties spring from ignorance. While there is certainly a significant correlation between creative invention and the "power of ignorance," and while as a general rule ESP operates best without rational information, we should not from this conclude that the "whole seeing" on which these faculties depend springs from a primitive state of total ignorance. The question is rather one of at what phase or state in the cycle of action the person is situated. We may diagram the cycle of action as shown in figure 2.

The learning cycle begins with unconscious *action* (1), and passes through unconscious *reaction* (2) to conscious reaction or *observation* (3). It is only at this point that the subject

6 WHICH WAY OUT?

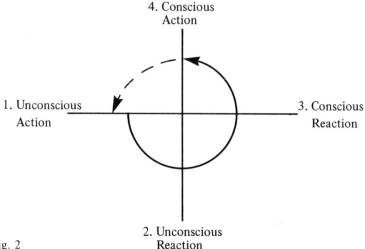

Fig. 2

knows, and is then able to apply this knowledge consciously (4), but this conscious action, while effective, is awkward and difficult. However, with continued practice, the controlled action becomes instinctive, and we come again to unconscious action (1), with the difference that some learning has been achieved.

"Unconscious action," then, is not necessarily or totally ignorant. It may be backed up by the entire learning cycle, as it is when the pianist has passed through his long training and is able, without conscious effort, to translate the written notes on the score into musical performance. Moreover, there may be many completions of the learning cycle which underlie a given spontaneous act. We must realize that the spontaneity or unconsciousness rests, as it were, in the concrete accomplishment of the entire cycle; it is "ignorant" because it is no longer concerned with the separation necessary for conscious observation and learning. It "forgets" what is no longer necessary to remember—the awkward self-consciousness and self-corrections involved in the learning process—and now acts "on instinct."

Such considerations apply also to animal instincts, which lead a wasp to bury its egg in the larva of another wasp, seeking this object out with no parental training. Again, there is the blackcap, a bird that annually migrates unerringly from Germany to Egypt—a case which has been carefully investigated.* It has been found, by placing the bird in a planetarium and displaying the stars as they would appear at different places if it got off course, that the bird always flies in such a direction as to put it back on course. The blackcap, in short, navigates by the stars: and since it does so without training by its parents, we are faced with a problem that transcends even the vast capabilities of the DNA blueprint. We need not insist on this here; suffice it only to say that the bird behaves as though it had learned a complex behavior pattern. We can say that the bird does so on instinct, but this explains nothing. It would seem, rather, that this instinct is the end product of a long learning process. All such automatic processes that we know from our own experience are based on what has been learned. An aviator's ability to navigate becomes instinctive only because it has been thoroughly mastered.

Having considered these examples of instinctive knowing insofar as they indicate that there is such a thing as unconscious intelligence, apparently spontaneous and not based on rational processes, let us return to the universe and what its finiteness means.

The infinite, let it be noted, is not an intellectual idea or concept, for all concepts are constructs having limits—containers, as it were, to which the occasion adds content. The infinite is the *denial* of limits. Nor is infinity a question of size: it is not larger than anything finite—or at least it is misleading to think of it this way, because in becoming larger a number gets no closer to being infinite. Similarly, any part of infinity is itself infinite.

It follows that if the universe were infinite, it would be impossible to assign measure to anything in it. We could not say: this is bigger than that, because each part of infinity is itself infinite. We can, therefore, conclude that because we can

---

*E. G. Sauer, "Celestial Navigation by Birds," *Scientific American* 199 (August 1958): 42–47.

measure things, the universe, which is made up of all things, is finite.

This argument is not altogether convincing, however, because it sidesteps the issue and does not come to grips with infinity itself. But then, since infinity is not a concept, how can we describe it at all? It is, in a way, the unscalable barrier that confines intellect.

How did intellect get caught in this trap? When we ask this question, we are asking how intellect came into existence, for the barrier to intellect is precisely intellect. Without the invention of limitation, there would be no intellect. There could not be, for intellect is definition, the use of limits. What was there before there were limits?

There must have been something, some stirring or moving that was to be limited. In physical terms this might be motion itself—but an unconfined motion, like water spilled out of a bucket—reaching out blindly, till it hit an obstruction, and then turning and groping again. In human terms, this would be emotion, hunger, longing.

Perhaps there was more than one longing, but there was no comparison of longing—for as soon as there was a comparison of one longing with another, there sprang up the notion of length, and with this limits and hence intellect. It might seem that we are taking advantage here of a pun, a coincidental word similarity; but as a matter of fact, the two words originate from the same Anglo-Saxon root, *"langian"*—to grow long, to long.

Was there something before longing? There must have been, even if we can only refer to it as a sleep, from which something was awakened by a longing. But sleep is not a satisfactory antecedent, which must rather be that which is asleep, or which is awakened by longing. Because it is before longing, it can have no extension, and we might liken it to a point (whose motion can create a line). But we need a richer image for the human equivalent. For this we might suggest the word "attention," which adds a dimension of potential direction to the otherwise barren point, and lays the basis for longing, which cannot arise except in some direction. That is to say, if we visualize longing as a pull toward something, or a motion toward something, the prior condition or antecedent is a stasis, with the option to move in a certain direction. Thus, the original condition may be likened to a point, having no dimensions but being free to turn in any direction.

In this state or condition, which is prior to the confines of intellect and the confines of desiring, there is stasis—pure being, complete and "perfect"—and the nature of this state resembles that which is attained in any cycle of action when it arrives at unconsciousness or spontaneous action, and is capable of an instinctual competence to perform what it has learned.

But how did we get *here?* We seem to have wandered away from our aim, which was to come to grips with infinity. Why have we not stuck to the subject? Recall that we started by asking how we could get away from the limits of intellect to comprehend infinity. To do so we traced our way *back* to the kind of being that did not know limits—and conjectured this to be a desiring, a longing, which because it could not measure itself was in darkness, absorbed only in its own state of being. Some will say that this is no answer, or that I've shifted the discourse to something that is different altogether from infinity or concept. This objection would affirm that while infinity is not a concept, neither is it an emotion—as I say, a desiring. I cannot answer this objection except to point out that at the outset we referred to the basic predisposition of human nature to associate a feeling of respect, of wonder, of awe with what is larger than ourselves. It is in this *relationship* that infinity exists; for it is when we don't know how *much* larger that we feel awe—as soon as size is measured, we can deal with it. Now, of course, awe and fear are emotions, and if my argument is in error, it is because I've equated infinity with the emotion that attends it. There is, however, a valid correlation here. If we describe intellect as dealing with the self-contained relationships of an object, and emotion as dealing with the relations of self to object, finiteness will by definition fall into the first category and infinity into the second. Infinity is thus a kind of acceptance, related to faith, credulity, awe, and wonder. And if this seems unsatisfactory, let us recall that since the word "definition" means to set limits, we cannot define infinity.*

A more fruitful line of inquiry is to recall that there was an antecedent to the desiring. We likened this to a point with no dimensions, but free to turn in any dirction. This means that there must be something beyond infinity, or before infinity, and leads us to say that that which is beyond infinity has no extension.

But there is another, more powerful argument that leads to the same answer; and it is one with which we are now better prepared to deal, because in our first encounter with it, we could only list it as some sort of incredible fact, a truth that was stranger than fiction.

This truth is that the ultimate essence of this universe, light—or, as it is now discovered to be, the photon of radiation—contains more energy the smaller it is. As noted before, when it is small enough, it contains the energy of a proton. When it is smaller still, it may contain the energy of a million protons, and *there is no limit*. We could have a single photon that contained the energy, and hence the potency, to materialize into an entire universe.

We have thus gathered from the world of science, from the findings of relativity and quantum theory, a garland of curious thoughts:

   1. The universe is finite.

   2. Our existence as parts of a universe is implied in and requires a finite universe.

   3. Before there was this finitude of the universe there was a deep longing, a bottomless pit of desire.

   4. And before that there was a dimensionless point, not infinitely small, but as small as might be for what it had to do, and its power was in its smallness and it was equal to a universe.

---

*When this essay was originally published in *The Journal for the Study of Consciousness* 5, no. 2 (1972–73), editor Charles Musès appended this footnote: "Jakob Böhme reaches much the same conclusion and identifies the infinite with 'the Wonder-Eye of the foundationless abyss (*Ungrund*),' out of which the magic of desiring arises."

Modern Idolatry

The cumulativeness that makes science the unique phenomenon it is and the glory of our age is misleading. It seems to imply that there is something called science which is an entity in its own right. This science with a capital S takes on an authority; it becomes a kind of institution. It is said that new discoveries are made by science, predictions are made by science, progress is made by science. And so on.

In point of fact, science never discovered anything. *People* invent microscopes, telescopes, dynamos, spectroscopes, interferometers. People discover electricity, gravity, calculus, new planets and new elements. But when they have done so (and very often it is years before the accomplishment is acknowledged), credit is given to "Science."

We don't speak of "Art" as having "discovered" impressionism or photography or motion pictures. Nor do we think of "Music" as having developed the pianoforte. But we even credit science with discovering things which were initially denied and rejected by the science they later "became."

People don't believe in ghosts any more. At least they say they don't. But everybody believes in science. *And science is a ghost.* Like most ghosts it is not only scary, but is used by some people to scare other people. Science, in the sense of discovering the things that make up the body of modern progress, does not exist, because it did not discover or invent anything.

We are concerned here with the nature of science as a component of our culture. In its role as custodian of the creative insights and discoveries of individual persons it deserves our admiration and support. Insofar as it channels the thinking of the many—whether these be the professional scientists as members of the order, or the public to whom the pronouncements of science have an authority that cannot be challenged—it is quite a different thing. It must be questioned in much the same way as the abuses of the church were questioned in previous centuries.

Science, ghost or not, does exist as the embodiment in communicable form of the findings of hundreds of creative persons whose individual accretions have built up an edifice of scientific tradition. It was this that enabled Newton to say, "I stood on the shoulders of giants." And it is this that enables even a college freshman to do calculus. As such, it is both a great contribution and also a great encumbrance. The more monumental science becomes, the more it constitutes a massive prejudice, or more correctly a blanket endorsement of the unthinking inertia of credulity. Science, once the avocation of a few pioneers, is now common agreement, not merely among the hundreds of thousands of personnel on research teams, the engineers, the white-collar class (who are actually no longer an elite financially, numerically, or hierarchically), but to the man in the street, who looks to science to tell him what to think and what to believe. Science has become the church of today. And like the Christian Church that it has replaced, it interprets the initial teachings for its own salvation and preservation.

We must, of course, see a merit in the stabilizing influence of science as a conservative body of opinion, and recognize that creativity needs a flywheel, that all new ideas cannot be indulged but must meet the test which conservative science imposes on innovation. Science as a body of agreement controls the machinery of production and of proliferation which can expand innovation, and innovation has no God-given right to these means.

Creativity, innovation, and invention per se are basically irresponsible and inconsistent, and recognize no vested interests. Nor do they make any concession to necessity or expedience. If the means of production and of dissemination were entrusted to the erratic muse of innovation nothing would be accomplished. Trains would not run, airplanes would fly on random schedules and to unexpected destinations. So the conservatism of science is on the whole a good thing, provided it does not completely blot out the innovation which supplies its continued renewal. In the larger sense, however, we must question this dogmatic aspect of science. I refer to its influence in our culture in replacing religion as an authoritarian doctrine to guide man's philosophy.

Now it is true that the view that science can tell us how to get out of our moral predicaments is not one in which scientists as such commonly indulge. If scientists take a stand on a moral issue, they can do so as humans but not as scientists. Even as psychologists their function is to show how the implications of one thing lead to another, rather than to say what ends man should pursue. Scientists usually disclaim any role as moral advisors, and with the exceptions of certain enfants terribles and their followers, whose opinions and zeal are their own, science is not trying to stage a takeover. Mostly, scientists find the elevation of "Science" into a religion for the layman an embarrassment.

How, then, can we disabuse the layman of the notion that science has the answer to his genuine hunger for moral truths and abiding values? The modesty of scientists on this subject only leads to the suspicion that they are holding back.

Perhaps the blatancy of advertising has developed in the layman a suspicion of anything that asserts that it has the answer, and makes him yet more certain that this lies where he's told it doesn't. Certainly the impressive progress of science (which we repeat was achieved by people and not by science) makes him feel that that which discovered penicillin and quasars can tell him the answer to such questions as whether or not God exists.

But Fleming, who discovered penicillin, would probably say he was as puzzled as anyone about these questions, which are, in any case, not within his competence as a scientist. In fact, when we do find a scientist setting forth his opinions on such subjects, he is more likely to lose his standing as a scientist than to gain acceptance as a philosopher. There are exceptions; many good scientists are also philosophers and have made contributions to the spiritual quest. But again, this is people, it is not science.

Our questions about man's ultimate nature may well be asked in the "spirit of scientific inquiry," but we might begin our inquiry by asking what restrictions the term "scientific" places on the questions we can ask.

It might be better just to ask.

Toward a Life Science

Science, having made prodigious advances in the last few decades, today confronts a new and greater challenge, commensurate with the prestige it has rightfully gained. It is no longer enough to provide better moon shots and build larger cyclotrons. It is now appropriate that science expand its scope to include questions which have hitherto been set aside as refractive to the scientific method.

Not that these questions will be welcome to science. Their difficulty has no doubt been responsible for their neglect. What is the nature of life? Does a man have a soul? If so, is it immortal? Such questions have for a long time been out of fashion. But the mood that thrust these issues aside is at present with equal insistence pulling them back; for they are questions which man must ultimately answer. He will inevitably be drawn to them when the fascination of gadgets palls and the present preoccupation with the bigger and better, the faster and cheaper, loses its charm and relevance. And man will insist on these questions being resolved when the very growth of technology and increase of population, poisons, and pollution put a finite limit on the habitability of the planet unless we change our ways in time.

Nor are these questions merely speculative. They are intensely practical, for if there is to be any valid human or social science, it must depend on our knowing what man is, and what the goal of evolution is.

NEW SCIENCES OR RENEWED SCIENCE

It is the spiritual nature of man that is involved. Although this has been considered outside the province of science, the fact that science has had greatness thrust upon it, and the authority formerly invested in religion transferred to it, leaves it with no choice. Science must expand its role. If it cannot do so, man will turn away from present science and transfer his attention to new sciences that can answer his most searching questions.

This takes us to the problem of the *self,* a subject which has no status in present science. The self does not share even the reality attributed to lesser entities, such as atoms and molecules. We know the weight of the electron to the fifth decimal place. Of the self, we do not even have scientific confirmation of its existence.

Psychology, by definition, is the knowledge of the psyche. But the psyche, which may be rendered either as the mind or the soul, is not given a status. Between behaviorism, which dismisses consciousness as a superfluous concept, and psychoanalysis, which attempts to trace the self's motivation, there are many schools, representing wide differences of approach. But all agree in venturing no definition, or even postulation, of selfhood. While the behaviorist deals with reaction to stimuli and the psychoanalyst with drives or motives, there is no postulation of an independent entity, leaving the way open for philosophers to deny that the self is any more than the harmonious functioning of the physical organism.

For information on the living organism, we turn to biology, the "knowledge of life" (*bios* = life; *logos* = knowledge, literally "word"). Here we find again an avoidance of the central issue—life itself. Although the basic characteristic of life, to build order against the general tendency to disorder, is contrary to a basic principle of science (the second law of thermodynamics), life is explained as merely a chemical activity.

Just as science has given formal status to mass, energy, force and so on, that which builds order against the general tendency to dissipate order must also be given formal status. However, in the reluctance to concede this formal status, there has been a certain rightness. Were we to give the self or life the character of a physical entity like an atom, we would be guilty of misplaced concreteness, to use Whitehead's phrase. The self is not a concrete particle like an atom, nor need it have the properties which provide the basis for scientific operations such as verify the atom and electron.

The reality of persons is the most taken-for-granted of all realities. The child endows the rag doll with a human personality. The novelist creates his Tom Jones or his Moll Flanders, drawing solely on this reality. It is only when we become the scientific observer and put a person on the operating table and dissect his body that the being of the person evaporates.

THE QUANTUM AND THE SELF

This takes us back to our opening theme: How can science verify the self to its own satisfaction and, having done so, develop a knowledge of self commensurate with its achievements in other areas?

Help with this problem comes from an unexpected quarter: quantum physics, which gives fundamental status to the unpredictability of individual particles. The layman has perhaps heard how the findings of modern physics have revolutionized the concepts on which classical physics was based. But there have been few attempts to uncover the repercussions of these new findings for concepts in general. Philosophers and biologists have shied away from an area that bristles with thorny formulae, and even physicists, habituated to the explanatory and predictive function of science, disguise the new element discovered by quantum physics as one that does no more than change the status of predictive laws from that of certainty (the classical view) to that of statistical probability (the quantum view).

It is our thesis that quantum theory provides the clue we need for a science of life. The new discoveries of physics deny classical determinism on principle, but this is only the negative side of the issue. What is important for our science is that action is established as more basic than the physical particles which interact. There is something other than the billiard balls which endows them with independent activity, an activity that is unpredictable.

Naturally we are tempted to call this energy. But to do so misses something of importance. For it is not energy but action that is quantified. It is action that comes in units of the same "size," whereas the unit of energy can be of any magnitude whatsoever; it can vary from zero on up. Moreover these units of action are unpredictable, and it is this concept that upset the classical physics which was so rational and comprehensible.

INTEGRITY, ACTION, AND DECISION

From the beginnings of rational thought, it has been natural to think of matter as coming in discrete pieces. Democritus held that atoms (units of matter) were at the basis of things, and this is what physics has found to be in fact the case. Atoms are exactly similar units of a given element, while protons and electrons are the even more basic units whose combination produces atoms. So the "atomicity" of protons and electrons is a reasonable explanation of matter.

But what else, it can be asked, comes in integral units, in wholes which cannot be divided? I have asked this question many times, and by the negative results must suppose the answer difficult. Yet, as with a conundrum, everyone recognizes the solution as simple when it is given. The answer is that the act of making a decision comes in wholes. One cannot decide to lean out a window one-and-a-half times. One cannot vote, sign a paper, decide to get out of bed, to stand up, or do any act one-and-a-half or 1.42 times. Decision comes in whole units, or not at all.

But is this human act of decision the same as the physicists' quantum of action? Some physicists tend to claim that Planck's constant refers to something quite technical which non-physicists can't understand. But, apart from the fact that Planck's constant is appropriate to the domain of physics and decisions to the human domain, I can see no essential difference, especially as both are uncertainties. Of course, any human decision to act—for example, to go to the dentist—can be described; but describing it is not doing it. Nor can even the specific decision to go to the dentist be predicted. The toothache is a cause for deciding to go to the dentist, but the decision to do so requires "making up one's mind." One might procrastinate indefinitely before doing so. It is this unpredictable human act of decision that parallels the physicists' quantum of action. And, just as the interactions of the world of the physicist can be traced to quanta of action, so the interactions of human life can be traced to acts of decision.

One might argue that one distinction between a human decision and the quantum of action is that the former has direction. Here, however, we must point out that the element of direction is just what the quantum of action, as radiation, *does* possess—for there is no radiation that does not have direction. Technically, direction is implicit in even the bare formulation of Planck's constant, which contains the factor 2π, and 2π is a reference to our complete uncertainty of the direction radiation may take. (In Eddington's apt phraseology, 2π is a dimension, the dimension of timing or phase—measured by an angle.)

But these are technical points which need not encumber what is of philosophical importance. This is that physics, starting on the solid ground of material particles and tracing them to their origin, has discovered an unpredictable entity, action, which comes in wholes (quanta). While this entity may be called physical (simply because it is part of physics), it has none of the properties normally associated with physical objects, such as rest, mass, charge, etc.

It is, rather, the unit into which the activity of objects can be resolved. Its function in physics accounts for the entire range of interactions of physical particles—the creation of matter, the radiation and the absorption of light by atoms, and the chemical activity of molecules.

SELFHOOD AND FREE WILL

We may now compare this entity to selfhood. It exists, yet it is not a material particle. Further, it can in principle, and does in fact, cause physical interactions, not only between atoms, but between molecules. Our thesis goes further, and states that we can invoke the quantum of action on categorical grounds to account not only for the negentropy (movement against entropy) in life, but also, at least in principle, for that primary function of selfhood, free will. This is not a new idea. It has been considered by philosophers before, but without the appropriate coordination of other relevant information.

Peter Caws, a contemporary philosopher, says, in discussing the pertinence of the Uncertainty Principle to free will, "Modern physics has often been thought to throw light on this question of choice, but it does not. After Heisenberg it was claimed that everybody was free; but of course it is absurd to pretend that my freedom depends on my manipulation of submicroscopic particles within the range of non-commuting variables p and q in such a way that $\Delta p \Delta q \geq h$. These values are much too small for my conscious endeavors to make the slightest difference to them."*

But this apparently honest scrutiny reveals a misconception of the nature of processes. It involves, first of all, a simple error of interpretation. The quantum of uncertainty is not restricted to subatomic particles. It has application to atoms and their interactions, and hence to molecules. Although the observation of an electron is the example usually chosen to illustrate the importance of the quantum of uncertainty (Planck's constant h), the principle applies to the electron as a bond, either in an atom or in a molecule.

For the molecule, the corresponding uncertainty involves an energy at least of the order of that contributed by its agitation due to temperature. This energy, at room temperature, is about $1/25$ of an electron volt, and its availability must account

---

The Philosophy of Science (New York: Van Nostrand Co., 1965), p. 305.

for the great sensitivity of vital processes to temperature. For the chemical changes necessary to life cannot occur at temperatures that are either too high (130°F) or too low (32°F). In terms of absolute (Kelvin) temperature, this is not a large range. (Absolute zero, 0°K, at which there is no molecular motion, is −453°F.)

So it would appear that far from it being out of the question that the Uncertainty Principle concerns life processes, the quantum of action or of uncertainty is of just the order of magnitude that is required by life. In fact, we might say that this uncertainty, whose energy is 1/25 of an electron volt, is just that which enables the high polymers and the proteins to build up—which is to say, to store—order, and hence move locally against the current of entropy that characterizes non-living matter.

To answer the objection raised by Caws—and similar arguments voiced by Cassirer, Waddington, and others in the period following Heisenberg's announcement—that the order of magnitude is too small for conscious processes, we must move from the level of the molecule to that of the cell.

Here we must recognize that one of the important characteristics of life is *organization*. The DNA molecule directs the building of the cell and, in the higher life forms, controls the cell division which makes possible growth into a coordinated multicellular organism. The full-grown organism may involve *trillions* of cells built into a structure which can achieve functions such as reproduction, self-repair, rejection of tissue not identical with itself, and innumerable others. Contrast this with the complexity of a large industrial corporation, such as General Motors, which is an organization of only some hundred thousand employees.

All of the organization of a living creature proceeds from the original fertilized cell, itself a carrier or vehicle for the marvelously potent DNA molecule, which stores the information and conducts the building process. And the operation of this DNA molecule resides in its molecular bonds. These bonds, as we have indicated, are "subatomic" (electronic) in nature, and involve the same quantum of uncertainty—which, we must keep in mind, is a quantum of *action*—the very thing that is primordial both in its status as the origin of matter, and in its essential spontaneity.

CHOICE AND CONSCIOUSNESS

In moving on to the more difficult question of choice, and hence of consciousness, we cannot draw on current science. Biology, however, shows that life as organization constitutes an organized hierarchy, a chain of command with the DNA molecule at the top, a sort of central office in which the quantum of action operates as top executive. We might also point out that the nervous system, the conveyor of the orders which control the motions of man or animal, is subject to what is known as the "all-or-nothing" law. This law states that for a nerve to operate, it must do so totally or not at all—it is either on or off. It cannot operate by gradual or continuous increments as can an electrical current.

Confronted with the all-or-nothing law of nerve currents, one is forced to reflect: What kind of electrical flow or current is of an all-or-nothing nature? Passing over the concept of an on-off switch as not pertinent to nervous excitation, because it merely transfers the question to the nerve that operates the switch, it is apparent that the "current" involved must be triggered by *single* electrons; otherwise it could be more or less rather than all-or-nothing.

The all-or-nothing nature of nerve action can thus be understood as simply another instance, in this case in higher organisms, of the quantum leap so firmly established in the atomic and molecular realms. Now, since control, conscious or unconscious, employs nerve action, and nerve action proceeds by quantum jumps, we may deduce that control in life processes proceeds by quantum jumps—and that the terminal point in the chain of command is the quantum of action. The identification of the quantum of action with choice, which first emerges in animals, is thus demonstrated.

The question of choice in animals—for example, their election, on the approach of an enemy, either to fight or to flee; or the decision to select this or that mate—is not answered by the claim that it results from the mere competition of stimuli, and that the reaction occurs in the direction of the strongest stimulus. The stimulus and the response follow *different nerve channels*. They must be received in, and emanate from, a central headquarters which possesses generalized sensitivity to several kinds of stimulus, and authority over a number of motor actions. Many reactions, it is true, are reflex—the stimulus produces a kick; the increase of light, a contraction of the iris—but where choice is involved, a "lower" reflex must often be suppressed and a different decision taken.*

A normal dog, faced with starvation, will eat food which has been rendered distasteful to it, whereas a dog with its higher brain disconnected cannot "override" its aversion: it will starve to death rather than eat the tainted food. A frog, if tickled on its right side, will scratch with its right foot. If this nerve reflex is severed, it will scratch by rubbing itself against a rock or the like.

These examples indicate the functioning of the central nervous system, a concept which has meaning only if there is indeed a "center." This self-center, in its initiation of action, cannot be distinguished from the quantum of action itself. Inasmuch as the latter is uncertainty, or free will, there is no going beyond it logically or casually.

Let us remind ourselves once more what this quantum of action is. First of all, as a quantum, it is a unit, but it is not a unit of inert matter, as is an ordinary atom, it is a unit of *action*. Its function in the scheme of things is the conveyance of the energy necessary to induce change. Like the spark that ignites a fire, it *initiates* a change, which may be uncontrolled as in an explosion, or controlled, as in life processes (or even as in an internal combustion engine).

Physicists would agree that this quantum of action is responsible for the interactions between fundamental particles, be they electrons, atoms, or molecules. We have extended the quantum's efficacy to life, and have endeavored to show that it meets the requirements of the "life spark," first at the cellular level in the synthesis of carbohydrates (energy stor-

---

*Modern biofeedback investigations have shown that choice can intervene at any level of unconscious process.

age), and again at the animal level, where choice between conflicting stimuli becomes a possibility. The quantum nature of such choice is indicated by the fact of the all-or-nothing law of nerve action, which can be accounted for only by the discontinuous shift of a single electron.

The fantastically small amounts of energy involved imply a high order of organization for there to be effective control of the comparatively immense living creature. There must be a central office that activates an electron that creates a molecular bond, in turn activating a nerve cell that triggers a muscle fiber. Such organization is the very basis of a living organism, whose billions of cells have been created by and developed in accordance with the blueprint in the DNA molecule of the original fertilized *single cell*.

At this point we can rest our case. Future discoveries will undoubtedly reveal ingenuities of means in biological and biochemical processes far beyond our present science, but I do not see how there can be found a better or more suitable protagonist for the central role of selfhood than the quantum of action, whose description as unitary, unpredictable, and without materiality identify it with what philosophers, speaking out of inner conviction and with no deference to an objectively limited science, have been wont to call the divine spark, or spirit. New discoveries will fill in the details of the connection between the quantum of action and the physical organism, and a life science worthy of the name will emerge.*

---

*Since this essay was written, I have read of several scientists, among them Brillouin, who insist that Maxwell's Demon—which my reference of the origin of free will to the quantum of action might suggest—has been exorcized. The exorcism is based on the fact that to get the requisite information about the speed of molecules, the demon would have to use up more energy than he would collect. This conclusion is not relevant to my view of the quantum, however, because as I have described the situation, the demon, like a jujitsu wrestler, *interacts* with each molecule and uses timing to control it. So Maxwell's Demon is alive and well. Evan H. Walker has latterly also published his theory of mind-body interaction, which assumes that quantum effects take place at the nerve synapse. This requires energy of an order of magnitude greater than is available from the random fluctuations due to room temperature. Moreover, it is my contention that we must look to a more elemental level, if not for consciousness, then for life process (to me synonymous).

Crossing the Psychic Sea

Science has been unable to account for the complex world of particles that precedes atoms. This is something of a surprise, for the steady progress of our understanding, from molecules to atoms, once encouraged us to think that things were getting simpler. Atoms are by now the most thoroughly understood of all entities. So it is a rude shock to find that "below" atoms there is a world more complex and less explicable.

The main purpose of science, of thinking about the world, is to increase our comprehension. This occurs through the interaction of reason and experience. Empirical testing produces results, which are interpreted and formulated. Such formulations, in turn, provide a basis for predictions, and for further empirical tests to verify the interpretations. If our predictions prove wrong, we are obliged to rethink, until we have really understood the empirical evidence. We can perhaps liken physics to a game of tennis. The scientist serves, Nature returns the ball, the scientist sends it back, and so on. Nature never misses, but for the sake of the metaphor, let us say that when the scientist can confirm a theory by test, he wins the point; when he cannot, he loses. For a long time the scientist had been winning pretty steadily, but recently he has been losing. Not only because recent findings in particle physics have forced him to invent a series of face-saving ad hoc hypotheses—quarks, partons, gluons, neutrinos, and so on—but because the whole picture is somehow getting out of focus.

Up until about twenty-five years ago things were going "well." All substances had been found to consist of molecules. The molecules were discovered to be composed of about one hundred kinds of atom. Then, as a final and triumphant climax, all atoms were found to be made up of two fundamental particles—proton and electron. The number of proton-electron pairs was pinpointed as the sole determining factor in the atom. It was as though someone had decided to try to explain the written language, first finding that all writing can be reduced to words, then that these words are composed of a limited number of letters, and finally that the letters themselves are composed of segments of straight and curved lines in various combinations.

There may have been biologists who felt that this still did not explain life, but for the most part, impressed by the manner in which the complexity of matter could be explained as the organization and combination of elementary units, they became convinced that this was the right way. When it was found that one DNA molecule could code sufficient information to determine the cellular pattern of an oak tree or an elephant, the biologists' work was cut out for them. It was only a matter of building an even more complex level of cellular organization on top of an already complex, but explicable, molecular organization.

As for the philosophers, they retired in confusion, venting their frustration in quibbling about the meanings of words. Even here there was mute deference to science. There emerged what came to be called the philosophy of science, largely consisting of freezing into dogma a more or less literal description of the so-called "scientific method" the scientists had used to achieve their success.

But with the reduction of protons and electrons, nature began to produce some surprises. First there were mesons, particles heavier than the electron yet lighter than protons. Then baryons, particles heavier than protons. Then anti-matter, a species of mirror-image particle duplicating all the others. Then neutrinos, which could not be detected, but were required to balance the books. There proliferated a host of new and unexplained particles that in number now exceed the number of atoms. More recently it has been proposed that there may be a simpler entity which will account for all these new particles. But such entities, called "quarks," are themselves becoming complicated. They carry fractional charge, they differ in "color" and "flavor," they require "gluons" to hold them together; in fact they rival in complexity the particles they are supposed to explain. Moreover, they have yet to be observed.

Most scientists would say, "Wait! Just be patient. We are making progress. Any day now all this will be accounted for." But there is evidence that it's not just a matter of a profusion of inexplicable new particles thumbing their noses at us. The whole new position of physics demands a new approach. There are basic experiments, such as the EPR experiment first proposed by Einstein, Podolsky, and Rosen, which indicate that we cannot any longer base our world view on the motion of particles, or even on the web of interrelationship between events. *These experiments imply that space itself is not fundamental.*

How do all these unaccounted-for particles fit into a larger scheme of things? Recall that their lifetimes are very short. None exceed a hundred millionth of a second. The particle physicist may retort that a hundred millionth of a second is a long time for a particle. Very well, but what has this particle, with its short lifetime, to do with anything else? Recall that to get evidence of them at all it was necessary to build huge atom smashers, cyclotrons, and bevatrons costing hundreds of millions of dollars, and even then to take hundreds of thousands of pictures to find their tracks in bubble chambers. What is perhaps more to the point, these unexplained particles have no part in building atoms, molecules and cells. They do not contribute to the levels of organization which make life possible.

Organization is composite, made of parts. What is on the

far side of the wall is not organized. That the subatomic world is not made of parts is confirmed by the fact that all the nuclear particles contain or involve one another. A baryon decays into a proton and a meson. The meson into a muon, the muon into a neutrino and an electron. These are not parts but stages in a process of transformation. It's as if nature, in the world of particles, were boiling up a huge stew, out of which it distills the proton and electron, the *only* particles which are permanent and can therefore build atoms, and hence molecules and cells. From one point of view, then, the whole of fundamental creation travaileth and groaneth to produce these elementary particles which endure and thus provide the building blocks for atoms, molecules, cells, and life.

The inferno of hydrogen burning in the sun is the preliminary condition to life on the planets, not only because it forges the elements (hydrogen turns to helium, helium to lithium, and so on) but because the sun provides just the right temperature to sustain the molecular combinations that contribute to life. But what is life *for?* Why should the solar furnace bubble and boil to turn this ephemeral existence which will expire when the sun burns out or blows up into a nova?

Let me return here to the metaphor of analyzing language into words, letters, and ultimately marks on paper. Is this a *complete* explanation? I think not. This analysis does not tell us what language is *for*. It tells us how, but not why. And with language, at any rate, we know that it is brought into existence by the need to communicate. It has a function—to convey the thought of one person to another, perhaps to build a science. It is not always adequate, but as we strive to learn from each other and from nature, language evolves and ultimately enables us to convey a great deal of thought and understanding. Thus we distinguish between the means and the end. The organization of language, even if it began with grunts, gradually evolved to the point where it could even describe things not known to common sense (a "wireless telegraph" for example could be described even though it seemed a contradiction in terms).

But the organization of language is the means, not the end.

The end is understanding, ultimately self-knowledge, because this is the ultimate end of knowing.

We can now return to nature. We say that the evolution of nuclear material produces protons and electrons, and these in turn produce atoms, molecules, and life. All of these are *forms* of organization, and because they are forms, and hence composite, they are temporary. They can be created and destroyed. But, like language, they *can* be the means for increasing understanding. As the old myths have it, God wants to know himself, so he created forms which the life force can inhabit—he breathes the breath of spirit into man.

Of course any scientific critic will say this is nonsense. I can only answer that he had better not be too sure about that. He can always claim that his reductive method will win out, that given enough time he will find how to explain all these particles. On the other hand, supposing he is wrong. How long would it take to convince him? I do not want to get further embroiled here because there is more to the story. I would like to complete the picture by showing that the evolution which distills the permanent particles out of nuclear substance and organizes these particles into life forms, does not simply leave these life forms like Christmas trees to be cut down and sold. The *life force reenters the psychic sea* which gave birth to nuclear particles.

Just as cellular life includes and transcends molecules, and molecules include and transcend atoms, there is another level of organization, animals, that includes and transcends vegetable (cellular) life. It includes it in the sense that the animal organism is cellular, and it transcends it in that it adds the power of motion which is lacking in plants.

Even if you think of the animal as Delgado or Skinner do, as motivated by *drives,* the animal can respond to drives in a way that the vegetable cannot. In fact I would insist it can and must choose between drives, as when an antelope, on the way to the watering hole, scents a lion.

These drives, which keep the animal on the move, searching for food, shelter, or a mate, or make it fight or flee when endangered, are the counterpart of the *forces* we encountered

30 WHICH WAY OUT?

with nuclear particles. What is important is that organization now no longer exists for its own sake. Evolution takes a further step and renders the organism manipulable by drives. Animals thus encounter and have to cope with the conditions which gave birth to nuclear particles and evolved proton and electron.

How can we compare a nuclear particle with an animal? Of course they are quite different, but what is important here is that both particles and animals are concerned with forces rather than organization. The animal differs from the particle in that the attractiveness (e.g., of the quarry in the case of a hunting animal) is endowed by the animal itself. This is an example of inversion between right and left sides of the arc of evolutionary development hypothesized in my book *The Reflexive Universe.** The particle (stage 2) is subject to forces; the animal (stage 6) *uses* the forces for its survival. Both are thus endowed with movement, and movement is linked with forces and with their value. On the left-hand side of the arc, movement in response to a force involves value, which is a constraint in one dimension. In the case of animals, on the right-hand side, movement is again important. It is not a constraint (being forced to move) but learning how to move.

To review the whole process once more, the evolution from particles to atoms involves an increase in constraint from one to two dimensions, and to three dimensions of constraint for molecules. With molecules we reach the world of determinism. At this bottom level, matter behaves according to law, and because it is determined it can be made use of. At the plant stage, life uses determinism to store energy and create cellular organisms. The evolutionary step to animals takes these cellular organisms and makes them depend on their own efforts to get food. To do so the organism *uses* drives in learning how to move. The arc is this expedition to acquire determinate means which can be used.

This explains what the traditional religions of almost all

---

*See Arthur M. Young, *The Reflexive Universe: Evolution of Consciousness* (New York: Delacorte Press, 1976).

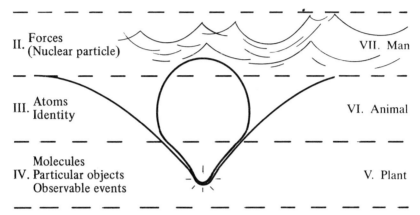

Fig. 3

cultures have referred to as the "Fall." It is a fall from the divine state of not knowing. In seeking to know, one must interact with matter and with otherness in order to learn. And in so doing, one learns the secret of life, and can eventually win back one's freedom. It is the same as the pattern of descent into matter, the same whether we are talking about the evolution of atoms and molecules, which must precede the evolution of life, or about the descent of man from the Garden of Eden, a descent which precedes the birth of the hero, or, in the Christian tradition, the birth of man as the son of God. Should I avoid religious concepts in a scientific discussion? No, because we are in a position to interpret this notion of the Fall as an expression of the same process which we have shown to prevail in the levels of organization.

We've been stuck too long in the limited focus of scientific objectivity, which blinds itself to deeper issues and evades basic questions. We should broaden the use of the discoveries of science to understand our own life and our own evolution. It is my contention that science itself is going to have to make some basic revisions of its view of reality or lose by default. In other words, we can now combine the factual discoveries of science and the insights of religious teachings to the great benefit of both. I opened with the point that we have two faculties for comprehending reality, reason and experience. These enable us to progress by using first one and then the other, like two feet. So let's now go back and fill in the rest of the picture.

What is more basic than nuclear particles? Quarks, partons, etc.? Forget it. There is no empirical evidence of these. Besides, even if we were to show that quarks were in fact responsible, we would want to know what gave rise to quarks? Well, it so happens we do have the answer to what is more basic than particles, for there exists the phenomenon of "pair creation" in which proton–anti-proton pairs and electron-positron pairs are created de novo, from light. Not only does light create particles—protons, baryons, mesons, electrons. It is the universal agency by which interactions between particles occur, whether this be the storage of energy by the chlorophyll of plants, the chemical energy of food, or communication through vision.

What can we say of the world of light? The ancients referred to it as the *intelligible* world (mundus intellectus) as opposed to the material world of sensation (mundus sensibilis). It is always in reference to light that we know. We use expressions like, "I see the light," "the light dawns," "see" for understand, "bright" for intelligent, "elucidate" (from *luce* = light). It is to the extent that we are enlightened that we have profited from our involvement with nature and natural law. This participation in the light, our ability to "see" how things work (which is because we see with the light of understanding) is testimony to the origin not only of matter, but of ourselves, in light. And it is in this sense again that we are children of the divine, sparks from the anvil of God.

But what has this to do with the chaotic sea of nuclear particles? Just this. That it is this same world of forces that we must transcend to reach the world of light. It is the psychic sea of Egypt which Osiris must cross in his boat. It is the

Bardo of Tibetan initiation. It is a world of demons, ghosts, all the forces that stand between us and salvation. Castaneda, in his four books about the teachings of Don Juan, gives very vivid first-hand accounts of how he learns to cope with witches, terrible animals, etc., to make them allies. And this is beautifully borne out in the world of nuclear particles. For when a particle and its anti-particle meet we get what is called pair annihilation, the opposite of pair creation. And this interaction causes a burst of light, which means that that which went forth has now returned. Again in psychoanalysis, when an emotional trauma is brought up into consciousness and understood, there is a burst of new energy that was formerly locked up in the trauma.

I have now given an overall survey of what I call the arc, which is the descent into matter followed by an ascent. The arc goes from light, at the first level, down to molecules at the fourth level. There it can stay, but there is an option here. The molecules can start to create life, which begins with cells, then vegetable life, then animals and finally man. (This is described in *The Reflexive Universe* in considerably greater detail than I'm able to give here.)

How, then do I explain this world of forces and nuclear particles? Well, the first step is to realize that there is a fundamental distinction between the world of particles and the world of atoms and molecules. The latter consists of separate entities which can be built into structures, the world of things that can come into being. And because they can come into being, they can also not be; things made of atoms and molecules can be destroyed. They are composite, but they are also separate. The world of particles, on the other hand, is interconnected, and obeys the laws of the conservation of energy and of mass. The creatures of this world cannot be destroyed, for they are not separate; they transform into one another.

This is very significant because it is a different picture from the one usually suggested by science. Science has been saying that the world is made out of fundamental particles, protons and electrons. But this cannot be true, because we've found that there is, so to speak, a factory manufacturing protons and electrons. It's as if to understand nature we made an analysis of houses and found them to be composed of fundamental units called bricks, and were so pleased with this discovery that we supposed from it that everything was composite. What about the people who live in the houses? They must be made of bricks, too!

But, thanks to science, we can now see that the bricks are not the whole story. There is a confused irrational ocean of stuff that underlies them, and this ocean is not only comparable to, but may even be, the psychic world. So to account for people we can now say, yes, to be sure their *bodies* are made out of bricks (atoms and molecules), but the *person* is something else, more comparable to a pitcher full of that *psychic sea,* containing all kinds of motions and forces. We are mistaken when we take the clothing for the reality, and dogmatically affirm that people are composed of molecules, and that there is no such thing as that creature of the psychic sea, the soul.

Where Is God?

Newton thought God was to be found in the *order* of the planetary motion, and science has found an even more intellectually satisfying order in atoms. The starting point was the spectrum with its mysterious lines, wave lengths of light in a precise order, different for each atom. Balmer found a mathematical expression which accounted for their exact values, and even predicted lines that had not yet been observed. Later Bohr explained how the lines were produced in hydrogen by the electron jumping from one orbit to another. Other scientists refined the theory to explain complicated atoms. Meanwhile, the concept that atoms were made up of an integral number of proton-electron pairs made steady progress; the gross discrepancies from exactness were traced to mixtures (e.g., chlorine was a mixture of atoms of two weights). Minor discrepancies were then found, due to the loss of a small fraction of the atoms' weight when packed together—thus four hydrogen atoms, with an atomic weight of 1.0008 each, weigh 4.000 when packed into helium. It was this loss of weight, or packing fraction, that was later found to account for the sun's enormous heat output: hydrogen being consumed to produce helium on the basis that energy equals mass times the velocity of light squared.

36   WHICH WAY OUT?

When I read about these earlier findings in the 1920s my enthusiasm was unbounded. It was so beautiful, so elegant. Democritus had, to be sure, imagined that the ultimate constituents of matter were atoms which were of "various shapes and sizes"—but how much more elegant was the truth uncovered by modern science. There are only so many kinds of atoms, each of which is an integral number of proton-electron pairs. One pair for hydrogen, two for helium, three for lithium and so on up to ninety-two for radium. The fact that additional neutrons are necessary to hold the nucleus together and so make necessary a different number for atomic *weight* (e.g., helium *4,* rather than *2*) does not impair the elegance of the picture, which traces the atom's shape, its chemical and other properties, to a pure number!

I felt then and still feel that this order is surpassingly beautiful. And there is much more. The same emphasis on economy and elegance recurs through all of chemistry. At college all I knew of was the combining ratios of salts, acids, and bases. All molecules are made of atoms combined in the ratio of simple whole numbers: water (H_2O), carbon tetrachloride ($CaCl_4$), calcium chloride ($CaCl_2$), and so on. More recently I've become aware of the marvels of organic chemistry—the methane series forming an alphabet from which thousands of kinds of organic molecules are compounded, much as the letters of the alphabet combine to form hundreds of thousands of words. Nor does the order stop with chemistry, for the DNA molecule provides the blueprint for entire organisms, and that, too, rests on the twenty-letter alphabet of amino acids.

Might we then not revere this spectacle of creation in its enormous variety: its millions of species of plants and animals all out of one alphabet of twenty amino acids, its endless yet ordered kinds of molecules, all made out of ninety-two kinds of atoms, its ninety-two kinds of atoms constructed of but two kinds of particles, protons and electrons? What sublime economy and eloquence! How could one describe such ingenuity, beauty, and exuberance except as divine, an order such as man strives for but which is far beyond man's capacity to achieve! No wonder that science became the new religion, with its godhead the sublime *order* that could by diligent effort be uncovered wherever one looked, whether downward toward infinitesimal molecules, or upward toward the stars and galaxies.

How could anything be said against such sublime ordering? But it was its very perfection that led to viewing order as self-sufficient—if such were the case, why was God necessary? In answer to Napoleon's question as to why he had not mentioned God in his *Celestial Mechanics,* Laplace replied, "Sire, I have no need of that hypothesis." Darwin, by accounting for the variety of life forms on the hypothesis that the principle of survival of the fittest, together with intrinsic variation, led to different species, effectively discredited literal interpretation of the Bible, and so lessened belief in revealed religion and strengthened the scientific explanation. Even though science could not account for life, it was—and still is—assumed that it was only a matter of time before it would "explain" that too. And so God gradually dropped out of fashion.

Then a new phase of development began. The sublime order, the divine perfection, developed a blemish. It was found that exact knowledge of small particles could *not* be obtained. And it was not a case where better instruments might provide this exact knowledge—the perfection of the tool interfered with the precision of the measurement. As Heisenberg proved, simultaneous knowledge of position and velocity is unobtainable in theory as well as in practice. Nor could the quest rightly esteem order even as an ideal truth, for uncertainty proved to be even more fundamental than Heisenberg had shown it to be. It was ontological rather than epistemological, and due to an inherent free range of action in the particle itself, not merely to the observer's probe.

This was ironic. It was as if nature, which had lured man to worship her symmetrical front, her conformity to classic ideals, now disrobed and presented her asymmetric derrière—a posture that called for a quite different assessment. Science still has not made the necessary change in point of view. It has, to be sure, shifted from emphasis on perfect order to an emphasis on statistics, but this compromise only accentuates the conflict between two camps, the one seeking the order it has found in other areas, and the other seeking in the name of uniformity to abolish all order. Einstein wanted to reject quantum theory on the grounds that, "God does not play dice with the universe." Others enshrine probability by allowing the possibility that the desk will fly out the window. One can hardly exaggerate the paradoxical and contradictory nature of the situation now existing. At the frontier of physics, we have had fifty years of efforts to accommodate the difficulties of admitting the fact of uncertainty, while behavioristic psychology, anthropology, and other pseudosciences march forward with complete confidence in the deterministic model. In the heart of logic, Gödel proved logic incomplete; meanwhile everywhere else there is a wild scramble to turn everything over to the computer, the supreme example of a logical tool.

But to get back to our thesis: the perfection of natural order, which led Laplace to dispense with God because there was no need for this hypothesis, gave way to the discovery of a fundamental "disorder," which Einstein could not accept. If Einstein could evoke God to oust disorder, how could Laplace use order to oust God? I think clarification is overdue. Let us therefore attempt it by listing the ideas involved:

1. The concept of order: Newtonian mechanics, atomic theory, and chemistry all display a superb elegance, a sublime order based on number itself.

2. The concept of chance or "disorder": in nuclear physics, subatomic particles reveal uncertainty and unpredictability; the electron is a "probability fog"; a radioactive atom has an average lifetime, not a predictable one.

3. Quantum theory: while it greatly extended the scope of order, in that it made possible the precise prediction of atomic phenomena, it did so at the expense of requiring theoretical sanction of "uncertainty" in the form of the quantum of action.

The quantum of action proved to be a divine imp, an atom of unpredictability at the core of predictability. While it explained microscopic phenomena, it could not itself be explained. It could create matter where there was none before. It was the agent of all changes, all interactions of matter with itself, yet because it could not be predicted and could not itself be explained, it could also be outside of the casual chain on which ordering is based. In other words the quantum of action can have no casual antecedent. *It is first cause!* This is why the quantum of action is indeterminate, why it is uncertainty, why it appears in such contrast to order, and why it seems to threaten order.

But it is not so much disorder as it is a spark of divine will which, as the myths declare, has descended into matter to stir it into activity. This gives to God the role of chance, a concept shocking both to theology and to science.

So what do we conclude? Is nature orderly, responding to law, or is it essentially subject to chance? We conclude that it is both. Is order basic or is chance basic? We conclude that chance, or first cause, is the more basic, but ineffective unless it is wedded to order.

It may be said that if we admit this uncertainty, this inconsistency, the whole structure of order will break down. As Bertrand Russell put it, "If $1 = 2$, then I am the pope."

But to admit first cause, which is a kind of inconsistency in that it has no antecedents, is not like admitting an inconsistency such as $1 = 2$. It is rather that first cause is analogous to an invention. An invention, in the definition given by the patent office, must be *novel,* and novelty is defined as that which could not have been thought of by one skilled in the art. In other words, novelty has no antecedents. But a novel invention doesn't mean that the laws of matter or the laws of logic break down. On the contrary, an invention is a novel application of the principles of chemistry, of physics, of mechanics. Inventions, in fact, only become possible when the laws and regularities of matter are sufficiently understood for it to be possible to put them to use.

We can thus reconcile the apparent contradiction between

the uncertainty of the quantum of action and the certainty of scientific law as not a conflict at all, but a partnership in which each factor is essential. The first is purpose, and the second is order. They are not the same, nor are they contraries. Order characterizes the means by which purpose achieves its goal.

Time to Go Home

Some Reactions to Arthur Koestler's *The Ghost in the Machine*

"All branches of science, except one, [have been] expanding at an unprecedented rate," says Arthur Koestler. "The one exception is psychology."* He might have added anthropology, sociology, biology (for the advances in biochemistry have not been matched by biology proper), economics, and even, as we look deeper, cosmology and logic. Without expecting agreement, I would also add mathematics.

Koestler is, however, quite right in his justification for beating the dead horse of nineteenth-century determinism. Though the animal as such is dead, its corpse is still around and full of life—i.e., of maggots, flies, and other forces of disintegration. But the speculative instinct that originally brought determinism into being, and which once was the source and inspiration of science, has ceased to function. The trouble with beating the dead horse in Koestler's case is that he has nothing to do it with. The Skinner behaviorist school of psychology, which avers that man is purely a mechanism, is holding the gun, and it is not a question of who is right, but of weapons.

Like the logical positivists, and the similarly oriented Ryle school of philosophy, the behaviorist school has invented a

---

*The Ghost in the Machine (New York: Macmillan Co., 1967), p. 5.

sophistical and deceptively simple weapon which can be used by anyone, the ignorant as well as the intelligent. It is a weapon that bypasses the subtleties of organic nature; it owes its success to the ego inflation it bestows on its adherents, giving them credit for toughmindedness while sparing them the trouble of genuine inquiry.

It is precisely the lack of an adequate weapon that is the weak point in Koestler's otherwise superbly executed inquiry into the fallacies of psychology—fallacies which, on further search, may be found in all the sciences. He uncovers the absurdities in the oppressor, but provides no corresponding help for the oppressed. He seems to believe the pretension of science that it is responsive to fact or to truth, whereas as Thomas S. Kuhn* points out, and as has been evident for some time, a scientific theory can only be displaced by another theory. Koestler's appeal is to common sense, but common sense is out of fashion. More than that, it has fled the field, a victim of fright, and may even be as dead as the horse. Common sense cannot function in the watery realm of emotion in which religious and scientific dogma thrive; its equipment requires a certain dryness. The spark of intuition can ignite only dry tinder.

Put in less metaphoric terms, the appeal to common sense cannot be effective in an atmosphere charged with emotion. This emotion arises because any question about foundations which challenges entrenched scientific dogma is a threat to the institutional setup which claims sole prerogative for the promulgation of science—much as in earlier times, the church claimed jurisdiction over truth and morality.

This brings us closer to the real issue. Organized science—and I'm not speaking only of psychology—is today the vehicle of a struggle for power. Only outwardly does it deal with the search for truth. Behind its claims of expanding the frontiers of knowledge are the more basic dynamics of power. Security depends on defense of position, and this, in turn, on weapons.

*The Structure of Scientific Revolutions (Chicago: University of Chicago Press, 1962).

The weapons, in this case, are first of all the usual ones which the party in power has always enjoyed: control of social institutions (and of the requirement for institutional processing: the college degree), and appeal to the instinctive fear of that which is outré, or doesn't conform to the established mores. But there is a subtler and more important weapon: the elevation of dogmatic prejudice (which is to say, belief and superstition) to the realm of respectability by calling it scientific method. So potent is this weapon that it deceives even its users, and endows them with that excess of zeal which in earlier times motivated the crusades, the Inquisition, and most of the wars that have characterized Western civilization.

In the face of this union of science and superstition, it is difficult to halt the "progress" of science, or of a civilization dominated and equipped by science, toward its own self-destruction. The court is packed, the police paid, the jury brainwashed by the party in power. This party is organized science, organized not only laudably on the necessary common denominator and the cumulative achievements of scientific discovery, but also unfortunately on the inevitably mediocre capabilities of many of the great number of practitioners needed for mass production, to whom participation in the profound mystery of existence is only made possible by the provison of wooden idols—the hardware of computers, electroencephalographs, and Geiger counters—to supplement the intangibles of true wisdom.

The Buddhists warn us not to mistake the finger which points for the thing pointed at. The religion of the scientific method, reversing this, makes the pointer reading the final reference, and demands that we disabuse ourselves of the notion that it means anything besides itself. Thus, religion is reversed, but it is *still religion,* and it gains adherents by sanctioning the illusions of matter, or, rather, the illusions of conceptualization.

This interpretation provides an additional clue to the fragmentation which characterizes modern science. This fragmentation into many separate disciplines in science, each of which is a world unto itself, also derives support from the search for security, for these many disciplines provide many more niches for territorial sway. As in a bureaucracy, each department gains by creating more departments under it, and the unitary thread that organizes this hierarchy is power rather than a cohesive theory. It becomes expedient for each discipline to have its own private language to prevent, rather than encourage, the unification which understanding would provide. Training in a discipline and, hence, qualification, then becomes ability to use that discipline's jargon, rather than to recognize in it the manifestations of a deeper generality.

In addition, the philosophy which makes the pointer reading the ultimate reference and denies the intangible to which the pointer refers, this philosophy, I say, supports fragmentation, for scientific ingenuity can devise many kinds of pointers which use different means for pointing to the same thing. We can have conventions of cosmologists discussing the red shift with never a query as to what the red shift means: velocity of recession need not be the only interpretation—it might be energy decay due to its billion-odd years of travel; it might be that in earlier times, atomic clocks, like most other things, moved more slowly; it might be that all of these things are equivalent; but such speculations are ruled out. The scientist has become the housewife. He watches his pointer readings because they are within bounds. He has found a way, so he believes, to make supremely comprehensive statements about the universe without leaving the confines of his lab.

I once met a nun who was obtaining her Ph.D. in behavioral science. I asked her whether this did not conflict with her religion. Her answer, "Like the bar-pressing rat, I learned what would get the rewards," indicates that she felt no conflict. The experiments to condition rats prove only that you can condition rats. Here is where the real blindness occurs: having denied that pointers point to something else, and having persuaded us that it is wrong to look for ghosts in machines, the exponents of the mechanism of conditioning, by a singular feat of self-contradiction, have convinced themselves that this means that conditioning accounts for all learning. Sure you can condition rats, and by a similar enforcement of cages and withheld rewards you can condition people, but what does this point to? That you can condition rats and people, but not that you have explained the learning process.

One can condition children to say "please" by withholding rewards until they do, but this does not explain learning. It is a condition imposed on a process already present but latent. In other words, it is a limitation of something already there. The point is that limitation does not supply content. A *quart* of milk does not explain or create milk, though it does limit it. This leads to the question of what it is that behaviorism purports to do: to predict behavior. But the behavior that can be predicted is only that which has been implanted. The child that has been conditioned to say "please" will probably continue to do so, but this won't educate the child, though it may fit him to please other people (conditioned in turn to expect "please"). This is a minor achievement, and neither particularly significant nor even new.

At the end of his book, Koestler makes a plea for developing a drug that will improve the rapport between emotion and intellect. Though I admire the book, it seems to me that, despite his perceptive and reasoned restoration of human dignity, Koestler turns traitor to it by thus pinning his hopes on some elixir from the drug store. I make this charge not because of the principle on which he draws (he argues well that much has been done and many sacred fixations broken by medicine, i.e., vaccination, etc.), but because he omits any reference to those intangible aspects of self which, in the first part of the book, he shows to be so important.

Put simply, Koestler's own psychology is insufficiently developed. He provides no theory to supplant the dead horse of determinism; and determinism, even while recognized as unsatisfactory, has become not only a habit of mind among psychologists, but a kind of "home safe" in the game of philosophic tag.

If we recognize the importance of theory to scientific method, we can understand why determinism still keeps its hold despite the deficiencies that Koestler has pointed out. To make good the gains which Koestler has won, *we need a new theory*. It is not sufficient to reemphasize teleological factors; one must give such factors legal status if they are not to be thrown out of court.

Here I can mention my own approach, which is directed toward such a theory, and one to which Koestler's facts give support. It is by virtue of this theory that I take exception to his later chapters, in which he indicates that the predicament of man is due to the rapid evolution of the cerebrum, that part of the brain capable of reasoned thought, the neocortex. "Evolution *superimposed a new, superior structure on an old one,* with partly overlapping functions, and *without providing the new with a clear-cut, hierarchic control over the old*—thus inviting confusion and conflict," Koestler observes (pp. 281–82). This dichotomy between limbic and neocortical areas (or between emotion and intellect), which he calls an evolutionary mistake, leads to his proposal in the final chapter for some means to integrate the two brains, perhaps a drug.

As I have said, it is in this conclusion that I feel Koestler lets us down, primarily because it is based on a misunderstanding of the problem. Under the subtitle, "The Physiology of Emotion," Koestler says,

The distinction between "knowing" and "feeling," between reason and emotion, goes back to the Greeks. Aristotle in *De Anima* pointed to visceral sensations as the *substance* of emotion and contrasted them with the *form*, i.e., the ideational content of the emotion.*

To me, this distinction between substance and form provides the key to the whole problem of feeling vs. knowing, and leads

---

***The Ghost in the Machine*, p. 274. Aristotle's distinction has been revived in the theory of hemispheric specialization. Despite the recent enthusiasm for attributing this distinction of function (form and content) to the left and right hemispheres, this cannot be the solution. As shown here and elsewhere in these essays, the content side requires something more basic, more like what is called soul. However, the alacrity with which the two hemispheres have been seized upon for this distinction shows that the need for something to which to attribute the difference is urgent. That the hemisphere explanation is inadequate becomes clear if we consider telepathy or distant viewing; how can the brain (right or left), a physical object with location in space, receive signals that don't depend on the inverse square law?

to a quite different answer from Koestler's. I am aware of the philosophical difficulties in the word substance, but the *distinction* is basic. A glass of water illustrates the duality: the water, which is the substance, needs the *form* as its vehicle. So, too, feelings or emotions are the *substance* for which the intellect provides the *form*.

Going further into the duality, we know that emotions are accompanied by and correlate with chemicals—for instance, adrenalin in the blood stream—whereas reasoning, concept formation, and rational thought occur on the pia mater, the *surface* of the forebrain.

On the basis of this elementary distinction, we can represent emotional state by a value, which is to say, by a single number, one dimension, whereas a mental concept is essentially a form, and requires two numbers, two dimensions (corresponding to the two dimensions of the surface, or plane, of the pia mater). This puts the distinction between emotion and intellect on a fundamental basis and leads to the question of why there are two aspects necessary to consciousness.

Here another image is helpful. Consider the factors involved in firing a gun. There is, first of all, the charge which propels the bullet, and has the nature of energy. Then there is the aim of the gun, which is not just simply a matter of pointing, but requires mental computation to allow for the movement of the target, the effects of the wind, and the trajectory of the bullet. The charge is analogous to emotion—it supplies the physical motivation that drives (emotions are referred to as drives; there is also a connection between motion, emotion, and animation)—and the computation necessary to aiming correlates with the rational mind. This latter function, while not physical, can be achieved by a computer whose circuitry is again two-dimensional (circuits are diagrams and essentially two-dimensional).

The analogy indicates the essential nature of emotion, for without emotion, brain work is without issue, has no effect. It also helps to indicate that it is not reason that controls emotion, but *purpose,* which for the gunner is the intention of hitting the target, and requires the driving power of emotion to be effective.

This interpretation of the roles of thought and emotion actually supports Koestler's plea (p. 336) for "a state of dynamic equilibrium in which thought and emotion are re-united, and hierarchic order is restored." What is this hierarchic order? Obviously, the one who aims must have the final say, and this affirms Koestler's repeated emphasis on the innate purposiveness of all life. To give this purposiveness formal sanction, we can correlate it with what might be called zero-dimensionality, the dimensionality which precedes and thereafter accompanies the one-dimensionality of emotion (of consciousness as it experiences the time sequence, or linear flow, of events).

Next in the hierarchic order (after emotion) is reason, which compares one emotional "longing" with another, and decides between them by measure, or ratio—the concept of "length." (It is of interest that "longing" and "length," with their apparently different meanings, have the same Anglo-Saxon root.) So thought is a later development than the more primitive emotion.

However, it would seem that emotion comes in again at a higher hierarchic level than reason, for emotion is not only the initial urge toward, but the ultimate repository for the fruits of mind. Again, this can be seen in the gun analogy; the computations have no other result than correctly to aim the bullet, which is to say, to direct the energy of the charge to reach the target.

To supplement the analogy with a specific example, we can instance the experiments with planaria worms that Koestler reported on some years ago in the London *Observer* (30 June 1961). Here it was found, first, that any part of the cut-up worm retained the conditioning that had been inculcated in the whole worm; second, that headless worms could not be taught. This implies that association (in this case, of an electric shock with light), while it does not require the brain in order to be retained in the creature's memory bank, does require the brain for the conditioning (the association of the shock with the light) to be registered.

We could go even further and instance psychic energy as it persists in the absence of brain activity either in the sleep state or the still discredited after-death state. The question of immortality is, perhaps, the most important question today, not only because of its bearing on values and on moral issues, but because it provides the touchstone or test for any serious hypothesis about the nature of the self. Admittedly, the position of science on this score is assumed to be negative, and sober thinkers, deferring to scientific respectability, bide their time.

However, the foreboding that civilization is on a collision course with self-destruction raises the question, "Have we time to wait for such inquiry to become academically respectable?" One cannot ponder philosophical issues at gun point (under the duress of circumstances), but one can use the gun point to break up the interminable game of pass the buck to which scientific respectability resorts when it answers inquiry with the excuse that it is "beyond my competence." Another such dodge, the more difficult because the user doesn't realize that it is a dodge, is to demand "criteria." "What are the *criteria* (for the soul)?"

The answer to this logical-positivist tactic is to turn it about, to toss the grenade back before it goes off, and ask, "What are the criteria for destructibility (i.e., death)?" Science recognizes the conservation of mass-energy (matter and energy *cannot be destroyed*). What kind of thing, what order of existence *can* be destroyed? Clearly, only that which can be constructed can be destroyed (for example, buildings, but more generally, relationship structures, organizations, etc.), which is to say, anything put together out of parts can be reduced to parts. The statement that the soul is immortal is equivalent, therefore, to saying that the soul is not something built out of parts.

To return, then, to Koestler's original point that the new mind does not have "a clear-cut, hierarchic control over the old, thus inviting confusion and conflict," it is our finding that there is no reason why the neocortex should control emotion.

Its role is rather that of instructing the more permanent emotional principle, and it is the latter, if anything, that is of a "higher" order. Value takes precedence over definition.

If neither the rational mind nor the emotions have final jurisdiction, what is it that does (or should)? The answer is implicit in Koestler's own thesis. In both *The Sleepwalkers* and *The Act of Creation*,* Koestler showed that scientific discovery and invention do not arise in the rational process, but from a deeper source. In *The Ghost in the Machine,* he refers (p. 180) to "spontaneous intuitions and hunches of unconscious origin," rather than rational and verbal processes. He seems to regard such hunches as a regression to the level of emotional imagery (Freud's primary process). He is quick to add, however:

It seems that between this very primary process, and the so-called secondary process, governed by the reality principle, we must interpolate several levels of mental activity which are not just mixtures of "primary" and "secondary," but are cognitive systems in their own right . . . (p. 181)

and goes on to confuse the issue with the paranoid delusion, the dream, the daydream, etc.

Here I must interrupt. Clearly there is an intuitive layer, but it is not *between* the primary (emotional) and secondary (rational), it is antecedent to both. It is the zero-dimensional unitary awareness itself, which normally accompanies the linear time dimension and the intellectual space dimension, but which, as the progenitor of both, can on occasion stand clear and *recognize*. It is that moment of clear seeing, the "Eureka!", the dawning of light, the blow on the head (which is the true reference in the fable of Newton's apple). The "pictures" (emotional imagery) are secondary to this. Our hierarchy is thus:

    I. Unitary consciousness, analogous to the point.

---

*The Sleepwalkers: A History of Man's Changing Vision of the Universe (New York: Macmillan Co., 1959; *The Act of Creation* (New York: Macmillan Co., 1964).

II. Time flow (emotional experience), analogous to the line.

III. Spatial relationship (rationality), analogous to the plane.

Now, if it is true that this higher function—intuition or purpose—must align the separate functions of mind and emotion in its service, then mind *versus* emotion is not the problem, and Koestler's prescription for a drug to unite the two is wrong, or springs from a partial view of the situation.

But Koestler is otherwise on the right track. Something must be done. But what do we do if we don't know what to do? I'm reconciled to doing something, provided one is able to turn things off if they produce a cure that is worse than the disease. The enthusiasm for the cure usually precludes any such caution, and by the time the results are in, it's too late. Almost every ecological reform has become a curse: the English sparrow, the starling, honeysuckle, and the water hyacinth were all introduced with the best of intentions. In fact, the problems man faces today are to a striking degree the result of his own attempts to improve on nature. Control of famines and diseases (by food transportation and hygiene) raised the population of India even in the nineteenth century to the point where no one had enough to eat. Africa is now experiencing the same fate. Control of disease and extension of lifespan is giving even the United States the problem of supporting a population of invalids. One could add air pollution, crop diseases, chemical poisoning, bacteria nourished by detergents, etc., as examples of side effects which are worse than the conditions which the reforms were intended to rectify.

But, as Koestler says, and most agree, the problem we face calls for immediate action. We have perhaps twenty years grace, perhaps less. I, too, agree, and my general answer is that there is nothing we can do that will avert disaster. Civilization is in a nose dive, and the momentum already exceeds our ability to halt it. So the world we know is doomed, and we should act accordingly. This does not mean we can do nothing. Rather, it is our crucial opportunity to find an answer to a problem that is basic to evolution itself—and if it is found, it will be worth a holocaust. We will take the answer into our immortal souls, and it will serve us, whether as a surviving remnant, or in the next rebirth of civilization, whenever that is, hundreds or thousands of years hence.*

Hence the importance of self-development rather than of more gadgets and cure-alls. And this requires an about-face, a reverse in our direction—not a return to former conditions, but a turn of a different sort, a turn away from fragmentation and specialization and toward integration and understanding. We have all the groceries we can carry. It is time to go home.

What I mean can be illustrated by the problem of the alcoholic or drug addict, whose appetite for self-transcending experience carries him from one intoxicant to another. He is led on and on, each bottle or fix only temporarily satisfying his craving. But his true need is not more alcohol or new stimulants. He is in need of self-examination, of pulling himself together and controlling his appetites, if necessary of starving out the craving that is destroying him.

Psychologists talk about a death wish, but I do not think the concept is sound. It is not so much that man has a death wish as that he has appetites which can destroy him, or at least can destroy his vehicle. Like any machine, he contains some form of motive power. An elaborate machine can suffer short circuits, an automobile can catch fire, an airplane can explode. These are not instances of the death wish in a machine, but of failure of control, failure which permits the motive power to get the upper hand. Suppose an automobile were rigged so that the spark plugs were placed in the gas tank. Then, when the ignition was turned on, instead of the engine turning over, the tank would explode. The tiger would get out of the tank—emotion getting out of control.

---

*The dinosaurs perished—but as a result of their failure, nature produced mammals, a new sort of creature that did not depend on size for efficient metabolism. Mammals increase self-control by employing a thermostat to maintain constant temperature. It would appear that the size of the dinosaur was a means of increasing body temperature to obtain higher metabolism, because the temperature of a cold-blooded animal is a function of volume (which determines heat production) divided by surface (which determines cooling).

This is, of course, an old idea. What I'm joining it to is the perception that the appetite for mechanical aids can also get out of hand and outgrow its function. This desire for organized structures to do our bidding is itself sound, and even laudable. It is perhaps the main force in evolution, for it accounts for our growth from a single-celled amoeba, which had volition of sorts, to an organism of some quadrillions of cells, a hierarchy of systems and subsystems with a central government. This central government (the self) has gained, through its employment of subsystems, a territorial sway that is far more extensive than is that of the amoeba. Through the device of civilization, it is achieving even greater sway.

But what we must not forget is that structures and mechanisms are still only means—means which the self engenders and uses to increase its scope and better attain its ends. What is the goal of man, or, rather, of the self, for man is only a stage in an evolutionary chain that stretches back beyond the amoeba to even more fundamental units?

We are all searching for an answer almost impossible to express in rational terms, since intellect itself is only one of the means the self employs, and a relatively recent one at that. But an indication of an answer may be obtained by surveying evolution, which is to say, the ladder of being of which man is a rung, (a most advanced one, in that he includes all the other rungs: his organization incorporates that of animals, plants, cells, molecules, and so on).

If man is a rung in the ladder of organizational forms, held together by a hierarchy of subsystems under a central government, then the goal of man is the extension of this ladder, which is to say, the more complete control of means.

We could pause to explain why the implied super-organism is not a government or state, but individuals. Certainly governments are organizations of persons, but civilization is an evolutionary device which enables the self to learn and to rehearse the organization within the self. The soldier learns first to obey, then to issue orders by playing the role of general in the army. To the self this is its opportunity to master the equivalent hierarchy in itself.

This brings us to the crucial point: the control of the intellectual appetite. This final act of control cannot itself be delegated. It cannot be supplanted by a mechanical, chemical, or even a psychological aid. (Progress from steering to power steering does not lead to the machine which steers itself; it must always be possible for the pilot to override the servo controls and return to manual.) This realization that intellect is not an end in itself, but must serve the self's ultimate goals, is the self's rebirth.

Individual resurrection amounts to a turning away from expenditure, exploration, and experiment toward the digestion of the materials the self has gathered in order to grow. There must occur a voluntary surrender of freedom of outer movement, a settling, a commitment. The hold-fast cell in algae clinging to the ocean floor, the blastocyst clinging to the wall of the womb, the seed falling to the ground, and the marriage contract are examples of self-limitation followed by growth, an amplification of self.

It would thus appear that since evolution predates modern science, the essential ingredient for further evolution is not more science, but the digestion of that science which we have. It is, of course, not possible arbitrarily to say at what point any one individual must turn, but turn he must if he is to go on. If the self is not reborn, it deteriorates into aggregate matter, the humus in which other seeds germinate.

SOCIAL EVOLUTION

But, my reader may say, our concern is not for the individual, who is not going to survive more than his appointed span anyway, but for the race. What is to happen to our children and our children's children? That this view has the picture the wrong way about (the individual monad by our tenets, will survive anyway, while civilizations always ultimately perish and give place to others) need not concern us here. We are addressing ourselves to the problem, as currently interpreted, that for one reason or another—overpopulation, pollution, civil strife, disposal of waste, atomic holocaust—the human race will be wiped out.

To deal with the question, we must first decide whether civilization exists for man, or man for civilization. My theory is the former, and it leads me to say categorically: *there are no social problems*. The problems we call social are those which arise in the individual, and which many individuals share. People pool their problems, and a social issue is the result. Civilization, with its human interactions—its wars, revolutions, conflicts of opposing dogma, is a theater in which individuals act out or exteriorize the drama within.

How does modern civilization differ from others? This question requires a measure by which to characterize a particular civilization. One measure is certainly the nature of its wars. Greece and Persia struggled for supremacy, and the Greeks prevailed. Rome and Greece struggled, and Rome conquered. Roman culture gave way to that of Northern Europe, not so much as a result of losing wars, as through the deterioration of its religion in the face of the more vigorous Christianity. And so on.

It is apparent that these wars, whether under the banner of nationalism or of religion, even up to World War II, were conflicts of man against man; they were *intra-human*. The threats we are now concerned with, and which it is anticipated will eliminate the human race, do not involve man against man, but are threats of a technology gone rampant, of Frankenstein forces that recognize no boundaries. Even the famine which overpopulation threatens will not stay within the borders of a given continent. Communication and humanitarian considerations would make famine universal, and drag us all down together. In other words, unless we lose even what civilization we have, a well-fed nation cannot watch another starve to death. So what is new and different is that man, who up to this point has used the therapy of conflict for his own development (and this applies to both games and wars), is now faced with a conflict that is no longer intra-human. It is now a matter of human nature versus technology.

It may be insisted that technology is still a problem only because man has not learned to use it; that it is man who must change, adapt to the wider horizons opened up by technology, and free himself from the inner conflict which vents itself in weapons. This, of course, is true, but I cannot see how it answers the real problem. It would require, ultimately, that everyone be elevated to an order of sanity hardly possessed by any of us at present. For suppose progress in nuclear science advances through general adoption of nuclear power to more and more, smaller and smaller nuclear power sources—first power plants for the generation of electricity, then for the propulsion of ships, aircraft, and private vehicles? It is inevitable that eventually anyone will be able to put enough fissionable material together to blow up a city, just as today anyone can make a homemade bomb to blow up a church.

Again, take the progress in technology which, for example, has made the automobile available to everyone, and air pollution a growing menace. Here the problem is not one of human conflict, of man against man, but the momentum of mass transportation, which in order to facilitate travel obliterates the countryside to provide roads. This expansion permits the factories to locate further from population centers, and this, in turn, requires still more and better roads. This is a development independent of man's inner harmony; it is due to the momentum of technological progress producing what may be called positive feedback, which amplifies instead of stabilizing a given adaptation, so that the disease feeds itself.

We cannot put the blame for overpopulation on an imbalance in man's inner harmony. This threat would be in no way reduced by universal peace. It exists because of improved technology—food supply is increased, overpopulated nations fed. Mass transportation makes it possible to avert local famine (as it did even in the nineteenth century), and then national famine becomes a threat. This predicament is relieved by improved agricultural methods, manufactured fertilizer, and transport of crop surpluses. The cure only feeds the disease, however, for it increases population, which was the cause of the trouble in the first place.

What, in fact, distinguishes our present civilization from all others is that it is not guided by the interests of any one section of the human race. It is not a matter of American versus

Russian, of communist versus capitalist, of dictatorship versus democracy, of white versus black. Ours is a civilization that derives its standards, its ultimate points of reference, from the needs of the technical devices man has invented and mass-produced for the conquest of his environment. Some of this technical proliferation is an outgrowth of the heady intoxication of the arms race, but even without this there prevails a partial or short-term view of human needs being solvable by improved technology. We can improve health by sanitation and sewage disposal, we can improve the yield of soil by fertilizer, we can improve living conditions by mass production of building materials, food and clothing. But while technology can win the battles, it cannot win the war. Like some of the evolutionary devices of prehistoric animals, it contains the germ of its own decease. The saber-toothed tiger eventually evolved a tooth so long that the animal couldn't open its mouth wide enough to use it. In the same way, our technology is outgrowing our true needs, by which I mean those long-term checks and balances by which, and only by which, a civilization can maintain itself.

We come, then, to the same conclusion whether we blame technology or the intellect that has produced it. Our predicament has come about because means have replaced ends; the self has fallen victim not to its passions, but to its central intelligence.

The Queen and Mr. Russell

What I would like to show in this essay is that mathematics, the queen of the sciences, has in the last century departed from its true path and become entangled in certain errors. The causes of this decline from grace are hard to determine; in a way they are manifestations of tendencies in the current culture and are shared by other sciences, but because mathematics represents the highest expression of abstract thought and cannot benefit from the correction afforded by practical experiment, these errors of the mind continue unchecked. Again, the prestige of mathematics being what it is, it is extremely difficult for corrections to gain a hearing.

The errors we will attempt to track down have to do with epistemology, itself a lost science, which, were it not bereft of its faculties, might assist in the queen's council. Again such errors might be traced to too great a reliance on technical devices, a failing to which all modern endeavors are subject, but which in a queen is most ill-fitting.

There must be a reason for mathematics to have come to be known as the queen of the sciences. Is it possibly because its inner workings depend on the highest degree of intuitive insight, mysterious and penetrating as woman herself? The attempt to logicize mathematics has been a conspicuous failure. Not only has it been without issue, but logic itself, as Gödel has proved, confesses its own incompleteness and challenges us to look to some other source for ultimate answers.

Mathematics, of course, rightfully claims a province of its own, one differing from that of physics and the other sciences. It deals in ideal forms, in the properties of numbers, in the performance of operations unencumbered with material necessity. In this inner sanctum mathematics reigns supreme, its deliberations unimpaired by the storms and broils of the physical world. But we must also recognize that the boundaries that separate disciplines do not separate the principles necessary to their performance. Nor can they isolate the parts of any whole endeavor. Two people who own separate properties, and conduct different businesses, do not differ in their organic constitution. Both are vertebrate mammals, have the same needs, organs, and senses, the same bones, flesh, and blood. The incompleteness discovered in logic is not peculiar to it, but characteristic of all organic systems. It is only an embarrassment to logic because it is inconsistent with logic's self-image, its postulates.

Thus the "ideal" world of mathematics is not exempt from the requirement that, in the course of abstracting the ingredients of pure meaning, it must omit none. Like an astronaut preparing for existence in outer space, its withdrawal from terra firma places it under even more severe requirements to include all the elements necessary for life. It is thus obliged to search out the parameters of existence even more thoroughly than the other sciences, which are content to work on the superstructure and need not question such absolutes as are provided gratis by life on Mother Earth. Of course, since Greek times, the emphasis in mathematics has been placed on pure forms. The circle takes on the perfection of a "Platonic idea." This ideal circle provides a more exact value for Pi than can the physical measurement of any circular object, however accurately made. So, too, the squares of mathematics are perfect squares, its lines are without thickness, its points dimensionless, its planes frictionless, and the terms of its infinite series unwearied in their passage.

Our intent, however, must not be interpreted as practical criticism. What I want to show is that mathematics has hidden potential which could greatly expand its scope. I would like to arouse curiosity and enthusiasm for an enlargement of mathematics. I do not call for defense of its borders, but for their expansion to meet even greater challenges and undertakings than its conquests of the past. With this introduction we may proceed with our task.

POSITIVE AND NEGATIVE NUMBERS

Important to mathematics has been the extension of the definition of number, first to zero, then to negative numbers and fractions, and beyond these to what are called real numbers (numbers that are the solution to equations), complex numbers (having a real and imaginary part), and transcendental numbers. Let us first consider negative numbers.

In real life the lack of something produces an imbalance which cannot be dismissed as nothing. If we run out of gas on the highway, we experience an acute need for gas. This need is a state which is positive in its own right. It is a disturbance of equilibrium that requires correction.

Carried over into mathematics, this need is translated as the minus sign, which mathematical custom has interpreted as creating a different kind of number, *negative number,* whose existence makes possible the extension and subtracting of a larger from a smaller number. Let us note that certain conventions have been established in so doing; for example, that the meaning of the negative sign has been expanded to denote *a species of number* different from the natural numbers. It follows that subtraction becomes equivalent to adding a negative number, and that subtraction of a negative number is equivalent to addition.

Here common sense might raise an objection and ask if it is indeed true that subtraction, which is a kind of operation, can transfer its functional significance to an essentially unchanging noumenal entity. Are numbers by which we count "things" changed by the direction in which an operation is made? For example: suppose *A* gives $10.00 to *B*. Then *A* is minus this amount and *B* is plus. But why should the *number* of dollars be either positive or negative? It is true that negative and positive are very real. *A* may feel the lack in a more real sense than *B* feels the gain, but what has this to do with a different kind of *number?* Moreover, since for *A* the amount is negative and for *B* it is positive, how can the *direction* of the transaction become permanently attached to the number 10, which is clearly objective and agreed upon by both parties, whereas its sign is different for each.

It might be objected that these questions do not pertain to mathematics, which is concerned only with keeping its own house in order. To do this mathematics has found it necessary to create this different type of number, the negative number, and prefers this alternative to that of confining the negative sign to the operation of subtraction. I was satisfied with this until I became aware that there was confusion within mathematics itself about *positive* numbers. What appeared to be the case was that the clarification necessary for negative numbers (i.e., to distinguish them from natural numbers) was not applied to positive numbers, which the mathematical dictionary described as "non-negative." This disturbed me and it was only with the greatest difficulty that I extracted from Church (who had written the dictionary) that it was important to distinguish positive numbers from natural (or signless) numbers.

In view of our objection to the creation of negative numbers, we can in fact ask, is there any reason for positive numbers? In other words, are numbers properly positive or negative? Are positive and negative not rather the direction in which an operation is taken, and numbers the count or measure of this operation?

The mathematician would probably say that this distinction does not concern him, because it makes no difference to the way he carries out his business. To this I reply, perhaps not in the ordinary run of business—but in some cases it can and does cause a lot of trouble.

Before we undertake to deal with this difficulty, let us first indicate how we would propose to fill the gap which removal of negative and positive numbers would leave. Actually, we do not imply that mathematics should not have a place for positive and negative. On the contrary, these signs indicate an important domain of mathematics not covered by the term quantitative. For if numbers are quantities, *signs are qualities*. In any case mathematics contains both—not merely because mathematics is comprehensive, but because quality and quantity are in ordinary usage inextricably intermixed. It should be the business of mathematics to elucidate the distinction, rather than fall victim to it. Assigning signs to qualities would be inconsistent with the fact that the quality of workmanship, accuracy of a machine part, hardness of steel, fineness of gold, etc., are *quantitatively* measured. On the other hand, we need to distinguish difference in a non-quantitative way—i.e., red is different from blue, and for this the wavelength of light will not do, because it does not give us the essentially two-dimensional display that is needed for mixing colors. Thus, red plus green = neutral, whereas red plus yellow = orange; red and green are opposites, red and yellow are at right angles. These distinctions are formal and abstract. They belong to the corpus of mathematics. In fact, they already exist there. It is only the obscurity in which logic has enveloped mathematics that prevents their general recognition.

If not as attached to numbers, how then should signs be used? Or, since their use is established in satisfactory fashion in addition and subtraction (with the exception of the troubles to which we have referred), perhaps it would be better to say, how described?

In creating ideals, or abstract entities, it seems to be implied that mathematics transcends the "physical" world, a world in which substance plays a necessary role. While in the physical world we can only exemplify a circle by making a wheel, or a disk of metal, or a figure in chalk, in mathematics we operate with ideal figures which need have no physical representation. We obtain the value of Pi to any number of decimal places far beyond what could be obtained by empirical measure. It would thus appear that mathematical science does not require the notion of substance. A triangle is a triangle irrespective of whether it is red or blue, gold or paper; its geometrical and mathematical properties are in no way influenced by the material of which it is formed. It was this argument, in fact, that led Bishop Berkeley to suggest that the notion of substance be abolished from philosophy, for, he said, the chemist had no need of it, since all his operations had to do only with ratios—acidity, density, solubility, etc. How much more true would this not be for mathematics, which needs no test tubes and no Bunsen burners and writes its equations on the tablets of mind?

True enough, but why then does mathematics see fit to speak of positive and negative quantities? If its sheep graze in mental pastures, how can they get lost?

Positive and negative imply a parameter—I hesitate to say a "dimension"—that is not encompassed by the dimensions of pure form. We may hazard a guess that the dimensions of mental space which forms require are those of a surface, requiring lateral extent but no thickness. The dimension of positive and negative, if it can be called a dimension, is not on this surface, because it does not concern the kinds of statement we would make about the properties of figures. Thus a square has four sides, and the number four in this case is neither positive nor negative as long as we remain in the world of form. If, however, we enclose a square field by a fence, we encounter positive and negative numbers. We require four positive lengths of fence, and are aware of a negative side if one has been removed or stolen. Thus, when we encounter a situation where we feel that something which *should* be there is lacking, we are often painfully aware of its absence. When, on the other hand, some expected thing is present, we do not celebrate its presence. (It is not in our nature to glorify the fact that the floor is still beneath our feet.) There is a basic difference in the way we relate to positive and negative values in our experience. My thesis is that this essential asymmetry between the presence and absence of some necessary thing must and does carry over into the abstract realm of mathematics.

The rational faculty rejects this notion categorically, but it is precisely for this reason that the case for it must be heard and judged without regard to rational bias. *Rational bias* is a contradictory phrase but it is indicative of the urgency of the obligation to deal with asymmetry. Reason purports to be symmetrical, but is contiguous with an asymmetry—whenever reasoning is involved, an important consideration is whether the reasoning is hindsight or foresight. This suggests that the symmetry that exists in logic is not present when the logic is applied. Here we have a reference on which to base an absolute asymmetry: time. I will assume the reader recognizes the problem posed by the direction of time. It is that it is impossible to "map" the direction of time. Space, on the other hand, is inherently mappable. In fact, we could equate spatial extension with mappability. Time, as one-dimensional extension, is also mappable. But the direction of time is not.

Again we hear the objection that the direction of time is a "reality condition" which does not concern mathematics. But, as we stated earlier, this view of mathematics is what is under scrutiny. In sum then, we are stating that there exists a prime asymmetry—of which the direction of time is an example—and that the duality of positive and negative is its mathematical correlate.

CANTOR, CONTINUITY AND INFINITY

We have indicated that our purpose is to increase the scope and power of mathematics, to expand the frontiers of the queen's domain. Now it so happens that the borders of her majesty's kingdom have for some time been in dispute, and I would like to show that the sources of this confusion lie in the same notion of positive and negative that we have been considering.

The border of the kingdom appropriately enough is infinity, and it might have been thought this was enough. But Cantor, in the late nineteenth century, introduced the notion of *many* infinities. The first is that of the natural numbers. Cantor devised a method to show that the set of all fractions could be put into one-to-one correspondence with the natural numbers and hence counted, and that the number of all fractions was therefore the same as the number of natural numbers—what he called "countably infinite." He then placed all fractions between 0 and 1 in sequence on a line segment and had the fractions designate points on this line segment, each fraction corresponding to a point. Since there are irrational numbers (e.g., ½√2, ½√3, etc.) not so designated, there are an infinity of gaps, each containing its own infinity of points. He later devised a technique for counting the irrationals, but the problem remained because this method would still not accommodate the transcendental numbers which would occupy further gaps. Cantor concluded that the countably infinite number of fractions and irrational numbers is exceeded by the uncountably infinite number of transcendentals. This latter number he called c, the number of continuum.

Aside from the question of whether the one-to-one correspondence between fractions and whole numbers (which is open-ended) is acceptable, we question its assumption that gaps contain points.

We also note that there are as many gaps as there are points (plus or minus one depending on whether 0 and 1 are counted as points), i.e., there are an infinity of gaps. Our effort to fill the initial gap between 0 and 1 has resulted in an infinity of gaps. We could insist that a point does not exist until it has been designated, in this case by a fraction. We can then say the gaps between fractions are *non*-points, since no construction has been provided to specify points thereon.

The conclusion is that far from achieving a continuum by filling the line segment with points, we have moved in the opposite direction; we have *interrupted* the continuum by the insertion of points. Each point is, in effect, a discontinuity. The gap between 0 and 1 was only a continuum before points were introduced. By introducing the points we have interrupted the continuum. This calls for some thought; since points *interrupt* the continuum we should categorically distinguish the points and the gaps (which are non-points), making the former positive and the latter negative. *Hence the number of the continuum is* − *1,* each constructed point being a plus one.

THE REAL NUMBERS

Another class of numbers as defined by mathematicians are the real numbers, defined as solutions to equations. In the case of equations involving square roots, our position is that since it requires two dimensions to construct these numbers they cannot properly be said to exist on the line segment. We find $\sqrt{2}$ as the diagonal of the unit square, $\sqrt{5}$ as the diagonal of the rectangle whose sides are 1 and $\sqrt{2}$, and so on. (In this connection Cantor's construction to show the denumerability of points on the plane is misleading. What is unique about the plane is not the construction of points upon it, but the construction of line intervals, such as $\sqrt{2}$, which cannot be constructed on the line.)

It is considered that Dedekind, by the device of the "cut," provided a method of locating an irrational on the line. But Dedekind's construction requires that we first obtain the square root of 2 on the plane. It is then transferred to the line by swinging a compass. But this operation is approximate; it does not provide a criterion by which the accuracy of the approximation can be improved upon without resorting again to the plane. Imagine two persons, one living on the line and another on the plane. The first knows all rational numbers. He is then told about $\sqrt{2}$, and is instructed to put it on the line. Where is he to put it? "Between 1.4 and 1.5," says the second who has measured the diagonal of the square whose sides are one. "Yes, but I want a more accurate value," says the first. The inhabitant of the plane measures the distance more accurately and gives him the value 1.41—which the inhabitant of the line can now place. But if he wants a still more accurate value he must again get it from the inhabitant of the plane. And so on. It is always the latter who determines the value of the $\sqrt{2}$. Thus the $\sqrt{2}$ cannot be said to be on the line between 1 and 2 because the inhabitant of the line can only learn its value from the inhabitant of the plane.

It is beside the point to say that this is a practical consideration. If we dispense with the plane and with physical measurement, we would then have to solve a quadratic equation which implies the two-dimensionality of the plane. This is a fundamental distinction in mathematics.

To conclude: we *cannot* include the irrational numbers on the line. Approximations yes, but to find their value, or even to refine the approximation, we must use the plane.

Even more simply, the "cut" (employed by Dedekind) is a line which by definition intersects the line segment; the two lines together define a plane. Hence the cut invokes the plane, since it requires another dimension perpendicular to the line.

TWO KINDS OF KNOWING

Let us see where we stand. We first stated that the intrusion of positive and negative signs into mathematics is an indication that this subject involves something in addition to, and quite different from, the world of ideal forms, something that is not form. To offset the tendency to dismiss positive and negative as merely a symmetric duality, we stressed its *asymmetry* and called attention to its connection with that aspect of existence pronounced untouchable by logic: time, handedness, value, absolute magnitude, even that anathema to reason, substance. Even one of the simplest notions involved, that of scale, provides an example of the difficulty involved. Scale is *measured* as a ratio, but its determination goes further, in that it must refer to some real object and thus cannot be defined solely by reference to ratio. For example, a map is marked "10 miles to the inch," and such a statement establishes the scale. But even if we go on to say a mile is 5,280 feet, and a foot is 12 inches, the true reference is still hidden. This is, that the inch referred to is present as a *physical object on the map*, not in the sense of representation, as are the distances, but in actuality.

We would become acutely aware of the importance of this concrete reference if it were absent, if the map were, say, televised to a recipient who had no previous knowledge of the value of the inch. This value can be transmitted only by a different kind of reference than that employed by the map. Normally, this is accomplished by a standard reference, say the Paris meter, the wavelength of a certain line in the spectrum of chromium light, or whatever—a physical referent which must be common to the experience of both parties. Thus the actual piece of paper on which the map is drawn, and which has no importance so far as the relationship structure depicted by the map is concerned, is essential in order to convey all the information necessary to know the true magnitude of all distances.

So, in order to know scale, by which we mean actual magnitude, we must have some physical object known to both parties. This is a different sort of knowledge from that which the map or drawing can depict. It is not a knowledge of form at all. The fact that such knowledge is necessary is an indication of the incompleteness not only of logic but of *any* abstract method of communication.

In order that this incompleteness may not limit mathematics, we need to draw out of mathematics such hints and promptings as can be found in its discomfort with the straight jacket prepared for it by those misguided enthusiasts who would "logicize" it.

We have seen that the positive and negative numbers are one symptom of its dependence on the physical world, but it would be desirable to show that this existence (positive and negative) can be generated *within* mathematics itself. How can negation be defined and created mathematically?

This was accomplished a long time ago when it was realized that one of the square roots of unity is minus one. Strictly speaking, *plus* one has its square roots plus one and minus one. This defines negation as an operation which, multiplied by itself, brings us back to the starting point or assertion. Just as the opposite of negation is assertion, so − 1 is *opposite* + 1. They can be represented as opposite ends of a line.

Fig. 4

Having obtained plus and minus in this way, independently of gain and loss, i.e., through pure mathematical formalism, how do we *assign* them to gain and loss? This question forces us to recognize that plus and minus are not interchangeable. Why? Because if we go one step further, and extract the square root of plus and of minus, we get different answers. The square root of plus one is minus one as before, but the square root of minus one is $\pm\sqrt{-1}$ which must be represented as a separate axis perpendicular to the axis plus and minus.

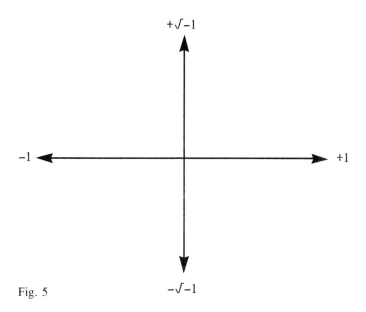

Fig. 5

In other words, + 1 and − 1 are *not* interchangeable. The square root of unity is its opposite, while the square root of its opposite is a median axis at right angles to the line joining the thing to its opposite. (This device is the celebrated domestication of imaginary numbers formulated by Gauss.) Note that we are getting several kinds of payoff:

    1. The generation of positive and negative signs from the natural or signless number.
    2. The generation of an asymmetry by a process which superficially might appear symmetrical.
    3. The generation of dimension out of no dimension. When we took the square root we generated a one-dimen-

sional line; when we took a second square root we generated a two-dimensional plane.)

These points are most important. We have stressed that what we wish to accomplish is an expansion of mathematics (not a reform) which will give it a more perfect jurisdiction in its domain. This domain is not set apart from the physical world but interpenetrates with it.

Normally mathematics is thought of as having nothing to do with the physical world. It is confined to a closed room in a tower; like a bewitched princess—or like a queen termite completely confined in a closed chamber—it is expected to hatch out its progeny, its "n dimensional" equations, without ever being allowed to see how many dimensions the actual world has. Centuries ago this termite queen gave birth to a monster—I refer, of course, to this same square root of minus, one—which the doctors in attendance could find no use for and labeled "imaginary." Some, such as Leibniz, speculated that this creature, which was neither positive or negative, corresponded to the *idea* of the object, but this suggestion has been ignored.

We are reviving it; but we are doing more, we are endeavoring to show that this and other ways of joining the world of mathematics and the actual world should be made use of. It is not just that the imaginary sign is that aspect of an object which we can call conceptual (just as the negative is that aspect we can call its value, or the *need* for the object), but that other aspects of the physical world are *generated* out of these pale enfoldments to be discovered in mathematical embryo. The issue is cosmological—it concerns the origin of the universe.

ORIGIN OF THE UNIVERSE

The scientist measures objects in space and time, and he is able to carry out his program of explanation when he can show that organisms are made up of molecules, and that molecules are made up of atoms, which are still spatio-temporal objects.

But when he presses further, his pursuit of ultimates takes him to protons and electrons. Here, to his great consternation, he finds that the principle of spatial location has evaporated! And if dissection is carried further, into pure radiation, the concept of time becomes meaningless as well. Mass, too, has evaporated into radiation. We are no longer in the world of space-time. This is an interesting journey, but the physicist is embarrassed by the disappearance of just those measures—mass, length, and time—with which he is at home. More than one physicist has told me he could not venture into such a sphere. The problem comes up with great urgency in the question of the disappearance of matter into Black Holes, the ultimate fate of superdense neutron stars. Predicted years ago by the mathematicians, these are now recognized as a reality by the astronomers.

We can now see the great importance that attaches to the fact that the successive square roots of unity *generate* just those dimensions which the physicist has seen vanish. It was a badly shaken fox that saw the rabbit (matter) disappear into a Black Hole! This was no ordinary rabbit hole—no hole in solid earth—this was a hole in space! I cannot go there, says Mr. Fox.

But let Mr. Fox, the physicist, relax. For the hole leads into a familiar place, the no-thing land of pure mathematics. This, in fact, is where we should *start*. Rather than deny the correspondence of mathematical entities to the aspects of real objects, we should examine the generation of the spatio-temporal world in the light of this correspondence—a world not just of positive and negative *signs,* but of positive and negative *objects,* protons and electrons, males and females, rabbits and foxes that chase them, a place where sheep can get lost. A land of things that has sprouted from the land of no-things.

This land of no-things *is* the queen's domain, and she is queen for a good reason. It is from her that the physical world descends. We can no more understand the world of objects without mathematics than we can understand communication by studying a typewriter.

For it is when man purposes to write that he makes a

typewriter. In fact, before he can even do that, he must make an alphabet. So, too, life, the supreme ineffable, articulates itself through DNA. But DNA is still a vehicle, a means, and if we are to understand life, we must recognize that the initial unity is purpose, which by dividing itself creates the things whose very incompleteness induces ceaseless activity in pursuit of completeness, striving to regain wholeness and unity.

Unity: that was where it all started and that is where it will all end. But if this is so, and if we admit that by this division things came to be, how is it that we have lost not only the original unity but even all consciousness of it, so that we no longer even believe in an integrity or purpose threading together the universe of things?

This too can be answered in terms of our thesis, the sickness of the queen, and it takes us to an even simpler critique, so elementary as to have totally escaped the philosophers of higher mathematics.

COUNTING AND MEASURING

We have shown that directions, or dimensions, can be generated by extracting roots—first the one-dimensional line, then the plane. Dr. Charles Musès has proposed still further dimensionalities, created by quartic, sextic, and octic equations, which in the present context are equivalent to roots.

Alternatively, we can say that a first application of the square root generates positive and negative numbers; a second, the imaginary numbers—or perhaps complex numbers—it being desirable to keep the real and imaginary axes as complementary, and hence together.

But there is another distinction which we have not yet considered, that between the natural numbers and fractions. Doubts as to the legitimacy of fractions as numbers are not likely to be raised today, it being assumed that it is only a matter of labels. One-half is as legitimate as two, in this view. Kronecker, of course, objected, saying, "God made the whole numbers, everything else was made by man." Our position is similar. Our present concern is not with the legitimacy of kinds of numbers, however, but to point out that the distinction between the natural numbers and the fractions, like that between signless numbers and numbers that are positive and negative, conceals an aspect of mathematics that deserves close attention, one that has implications of the profoundest sort. This is that the whole numbers are used to *count* and fractions to *measure*.

I believe that the fact that this distinction has been forgotten has had serious consequences—it has led to misunderstandings about the interpretation of science. The failure to distinguish measure from counting, added to the great prestige of science, in which measure is equally or more important than counting ("anybody can count"), has led to greater prominence being given to the properties of things than to things in themselves.

This leads to a subtle undermining of our sense of the integrity of persons, and to the reductionism which has come to be thought of as scientific, but is just an opiate for the rational ego, rendering it incapable of open challenge.

The child is taught that ½ + ½ = 1. This is true for measure, but if we cut his toy engine or his pet turtle in two pieces and gave them back to him, he would be most unhappy. In this case two halves do not make a whole. In the pressure to learn arithmetic, this truth is thrust aside. Later the child becomes a scientist, and may permanently blind himself to wholeness in general, replacing this vision with an interest in measure. Under this regimen, an amoeba is a bunch of molecules, as is a man.

It is frequently stated that science has reduced all things to mass, length, and time; one philosopher has even praised science for its modesty in not trying to go further. I speak feelingly on this topic because in my own researches I have expended a great deal of effort trying to find a fourth "measure." Harlow Shapley has commented* on the need for an entity in addition to space, time, matter, and energy (overlooking the fact that energy can be reduced to the other three). For this additional entity he suggests, "Drive, Direction, Original Breath of Life (administered by the Almighty) or cosmic evolution." Coming from as strict a custodian of science as Dr. Shapley, notorious for his treatment of Velikovsky, this is

---

*Zygon, September 1966.

startling. So I am not alone in my quest for an additional factor.

This factor, like Aristotle's final cause, has to be purposive; and like Shapley's Drive, it has to be progressive (to account for the ongoingness of evolution). My search has led me to identify it as *action*. Action, as Planck discovered, comes in wholes, or quanta, and it was this discovery that gave rise to quantum theory and caused the great revolution in physics. Now it is obvious to common sense that action comes in wholes; one cannot, as Veblen used to say, "lean out a window one-and-a-half times." We can drink half a glass of water, but that's one drink. If absolute clarity is demanded we can put the stress on *decision to act* and a partially successful outcome need not obscure the fact that *an act* was initiated.

It is necessary to go into this detail because the significance of Planck's discovery, while it has affected science more profoundly perhaps than has any other discovery in the last 300 years, has not percolated through to the philosophy of science. The indeterminancy of quantum theory, which gives to each individual particle and subsystem a dimension of freedom that invests it with a generative function (final cause) is still shoved into the closet like a family skeleton. Why?

I think the reason can be traced to the fact that we have forgotten the importance of wholes. We have "learned" to grovel before measurement, to think of the doubly arbitrary result of measuring as sacred (doubly arbitrary because both the magnitude and nature of the unit are arbitrary). This acceptance of arbitrariness overflows and infects our reason, we begin to *try* not to see wholes, and take refuge in reductionism.

The blame for this failure lies with man himself—or, rather, with that faculty he calls reason, which as *rational* deals in ratios, in the relationship of one thing to another, and hence not in wholes. This is sanctioned by mathematics, in the sense that whole numbers are regarded as "merely" a subset of rational numbers, and I believe that mathematics itself suffers from this to a greater degree even than in its failure to correctly understand positive and negative.

To demonstrate this, I must once more go outside the limits

THE QUEEN AND MR. RUSSELL 75

that mathematics erroneously places on itself. I refer again to the notion that mathematics touches on only one aspect of reality, which deals in relationships. We have already shown that this is not true in regard to the intrusion into mathematics of certain real-world or substance factors, characterized by asymmetry, which crop up in mathematics as positive and negative signs.

Now we want to show it is not true in regard to number theory itself—for this includes an aspect of existence, mathematical existence we can say, that logic or reason cannot possibly deal with and still retain even a semblance of itself. This is important. While it accounts for the failure to logicize mathematics, it also brings into view the larger scope of mathematical truth, and the possibility of the formal recognition of just those aspects of living and even of science that logic, like the medieval Inquisition, is currently trying to stamp out.

Mathematics deals with numbers, and I would like to stress especially that it deals with whole numbers. When one goes back to the speculations of Fermat, Gauss, Euler and other early giants, one has a sense of a peculiar sort of mathematical insight, or intuition, that is altogether different from reason. It must have been this that gave birth to such truths as the assertion of Fermat that every integer is the sum of four integer squares.

As an example of mathematical endeavor where rationality fails, let me cite the study of prime numbers. If we note the occurrence of primes in the sequence of natural numbers—*1, 2, 3,* 4, *5,* 6, *7,* 8, 9, 10, *11,* 12, *13,* 14, 15, 16, *17,* 18, *19*—and make a table of all even numbers, all numbers divisible by three, etc., continuing in this way with all primes

	2	4	6	8	10	
	3			9	*12*	
	5					
			7			14

we see that a prime, say 5, is succeeded by an infinite succession of multiples of itself—10, 15, 20 and so on—and that members of this family are *rational* to one another, i.e., $^{10}/_{15} = ^{2}/_{3}$. But the first number is not, it is the first introduction of the peculiar quality of fiveness. Thus "prime" numbers have a certain priority to non-primes because they introduce what never has occurred before in the sequence of numbers.

Think of it—the universe counting away and every now and then producing a new prime, a number not anticipated by anything before it, something absolutely new and novel. This is the direct antithesis of the rational machine of the determinists, whose cyclings theoretically repeat the same thing over and over.

Here at the highest level of abstraction, we have the occurrence of creative novelty; from the highest authority, mathematics, examples of first cause. Where the numbers themselves, the barest bones of nature, proclaim divine novelty, there is no excuse for reducing the universe to a mere population of senseless puppets.

THE DEFINITION OF NUMBER

The reader may wonder how such misunderstanding within mathematics of its own content is possible, and may prefer to believe me mistaken rather than to suppose there to be such confusion in high places.

Our confidence in the beauty, profundity, and wonder of mathematics is right and proper, but there is a difference between the magic of its natural estate and the self-conscious explanation of its own processes. All living things possess their own natural magic; the flower unfolds its blossoms, the embryo builds the complex organism of a living creature, even the digestion of a meal requires a complexity of chemical process that far outstrips even the present knowledge of chemistry. The centipede in the fable, asked which foot it moved first, was not only unable to answer, but was rendered incapable of further movement. Man himself is a masterpiece of chemical, hydraulic, electrical, and mechanical engineering, yet when he attempts to describe himself can only create a scarecrow that hardly suffices to deceive the birds.

THE QUEEN AND MR. RUSSELL 77

So mathematics, like an electric light bulb, gives forth its radiance when connected with a battery such as Euler or Fermat, and can throw light on everything about it, but cannot illuminate itself. In fact, the attempt to define mathematics did not really begin until the efforts of Frege, Peano, and Russell at the beginning of this century. Hence, what we are calling errors in mathematics are more correctly first steps of a relatively young and untried science, the self-consciousness of mathematics.

If we attend to the proposals of these innovators, we need have no difficulty in discovering where the trouble arose. Their erroneous attempts should not be charged to mathematics but to the limitations of reason—to the assumption that logic has a right to say what mathematics consists of. We encounter this difficulty in the definition of number.

What is number? This question was the beginning of mathematical self-consciousness. Peano proposed to reduce number to two undefined terms—"one" and "successor." Thus "two" could be defined as the successor of one, "three" as the successor of two, and so on. Bertrand Russell endorsed this definition, and is probably mainly responsible for putting it before the public.

Now I admire Russell; he was witty, forthright, and energetic, and is to be credited with inventing mathematical philosophy. He has therefore made it possible for me to write on this subject. Nevertheless, I think many of his ideas are erroneous. This does not make them less interesting—they have a diabolic flavor that is in contrast to the former vacuum.

So let us stop to consider this notion that number is to be defined as based on one and its successor. "What's one and one and one and one and one?" the White Queen asked Alice, and Alice, who had lost count, couldn't say. Ever since I learned of Peano's definition, it has struck me as missing the point. It transfers the difficulty to how many times the successor is added, and so does not define number at all. To really grapple with the problem we must recognize that number involves the coexistence of multiplicity and unity. Thus 27 is both 27 units and one unit of 27 parts. The number 27 would be more correctly defined as the *division* of a whole into 27 parts. This method does not offer the advantage of reducing number to two undefined items, as does Peano's, but I don't think number can be "reduced." *It* is already the simplest way of dealing with enumeration. The question is whether the reductive definition of Peano catches the important essence with which number deals.

Another answer is to dismiss the whole question of the "definition" of number as a pointless exercise. Certainly nothing has come of it, and most mathematicians don't pay any attention to the logicians who are attempting to take over. They continue with mathematics as if Russell and Peano had never existed.

But, however appropriate it might be, I cannot adopt this indifference. I consider the kind of error that has led to this simplistic approach important, because it has become widespread and affects thinking in other areas. Let me make one more attempt to catch the error. "Number," the definition says in effect, "consists of successive additions of unity to itself." We have the right to ask, "If so, what is the difference between one number and another, since both are achieved by this process?" "Well, m is less than n if, in the course of adding successors, m is reached before n." But that answer doesn't mean anything different from counting, which is what we are supposed to be defining. In other words, this definition of number does not define counting—it only replaces it with an equivalent process.

Since the method we propose, of defining by division, would require that every new number, or at least every prime, be defined over again, we might ask whether it is possible that number *cannot* be defined (in any manner other than by learning to count)?

I believe this gets to the heart of the question. In fact, the subject now starts to become interesting. It has always been the presumption of reason or intellect that it carries a special authority that permits it to deal with any and all particulars without individual consideration: "All particulars may be generalized under some law. This particular is one of them. Off with its head."

All well and good. But it would appear that this principle, which goes unchallenged in ordinary affairs, including science and government, has come up against its nemesis in trying to define number so as to logicize mathematics. For the difference between 12 and 13, or between any two numbers, which is the essence and reason for number, is *not* caught by the device of one and its successors.

LOGICAL TYPES

In other words, every number is *unique,* eleven is different from ten, and ten is different from nine. Number is a particular quality possessed by a set of objects, and this quality is not definable in terms other than itself. Eleven may be the successor of ten, but that does not describe its unique essence, which is not anticipated by the notion of ten plus a successor. Again because there is only one (immediate) successor to ten, it is unique, and cannot be a logical entity like a proposition. (By its own rules, all the propositions of logic are generalities; i.e., all squares have four sides.) Thus the *class* of elephants is a generality. It has as members all particular elephants, whether they be in jungles or in zoos. But the elephantine property, which is of a yet higher order than the "class" it creates, is *particular.* It is true that the class of elephants is a subclass of the class of mammals, while mammals are a subclass of the vertebrates, etc., but such increasingly general classification takes us in a different direction. *It* seizes on and deals with only a property shared by elephants with other mammals, warm-bloodedness for example, but neglects the elephantine property that all elephants share.

We are thus led to the problem of logical types, which was left unfinished by Russell. Here too Russell made a first step, for it was he who noted with his characteristic flair for epigrams that, "The class of elephants is not an elephant."

This statement established that the class of elephants was of a different logical type from a particular elephant, and introduced the possibility of a hierarchy of logical types, beginning with actual elephants (for example) and ascending to the class elephant (a conceptual or logical entity) and on. But the "and so on" was not completed by Russell, nor have I been able to discover any further elucidation of the idea in the literature. The general impression seems to be that there are an infinity, or at least an indefinite number, of higher classifications—i.e., mammals, vertebrates, animals, life forms, objects, etc.—and that these higher classifications constitute successively higher logical types.

But these classifications are still classifications. They are all conceptual entities, not a different logical type. To get to a different type we have to go outside of logic, get away from the conceptual. Taking a clue from number we can form some notion of this extra-logical existence. It is as in the case of number, a particular *quality*. With number it is the quality common to all sets which have that number of members. Thus all pairs share the property of twoness, and while pairs are general, twoness is a particular quality. This is the next logical type after classes. The collection of all such qualities is the set of all numbers 1, 2, 3, 4, and so on to infinity. There is *only one such set* (of all numbers) and this means it is particular. By this ascent the exuberant variety of natural objects leads to unity. A medieval philosopher would have seen it as evidence for the existence of God. In any case it is an ascent, from the alleged "objects" we encounter (ultimately the probability fogs of the physicist) to the defined concepts with which we label them, thence to the qualities (of which numbers are an example), thence to the ordering of all qualities (of which the set of all natural numbers is an example).

This, in fact, is precisely Bertrand Russell's own definition of number. "Number is the class (3) of all classes (2) of classes (1) of things." The first class "things" (1) being the sets into which we group things, i.e., into pairs, or sets of three; the second class (2) being the quality specific to each set, "two" for pairs, "three" for triples; and the third class (3) being the set of all numbers.

But why did Russell not see that in this definition of number he had a clue to the resolution of the problem of logical types? I don't know. When I had the opportunity to ask him this question he was noncommittal and went on to say that he had transferred his interest to women. I suspect that having realized the futility of logicizing mathematics, he was seeking a more direct route to the source of intuition.

Seriously, however, there is something to be said on this score for a psychoanalysis of mathematics. If the overemphasis on intellect at the expense of sensation, emotion, and intuition can cause an imbalance in individuals, the same overemphasis can distort an area of knowledge or a discipline. True discipline is not attained by shutting out one function at the expense of others.

Of course, it is generally supposed that empirical science specializes in physical fact, poetry and the arts in emotion and intuition, and logic in intellectual exercise. But pushed too far, and fortified by self-righteous policy, each of these delimitations becomes absurd. The discovery of fact requires intuition, and its interpretation depends on intellect. The emotions, which are the content of art, require sensation for their transmission and intellect for their understanding. The propositions of logic are dependent not only on an initial hypothesis whose validity rests on belief and hence is an expression of emotion, the implication of one proposition by another depends ultimately on intuition. The conclusion of a logical argument must be capable of verification by fact for the initial hypothesis to be valid.

Thus, whatever the subject, there is an interdependence of one function on another, and if mathematics is to preside as queen of the sciences, she must not become the victim of one department. Mathematics cannot be made the creature of formalized logic. I would even go a step further, and stress that intuition, being the seed principle from which the other functions derive, should have the primary role. Formal logic, along with the other functions, should be intuition's agent. Mathematics might thus sanction the art of good judgment, all but banished from current affairs.

Rotation, the Neglected Invariant

The theory of relativity is both a scientific theory and a philosophical credo that has caught the public fancy. Everyone has heard of measuring rods that shrink, the impossibility of observers at different places synchronizing their watches, and the anecdote about the kitten transported at the speed of light without aging. The influence of relativity on the thought of our time, surpassing even that of Darwin's theory of the descent of man from apelike progenitors, has been widespread. It has been most influential in the social sciences, where the concept that values are relative unfortunately enters the public mind as a scientific truth.

There are two errors here. In the first place, the term relativity is a misnomer. The theory is a method for deriving from measurements, themselves relative, a description in terms of invariants, which are not relative. As Stanley L. Jaki, a writer on the history of science, says: "His [Einstein's] two theories were in a sense mislabeled with the word *relative,* because both the special and general theories of relativity were more absolutist in character and content than any other scientific theory."*

The Road of Science and the Ways to God (Chicago: University of Chicago Press, 1978), p. 188.

ROTATION, THE NEGLECTED INVARIANT 83

In the second place, the important invariant, gravitation, is attributed to the curvature of space-time by the general theory of relativity. This has been hailed as a great contribution, because it dispenses with the "elusive notion of force." Force has always been an embarrassment to science, as is evident from the current vogue for explaining a force as a shower of "virtual" (unobservable) photons. To call a photon a particle is absurd, as it has no rest mass, charge, or position, and can be observed only once. Similarly, gravity is attributed to "gravitons," which have not been observed at all.

But this way of dealing with force is not the one used by relativity. I mention it because it indicates the predicament in which science is placed in dealing with force. Relativity deals with this problem by attributing forces to the curvature of space-time "lines," thus making it possible to say, for example, that the moon orbits the earth because the earth's gravity bends space-time. Newton would have said that the moon moves as it does because the centrifugal force just balances the gravitational force.

This brings us to the second objection to the term relativity. The curvature of space-time is equivalent to the acceleration of one body in the presence of another. Acceleration is the rate of change, or derivative, of velocity, which in turn is the rate of change, or derivative, of position. In other words, position, which can be directly observed, and which is relative (what is to your left may be to my right), and velocity, which is also relative (the passengers in a moving airplane are at rest with respect to one another), by derivation yield acceleration. This is the invariant that relativity sets out to obtain.

As shown in *The Geometry of Meaning** and in other essays in this volume, velocity and position are objective; acceleration is not. It is what I call *projective,* internally felt (we *feel* force). Thus the misconception arising from relativity is that invariants are objective, whereas in fact they are projective. Or, if it be insisted that invariance is equivalent to objectivity, then it follows that the sense data which the scientist directly

---

*See Arthur M. Young, *The Geometry of Meaning* (New York: Delacorte Press, 1976).

observes, not being invariant, are not objective. In either case, the implications are contrary to the usual assumption that science is dealing with what is objective.

The point may be evident even without my argument in *The Geometry of Meaning* and *The Reflexive Universe*, in that as J. B. Watson, the father of behaviorism, pointed out, the only innate instinct of the newborn human is the fear of falling, a sensitivity to acceleration. Position, which depends on vision, has to be learned, and depends on the coordination of eye movement (parallax) with the experience of depth. The fear of falling is a value judgment and, since falling is synonymous with acceleration, which is an invariant, value judgment is an invariant and the interpretation of relativity as a warrant to say that values are relative is a misconception—in fact, an inversion.

Unfortunate as this misinterpretation of relativity has been—and it has had a profound influence on our culture—there is another, even more serious charge which can be laid at the door of relativity. This is not so much a fault of interpretation as an oversight: its failure to perceive that rotation is an invariant.

I am not alone in maintaining the invariance of rotation. Percy Bridgman discusses the subject at some length.* Bridgman pictures two worlds rotating in empty space, surrounded by impenetrable clouds and each provided with a Foucault pendulum. (Foucault was the first person to use a pendulum to demonstrate the rotation of the earth.) Unlike the case with velocity, in which observers have an equal right to consider themselves at rest, the scientists on the rotating worlds can measure their absolute rotation by reference to the Foucault pendulum. "This demands that we give up our physical hypothesis of the possibility of isolating a system," Bridgman observes.

He goes on to consider an extreme case. Suppose we were to consider the proton at rest and the universe rotating. This would require that the distant stars move at 10^{40} times the

---

*The Logic of Modern Physics (New York: Macmillan Co., 1927).

speed of light. More simply put, if in getting up from your chair and turning around through a complete rotation, you were to insist that the universe and not you had rotated, your interpretation would require that the nearest stars make a circuit of the heavens at a velocity of forty light years a second, which exceeds the speed of light by a factor of 400,000,000. More distant stars would have to move even faster. Since the theory of relativity requires that the velocity of matter not exceed the speed of light, it is evident that relativity itself makes rotation absolute.

What if there were no stars? In that case, the question of absolute rotation would have no meaning, for, according to Mach, there would be no centrifugal force, and the Foucault pendulum would not keep to a fixed plane. However, let us consider systems composed of a limited number of bodies, where we want to know which are rotating.

Let us assume that we do not measure centrifugal force or use Foucault pendulums, and that there are only two bodies revolving around one another. How would we know their rates of rotation? In this case, the difference between rotation and revolution cannot be established. Each would have a right to consider the other revolving around it, much as primitive people considered the vault of the heavens to be revolving once in twenty-four hours.

Now suppose that there are three bodies. While they might still be revolving around one another, it is clear that the further away one of the bodies is from the other two, the more right we have to consider it at rest (because the further away it is, the greater velocity it will have to have). Thus, if we were in a system consisting of the earth, sun, and fixed stars, the earth's rotation with respect to the fixed stars would override its rotation with respect to the sun in the determination of its absolute rotation. An observer on the moon, which always keeps the same face to the earth, seeing the distant stars rise and set, would favor the stars (over the earth) as the point of reference—not because there are more stars, but because the stars are further away by a large factor.

Hence we come up with a rather surprising principle—the

"weight" of a body (in the metaphorical sense of the word) as a factor in determining the rate of rotation *increases* with the distance of the body. This is surprising, because physical influences such as gravity *decrease* with the distance, in fact as the square of the distance. This is important because it illustrates how the remote universe has an influence on what we would normally think of as local phenomena. The invariance of rotation is thus not only measured by reference to the most remote bodies, it is determined by these bodies.

Let us now examine rotation itself. What is its significance? Recall that rotation is inextricably linked with angular momentum. As angular momentum, it is also Planck's constant, the quantum of action, and thus the most fundamental of all measures. The quantum of action is the universal constituent of all interactions of particles. When an atom radiates or absorbs light what actually occurs is that a quantum of action is given off or incorporated into the atom by a shift of the electron to a lower or higher orbit. If the energy of the photon is sufficient the electron may escape from the atom (photoelectric effect). Similarly with a molecule, the chlorophyll molecule can store photons in the form of energy available for plant growth.

Thus the photon is the universal conveyer of energy, and accounts for "action at a distance." How may this be explained? It may best be understood if we think of the quantum of action under its other alias—*angular momentum*. Momentum is the product of mass times velocity, like the momentum of a car hitting a telegraph pole. Angular momentum is momentum in rotation. We may think of it as two weights tied together and spinning rapidly. The spinning weights can store energy, any amount of energy, by being pulled together but not changing the angular momentum, and the energy stored in this spinning can travel from one point to another. The amount of energy so carried depends on the rate of spin rather than, as with ordinary momentum, on its velocity. With actual weights there is a limit to how close together the weights can be, but with the photon there is no such limit, and the diameter of the circle has no lower limit. The size of the circle is proportional to the wavelength of the photon, and is approximately 10^{-13} or one ten-trillionth of a centimeter for a photon which can create a proton. Such a photon "spins" 10^{22} times per second. The size of a photon having the energy of a molecule at room temperature is approximately one thousandth of a centimeter.

To portray our uncertainty about the photon, we may think of the wire holding the spinning weights as suddenly cut; spinning at this prodigious rate, we would have no idea which way the weights would go (unless our consciousness could discriminate time intervals shorter than the period of rotation).

So this rotation of the quantum of action is important and needs to be emphasized in the present context, for since the quantum of action is an invariant and involves rotation, we see that not only is rotation invariant by the criterion of relativity, it is invariant in that it is an aspect of Planck's constant.

But does this rotational aspect have a more general significance? I believe it does. One way to show this is to draw on what we have elsewhere indicated, that the quantum of action is decision or *first cause*. This is a major thesis of *The Reflexive Universe,* where it is shown that the evolutionary process that produces increasingly complex levels of organization, namely atoms, molecules, cells, plants, animals, and man, has its origin in the quantum of action or photon.

The photon initiates the process (the fall into matter) by creating particles (pair production) which in turn become organized into atoms. Atoms combine into molecules, molecules are organized into cells, cells into plants and animals. The process is depicted as a V-shaped arc:

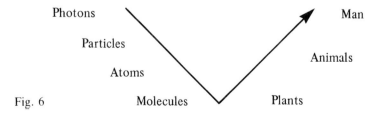

Fig. 6

The quantum of action is the dynamic which initiates this process and pushes it through these stages of increasing complexity. In particles and atoms it manifests as uncertainty (in position and energy respectively), in molecules as the making and breaking of bonds, in plants as growth (cell division), in animals as voluntary movement (choice), and in man as free will (option to elect remote goals).

Since the quantum of action has this central role in evolution, it is essential that it be an invariant. To meet the requirement of storing energy (negative entropy), the quantum of action must have the ability to move against the flow of entropy, that is, to act rather than only to be acted upon. To meet the requirement of choice of goals (as in animals), the quantum of action must be able to act in a chosen direction. To meet the requirement of free will, this choice of goals must include long-term goals—that is, goals envisioned rather than already present in sense experience.

All these requirements involve the rotational aspect of the quantum of action, which may be explained as follows: Given a consciousness which can discriminate time intervals within the period of the rotation (timing), control is the ability to time the action, or in the absence of this ability, to learn by trial and error when to time the action (the learning cycle involves this same "rotation"). Thus the freedom to choose one direction from the totality of possible directions, a circle of 360°, is provided by the rotation of the quantum of action. On the other hand, to an observer this circle is the observer's uncertainty as to what will occur, just as the roulette player is uncertain as to which way the ball will roll.

In sum, the basis for consciousness and free will must not only be invariant, it must contain the extra dimension of rotation, which in "Constraint and Freedom" (see below) I call zero-dimensionality. This is precisely what the quantum of action provides, and I see no impediment to assigning consciousness to the quantum of action. We can thus provide the bridge between consciousness and physical matter; in fact, since the quantum of action is responsible for all changes in physical matter, to equate consciousness with the quantum of

ROTATION, THE NEGLECTED INVARIANT 89

action, or more correctly to one phase of it (action can be unconscious as well as conscious), puts upon consciousness the sole requirement that it be able to discriminate time intervals within a pertinent period of rotation.

It might be objected that this explanation makes consciousness a physical phenomenon. But does it? The photon is not objective because it cannot be "observed"; its detection is its annihilation. Nor is it physical in any meaningful sense, since it has no charge, no rest mass, no position, and time does not exist for it.

I think we are now ready for a rather mind-boggling statement that Sir Arthur Eddington makes on page 137 of his *The Mathematical Theory of Relativity* (published in 1923 before Heisenberg announced the Uncertainty Principle). After establishing that action is invariant, Eddington says:

> From its first introduction, action has always been looked upon as something whose sole raison d'être is to be varied and, moreover, *varied in such a way as to defy the laws of nature!*

Which brings us to an interesting enigma. We have shown that rotation is an invariant, and we have shown that the rotation implicit in the quantum of action can be correlated with consciousness and choice, or free will. But how can choice be an invariant? We have arrived at a paradox. We started by defining an invariant as something to which all observers would give the same measure or value. How can choice be an invariant when I am free to change it? Surely an invariant is something that does not change? An invariant that can be changed is self-contradictory, and therefore meaningless.

This is where the rational mind, being confined to hindsight, cannot be trusted. For one cannot give a reason for first cause, or isolate oneself from first cause. First cause is what it says it is. It initiates. It decides on the goal, and in so doing sets up the criteria by which subsequent supportive decisions will be made. If we change the primary decision, it becomes a whole new game—nothing is as it was. In other words, choice, *once made,* is an invariant, and remains so as long as it is not cancelled by another choice. It is an invariant not because observers agree, but because it directs subsequent action. Choice has thus an even higher status than an invariant in that it creates the criteria for invariance.

We can illustrate this point by reference to the instinctive fear of falling to which we alluded earlier in this essay. It was pointed out that this awareness of acceleration has for the infant the status of an invariant. But as the child gets older, he may learn to enjoy the same feeling on a roller coaster. Here the choice to experience a thrill reverses the situation and puts a positive value on falling. This is a special, anthropomorphic case which may have no bearing on physical science; the child's choice does not reverse gravity. It does, however, illustrate the priority of choice, which can turn almost any reality condition—no matter how overtly unfavorable—into an asset.

A Formalism for Philosophy

I.

COSMOGONY VERSUS ONTOLOGY

Cosmogony (the body of theories dealing with the origin of the universe) must be based on the procedures of experimental physics, and must be stated in terms of objective science. We measure objects, observe interrelationships and arrive at fundamental concepts regarding the "ingredients" which account for the universe (i.e., atoms, molecules, motion, mass, acceleration, etc.). These basic ingredients and the operations that lead to them constitute an intellectually satisfying scientific description. It becomes even more satisfying and takes on a philosophical appeal when further analysis shows that the number of ingredients may be further reduced, or accounted for in terms of one another, as has occurred many times in the history of science.

But there is an important sense in which scientific reduction does not meet the requirements of philosophy. We must realize that, however basic these scientific ingredients are, they are known only indirectly; nor are they, despite being obtained in a very special way, free of the charge of anthropomorphic contamination. What, for example, is length? How do we *know* it? The indirect nature of our knowledge of such physical constants disqualifies them as philosophic realities, not only because they are remote from direct experience, but because their exclusively objective nature leaves the question, "What are they objective to?" entirely unanswered. And if it be insisted that the philosophical subjectivism which condones such a question is outmoded, we need only refer to the Heisenberg Uncertainty Principle, which reveals that the act of observation cannot, even in theory, be performed without disturbing the small system it observes, and thus rendering false the data it obtains; purely objective information about individuals is theoretically impossible, and the observer is a necessary part of the universe.

It might be proposed that our final knowledge can only be statistical. We should not expect to predict precisely; the electron is not a thing but a probability fog, etc. Such admonitions strike me as missing the point. The fact that the insurance company cannot predict who will die on a certain day does not suspend the law of gravity, which will continue to operate with great precision if I predict an eclipse or jump out of the window. There is, in short, a proper sphere for exact science; there is a sphere for statistical science; and there may even be a sphere for no science at all. It is the job of philosophy to delineate those spheres and assign their responsibilities. The scientist, like the motorcycle cop chasing the Dusenberg across the state line back in 1930, may be happy to have the issue taken up by another department.

Philosophers, then, deal with ontology: "The science of being, the doctrine of the universal and necessary characteristics of existence" (*Funk & Wagnalls' Dictionary*). Here we are plunged into difficulties almost worse than before. The long history of philosophy is replete with endless dissertations, distinctions beyond reach of empirical verification, into which one cannot enter without recreating a context that has long ago been obliterated. We also miss the clearly defined categorical distinctions of science, and are deprived of empirical findings of fact. We particularly miss science's formal and operational procedure, so much less subject to semantic and subjective misinterpretation. How can we cut through this confusion of terms and interpretations, which so encumbers metaphysics that it has become common parlance to use the term "metaphysical" as a derogatory epithet? The true purpose of metaphysics is to clear the air and guide us to what is and is not pertinent, and it is to be regretted that modern philosophy has abandoned this role.

CAN PHILOSOPHY PROFIT FROM THE "SCIENTIFIC METHOD"?

Why the confusion? Why has philosophy not "progressed" as has science? Is it because science has abandoned difficult or unanswerable questions, leaving them to philosophy? Is it because, as is charged by the logical positivists, the questions of philosophy are meaningless? Is it because as soon as a technique is found for answering a question it is removed from the philosophical to the scientific domain? Is it because, as my college professor said, it is the business of philosophy to ask questions, not to answer them?

Let me take time to consider these points rather briefly. That science has answered only the easy questions is not a fair statement. It is only now, with the benefit of hindsight and the genius of Kepler and Newton, that we can dismiss the rather simple laws of planetary motion as easy. They were not easily come by, even though, now that the job is done, we look back and say, "How simple!" Furthermore, the progress of science has brought it into increasingly difficult areas, and the questions science has encountered and answered in recent years, such as those of quantum physics and of the nature of the chemicals that comprise the cell nucleus, are supremely difficult. (It would now appear that the questions of nuclear physics are even more so.)

As to whether the questions of philosophy are meaningless, as it has been the fashion of late to maintain, it seems that this, too, is a matter of hindsight. The meaning of a question emerges after we have found out how to answer it. Sometimes we have to do a lot of work to prove that a question is meaningless, as was found in trying to measure the ether drift.

And the charge of "meaningless" itself has become a stereotype, an epithet which discourages the very curiosity that has brought science into being.

The next point, that when a technique is found for answering a question, it is removed from the domain of philosophy and transferred to science, seems to have merit. This charge does not try to upgrade one profession at the expense of the other. Bacon said, "It is the business of God to conceal a thing, and the business of man to find it out." This gives everyone a job, including the philosopher, who has the task of rooting for the truffles, while science lifts them up.

But there seems to me to be something else, some lesson that philosophy can learn from science. This can hardly be the experimental method, which, though it has plenty of advocates in fields outside hard science, and has often been applied in the hope of making a subject more "scientific," and hence respectable, is not appropriate to philosophy. I rather think that it is something on the order of grammar—a language, or vocabulary, which, whether it be like that of science or not, is necessary for philosophy so that its conclusions won't continually melt or shift as fast as they are deduced. This seems to me to be the lesson that philosophy can learn from science: *Get a language.* Or perhaps it would be better to say, find a way to formalize philosophical insights so that, having established one point, we can retain it and move on.

The problem is rather similar to that of adding columns of figures. It is easy enough if you have a method of carrying over from one column to the next. This, incidentally, was the great contribution of zero in arithmetic; it made it possible to cycle the numbers, so that when you get to ten you just use the digits one to nine over again, shifting the decimal point by adding a zero. Imagine the problem of multiplying 37×89 in Roman numerals!

Suppose we have a table before us. We observe that it presents a continuous hard surface and conclude this to be the character of tables. Then the scientist tells us that the table is not continuous or hard, that this is our impression only, that in actuality it is made up of discrete molecules which have no intrinsic hardness; that in fact it is 99.9 percent pure space.

Now what has become of the quality of continuity and hardness which we observed? To say that these are illusions does not account for their origin. To say they originate in the self begs the question, since "the self" is not defined and has no scientific status. Similarly the word "illusion," which implies the existence of a valid or non-illusory way of knowing, itself turns up at the end of the line in the scientist's description, for how else can we interpret the final ingredient of scientific description, the electron, which is not a particle but a "probability fog"?

So it would appear that science leads us up the garden path with a promise it cannot fulfill. It would be better to begin with that which is directly apprehended, the fact of consciousness. Understand that we are not disputing the operational effectiveness of science, whose investment in "electronic theory" or other constructs is productive of technical achievement; but such devices leave science forever a third party to the marriage of its components. As philosophers, we want to taste the joys of the marriage bed and experience its fruits. Or perhaps it would be better to say that we want to know life without renouncing forever our participation in it.

What is it that leads to this renunciation of participation which the scientific method requires? One suspects that it is the use of abstract concepts, that it lies in the "objectivity" of science. Recognizing the value of objectivity, but realizing that it has its place—that of removing us from the very ground of our being, which it is our purpose as philosophers or ontologists to study—how can we establish the legitimacy of our status as participants?

We have implied that participation is prior to objectification, and it is certainly true that the constructs of scientific objectivity are fabricated out of the content of experience, and continue to retain this "contamination," to the embarrassment of pure science, which, as Eddington put it, finds a footprint on the sands of time and "Lo, it is our own." We should, therefore, look for our formalization in the neglected area that is prior to scientific objectification. What is it we do before we create a scientific construct? How may we describe the immediate fact of experience?

THE PRIMORDIAL UNITY OF CONSCIOUSNESS
(The First Level)

Our primary datum is the fact of consciousness itself. If we question it and demand that it be defined, we cannot expect satisfaction. We could, of course, take our cue from geometry and list it as a primary and undefined term, much as the "point" and the "line" are undefined terms, but here we do have more positive grounds: that it is the prior condition to any consideration of entities. Even if we decree that all objects of consciousness or of sense are false or illusory, the fact of consciousness itself remains and cannot be removed. Such priority has a sanction more basic than "space," or whatever we call the matrix of relationships in terms of which the objective properties of things are described, because such a matrix is secondary to the consciousness that views or projects it. The objective, and all objects, "stand against" this projection as the word "*ob*-ject" connotes.

This, then, must be the initial and basic unity from which all else derives. We might postulate other existences, and agree to adopt them "if useful" or "if productive," but such postulations ultimately revert to a knower who postulates, and in relation to that initial knowing must stand as derived principles. Having recognized this—and it is in no sense a novel idea, for it occurs in many philosophies—let us give it formal expression, that is, incorporate it into our structure. (Recall that it was the need for such bookkeeping that we have set about to fulfill in our "grammar" of philosophy.) From the formal unity of consciousness that underlies experience, postulations of whatever sort must derive—and therefore stand to this unity in the relation of parts to a whole. This unitary level we call the first or primary level.

DIVISION OF THE FIELD OF CONSCIOUSNESS
(The Second Level)

Proceeding from this, we might divide the "field of consciousness" into two parts: what is known and what is unknown. This, it turns out, is not primitive enough, for it would require a criterion for knowing versus not knowing, which has not been established. Better would be to divide the field into events which have happened and events which have not happened. We would thus have a sequence of events which have happened trailing into the past, and events which have not happened projecting into the future. We must necessarily begin with an undefined term, in this case "event," but need make no distinction at this point between interior and exterior events, or between false and true. Definition is not involved, because we assume that happening or having happened is the only criterion. By "happening" we here imply only the impression—we would *not* distinguish between "I saw Jones yesterday," and "I think I saw Jones yesterday," but we would distinguish between "I saw Jones yesterday," and "I will see Jones again tomorrow."

This distinction honors time flow as next in priority to consciousness, and describes this time flow as a linear sequence of point events divided into those which have happened and those which have not. We add for the sake of our formalism that this time flow can only arise from what has already been assumed; it must be engendered by a *division* of consciousness, or of the field of consciousness. How then does time divide consciousness? Since we have mentioned only that past and future are distinguished, we might suppose it to be a twofold division.

But this leaves out the *present* or that which separates past and future. We cannot say either that the present is part of that which has happened or that it is part of that which has not happened. It has its own character, which renders it distinct from both. Moreover, it has the function of separating past from future. Again, in a different sense, it has some of the status of the original whole. For the present (as for example, the "present day" or the "present moment") has its parts: its morning and evening, its past and future. So we have a *threefold* division of the whole, or of consciousness. We call this the second level. This threefold division includes two new parts plus the uniting principle of the first level.

Nature of the Threefold Division

One interesting feature of this division is that *it does not provide for definition.* We are prior to even the possibility of a definition. We are not yet defining individual events; we are simply listing them in two categories: events that have happened and events that have not happened. Defining them would require our extracting from them some property—creating some abstraction not itself an event—and this is not yet possible because we have provided only for the distinction between what has and what has not happened, with the addition of something separating the two categories. That these categories, while distinct, are not *defined* in any proper sense may be seen merely by trying to define them. Suppose we define the past as all those events prior to 2:15 P.M., standard time, 28 March 1965. Such a definition is no definition, for it becomes false at 2:16 P.M. and continues false. For a definition of the past, time flow would have to cease.

But since it is the flow of time that is directly apprehended or known, any cessation of time flow is a postulation, and cannot be entertained at this level (of the threefold distinction). It will be appreciated that in this concentrating on time flow as the next-most-basic order of existence to the unitary consciousness, we are not necessarily discovering the only manifestation of a secondary (next-most-basic) order. Later we will return to this question with better equipment. For the present, we might note that, in describing God as the Holy Trinity, or God in three persons—Father, Son, and Holy Ghost—the medieval mind was dealing with the same question that is here being considered.

Alternatively we might describe this flow of time by asserting that the content of consciousness is changing. Let us examine this. We at once encounter a difficulty. Changing with respect to what? We cannot evoke a standard; this would be a construct of another order, and would be excluded on the same grounds that we used to exclude definition. We could, however, say that the past is changing with respect to the future. More directly, we could say that the past continually incorporates what was the future. This would be verbal juggling, but it leads to an important idea; viz., that there is a one-way passage of future events into the past; the past continually incorporates the future, and *this change is irreversible*. It betokens one-wayness, an asymmetry that is inherent in the motion we describe as a threefold division of the initial unity.

Reversible motion cannot emerge at this level, which we call secondary. We are not banning reversible motion; we are merely pointing out that it does not belong at this level. Here we are committed to time flow, and it will be apparent that the sequence of events in which every event is either before or after every other event—but not *both*—forbids a reversible time flow.

Positive Contribution of the Second Level

This second level is the sequence of events that impinge on any given consciousness and constitute its experience. In formal terms, it is an infinite line created by a moving point, but this statement hardly suffices to suggest its positive contribution. For here we have the *substance* of experience. Here are pain and pleasure. Here are hunger and love, hate and suffering. Here are all those *feelings* which provide content: the stimulus to action, the action, and the result of action (and this triplicity—stimulus, action, and result—is another expression of the threefold division which is appropriate at this level).

What more can be said? That experience is of this nature we know, because we experience. No use of words can transfer this content. We have already said that it cannot be defined, and hence cannot be communicated. So it may be asked, why talk about it? Why introduce it into our science?

To begin with, it may be objected that the area of inquiry is itself unscientific. This, of course, stems (ostensibly) from the impossibility of "defining" it. Presumably, to be scientific, we should construct our description in terms of relationships or structural correspondence, even though the members between which relationship is set up may be forever beyond scientific description. Thus a map is a relationship structure between points and lines on a plane, these points and lines representing cities, boundaries, coast lines, etc. The map does not purport to describe the (ultimate) nature of the cities, seashores, etc., but just where they are in relation to each other. As we shall see shortly, such relationship structure becomes possible at the third level, but here, at the second level, we have the experience of the traveler in passing through these points by moving about the country.

In order to make a map, surveyors must cut their way through the underbrush, ford the streams, and drive stakes into the earth. True, the measures which result from their efforts, set down in distances and directions, contain no trace of mud or clay and could even have been obtained by aerial survey without encountering such experiential content. Still the final emergence of survey maps brings no authorization to deny the reality that was its basis. This reality, like the feeling of hardness of the surface of the table, is subjective, projected by the knower upon the object. But it supplies the knower with the substance of his experience.

To touch on another aspect of the threefold, let us suppose we perform an experiment on an animal wherein we associate a reward (pleasure) with one object and pain with another. Eventually, the animal will learn to select that which is associated with pleasure. Why not immediately? It is apparent that the animal does not at first have an awareness of the character of the objects which the scientist matches with the pleasurable stimulus. But the animal does immediately experience the pleasure or the pain and doubts not which is which. This part does not have to be learned; it is given. In fact, it is the basis for the conditioning or association.

From here we can take a further step. This innate experience of pleasure or pain involves value, and hence leads to choice. Value, then, is involved at this second level. It would at first seem to be twofold, but we must reckon with that which chooses, which is not itself a value. It is a third factor, distinct from the values it chooses. If a correspondence of past, present, and future were set up with the three, we might correlate choice with the present, since it constitutes an action. Moreover, we cannot be content to make the two values—pleasure and pain—symmetrical. Their essential nature includes a bias of action toward one and away from the other. This echoes the asymmetry of the line, which always moves one way.

We should include *memory* as having existence here, and note that memory cannot be described except by reference to three ingredients: the experience of at least two events in sequence, and a third element which permits the comparison of the second event with the first. To illustrate: suppose an external event is experienced which is associated with pain. The second time the same event is experienced, there will occur some memory of the pain associated with the first experience. There are thus three irreducible factors: the two experiences and the remembered pain (or pleasure). This association occurs only one way (memory applies only one way); one cannot remember the future. Still further, memory cannot be negative; it may be absent, weak, or unpleasant, but it has no contrary. Memory is the incorporation of substance.

So we have a number of important contributing principles operating in this second realm: experience, value, memory. Each exhibits a threefoldness; each is asymmetric; each accompanies the experience of events in time, and involves linear sequence.

Résumé of the Threefold, or Second Level

Our formalism consists, therefore, in the division of the primary unity of consciousness (level one) into a linear sequence of events, concerning which we make the following observations:

  1. Every event has a unique position in this linear sequence. It has a successor and a predecessor, and of any two events, one is before and the other after.
  2. The events are distinguished in threefold fashion, as events which have happened, events which have not happened, and a mediating "transition event" which separates

the former two categories. (It is desirable to retain a more general expression of the threefold division than past, present, and future, which may be considered a special case. . . but we will not attempt to generalize now.)

3. This threefold division is to be regarded as a dynamic or changing relation, no definition being possible. It is characterized by a continuous incorporation of present into past, which moves one way and is, therefore, asymmetrical.

4. The formal model for this threefold division is a continuous and unending line, characterized as unbounded, one-dimensional, and asymmetric (time's arrow). Its division into two sectors (past and future) by a moving point (the present) will be referred to hereafter as a threefold mapping, or *threefold topology*.

We thus assign to the "stream of consciousness" the threefold topology of a line or sequence of points, not permitting the intrusion (at this level) of any further abstraction (such as definition or measure). There is nothing said of this line that dictates that it be a "straight line," nor do we have any criteria for straightness. Our sole rule is that each point on the line is uniquely determined and non-recurrent.

THE FOURFOLD DIVISION
(The Third Level)

The next step is to ask what occurs when the present—a sort of focus of awareness—seeks to know again one of its experiences, tries to examine some part of its past. It cannot do this by doubling back and running into itself, "looking at itself," for in so doing it continues to exist in its present, for the observation of a past incident or experience is not that past event, but a new event. Each event is a new event and stays on the line. But the *conscious* agent must have a way of equating one memory with another—a way of seeing them both together. It must create a *space* for comparison, and this space must be independent of the linear dimension of experience. The points of this space, *unlike the points of the line*, must exist in simultaneity.

Fig. 7

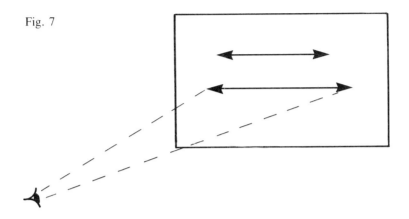

The "knower" must be able to apprehend both experiences at once, or at least to switch from A to B while retaining a memory of A. It follows that both A and B are, in this case, simulacra; they are only re-presenting the original experience. I recall that it was hot yesterday and that it was hot the day before. Which experience was hotter? To make the comparison, I have to cause their simulacra to coexist. Even if I am comparing yesterday with today, for the purpose of comparison I must objectify today's experience, that is, be apart from it, if I am to be fair to yesterday's experience.

It also follows that two dimensions are necessary. One dimension "measures" the intensity of each experience (hotness or the like); the other dimension must be available in order to put the two side by side and yet separate.

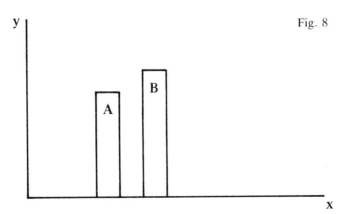

Fig. 8

For unless we can distinguish A and B in some way (A was "before" B, for instance) we could not keep them separate. The x dimension above distinguishes A and B (A was before B). The y dimension measures their magnitude (B is greater than A).

A third important feature is that neither of these dimensions (the x or the y) is the same as the time sequence in which the experiences A and B originally occurred. This time sequence was such that we could not "get off" of it. In fact we, the knower, are still "on it" when the xy plane is invoked to compare A and B. (It would be more rigorous to say that the knower, in the act of comparing A and B, is having a third experience. We are thus involved in three dimensions, one of these being a time or timelike dimension which is *necessarily* linear (only one point at a time), while the other two, which occur together, are *necessarily* planar, or two-dimensional.

We have now laid the basis for something quite important, the distinction between two "kinds of knowing": one experiential (subjective) and the other scientific (objective). The first, or prescientific kind, *involves* the direct impact of the varying content of experience; it occurs in linear fashion—one experience at a time—in an inevitable sequence; every experience is either before or after every other experience. The second involves the comparison of one experience with another and is necessarily divorced from the time flow of the first. It involves two dimensions or two axes, one being assigned to the separation of experiences and the other to their measure—and these two dimensions coexist. We may further add that these two simultaneous dimensions are *fourfold,* meaning by this that there are four distinctions involved (as compared to the three distinctions of known, knower, and not known; or past, present, and future).

These four distinctions consist of motion along each of two axes in either of two directions, i.e., four directions altogether. Each direction can be a positive meaning—but is the negative of its opposite. Thus, cold is opposite hot, but either may be considered positive. With the two-dimensional plane, reversibility is now for the first time possible. This is of great significance, for it leads to the possibility of transposing coordinates, and for the first time makes possible objectivity, or "thingness."

Another way to demonstrate the fourness of the plane is to recognize that any surface can be mapped in such a way as to require four colors to distinguish contiguous areas.

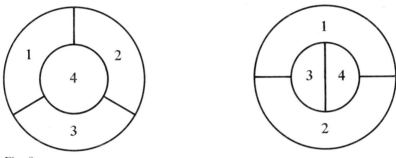

Fig. 9

Four colors are four distinctions. We say, therefore, that the surface is fourfold.

In order to show that the two-dimensional plane is the necessary and sufficient matrix for concepts, and hence for definition and measure, we may take an example that apparently involves only one dimension. Consider a unit length, say a "foot." While this is one-dimensional (and does not appear to require two dimensions for its measure),

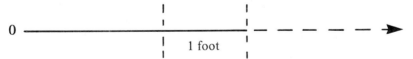

Fig. 10

we could not describe its boundaries without recourse to a motion that "crosses" the dimension of its extension. Thus, the vertical lines A and B, which cut across the unending line $O \blacklozenge \infty$, require for their existence a dimension perpendicular to the unending line. The finiteness of the unit length can only be stated in two dimensions. Furthermore, in order to use this length as a measure or *standard,* we have to be able to transport it to some other part of the infinite line. This means that it must exist at the same time that other parts of the line exist, which is impossible with only one dimension, because we have no basis for simultaneity. The points of the line being in a linear time sequence, no two can coexist. It is, therefore, clear that to move the line parallel to itself there must be the possibility of simultaneity or multiple tracks. To take a visual example, suppose we had a freight train on an endless track, and we wanted to compare the lengths of the different freight cars by selecting one car as a standard and moving it into juxtaposition with others. We would find we could not do this without another track and means for

Fig. 11

sighting perpendicular to the track's motion.

Furthermore, recalling that our original $O \rightarrow \infty$ line consists of moments in a linear or time sequence, no two of which coexist, we see that to compare even one point on the line with another, we must "freeze" the whole line, that is, view it in simultaneity. In other words, both the dimension of the line and the dimension perpendicular to the line must coexist in order to compare, measure, or define "length." So we must create a new space of two dimensions, neither of which existed before. This we call mental space, the fourfold, or third level.

Is this third level more "real" than the second? The answer depends on the definition of real. It would be better to say it is objective. The second level is not objective—but it is *tangible*.

Application to a Classic Philosophical Problem

We are now in a position to put our method to use. Consider the old question of *substance*. Bishop Berkeley perceived the illusory nature of substance, which he insisted we project upon objects. As Hocking sums up Berkeley's argument:

SECOND AND THIRD LEVELS COMPARED

2nd Level	3rd Level
One-dimensional and unbounded (infinite)*	two-dimensional and bounded (finite)
Timelike	Spacelike
Dynamic	Static
Projective and tangible	Objective and intangible
The matrix of feelings, participation, value	The matrix of concept, deinition, measure
Experience	Intellect
No definition possible	No participation possible
Has three distinctions (Past, present, future, etc.)	Has four distinctions (Up, down, right, left, etc.)
Topology (like line) is three-fold	Topology (like plane) is four-fold
Origin is a point with changing past and future	Origin can be anywhere†
Observer is unique	No unique observer
Impressions and illusions	Concepts (objective), Ratio, rational thinking

*"Infinite" may be held to be more basic than "finite" on the grounds that infinity is not defined.

†This is important and is implied in the "objectivity" of the plane. Thus we define a square in such a way that the origin can be anywhere. We make its properties invariant, that is, independent of a coordinate system and hence the same for different persons, i.e., objective.

The chemist can always determine whether the object before him is gold; but he never does so by inspecting its "substance," he reaches his conclusions solely on the basis of its properties—its solubility in different acids, its combining proportions and weights: these are all he has to work with, and they are all he needs. Is not the "substance" of gold a mere name for the fact of experience that these properties belong together?*

Berkeley used this argument to show that we impute substance to things but do not find it in the objective world, and that we may therefore omit it from our metaphysics.

*William E. Hocking, *Types of Philosophy* (1929), 3rd ed. (New York: Charles Scribner's Sons, 1959), p. 162.

We agree with the first point and not with the second. In the light of our discussion (of the experiential threefold and the mental fourfold levels), we may assert that the laws and procedures which the chemist employs to identify gold belong to the mental realm, and that the substantial nature of gold which, as Berkeley said, we project, belongs to the experiential realm. That the latter is illusory does not suffice for its dismissal. We can, in fact, say that it has a valid reality of its own, however useless this may be for objective knowledge. Without substance the experiments of the chemist could not be performed.

THE CONJUNCTION OF THREE- AND FOURFOLD
(The Fourth Level)

This level corresponds to the "physical world" which is customarily referred to as three-dimensional; thus a room has length, breadth, and height. We will take the position, for a reason that will become apparent, that these three dimensions are a combination of the one dimension of experience and the two dimensions of intellect (mind space).

One justification for this view is an appeal to practical experience: in any operation involving measure or other conceptual activity we see the world as though it were projected on a screen (the retina of the eye, for example). The objects on this screen have an apparent extent depending on their size and distance away from us. However, what is immediately given is their *angular extent,* which we measure by eye motions in two directions: up and down, right and left. To determine depth or distance of objects, we must:

1. Move toward them and detect their distance by change of size, vis-à-vis the velocity of our approach; or,
2. Estimate their distance from us by the difference in angle of each eye (parallax); or,
3. Send a signal and measure the time elapsed for it to return (a method used by bats and dolphins).

In any case the way we measure two of the dimensions (width and height) is of a different nature than the way we measure the third (depth). The measure of extent, as of the two-dimensional screen, is achieved in simultaneity. The measure of the single dimension of depth requires time (not being simultaneous); we must either move toward the objects or send a signal toward them. (The case of parallax involves comparison of two non-simultaneous determinations.) The distinction between the two kinds of measurement is illustrated by photographs: one is sufficient to show angular position, but two are necessary to show depth (stereoptograph).

Another justification for considering the "physical world" as one- plus two-dimensional is that this interpretation introduces no new postulate, no new factors; instead of having to say that three-space is a "different sort of space" from the time dimension of experience and the dimensional space of intellect (mental space), we can say that three-space is the *combination* of the time dimension of experience and the space dimension of mind.

Two objections present themselves. First, will this formalism deprive us of the freedom to express motion in terms of three dimensions of space and one of time? I do not believe so. For when physical phenomena are so formulated, the time dimension is always inextricably linked with one of the space dimensions, as is obvious in the case of a moving body. Even in the case of an observer at rest making determinations about objects in "three-space," the fact that light signals travel with a finite velocity in the depth dimension renders the concept of simultaneity, or absolute time, false in theory as well as in practice, as is recognized in relativity theory. This provides a stronger argument than we need. Our requirement is only to show that simultaneity in a radial direction from the observer is impossible to establish. But we would accept and expect simultaneity at points equidistant from the observer, and I believe relativity would concur.

The second objection is that our postulate violates the self-evident isotropism of space. (There is nothing to distinguish one dimension from another except the position of the observer; there is nothing to prevent our switching dimensions about, i.e., making the plan view an end view, etc.) This we would not deny, but would point out that in switching positions, we are switching observers. There is always an observer, in theory as well as in practice, and this necessary factor establishes the uniqueness of the depth or radial dimension. This does not mean that "a cube" is not objective. It is just as objective as the square in two dimensions. We mean that when we define the cube or, in general, any three-dimensional figure, we use a series of views, each of which is two-dimensional. In a typical instance we require three such views: plan view, end view, and side view. But these three views cannot be seen or even visualized in simultaneity. We can perceive only one view at a time and must observe some convention in their sequence. This sequence is the time dimension; it is independent of the two dimensions required for each view.

Other technical evidence for the sufficiency of the one- plus two-dimensional interpretation could be cited; for example, the diagonal of a cube is of the same form (square root of the sum of the squares of the sides) as is the diagonal of a square. Again, rotation establishes one unique dimension (the axis of rotation) and two dimensions which come into existence together (the plane established by the movement of any point not on the axis).

The Topology of Three-Space

We now come to the crux of our argument and can put into operation what we have been describing as our formalism. We have noted that the "topology" of the space of experience is threefold, while that of mental space is fourfold. So far there has been no advantage in this correlation with topology, since it only tells us what we had already invested in our postulates—that time is threefold and plane space or the flat surface is fourfold.

With three-dimensional space, something quite new and logically unexpected emerges. This is revealed in its topology. We have, of course, volume. But we have noted that we may only be aware of a volume, or "know" a volume, in terms of three (or more) two-dimensional views. We should now add that these views must be taken in a certain order or according to a certain convention; otherwise, we run the risk of con-

A FORMALISM FOR PHILOSOPHY   111

structing the mirror image of an object, e.g., a left-handed golf club instead of a right-handed one. The ordering which distinguishes a right from a left hand is employed in the description of the direction of a magnetic field caused by the motion of a charge. This is the "right-hand rule," which invokes an anthropomorphic reference to supply information that cannot be stated in a purely logical or objective way. In other words, we must use both the threefold (observer-oriented) and the fourfold (or objective) "way of knowing" to account for objects in three-dimensional space.

How can we give formal expression to this new complication? Here, our investment in topology pays off. For if we inquire, "What topology does a surface in three dimensions have?" we find there are several answers. The surface of a sphere or, in fact, the surface of any great class of three-dimensional shapes, requires but four colors to map. In the present reference or context we should perhaps say that this surface is the *interior* of a sphere (the maximum sweep of angular position from a center of observation).

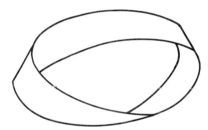

Fig. 12

But we also know that other topologies are possible, like that of the Moebius strip, which is a ribbonlike sheet that rejoins itself, but is twisted half a turn so that its lower edge joins its upper edge. There results a surface such that, starting on one side, one can reach the other by continuous transport, and if one tried to paint one side, before one had reached the end the other would be painted also. On such a surface, it is possible to make a map which requires six colors, so that a Moebius strip is topologically quite different from the class of surfaces represented by a sphere (which requires but four colors for even the most complicated map). Even more interesting, for reasons which we will develop later, is the topology of the surface of a torus, or doughnut shape, on which a map requiring seven colors may be drawn.

Recalling that the surface of a sphere is fourfold, and that the fourfold correlates with mental concepts, we may equate spherical surfaces to structure (understood here as a conceptual compound of relationships existing in simultaneity, analyzable and definable in terms of surface mappings, i.e., blueprints).

Since at the fourth level we have topologies beyond that of the sphere (and all other simply connected surfaces) there must be *something beyond structure* possible in three-space. We will return to this later after we have considered an important characteristic which distinguishes the physical world from the realms that precede it.

MULTIPLICITY OF SPACES

We have observed that our third level, which has a fourfold topology, corresponds to mental space. Up to this point, we have not said how many "mental spaces" there are, though we have noted that each analysis or "view" of an object or a situation requires the creation of a mental space or screen for its projection, and that there must, therefore, be many such spaces.* Kant discusses this point in his dissertation of plural spaces, calling attention to the fact that each picture on the wall erects its own space. And we cannot in general mix these spaces by asking how far the chair in picture 1 is from the table in picture 2. Nor can we say how far the tree in a picture is from the tree outside the window. Each picture has its own space, and none of these spaces is the space in which the pictures are hung.

A similar, if not greater, latitude must be extended to the *linear* spaces of the experience of different persons. As the

---

*See too, William E. Hocking, *The Meaning of Immortality in Human Experience* (1957), rev. ed. (Westport, Conn.: Greenwood Press, 1973), pp. 26–28. Hocking takes up this question in a most interesting fashion. I have also found the same thought in Heraclitus.

theory of relativity makes explicit, observers moving with velocities with respect to each other cannot, even in theory, adjust their clocks to read the same time. In a more direct sense, we each experience a unique time flow and burn our candles at our own rates. Each person has his own experience space. So we may say both of the second and third level (one-dimensional and two-dimensional, respectively) that their spaces are plural: there are a multiplicity of them.

THE UNIQUENESS OF THE PHYSICAL WORLD

In sharp contrast to this multiplicity is three-dimensional space. *There is but one such space.* This makes it impossible for two objects to occupy the same position in this space at the same time. But this, too, was recognized by Kant. Why, then, do we bring it up?

It is because it provides the simplest and perhaps most direct reason why there are no higher spaces, no fourth-, fifth-, or sixth-dimensional space, such as it has been the fashion to refer to in popular treatises on relativity.*

We started, at the first level, with the unity of consciousness. We divided this in threefold fashion to construct time. We divided it in fourfold fashion to construct mental space. Applying both divisions, we get a new unity! Nor is this unity of consciousness. Here consciousness is divided indefinitely. This, then, is the floor; consciousness has been spilled upon it, infinitely dispersed. And we must recognize this floor, because it is "there."

To sum up, three-dimensional space constitutes a sort of terminus to the fragmentation of the unity of consciousness. Due to its own unity, it presents a barrier to the proliferation

---

*We have already defined the position that one- plus two-dimensional space suffices to cover time and the three spatial dimensions. Time only exists in association with motion of some sort. There is always a spatial movement in any time passage, even in a clock. And it is interesting that the theory of relativity, while setting forth its dimensional requirements by measuring one time and three space displacements in the formula for interval, concludes with a formula for the volume of the universe in which no time dimension appears. Sir Arthur Eddington makes the most pertinent remarks on the subject of extra dimensions in his *New Pathways in Science* (1935), where he explains that when more than three dimensions are introduced they must, of necessity, have imaginary coefficients.

of further spaces, and enforces the reality of otherness. It is only here that the tangible *object* prevails. The freedom that permits a generality or a concept to have many instances does not exist. "A chair" becomes "this chair" or rather "this object"—for "chairness" is third level. The object here exists in its own right apart from the name it is given and the attributes projected upon it. Substance and form coexist. It is here that determinism applies. Laws here have to do with more than pattern or definition of things, as is their role on the third level. (Every square has four equal sides and four equal angles.) They *determine* specific results. Two aircraft collide because their flight paths intersect at the same time. A man fails to make a business appointment because there is a traffic jam. We thus encounter inevitability—the "laws of nature" make themselves known. Such laws have special interest not only because they enforce their recognition, but also because it now becomes possible for conscious intelligence to employ them to achieve results, to achieve its own ends.

II.

FEEDBACK: LAW AND PURPOSIVE INTELLIGENCE

Suppose I discover by experiment that heating food by fire (cooking) improves it. I can say that the fire is the cause of the cooking, and hence the improved flavor. Then I build a fire in order to cook. What is the cause of the fire? Clearly, the cooking—for it is in order to cook that I build the fire. There are thus two "cookings"—one (imagined) causes us to build the fire, and one (actual) is caused by the fire. But analytically we can only say that fire is associated with cooking. The analysis, which lacks the time dimension, can be read in either direction, and there is an important difference between them. One direction, by which law is discovered to operate, reads the "fire is the cause of the cooking." The other direction, by which the law, "fire causes cooking," is put to use, starts with an envisioned result, and *then* creates the fire. If you want to cook, you must build a fire. A complete picture starts with the

A FORMALISM FOR PHILOSOPHY

first ordering and ends with the second; we must learn the law before we can use it.

Let us now apply this to our two topologies. We have the line of experience, depicted by an arrow (which refers to the observer), and the plane, which portrays the conceptual statement of the law.

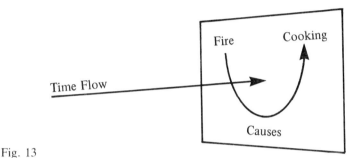

Fig. 13

Let us assume that the law reads that "fire causes cooking." Then the experience of the law will be represented by the movement of the arrow through the plane, and the order, fire causes cooking, by a circle (shown as curved arrow, counter-clockwise on this plane). How may we represent the "inversion," that is, the *use* of law? Since the use of law, as we've just seen, inverts cause and effect, we need to have the circle move clockwise. We may do this by making our arrow turn back and pierce the plane from the other side, thus reversing the order, so that cooking causes fire.

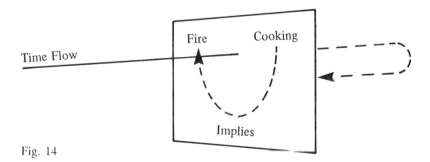

Fig. 14

> There was a man in our town
> And he was wondrous wise.
> He ran into a bramble bush
> And scratched out both his eyes,
> And when he saw his eyes were out
> He ran into another bush
> And scratched them in again!

While there are alternative ways of describing the necessary inversion, it is important only that the motion of the point (the moving observer) with respect to the plane be changed. The effect is to change from a right- to a left-hand spiral or screw. It is not the same as reversing a screw; this would simply back up or reverse the first penetration. The approach from behind makes the mapping of the plane a mirror image of what it was before. One suspects that the story of Perseus slaying Medusa while looking at her reflection in the shield of Athene describes this conquering of nature by inversion.

THE TOPOLOGY OF INVERSION

We can draw on the investment we have made in topology to deal with cases in which a structure (for example, an electric motor) can be used in two alternative ways. We can drive the motor from a battery and use it to lift a weight, or we can use the potential energy of the weight to drive the motor as a dynamo and charge the battery. The situation involves inversion, plus a free agent with the option of reversing the order. If the device were on board a pitching ship, where the weight changed, we could provide a sensing device to flip a switch so as to draw energy from the environment and charge the battery. Another example would be the self-winding wrist watch.

To formally describe such situations, we need a deductive principle, which can be supplied by our topology. In particular we want to know how many separate distinctions there are. We are interested in the three-dimensional space of the physical world (which we have analyzed as one- plus two-dimensional with a reversible relationship between the one- and the two-dimensional). What other topologies besides that of the plane surface are possible in three dimensions? As was shown earlier there are two: the Moebius strip (sixfold) and the torus (sevenfold). Of these, the latter is of special interest. For reasons which we develop more fully elsewhere, we have come to suspect that the seven distinctions required to map the surface of a torus must correlate to a sevenfold distinction in some occurrence possible in nature.

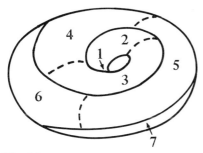

Fig. 15

What occurrences are these? We know they involve more than structure, since structure is accounted for in terms of four distinctions. We also know that they must involve time, because both the learning of cause and effect and the employment of cause and effect to produce an effect involve sequence in time. (It is this ordering of cause and effect that cannot be stated in conceptual, i.e., fourfold, terms.) So we can call such an occurrence a time structure or, more simply, a *process*. We have already given an example of a "process" in the purposive use of fire to cook food, but we need to draw out more of the inherent nature of process than this affords. We need to establish that,

1. Process consists of *stages*.
2. There are a distinct number of stages.

We are aided in the first case by the dictionary definition: "Process—a series of actions or operations definitely conducing to an end" *(Webster's New Collegiate Dictionary)*, if we can take the liberty of calling the "actions or operations" stages. For the second, we call on analytic description, which, while inadequate to process, can at least establish a minimum of four distinctions.

There are many ways of showing the four distinctions involved in operations capable of analytic description, such as cooking. One is to draw on Aristotle's classic four "causes," the *material,* the *formal,* the *efficient,* and the *final.* In the case of cooking, the material cause is the raw food, the formal cause the plan or procedure to be followed, the efficient cause the actual work of building the fire (application of the plan to the material), and the final cause the purpose of making the food palatable. Interestingly enough, the final cause is absent in the situation where something is burnt by accident, although when someone first *recognized* the improvement that comes with cooking final cause came into existence. Charles Lamb tells us that long ago, before fire was used, barns struck by lightning burned down, sometimes with the animals in them. Then someone noticed that burnt pig had an improved flavor. The idea of burning barns to get roast pig got around. Eventually others discovered simpler ways to make a "bonfire."

But, recalling that analysis cannot distinguish between the purpose to cook and the goal of the cooking (both of which are final cause), let us take another example in order to discover the anatomy of process—say the purchase of stock. Stock is bought at a certain price and later sold at a higher price. The purpose which initiates the transaction is to make a profit (the final cause). But we must also designate the realization of profit as a final cause. The price paid is the going market price established by supply and demand. We may designate this as the *formal* aspect. Also involved are the money, and the stock itself—a share in the company, in effect ownership of a piece of real property, land, machinery, or the like.

To which of the latter are we to assign the *material* and to which the *efficient* cause? Both are material in a sense in which neither the present nor the future price are (since price is not a tangible thing but a ratio between tangibles, i.e., so many dollars per share). A clue comes from a closer look at what happens to make a stock go up. Someone has to do the *work:* the company is making a new product for which there is a wide market; there is a change in management; the investor himself may intercede and bring about management reforms. Such work is the efficient cause and correlates to the stock ownership. We are left with money as the material cause; it is the substance of the entire transaction. So we have

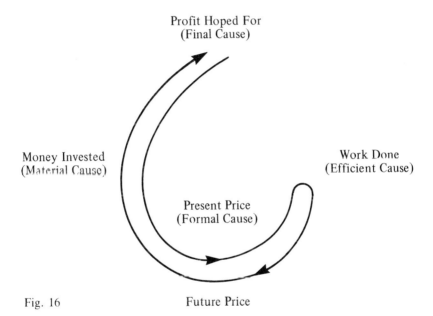

Fig. 16 Future Price

This representation deals with the four factors, or aspects, involved, and hence can be called the analytic view. It is the view that would be projected on the two-dimensional screen of mind. But we may now expand this example with three dimensions, and see that as process seven stages are involved.

Note that in the complete process the seven factors are *independent.*

The anticipated profit involves the investment of money in

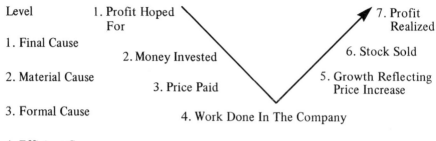

Fig. 17

a company. The *accomplishment* of this hoped-for profit requires a manipulation of factors at the practical (fourth) level—some control, work, or adjustment in material things—before the growth is possible. The growth results in a price rise (or even literally as growth in "stock splits") (stage 5), and makes possible the sale of the stock (stage 6) for more money than was put in, resulting in the hoped-for profit (stage 7).

Another example is communication: Here we learn a language in order to translate thoughts into words. As before, each of these operations is fourfold. There is the material cause (sounds or marks on paper), the formal cause (the letters of the alphabet and rules of grammar), the efficient cause (the construction of the message in words—again, this is the molding of the raw material according to the plan), and the final cause (the thought to be communicated).

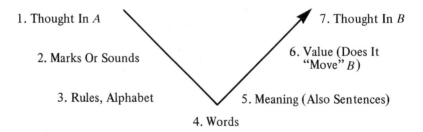

Fig. 18

CREATURES (ENTITIES) OF THE THREE-DIMENSIONAL REALM
(Fourth Level)

What we have shown is that the fourfold (i.e., structures, concept relationships, and "mechanisms") can be "inverted" and hence used either way. The key that locks the prison door also unlocks it—providing it is turned the other way. (The drawing of the key cannot tell us which way to turn it! For even if there is a legend on the drawing which says, "To open turn to the right," the visitor from Mars would not know which was "to the right," though he could make the key correctly from the drawing. Nor, for that matter, do we know whether the top or the bottom of the key is to be moved to the right.)

To completely account for the "aspects" of the fourfold entity in three-space, we do not need eight (which would be the four aspects plus their inversions) but seven, because one of the aspects is the pivot. We may diagram this as a V-shape.

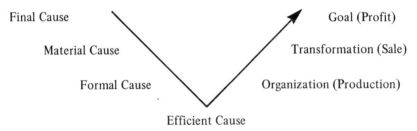

Fig. 19

TOPOLOGY OF THE FOURTH LEVEL
(Three Dimensions)

Since there are seven aspects or stages of the metastructure which we are now ready to call process, we may ask, "Does this sevenfold process correlate with the sevenfold topology of a surface such as that of a torus?"

The connection may seem farfetched, but as an aid to its recognition we can think of ourselves starting from a point and going out in every direction to gain experience. In thus creating a "sphere" of experience, we encompass everything within a given orbit. So doing, we encounter a great diversity of experience, represented by the sphere, but this material must be gathered and integrated, and then incorporated into ourselves. This incorporation can be represented as a drawing back together of the diversity to the central point of the self. But this return to source requires a second or vertical circularity, one different from the horizontal sweep of going out to gain experience.

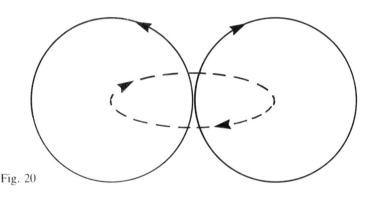

Fig. 20

There are thus two circularities involved: a horizontal circle (a sort of horizon), and a second vertical circling, which at first is expansive, carrying the self out and away from center, but which then draws back to center and internalizes the experience back into ourself (at the center). The torus represents this going out and return of process. But is there anything else in nature which behaves in this way (going out and returning)?

The answer we would like to propose is "life." That is, we can view life as the emergence of entities which acquire the ability to manipulate matter and store energy to build up order against entropy. We do know of the most primitive life forms (say unicellular plants) that they extract energy from sunlight to store it within their own boundaries and use it for fueling life processes (i.e., cell division, growth, metabolism, etc.). We would also suspect that the virus, an even more primitive life form (a molecule), manages to impose its own kind of order on the cell it invades by causing that cell to replicate more virus. It is thus itself a chemist. We may suppose that it accomplishes this function by using the same *laws* of chemistry that the chemist does when he synthesizes other chemicals.

The existence of such an agent, far from being excluded by determinism, requires it; the agent can be free, even if the laws are not. Is this not an unjustified assumption? Not in the formalism we are presenting, which does not exclude free agents. The situation is the reverse of what it is for objective science, which starts with objects and arrives at laws, thereby, like the man thoughtlessly painting the floor, hemming itself in. We, on the contrary, *started* with the self (or rather consciousness), and we can simply include volition as one of its attributes.

But we need not do this in quite so peremptory a manner. We have some quite sophisticated confirmations for this assumption of volition at the roots of cosmology. Here both quantum theory and relativity affirm a "phase" dimension of 2π. Quantum theory finds it in the quantum of action and hence applicable to individuals, relativity in the volume of the hypersphere and hence applicable to a universe. This extra dimension, which to an observer is uncertainty, and which may be interpreted as control of timing, provides the necessary condition for a free agent. More simply it *is* the free agent.

However, proof of this from physics is not within the scope of this paper. Its inherent validity must rest on its appeal to ontogony and on derivation of the parameters of the universe from a fundamental unity itself not defined.

THE SIGNIFICANCE OF TOPOLOGY

It might be asked, "Why is it that topology, the mapping of lines and surfaces, has been so emphasized in the present essay?" In answering this, we should first note that in an account of a universe, we expect to give special heed both to number and to geometry. The exact sciences bear this out in a convincing way. We have only to recall the discoveries of Pythagoras on the relation of musical harmony to the ratios of whole numbers, or the laws of planetary motion based on the solution of quadratic equations. Again, the account of the properties of atoms based on the number of orbiting electrons—an account which unfailingly traces the origin of qualities to numerical quantity—reminds us that here science is supreme.

But such tools do not suffice for all problems. The limit to the application of exact measure becomes important when a whole has to be dealt with. For example, in mapping the surface of the earth, local maps of a country or a state present no problem. But when the whole of the earth's surface is mapped on a plane, the measure changes from point to point. We can no longer use straight lines and equal intervals. (Mercator's projection shows the arctic regions too large.) The theory of relativity, in an even more significant sense, reveals how the length of bodies moving near the speed of light shortens and clocks slow down.

So we have to surrender the simplicity of Euclidean geometry and deal with deformation. Such deformation causes the square grid of longitude and latitude to give place to trapezoids at high latitudes and finally to triangles at the poles.

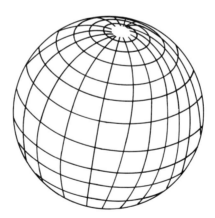

Fig. 21

Is there something even more basic than deformation? Mathematics tells us that there is (and we can verify it by reference to simple experiments). We can deform a plane surface into a sphere, or press a sheet of steel to the shape of a bowl, but we cannot deform the surface of a sphere into the

A FORMALISM FOR PHILOSOPHY   125

surface of an inner tube. This leads to the concept of the classification of surfaces based on topology. As Courant says, "Topological properties of figures are of the greatest interest and importance in mathematical investigations. They are in a sense the deepest and most fundamental of all geometrical properties, since they persist under the most drastic changes of shape."*

This fact (that the topology of a surface is so fundamental that it cannot be affected by deformation) makes topology significant for cosmology. Science pushes the principle of measurement to its utmost limits in relativistic calculations. But these limits do not suffice to reach into the areas opened to us if we introduce the torus, whose "doubly connected" surface is of a *different order* than that of the sphere of the plane.

What we have shown, and the reason for our emphasis on topology, is that we may use the more complex topology of the torus as a model for a universe which includes life—for the reflexivity (or feedback) which is necessary to account for life (or for purposive action by a free agent). This gives a totally different aspect to cosmology; it establishes a categorical basis for evolution and elevates life (often referred to as a mere accident or as a "green scum on the surface of a minor planet") to a cosmological principle of first order.

In this connection we can recall that the concept of *vortical motion* as a basic principle has been known through the ages. Leucippus traced the origin of things to vortical motion in the sixth century B.C., and later Descartes attempted to account for the generation of the sun and planets by a theory of vortices. But these theories did not recognize that the topology of the vortex is toroidal, a finding which we can now emphasize.

CONCLUSION

We opened our essay with the argument that philosophy would benefit, as science has, from some method of structuring

---

*Richard Courant, *What Is Mathematics? An Elementary Approach to Ideas and Methods* (London and New York: Oxford University Press, 1941), p. 234.

its development, so that it can progress in a cumulative manner and not have its earlier conclusions slip back into confusion as soon as it turns its attention to new problems. We called the proposed method of structuring a *formalism,* and described it as arising from an initial unity, from which was derived a threefold level (substance or time), then a fourfold level (form, a concept, or space), and finally a level which involved the combination of threefold and fourfold, the world of formed substance or *objects*.

This can be regarded as completing the formalism. However, in part 2, we went on to show that unlike a science based on objects, this formalism, descending as it does from an initial consciousness, is not barred from the inclusion of purposive intelligence. On the contrary the same formalism which depicts the descent from consciousness into a world of objects can describe the onset and development of life. This "going out" (to obtain determinate means) and returning (to reach a goal) can be shown to correlate to the compound "motion" in the *torus*.

The use of the torus may tax the reader unfamiliar with topology, but it is well worth the investment. The extra dimension it provides models our innate capacity to *be* cause. It can thus express the active as well as the passive role in experience. We are so used to having laws drummed into us that we forget that we can turn determinism around and *use* it. The structural model implied by science swallows itself and comes to life.

The Mind-Body Problem

*"The analysis of consciousness into parts
presents the same problem as the
analysis of the physical universe into parts."*

SIR ARTHUR EDDINGTON

I.

What is overlooked in the mind/body problem is that mind ⬌ body is not a valid dichotomy, but *two* dichotomies. Body is the dichotomy between emotion and sensation, between the need for something and its supply. Mind is the dichotomy between curiosity (the need for concept) and knowledge (the concept).

The first dichotomy, need and its fulfillment, is physical. The second dichotomy, search and its answer, is non-physical.

Moreover, with this scheme we can give formal expression to the aspect of mind not accounted for by the computer analogy—its higher or intuitive function as opposed to the function supplied by the brain, that of making generalities out of sense data.

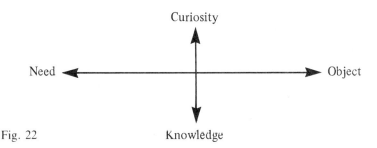

Fig. 22

Consciousness is the integration of these four aspects; it operates to attain their mutual equilibrium or cancellation. If we are driving along the highway and run out of gas, we become acutely aware of gas. If we are without certain knowledge, such as the direction of our destination, we ask for information.

Thus consciousness includes the two things that science doesn't: the negative object (need or value) and the negative concept (teleological purpose).

Note that the non-*object* is a double negative and returns us to the "ject" (or throw) which produces the object. But it is the non-object and especially the non-concept that are the true positive terms. The latter, as purpose, is the dynamic that starts action; the former is the need that sustains it. Their neglect by classical science was what made quantum physics such a revolutionary development, for the latter stresses the action involved in both radiation (Planck) and in the act of observation (Heisenberg).

A layman, upon hearing of the "mind-body" problem, might suppose it to be a special aspect of the "mind-matter" problem which has been a fundamental issue ever since philosophical problems were first debated. The question involved in the mind-matter issue was: Is reality primarily constructed of material objects, such as atoms, whose combination and interaction create the world of effects in which we live and have our being? Or does it primarily consist of and depend upon some intangible mental "essence" underlying the material objects?

This dichotomy, however, appears to have dropped out of fashion and has been replaced by another—the "mind-body" problem, which has increasingly come into prominence. Prompted by new scientific discoveries about the brain and by the progress of computer electronics, the debates have taken a new tack. The issue is not one of contrasting mind-as-computer versus body-as-bulldozer, since now both computers and bulldozers are made in factories. It is, rather, mind-as-subjective versus body-as-objective. For example, if I feel a pain in my arm, how is the direct experience of this pain related to the neuroelectrical activity or chemical changes that are observable through instruments?

Perhaps this recent interpretation of the mind-body problem has also been given impetus by the fact that the disciplines of science themselves differ as to their points of departure and their methods. Psychoanalysis takes its departure from the mind (the psyche) whereas biology and chemistry deal in objectively observable entities such as cells and nerves. The latter approach is also that of the behaviorists, for whom "mind" is a superfluous concept.

So present-day philosophy asks: how are cognitions, value judgments, intentions, and sensations—which are presumably mental—related to what the scientists observe through their techniques—the structure of nerves, their electrical disturbances, and their chemical changes, which are assertedly physical?

Stated this way, the question seems valid, until we realize that the scientist can only reach his conclusions through sensations and cognitions. This is not to deny that there are objective phenomena in the universe which have an existence apart from the scientist himself, but rather to emphasize the difficulty we encounter when we try to establish what the objective phenomena are. What is force? According to the relativistic interpretation, it is a bending of the space-time matrix. Does this answer make force mental or physical (body-like)? Or, what is a square? Of course it is a geometric figure made up of four points or four lines which are themselves physical and known through sensation. But the squareness of the points or lines resides in the relationship between them, which is a mental perception: the observer's projection of a concept of an object.

And when the scientist "observes" a nerve current, does he, in fact, do so? No, he does not. Rather he observes certain instruments which indicate that the nerve is stimulated.

To be sure, the scientist uses instruments to aid his senses. But reading an instrument, such as an ammeter, to measure whether a current is flowing is, if anything, more mental than the direct experience of the electricity which one would feel by touching the wire. Further, the reading of instruments and similar indirect scientific procedures have *meaning* only in so far as they lead to conceptual constructs or cognitions—in short, the mental category.

Moreover, the analogy which compares the brain to a machine does not show proper comprehension even of a machine. A machine is not just a number of parts—the hardware—but parts disposed according to a plan (a mental construct) even before they exist. The parts are, in fact, constructed to fit the plan. Plan and parts are thus two aspects of one whole. Another aspect is the machine's need for motive power. It must be driven by a source of energy—an electric battery, a falling weight, a coiled spring, nuclear energy, or some other fuel. A fourth (and actually primary) aspect is that the machine is made for a purpose. If it is a vehicle, it is made for travel; if a crane, for lifting; if a furnace, for heating; if a clock, for keeping time. It is impossible for a thing to be a machine and not have a function.

So anyone who equates the human organism with a machine and sees "mind" as a superfluous concept should realize that a machine involves more than hardware or even plan. When we look at a machine, all we *see* are nuts, bolts, gears, etc. But for the machine to work, the parts must be in proper order (the plan) and there must be motive power, which is itself not visible. Nor can we see, in the sense of having a retinal image of it, the purpose of a machine. In fact, the purpose of a machine cannot be discovered by physical examination apart from the context of its use. For example, in order to charge a battery, we might connect it to a dynamo, thus obtaining electricity from a mechanical energy source. But if we connect a battery to a motor, we can obtain mechanical energy from electricity. The fact that a dynamo can function as a motor indicates how difficult it is to ascertain the purpose of a machine from its description. So there are four aspects to a machine: hardware, plan, motor power, and purpose.

If we now turn to "mind-states," we discover that here, too, are four kinds or categories. Concepts are not the same as values. Intentions are not the same as sensations. We have here another multiplicity of aspects. And we owe it to ourselves to clarify the distinction between them, and to look for a possible correlation with those aspects already discussed.

It is my thesis that we can analyze the categories of mental states in much the same way that we can analyze the aspects of the machine. When that is done, we find purpose to be an aspect of both machine and mind. We find motivation in mental states just as we find motive power in the machine, and in concepts of the mind we have a correlation with the configurational or ordering aspect of plan in the machine.

Let us recognize that the word "mind" means at least two things. In the sentence, "Have you a mind to go to the opera?" it means "purpose." In the sentence, "Visualize in your mind a figure in the form of a square," it means "concept formation." Both purposes and concepts are thus included in the broader sense of the word "mind."

Further confusion arises when values (such as pain and pleasure) and sensations (such as color, texture, sound, smell, and taste) are included as aspects of mind itself, or when values and sensations are themselves confused with one another. This occurs, for example, in the word "feeling," which may mean sensation ("The room feels cold") or value ("How do you feel about LSD?"). The sensation of sweetness would be accepted as contributing to the taste of a dessert, but added to our soup it would be cause for rejection. The smell of perfume may enhance a beautiful woman, but if it emanates from our morning cup of coffee we might assume that the scented detergent was not washed from the cup. Even the fear of falling, which Watson found instinctive in infants, can be turned into a thrill in a roller coaster. Here it is not the sensation that has changed, but the valuation we place on it. In this connection, it is interesting to note the testimony of a person who had once had a very keen sense of smell—as sensitive as that of many animals—and had subsequently lost this power. He was asked the difference to him between his former keen sensitivity and his later ordinary sensitivity. "Formerly," he replied, "all smells were interesting."

This is sensation at its best: a purely informative function without valuation or judgment. For most humans, smell is a judgment (i.e., value) word. "Such a proposition has a bad smell" or "Our offense is rank; it smells to heaven."

Thus there is a distinction between values and sensations, especially when we are careful to remove from sensation the value that we may unconsciously place upon it. We should also note that whereas sensation, per se, supplies objective information about our environment, value supplies the motivating force in human activity. We are motivated by the pleasure-pain principle in terms of hunger, fear, desires for love, wealth, etc. To the extent that we are motivated by sensations, such as warm or cold, hard or soft, rough or smooth, that motivation is in terms of values. We seek out what is desirable or comfortable. If the room is too cold or too hot, we adjust the thermostat to establish that *value* of temperature we prefer. Neither hot nor cold, which are sensations, have value in themselves. We prefer our coffee hot and our orange juice cold, etc. Jung gave to emotion the function of evaluating. The four functions referred to by Jung—intuition, emotion, intellect, and sensation—correlate with the four aspects we are discussing: intuition (purpose), emotion (value), intellect (concepts), and sensation. However, our fourfold diagram is different from Jung's. He places sensation opposite intellect, while we place it opposite emotion. Figure 23 gives a diagrammatic presentation of the four aspects of mind that we have been considering.

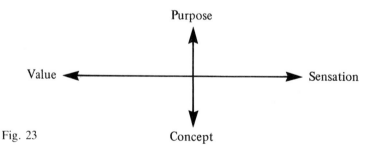

Fig. 23

We have placed purpose and concept on the vertical axis, and value (which includes emotion) and sensation on a horizontal axis. The horizontal categories are functions or faculties that are essentially supplied by the physical body. Values—fear of falling, for example—can be supplied by internal stimuli, including glandular secretions; sensations by external stimuli to nerve ends by objects in the environment.

The categories on the vertical axis, purpose and concept, are not physical. They do not arise directly from bodily states. For example, the concept of squareness is reached only through comparison of different observations. Of course, most people can immediately recognize the difference between a square and a triangle, but this obviousness is not present in polygons with a greater number of sides. When we reach eight or nine sides we have to stop and count the sides. This counting is a mental operation, in contrast to the more direct apprehension of sensation. That even squareness is not apprehended without long training has been shown in the case of people born blind who have gained sight by an operation.* They see as well as persons with normal vision do (in the sense of the mechanical functioning of the eye), but not until after several months of practice are they able to distinguish a square from a triangle.

The probable reason for this phenomenon is that to detect a square, the eye (or perhaps the mind's eye) must follow the outline and keep a record of where it has been. This activity is not sensation. It is rather the integration of a number of sensations, as a line is the integration of a number of points. The function of making this integration and coming up with the appropriate name, or the classification into which the object should fall, is one we may assign to mind.

Thus we may define "mind" as the faculty that integrates and classifies sensations or groupings of sensations. This faculty normally operates so quickly as to make it seem that it is directly perceiving reality. But it is not. Rather, the faculty of mind interprets, and may prejudge (or preconceive) and thus misevaluate. For example, an acquaintance confronted by a

---

*M. von Senden, *Raum- und Gestaltauffassung bei operierten Blindgeborenen vor und nach der Operation* (Leipzig: Barth, 1932).

substance he thought was milk was nauseated because it contained "horrid lumps." On being told it was oyster stew, he ate it with gusto.

Placing purpose on the vertical axis (which is independent of the horizontal axis) is justified by its non-physical nature. Neither in ourselves nor in other people nor in machines do we detect purpose through physical means, whether sense data or feelings. We must infer it (for others) or create it (for ourselves).

In sum, therefore, we can assign the vertical axis to mind and the horizontal axis to body. This analysis can be applied to any situation, whether it is approached internally (subjectively) or externally (objectively). The factors in the first case are sensation, concept, value, and purpose. In the second case, they are hardware, plan, power supply, and function.

II.

Let us now consider the scientist's observations by instrumentation as compared with the cognitions of a man observing the world from an armchair. In both cases, the direct data is obtained by physical means. In the armchair situation, data comes from the nerve endings; in that of the scientist, from instrument readings. Motivation for the human situation is supplied by the fuel of desire, say; the motive power of the machine is ultimately due to the attraction ("desire") of opposite charges (electricity). In the human armchair situation, purpose is created; for the scientific observer it is inferred. (Here we should note that for a more complete account of purpose in objects we must go to the inventor of a machine. Purpose provides the causative dynamic that brings the machine into being. The Wright Brothers built the airplane *in order to* fly. Necessity may be the mother of invention, but purpose is the father.)

Finally, in the mental process of comparison—whether a man is comparing the specifications of different automobiles or a scientist is looking at brain cells or the circuitry of a computer—it is the *relationship structure* that counts. This is

THE MIND-BODY PROBLEM 135

not a direct sense datum. For the man or the scientist, it is the product of both comparison and integration of sense data.

We conclude, therefore, that it is an error to speak of the contrast between a thinker reviewing his states of mind and the scientist looking at brain structure as the mind-body problem, since both the thinker and the scientist are dealing with the same four kinds of information. One kind is directly given. Others are inferred or computed. In fact, to say that the scientist deals only with what he perceives through his instruments, and not permit him to formulate and integrate this data, would be to confuse the scientist with his instrumentation, and thereby to reduce him to an instrument dial.

What other approach is indicated? Since the mind-body problem arises from the difference between the introspective and the scientific approaches, we may note that there are two *sequences* in which the aspects we have discussed can be apprehended. In the introspective approach, pleasure and pain are immediate and *direct:* the armchair human (or even an infant) recognizes discomfort without being taught—the discomfort *is* his experience. Concepts, on the other hand, have to be learned, and are thus *indirect*.

In employing the scientific method, however, the capacity to make observations and form concepts already exists. Indeed, it has produced the laboratory equipment. This capacity and the operations made possible by it are fundamental and direct in scientific method. Thus the scientist determines through his instrumentation that the invisible electricity is flowing in the wire. Similarly, he observes by outward signs that a laboratory animal is undergoing stimulus of the pleasure center. Neither the current nor the pleasure is experienced directly by the scientist. The current is inferred from the indication of the ammeter, and the pleasure from the animal's repeated self-stimulation. Nor can we say the pleasure is the animal act of stimulation, any more than we can say that the current is the motion of the meter. What the infant experiences directly, the scientist experiences indirectly, and there is a reversal of sequence.

This difference in the order or temporal succession in which the aspects of a situation occur (i.e., whether their effect is immediate, as with the infant, or inferred, as with the scientist) provides a hint of a different approach to the problem, one which takes *time* into account.

SELF AND NON-SELF

When a mechanic says, "Your engine is fouled up and you need a new set of spark plugs," is he talking about self? Obviously your automobile is not you. When the doctor says, "Your heart is overworked," this *is* you, but if so, what happens if you get a transplant? Another example: *you* (1) are brought to court for violation of a traffic law, but *you* (2) get a good lawyer who proves that *you* (1) were violently upset, and *you* (3) are released with a light sentence. The (1), (2), and (3) represent successive states of you. That a difficulty can result from a failure to recognize the sequence in time of these terms is illustrated in what is known as the Cretan paradox: X says, "All Cretans are liars," but X is a Cretan. The paradox is resolved by the recognition that X can only qualify as true or false statements *already* made, and not the statement he is in process of making.

It is the distinction introduced by succession that gives rise to the problem which has come to be called the mind-body problem. This distinction should be comprehended in its most general form, which may well include such apparently unrelated problems as the proof of continuity (given any two points, there exists a third point between them), the legal status of a will (it is impossible to make a will whose legality may not later be questioned), and other instances that involve succession.

As an example demonstrating the unending nature of succession, let us suppose that Smith designates a point exactly midway between Chicago and New York and that Jones is to locate this point. Here there is no vagueness of definition. We know such a point exists. In practice, however, difficulties develop.

Let us suppose that a preliminary survey has been made,

THE MIND-BODY PROBLEM   137

and that Jones has located the point. Smith then follows with a check and finds Jones's result to be in error. Jones now makes a more exact survey, and this time qualifies his result by adding that it is accurate only to plus or minus fifty feet. It is to be supposed that however small an error is permitted by Smith, a technique can be developed by Jones to locate the point within less than that error. But this implies that the tolerance permitted by Smith remains unchanged during Jones's survey. It is quite possible that Smith will improve his inspection technique as fast as Jones improves his survey, and will be able to show that the point designated by Jones is not, in fact, the exact midpoint required. We thus reach the equivalent of an infinite regression similar to the case in which we try to discover the "non-objective" self.

In order to provide a formal method for dealing with endless regression, let us take the familiar example of the picture that depicts itself, like the salt box that bears a picture of a girl holding the same salt box, and so on:

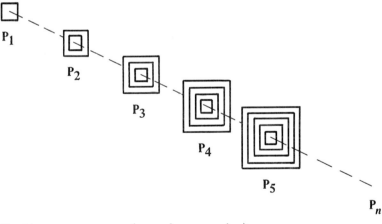

Fig. 24   (*n* equals any number)

The successive pictures in the above diagram comprise a sequence, $P_1, P_2, P_3 \ldots P_n \ldots$, etc. The P_n picture depicts within itself the P_{n-1}, which in turn includes the P_{n-2}, etc. For every P, there is a successor and an antecedent, and we can place the sequence of P's in one-to-one correspondence with the cardinal numbers. While each picture is two-dimensional, the sequence of pictures is one-dimensional:

$$1, 2, 3, 4 \ldots n \ldots$$

There are two important points here:

1. Whether we are talking about the series of pictures or the series of measurements or the steps in any example of infinite regression, each stage has a definite and unique position in the sequence. In the series of cardinal numbers, each number has a unique position.

2. Any particular term in either sequence includes all the preceding terms. For the set of cardinal numbers, this inclusion of predecessors reduces to the relationship "is greater than." We must appreciate that the mathematical analogue is symbolic; "is greater than" corresponds to more subtle inclusions in the other examples, such as in the case of successive court actions on a will in which each court action must take into consideration the findings of former court actions. Similarly, each successive approximation of the midpoint of a line must be an improvement in accuracy over preceding determinations.

To emphasize more strongly the second property, that each term includes or is greater than all its predecessors, we will say that the infinite series, or series of terms each of which includes its predecessors, is cumulative and hence asymmetrical. This helps to bring out what is important for our discussion: *that all such series are asymmetrical.* We thus have the general principle that cumulative phenomena possess asymmetry. This principle establishes that we are not dealing with the analysis of an "objective" ordering. In any objective consideration of order (i.e., apart from the process that created the order), say of the points on a line, or the houses along a street, we may arbitrarily begin at one end or the other. Such a spatial ordering is symmetrical. In fact, this symmetry is

THE MIND-BODY PROBLEM 139

equivalent to objectivity because it is independent of the observer. The elements of an object are related to one another, rather than to the observer.

But by hypothesis, the series of pictures is an unending line. One of its ends cannot be displayed. We may write down as many terms as we wish, but we cannot ever come to an end. Therefore, it is not, as the logician would say, "well-formed," and cannot be an object. Nor can we count the terms in inverse order, since the upper terms are not available. Even if we confine ourselves to a subset of the whole array—say the terms that have already been written down—the asymmetry still exists within the subset, because the nth term includes the nth-1 but not the reverse. If one insists that the relationship "is greater than" can be reversed and replaced by "is less than," hence gaining symmetry, the essential meaning of the succession, which portrays a hierarchy rather than a symmetrical relationship between objects, would be obliterated.

In the case of a will, the will assigns one set of objects (possessions) to another set of objects (persons). This is objective. But the question of the *legality* of the will with which the sequence of court actions deals is not a relationship between objects. Rather, it is *valuational.* In similar fashion, the relation between successive determinations of the midpoint of the line is not itself a determination, but is of a different relational or logical type.

Let us now return to our pictures. We notice that as we draw more and more pictures, each of which must include all the ones that have preceded it, they must continually increase in size. More precisely, they must extend over a greater angle, a greater proportion of the field in view. We could draw them, for example, on larger and larger sheets of paper, but the limit would not be an infinite plane, but the interior of a sphere with the eye at the center. More simply, *we can call this central point the "origin" or the self.*

We may well pause to explore what is involved. Since we are setting up a model for self and its field, we may be excused for employing the greater pictorial suggestiveness of an eye rather than a point, because this use of an eye is a way of

WHICH WAY OUT?

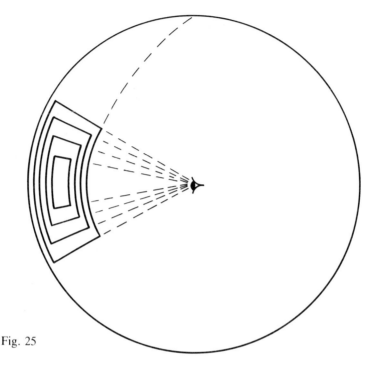

Fig. 25

describing the function of this point of origin. What is the nature or parameter of this ultimate "point"? Better still, what is its measure of *dimensional* character? For it is by dimensionality that we are able to distinguish between the two-dimensionality of each picture and the one-dimensional sequence of the picture.

While it is a truism that "a point has no dimensions," the question of angle must still be considered. The point, whatever it is, must be invested with *angular freedom* if a radius is to be projected from it. This angular freedom may be called orientation (the freedom of a radius to be rotated around the point of origin in any direction). What is the significance of angular measure in the context of the problem of the self? Here we would propose that the relevant concept is attention. The self can attend to any point within the sphere of perception. In attention we can include other notions that involve angular measure: "aim," "intention," "purpose," and "focus," for example. These ordinarily subjective words have now been

THE MIND-BODY PROBLEM

given status in terms of measure—in this case, in terms of angle—just as we give "value" status as a one-dimensional measure (scale), and "concept" status as a two-dimensional measure.

In our discussion, we have traced consciousness through four sequential aspects or levels, beginning at the level of *physical objects* having position in space. Such objects are three-dimensional and known through sensation.

We then make *statements* in the form of maps, charts, or theories relating the objects. Such statements or maps are two-dimensional and are known through the intellect. These statements have temporal or one-dimensional order. Or, if *evaluated*, we place them in an order of merit. Their ordering forms a sequence which is one-dimensional, known through feeling.

Finally we reach the *self*, which is a point of attention and essentially without dimension.

Constraint and Freedom

An Ontology Based on the Study of Dimensions

I.

Are there two worlds, an ideal world of mathematical relationship and a "real" world of actual objects? A consideration of dimensions, which link the two worlds, brings the question into focus.

Outwardly there are reasons to believe in the duality. Not only is it possible mathematically to construct spaces which may have any number of dimensions, and thus differ from the space of the physical universe with its three-dimensionality, but we may in mathematics have "imaginary" dimensions—a designation which, at least at first glance, indicates an unreality, or a reality of a different order than we encounter in the measurement of actual space.

The duality is further supported by the general trend toward specialization in scientific disciplines. This encourages the invention of mathematical fictions for their own sake and without regard to practical application; and, while what was thought to be a pure abstraction has often turned out to have practical application, there is always a feeling that if the criterion of usefulness were made the sole justification for mathematics, creativity would suffer. The finest blossoms would be moved down, and the garden of the queen of the sciences reduced to supplying the kitchen with "useful" vegetables.

But here we must reckon with another consideration. Mathematics, even of the purest sort, is fostered and even inspired by the necessity of having to deal with the physical world. In fact, the birth of probability theory is credited to a gambler who engaged Fermat to work out the odds in his gaming activities. Again, we have the more recent example of the theory of relativity, which, prodded by the negative results of the Michelson-Morley experiment, and the practical impossibility of synchronized clocks for separated observers, established categorical implications about measurement. Some of the advocates of pure mathematics would claim this was only a "reality condition," leading one to select from the larder of pure mathematics but one of many possibilities, but for the present inquiry, this claim begs the question, since it is the *nature* of measure that is under scrutiny.

We come back, therefore, to the question, "what is dimension?" Perhaps the most general definition is that it is the range of a parameter—that is, of a variable which can take on different values. Under this definition it is possible to have any number of dimensions. But when we use dimension in the more limited sense of its applicability to a universe, particularly to measurement in space, we are faced with certain facts about reality, or real space. It has three dimensions, not four or five, and, even if we call time another dimension, or assign dimension to rotation, we do not efface the obvious and sensible character of length, breadth, and height, and their intrinsic difference from time and rotation.

There are a number of ways in which the concept of dimension, as we are forced to employ it in dealing with the physical world, takes on a character that it does not appear to inherit from the realm of mathematics. The difference between time and space, the difference between position and rotation, and, indeed, the very notion of a world of precisely three dimensions, seem foreign to the mathematical mode of thought, to which the nature and number of dimensions is considered arbitrary.

We will endeavor to show that this view is actually not even valid within mathematics itself, but before we can do so, we must first examine the concept of dimension as it applies to measurement of the so-called physical universe.

CARTESIAN VS. SPHERICAL COORDINATES

Ordinarily we think of space as measured by rectangular or Cartesian coordinates. This custom seems so completely natural that it goes unchallenged (fig. 26).

Cartesian coordinates are thought of as extending in three mutually perpendicular directions from an arbitrarily chosen origin, O: an x coordinate extending from left to right, a y coordinate perpendicular to the plane of the paper (shown below at an oblique angle) and a z coordinate extending up and down. Such coordinates are understood to extend indefinitely in all directions. With them, one can describe the proportions of a house, for example.

Conventional and natural as is this coordinate system invented by Descartes, there are two important respects in which it is deficient, and has led to a false notion of reality.

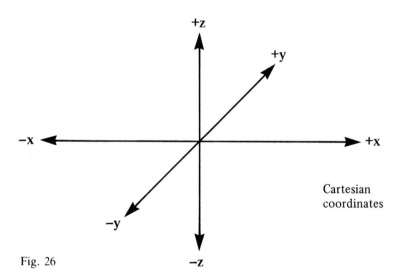

Cartesian coordinates

Fig. 26

CONSTRAINT AND FREEDOM 145

The first deficiency is that it provides us with no notion of scale, or the size of the object. While scale may be indicated on a drawing as so many feet to the inch, this only begs the question, since the feet which the inch represents are in a different logical category, as it were, from the measurements on the paper; they require experience for their proper appreciation. One must know what a foot is, and this knowledge is of a different order to that involved in the drawing. We may come to an appreciation of this if we think of the drawing as though it were on a transparent sheet placed in the line of view and coinciding with the outline of the actual object drawn (see fig. 27). The scale of the drawing is equivalent to the ratio of the distance of the drawing from the observer (or point of observation) to the distance of the actual object depicted. But to know what this means, the observer must operate in real space with the actual distances involved. He must *experience* the unit distance by some flesh and blood action, such as pacing the distance off, or perhaps even living in the house. The size of the door must be such that he can walk through it.

The second deficiency is that rectangular or Cartesian coordinates give no account of *orientation*. In ordinary engineering practice, or when we hang paintings on a wall, this occasions no difficulty. Gravity is always present, and we hang our pictures "right side up" and set up buildings with the floor down and the roof above it. The mason uses his plumb bob and level as much as his trowel and mortar board.

But it is not always so simple. In using a map it is very important to know how to orient the map with respect to north. In surveying, it is just as important to record the angles of the boundaries with respect to north as it is to measure their length. In the assembly of a machine we always require not just the drawing of the parts, but the general layout, a larger drawing that shows how the parts are put together, how they are oriented with respect to the whole. Otherwise the carburetor may be installed upside down, or the controls rigged backwards.

How do we show orientation on a map, or on a drawing? The drawing in Cartesian coordinates takes care of this by the convention that its bottom edge is intended to be parallel with the earth's surface. Similarly, the map carries an arrow to be brought into coincidence with a compass needle, or it shows parallels of longitude and latitude which ultimately refer to the earth's axis.

But such references—like that of indicating scale as so many feet to the inch—leave the crucial question of orientation to be resolved by methods outside the system. The coordinates of the map or drawing are incomplete. They omit or bypass the necessity of an observer of the map or drawing who will have the task of correctly orienting it.

SPHERICAL COORDINATES

This deficiency, as well as that of scale, are taken care of in spherical coordinates, which we will now consider (fig. 27). Spherical coordinates, like Cartesian coordinates, comprise three mutually perpendicular axes. But in this case, the origin, O, is assumed to coincide with the observer. Orientation is referred to this observer, and the object is seen as though on the inner surface of a hollow sphere. The distance of the object is given by the radius, R, of the hollow sphere, and the direction in which it lies and its shape are given by *angular measurements* of longitude and latitude.* The latter measures correspond to the horizontal and vertical dimensions of a drawing, while the distance, or the radius, is equivalent to the scale of Cartesian coordinates. Orientation is taken care of because the observer himself provides the base with reference to which angles are measured.

It is obvious, therefore, that the spherical coordinate system is more complete, for it includes both scale and orientation. Furthermore, by the fact that the measurements of the shape of the object are in terms of angle instead of distance, we are

---

*More correctly, right ascension and declination, but longitude and latitude are the more familiar terms.

CONSTRAINT AND FREEDOM   147

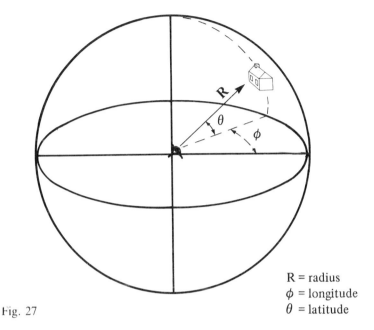

R = radius
ϕ = longitude
θ = latitude

Fig. 27

reminded that the distances of Cartesian coordinates are not distances at all, but *proportions*. The drawing of the box tells us that it is twice as long as it is high, or the like; the drawing cannot give its actual size, except through the *reference* of scale. Hence it is not the dimensions of the object but their proportions that the drawing depicts.

There still remains an important point. Since Cartesian coordinates do not include scale and orientation, how are these aspects established?

Let us assume we are looking at an architect's drawing of a house. How are scale and orientation established? The apparent size of the drawing varies as the eye of the observer moves toward or away from it, and there is a precise distance where the drawing may be held at which the building it depicts is the same apparent size as the actual building. As mentioned before, this distance would be such that the drawing would exactly cover the outline of the actual object (fig. 28).

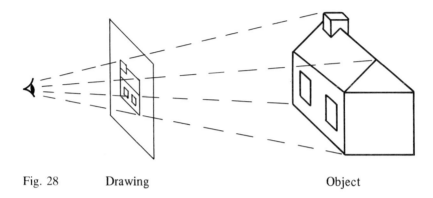

Fig. 28   Drawing   Object

The ratio of the two distances is the *scale* of the drawing.

$$\text{Scale of drawing} = \frac{\text{drawing dimensions}}{\text{actual dimensions}}$$

Similar considerations apply to orientation. Both the drawing and the building require the reference of a level or plumb bob which denote the horizontal and vertical directions of the actual world. For the *observer,* living and acting in this world, the horizontal and vertical directions are a prime consideration, indeed, a prime *condition* of his existence, for the mere act of standing requires continual muscular adjustments to maintain an upright position. This control is basic to being alive, so much so that the outward sign of death or illness is to fall down.

We conclude that the drawing or map in Cartesian coordinates, and, in fact, the Cartesian coordinate system itself, is included in and must be completed by a spherical coordinate system whose origin is the observer, and in which distance from this origin is the reference for scale. Orientation of the object is ultimately resolved by reference to the coordinates of this origin, which is the observer himself.

THE RELEVANCE TO SCIENCE OF THE NON-OBJECTIVE

At this point, the scientific reader is likely to object that such considerations are irrelevant to scientific objectivity, which is concerned only with such description of the object as would be common to all observers. Science, he points out, is careful to eliminate just that aspect of observation that we are reinstating by emphasis on the spherical coordinates of the observer. Apparent size and orientation are not objective realities and have no significance for scientific inquiry.

We could answer this criticism by pointing out that, since the conditions we are emphasizing are present in all scientific observations, they are as much a part of reality as are the objective determinations themselves, not because they concern the object but because they are an inevitable part of the act of observation. Support of the importance of the relation between observer and observed in a world scheme comes both from Heisenberg's Uncertainty Principle and from the theory of relativity. The former calls attention to the inevitable energy exchange involved in observation of individual particles such as electrons, an exchange which makes complete predictability and, hence, objective determination impossible in theory as well as in fact. The latter calls attention to the impossibility of establishing simultaneity, and, hence, of the impossibility of identical world views.

Since science is based on observation, and observation involves dimensionalities which are not necessarily objective, we must give attention to just those aspects of the act of measurement which underlie or precede objectivity.

What does the spherical coordinate system teach in this regard? We may at once note that the R dimension is of an altogether different nature than the two angular dimensions $d\theta$ and $d\phi$ (or the x and y of Cartesian coordinates).

In the first place, the two angular dimensions occur together, and are inseparable from one another. To speak simply, they are the dimensions of a *surface*. The radius R, which is the distance away of this surface, is a *linear* dimension. The surface dimension involves two numbers, or measurements; the linear dimension, only one.

In the second place, the dimensions that describe the surface may be read either way. Longitude may be measured east or west, latitude either north or south. In format terms, the numbers which measure a surface are positive or negative,

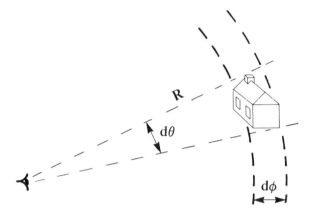

Fig. 29

$d\theta$ = small vertical angle
$d\phi$ = small horizontal angle

that is to say, they are numbers *associated with signs*. In fact, in some applications, it is convenient to assign to the vertical axis (or latitude) what are called imaginary numbers. These, we will endeavor to show, are not so much numbers as *signs*, and designate *directions* perpendicular to those indicated by the plus or minus sign. In other words, the two-dimensional plane creates the necessity of four kinds of signs to describe direction.

Returning now to the linear dimension, the radius R, let us note that it is both empirically and logically prior to the two dimensions that measure position and shape on the interior surface of the sphere. We now assert that, in itself, R is without sign. We speak of an object as being two miles away, but it would make no sense to say that an object is minus two miles away. This might lead to our saying that R is only positive, but positive has meaning only when contrasted with negative; it is meaningless to speak of R as positive when it cannot be negative. Hence we say R is a signless number. This implies that this R dimension is one-way (like time). It could be objected that we can measure R either way, but the answer is that when we measure R "backwards," we are measuring from the object, not from the origin.

CONSTRAINT AND FREEDOM 151

It is very difficult to accept this notion (that R can be measured only one way) because we have become accustomed to viewing a linear dimension as one we can measure either way—we add and subtract distances on a ruler or sums in a bank account. But this measuring cannot be done until we have the two simultaneous dimensions of the surface as the matrix for our operations of transporting measuring rods and reading intervals of length by their coincidence with markings on the ruler. In other words, the linear dimension we measure in this way is part of the surface perpendicular to the line of view.

This may be seen when we "measure" R. To do so, the observer must shift his position and place himself at right angles to R, and view it as extending across his line of vision instead of along it. This implies he is no longer at the origin from which R was initially struck. We have taken a new viewpoint.

In this way the linear R dimension eludes objectivity; for when we stand on one side to "measure" R (thereby making it objective), we create a new origin and a new R, which is not objective (fig. 30).

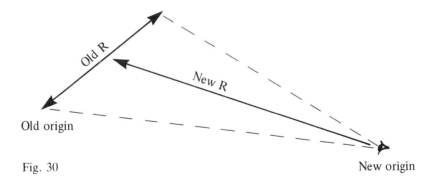

Fig. 30

So there must always be a non-objective R in an objective observation. To avoid the limited connotations of the word "subjective," we may call R *projective*. To reiterate our point, what is important for the question of dimensionality is that this one dimensionality of R is prior to objective measure, and that it is a dimension without sign.

There is another aspect of the R dimension that, while it might seem to impose a loss of generality, so enriches its meaning that it should be mentioned. This is that *the R dimension correlates with time.* Time here is as it is experienced, and not that abstraction to which we refer when we *measure* time, in which case we have to "view" time from a vantage point outside of it, just as to measure R we have to view it from an origin that is separate from it. *Further,* the correlation of R with time does not exclude space. R is, rather, both time and space together before their separation by analysis. Thus, when we say something is four miles away, we might also say it is an hour's walk away, recognizing that time and space are both involved. Here again, we can recognize that whereas measurement across our line of view is accomplished independently of time (we can estimate angular distances of remote objects instantaneously), radial distance, even when measured at the speed of light, requires time (a signal to the nearest star, for example, requires a time interval of ten years to go out and return).

We can now confirm an earlier statement. We said that R cannot be measured backwards. Suppose we extended R, say, by a signal to a distant star. At the speed of light this requires some five years. We now have the signal return; the time, unlike the distance, is not reduced as the signal returns; it continues to *add,* and when the signal reaches the starting point, the time elapsed is double what it was when it reached the object. The example gives reality to the notion of signless dimension. But we should not limit its application to time. There are a host of quantitative measures that are signless, viz., absolute temperature, mass, density, frequency, and so on, as well as "distance away."

FINITE AND INFINITE

Dimensionality, then, is of at least two kinds: *one-dimensionality,* which applies to the measure of how far away something is, and corresponds to the radius in spherical coordinates, and *two-dimensionality,* which applies to the measure of shape or position, and corresponds to latitude and longitude. The former gives the *scale,* and is a measure of the "bigness" of an object; the latter is independent of scale, and yields the *proportions* and direction of an object.

The former is in the "line of view" of the observer, and the latter is perpendicular to this line of view, a screen, as it were, intersecting the line of view (fig. 31).

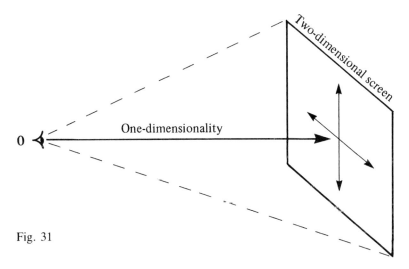

Fig. 31

We may now note an interesting point: the R dimension (determined by the time required for light signals to reach the observer from the object), *has no upper limit,* and may extend indefinitely, whereas the two-dimensional screen has, for any fixed R, a *finite extent.* For, as we enlarge this screen, we see that, if R is to remain the same, it must curve around the observer (fig. 32).

The screen can become no larger than would be required to fill the entire field of view, and becomes the interior surface of a sphere, with the observer at the center. The measure of this field of view is in terms of angle; it is the maximum amount of rotation sideways and up and down, the maximum *angle* through which the observer can turn his eye or his telescope. This maximum angle is 2π, or 360° in either direction. Hence the dimensionality of the screen which extends

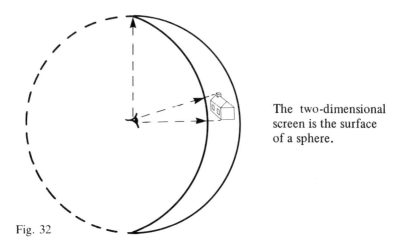

The two-dimensional screen is the surface of a sphere.

Fig. 32

across the line of vision is quite different from the dimensionality of the "distance away," the R dimension (fig. 33). Not only is it *angular* whereas R is measured in terms of length, but in its maximum extent it is finite, whereas R has no *upper* limit.

This is not to say that the interior of the sphere seen by the observer may not be as large as you please, but such size is contributed by the radius, and not by the angle.

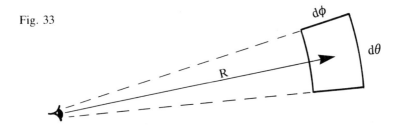

Fig. 33

POSITION AND ROTATION

We are ready for a further point: the distinction between the position of a body in space and its rotation around its own axis. This might appear to carry us into refinements only necessary to an astronomer, and without significance for the notion of dimension in general, but investigation will show that the concept of rotation is very important to and actually helps resolve the problem of dimension and extends its significance.

In a typical instance, an observer is situated on the earth and looks out at, say, a planet, itself rotating (fig. 34). The observer makes two sets of observations:

1. The set that determines the position of the planet in terms of the latitude, longitude, and distance away with respect to the observer.
2. The set that determines the rotation of the planet, which includes the direction of its axis, given in terms of two angles, one lateral and one vertical, and the number of rotations in a unit of time.

Note first that both of these determinations reiterate our distinction between the two-dimensionality of angle, and a one-dimensionality which, for position, is distance away, and for rotation, a number divided by time (e.g., revolutions per minute).

In the last analysis, all these determinations depend on

Fig. 34

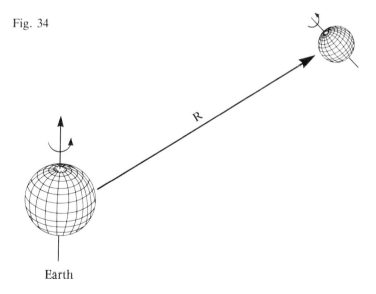

Earth

time. This is evident if we recall that the observer is himself rotating, and determines the longitude of the planet he observes by the time at which it rises, sets, or passes the meridian (directly overhead).

Since this rotation of the observer is the basis for regulating and setting his clock, it is also basic to his measure of the planet's period of rotation. Again, rotation establishes the direction of the equator and, hence, fixes the base for both the latitude of the planet's position and its axis of rotation.

Finally, the distance away of the planet is measured by the time taken for a signal to reach it, and this time, in turn, is measured by the earth's rotation. So we have all the parameters required to describe the planet's motion depending on time, and, hence, on the observer's rate of rotation.

This is not to say that these measures could not be determined by an observer on another planet independently of the earth's rotation; what is implied is that this other observer, or any observer, would find himself in the same situation, all his observations stemming back to his own rotation.

We have seen the importance of rotation in establishing these measures, both because it is the basis for setting clocks and because the axis of rotation establishes latitude. We must therefore accord rotation the status of a dimension. As it is prior to the one-dimensional extension of the radius, we may call it a *zero dimension,* or *zero*-dimensionality. In addition to providing the basis for the other dimensions, we may here note that it establishes *orientation,* a factor, as noted earlier, not included within the Cartesian system. (To the extent that it occurs in Cartesian coordinates, it is accidental, a supernumerary factor which has to be supported by references outside the system.)

Since the zero-dimensionality of rotation is the basis for the dimensionalities of the line and the surface, we may correlate zero-dimensionality to the point, and thus complete the analogy which portrays one-dimensionality as a line, and two-dimensionality as a surface. But in doing so we must recognize that the point in this case is associated with a number, i.e., a frequency. Why, we might ask, does this numerical measure not make zero-dimensionality like the one dimension of scale, which also has one numerical measure? The tentative answer is that the rotation of zero-dimensionality is a "pure number," devoid even of the property of extension, either in time or space, which characterizes one dimension. The point is devoid of extension, and we should keep in mind that, while we can call its number "frequency," we have already correlated time with one-dimensionality, and to define zero-dimensionality as frequency or inverse time would be to invoke a circular definition. (Frequency is inverse time because it is a number divided by the time required for that many rotations.)

The ambiguity can be settled by making the number of zero-dimensionality an *undefined term,* and hence fundamental. What we call time, and correlate with one-dimensionality, we can then define as the inverse of this undefined term. This is not a dodge, it is rather a matter of getting first things first. We must expect undefined terms, and it is proper that they should be the originating term in an ontological declension.

As we expect to clarify these issues in what follows, we can postpone this question for the present. We emphasize, however, that our study of dimension has indicated that at the root of the time and space measurement there is a number which counts rotation or frequency. Itself "dimensionless," it is the origin of dimension. We call it zero-dimensionality.

THREE-DIMENSIONALITY

None of the dimensionalities discussed are appropriate for the "three-dimensional" world of physical objects, but we may adapt them to this purpose by saying that in the "three-dimensional" physical universe, the one- and two-dimensionalities of spherical coordinates are combined. I incline to the view that a different dimensionality is not required. Despite this assurance, one is habituated to thinking of the three dimensions of space as coexisting and objective, even though the act of measuring them always involves the procedures of spherical coordinates, requiring two kinds of measure: the one of scale (radius), and the other of form (latitude and longitude).

There may be some sense in which the dimensionality of physical objects is more than a combination of one- and two-dimensionalities, but for the purpose of this analysis—which, as will be seen shortly, draws significance from the dimensions *not* measured as well as from the minimum number of dimensions that will suffice for measurement—it is not necessary to endow the dimensionality of physical objects with any properties which would not flow from the combination of one and two dimensions.

A further point about physical space (characterized by three-dimensionality or one- plus two-dimensionality) is that, as Kant said, the universe of physical objects is unique; there is but one physical world. On the other hand, every concept, every picture has its *own* space. There is no point in asking how far the tree in one picture is from the house in another* because they have different space. The significance of this we now propose to discuss. It suggests that with the three dimensions of physical objects we reach an end to dimension.

FREEDOM VS. CONSTRAINT

The two-dimensional surface on which the position of a body is depicted in terms of latitude and longitude, and on which the shape of a body is described, may be at any distance. We look out the window and see the moon. It is 232,000 miles away. At the time of an eclipse it will just cover the sun, which is 400 times further away, and an equal number of times larger. There is no limit to the number of objects that could, like the moon, be between us and the sun, and just cover its image, provided they were smaller in the same proportions as they were nearer to the observer, and placed squarely in the line of view to the sun.

This is equivalent to saying that an object whose latitude and longitude are expressed, but whose distance is not, has a "degree of freedom" not possessed by an object for which the radius is given as well as the latitude and longitude. Similarly,

---

*See William E. Hocking, *The Meaning of Immortality in Human Experience* (1957), rev. ed. (Westport, Conn.: Greenwood Press, 1973), pp. 26–28.

CONSTRAINT AND FREEDOM   159

if we specify distance away of an object but not its latitude and longitude, such an object has two degrees of freedom. It may be anywhere on the surface of a sphere whose radius is the specified distance. Further, if none of the three dimensions—longitude, latitude, and radius—are expressed, it has three degrees of freedom, or zero constraint. Thus we have:

For zero- dimensionality, 3 degrees of freedom, no constraint(s)
For one- dimensionality, 2 degrees of freedom, one constraint
For two- dimensionality, 1 degree of freedom, two constraints
For three- dimensionality, 0 degrees of freedom, three constraints

We may now note two points:

1. That dimensionalities are constraints.
2. That the above relation between freedoms and constraints applies only to objects or to phenomena in the actual world.

This second point is necessary to meet the criticism of conventional mathematics (which insists that we may have any number of dimensions, with or without freedoms) that a dimensionality can be considered or devised that need not have this limitation. Here we would offer that such dimensionalities are fictitious. Not only is the observed physical world of objects three-dimensional and not four- or five- or six-dimensional, but even the postulated spaces—such as the flatland of two-dimensional creatures which was evoked to explain relativity in the 1920s—are fictitious. There are two-dimensional entities which "exist," but such entities are possessed of one degree of freedom. For example, a circle is a two-dimensional entity, but because its scale cannot be specified except by reference to the physical world, neither the concept of a circle nor its description in a drawing go beyond stating its shape, and hence leave scale as a freedom. (Scale in a drawing requires experience in physical space as its reference.)

Similarly, there are one-dimensional "entities"—for example, velocity or momentum—but any actual velocity or momentum has *direction,* which requires two degrees of freedom to specify (two angular measures would be required). Other presumably one-dimensional "entities" such as temperature, population density, barometric pressure, etc., are associated with degrees of freedom: temperature with position in space, population with an area, barometric pressure with location in a volume.

But proof of the thesis involves questions of semantics. What is important is to recognize that postulations of "*n*-dimensional"* existence are either dealing in fictions or with actualities which have $3-n$ freedoms usually overlooked or neglected.

II.

To the reader who may view an excursion into science as tedious or demanding, let us explain that our ultimate goal is to apply this ontology to *self* and to achieve definitions of aspects of consciousness. We have already noted the correspondence of the third level to concept, and hence to the reasoning mind, and of the second to values (to which feelings inevitably attach). We can add that the first correlates with purpose and volition. Thus we have:

1. Creation of an origin and direction of coordinate axes = orientation, hence *purpose.*
2. Extension of a radius = absolute magnitude, hence *value.*
3. Simultaneous measure of angle in two directions = form, hence *concept.*
4. The combination of the foregoing to provide objective measurement.

Thus it establishes the necessary orders in the process of generation and has significance for ontology (the "universal

---

*Any arbitrary number of dimensions.

and necessary character of existence"). Ontology is general; it can be applied to many areas. In what follows we propose to show its pertinence to the basic entities of physics, an exercise which provides valuable feedback to refine the theory.

CORRELATION OF DIMENSIONALITIES TO ACTUAL ENTITIES

We now take up the principal theme of our discussion, for which the consideration of dimension has been a preparation. This is that there is a correspondence between dimensionalities and the basic entities of physics. We have already hinted at such a correspondence with the suggestion that the world of concept is two-dimensional, and the world of scale, one-dimensional, as well as in the question at the beginning of this essay, "Are there two worlds, the ideal world of mathematical relationship and a real world of actual objects?"

We may now generalize. We have seen that two dimensions are the basis for depicting form, and hence of objective definition and concept. We should recognize a similar inclusiveness in one dimension, and note that scale, in its dependence on an involvement with the object, is a special case of *value*. Thus value, like scale, is one-dimensional; in fact, it can only be indicated by a measure having a one-dimensional range. Otherwise, we could not say that one thing was more valuable than another. There is thus a correlation between dimensions and psychological functions (if we may so entitle the sense of value and the capacity to form concepts).

But the correspondence goes deeper, for when applied to the universe of physics, we can deduce the existence of entities which not only resemble the entities of physics, but predict their properties and behavior, and establish categorical distinctions between them.

THE ZERO-DIMENSIONALITY OF PHOTONS
(Three Degrees of Freedom)

We may begin with electromagnetic radiation, whose ubiquitous presence is most familiar to us as visible light. Can one imagine a universe without light? The sun gone out, the stars extinguished? Hardly, for not only does the sun supply the energy for plant growth, and thence for all animal life, but without light we could observe nothing—perhaps not even think. And if this plea for the importance of light is insufficient, we can instance the generation of matter from photons, for a photon of appropriate energy can condense into a nuclear particle. And photons are also the end point in that ultimate destruction of matter which science is discovering in what is known as "gravitational collapse," the final disappearance of matter which occurs when a star becomes so dense that its nuclear material collapses into a "singularity" (how effectively this word conveys its meaning without benefit of definition). Whatever this singularity is, it is not matter; it is something that can contain energy in less space than can matter, and this must be photons, which *increase in energy as they decrease in size, or wavelength.*

In short, the beginning of all things is light, by which we mean photons of all frequencies or wavelengths. (At a wavelength of 10^{-13} centimeters a photon can produce a nuclear particle.)

Thus, in its smallness of dimension and in its possession of frequency, the photon conforms to the description of zero-dimensionality, as a point with an associated frequency. Further, the photon conforms to a zero-dimensional entity in that it has three degrees of freedom, because its free passage in any direction involves two angular measures (two degrees of freedom), and its velocity of 186,000 miles per second constitutes a third degree of freedom, of distance, or extension. In answer to the response that this is a "limited velocity," and hence a constraint, we must insist that the velocity of light is not a limiting velocity as is the "speed limit" on a highway—if the police were to chase a photon in a space ship, not only could they never overtake it, but however much power they exerted, and however closely they themselves approached the speed of light, they would still measure the speed of the photon away from them at 186,000 miles per second.

Thus our definition of zero-dimensionality deduces that the

primordial entity of nature is point-like, is associated with frequency, and has three degrees of freedom.

It is now in order to make a turn about and obtain information from the photon to implement our definition of zero-dimensionality. It will be recalled that we had to accept that the number associated with zero-dimensionality could only be distinguished from the number associated with one-dimensionality by the fact that it was a "pure number," devoid of the property of extension. We can now draw on the known character of the photon to sharpen this distinction, as follows.

We know that the photon may be defined as a unit of action equal to Planck's constant h. We also know that action has the measure formula ML^2/T, that is, it is energy (ML^2/T^2) times time (T), or momentum (ML/T) times length (L), or other combinations of these basic parameters of mass, length, and time. We now assert that such divisions of the photon are analytic, in a way, fictions; they are *aspects* of the totality, action, ways of dividing it into parts to which sense experience has access.

We earlier said that time could be defined as the inverse of the initial undefined term. This statement requires modification. The initial undefined term is *action*, but we can say that time is one of its components. Action is the product of time and energy. If we extract time (divide by time), we obtain energy!

The implication is twofold. In the first place, we can now say the initial photon, a zero-dimensional entity, has a *constant* as its measure (Action = Planck's constant h). This avoids the difficulty of having to call it a pure number. We have traced the dimensions of cosmological genesis to a dimensionless constant.

In the second place, this cosmological goose egg, Planck's constant, when it hatches out into one-dimensionality and produces time, *also* engenders energy, or mass and length. The same womb that produces number and dimension produces mass and energy. This conclusion is of great significance, for it demonstrates how "physicality" emerges from ideality, how physics unfolds from mathematics.

THE ONE-DIMENSIONALITY OF NUCLEAR PARTICLES
(Two Degrees of Freedom)

In terms of dimensions, the line is extension and the birth of time. In terms of the entities of physics, this extension manifests itself as nuclear particles, protons and electrons, whose rest mass makes temporal existence possible. This rest mass is the permanent investment that makes the physicality of the universe possible, for like "real estate" it has tangible reality. That is, a nuclear particle interacts with other particles without loss of its rest mass, whereas the photon is completely transformed in its interaction with matter. Even if the photon makes a partial expenditure of its energy, as in the photoelectric effect, what emerges from the encounter is not the same photon, but a new one.

The rest mass of the nuclear particle, by its permanence, contributes both substance and duration. We might say that it is this persistence that creates time, or that the creation of time requires permanence as a referrent. But time and substance are not enough; a third element is necessary. The need for a third element springs from the essential one-wayness of time. The notion of succession requires two things, plus an indicator of which comes first. This deduction is supported by reference to nuclear particles. As we were previously able to obtain information about zero-dimensionality from the photon, so we can obtain information from the proton and electron about linear one-dimensionality.

The third element is charge, and its asymmetrical association with mass (the positive proton has a mass 1,800 times that of the negative electron) establishes the basis for such other asymmetries as time direction, attraction and repulsion, etc., and accounts for the distinction between value and scale. We said previously that scale was a special case of value. We can now add meaning to this: the "desirability" factor in value (not present when, for example, we are talking about numerical "value") has its ontological basis in the asymmetry of nuclear particles: the electron revolves around the proton, not the contrary.

CONSTRAINT AND FREEDOM   165

To sum up, then, there is at this second level of generation an essential threeness, whether we call it time, motion, and change, or persistence, change, and interrelation. Nor is this richness only present in nuclear particles. Once alerted to it, we discover it in dimension, for the asymmetry of time is one of its essential and irreducible attributes; past and future are not interchangeable, logic to the contrary.

With this assurance (that the one dimension of time implies a threefold syndrome), we can go on to designate nuclear particles as *entities with one dimension of constraint and two dimensions of freedom*. Their constraint is persistence in time, but for the purpose of this analysis, the nuclear particle results when a portion of the original energy of the photon is precipitated into rest mass. This amount of energy (1 BEV), formerly unpredictable, becomes exact and known; the remainder continues as the product of two uncertainties, that of position and that of momentum $(L \times ML/T = ML^2/T)$.

Thus, a nuclear particle is a fixed amount of energy (in the form of rest mass) of uncertain position and momentum. Admittedly, this view of a nuclear particle omits Heisenberg's observer who "cannot determine its position without changing its momentum, and cannot determine its momentum without a time interval during which its position changes." The view here taken draws upon mathematical usage in which the specification of one of three dimensions leaves the other two unspecified, or "free." The present view goes further only in that it invests such a mathematical entity with real existence, citing a nuclear particle as a *linear entity*.

THE TWO-DIMENSIONALITY OF ATOMS
(One Degree of Freedom)

This brings us to atoms, which interchange the roles that constraint and freedom play for nuclear particles. Atoms at absolute zero temperature are without self movement; their *position can be fixed and determined,* but *their energy cannot!* An atom, even at absolute zero, is free to emit or absorb energy (in the form of photons). Thus, we can say that an atom possesses two-dimensional constraint and one-dimensional freedom.

We may also view the atom as having a two-dimensional nature, in that its orbital electrons establish a plane of symmetry, a two-dimensional matrix in which the form principle operates to create the precisely determined electron configurations known as rings or shells (according to the Pauli Exclusion Principle). It is these configurations which dictate the chemical and other properties of the over 100 kinds of atoms.

Note the important philosophical principle that properties are constraints. The more qualifying adjectives and phrases are added to the description of an entity, the narrower (more constrained) it is. "A house" could apply to almost any habitation, "a six-room house" is a more limited category. The atom, being the bearer of "properties," is more constrained than the nuclear particle, whose properties are limited only to its mass and charge. So we see the dimensional concept applies metaphorically as well as literally; the dimensional constraints of an atom correlate with its possession of definite properties. In fact, it is two-dimensional constraint that makes *form* possible. There can be no "form" in one dimension. Similarly, substance has no "properties" save that of substantiality.

THE THREE-DIMENSIONALITY OF MOLECULES AND MOLAR MATTER
(Zero Degrees of Freedom)

The fourth category of three-dimensional constraint correlates with molecules and molar matter. While it is true that molecules have an inherent motion at ordinary temperatures even when locked in crystals, and are only completely inert at absolute zero, this random motion is *not an inherent motion;* it is imparted by impact with other molecules which are agitated by temperature.

Even this motion due to temperature disappears for aggregates of molecules, the so-called molar objects, which are the physical objects we see and touch, the world of matter which is subject to the laws of determinism, and whose behavior was judged by nineteenth-century science to be typical of all objects, including the ultimate particles like atoms and electrons.

The view that we are here expounding—that determinism is a special case—is that of contemporary science, though it must be admitted that science today does not give a positive meaning to indeterminacy.

To a great extent, this lack of emphasis is due to science's ingrained respect for law. This "respect" is essentially aesthetic; the rational mind has difficulty accepting indeterminacy, which cannot be distinguished from chance or accident. As Einstein said, appropos of his reluctance to accept the quantum theory, "God does not play dice with the universe."

This remark of Einstein's bears some scrutiny. It is not, as would first appear, a religious statement. The God of the Old Testament, for example, did indeed take a chance in giving man free will. Nor can we allow the notion that chance and free will are essentially different, since to an observer there is no way to distinguish between them. If Einstein was sincere, we must interpret his sincerity as naive, like that of the soldier who believes that he has God on his side. The side, in this case, is that of the rationalist who cannot, or does not wish to, conceive that there is anything in the universe not subject to law.

But here, with molar matter, the factor of chance is minimal; the planets do obey exact laws, and inert objects like paperweights do not exhibit self-caused or spontaneous action. The progressive diminution of freedom which characterized photons, electrons, and atoms is in molar matter complete; constraint has taken over entirely. We have reached the end of the spectrum that runs from complete freedom to complete restraint.

HIGHER DIMENSIONS

Let us recapitulate the main points so far. By reference to the actual operations necessary to measure objects in space, we first established that dimensions involve much more than "pure" number. They exhibit a *difference in kind* which provides the basis not only for different *aspects* of reality—viz., its scale or "value" aspect as well as its conceptual or measure aspect—but different *orders* of "reality," viz., nuclear particles, atoms, etc.

Going further, we discern that each order, or aspect of "reality," is characterized by constraint and freedom in varying proportion, but of constant sum, being such that the total dimensionality is always three. The significance of this is that it establishes, albeit empirically, that there are no more than three dimensions—or four dimensionalities, for we must include the case where there are zero dimensions (no constraint). This would appear to indicate the impossibility of higher dimensions—four, five, and six dimensions.

Such higher dimensions have often been hypothesized. It was argued that just as a plane moved perpendicularly to itself produces a volume, so a volume moved at right angles to itself will produce a four-dimensional figure.

The implication was that because we can know both a plane and a volume, there is no reason to stop there; there must be higher dimensional figures. In the light of our present analysis, however, a plane is not so much a two-dimensional figure as it is the specification of two out of three possible dimensions, and hence can be moved to create a volume; whereas a volume, in which three dimensions are specified, has no further freedom into which to move, so it does not follow that four dimensions are possible.

We may, of course, take the position that dimensions beyond the three physical ones are of a different sort, i.e., that they are not accessible to the physical senses. This cuts us off from experimental verification, and exposes us to the same criticism that has been leveled at the ancients, whose doctrine of epicycles to explain the motion of the planets created a new hypothecation for each new factor.

However, we do have sanction from nature for a different kind of dimensionality. The formula for the interval between two events (interval is distance plus time) is expressed in Pythagorean fashion as the square root of the sum of the squares: $I = \sqrt{dx^2 + dy^2 + dz^2 - dt^2}$.

In this expression, x, y, and z are the dimensions of space, and t is time. Unlike the x, y, and z dimensions, time has a negative coefficient. Since the negative sign is associated with time squared, time itself must have the square root of negative unity as its coefficient. This, of course, is i, the imaginary. It is essentially different from space.

In space I can measure a step forward, a step to the right, and a step up; each direction is perpendicular to the one before. But if I am to take another "step" in a direction perpendicular to all three, it cannot be spacelike. It has to be a different *time* (later or earlier).

Eddington, in *New Pathways in Science,* makes this categorical distinction between time and space quite clear. He does so, however, in connection with his E-operators, which are beyond the scope of the present article.

Thus the distinction between space and time is already foretold in the structure of the set of E-operators. Space can have only three dimensions, because no more than three operators fulfil the necessary relationship of perpendicular displacement. A fourth displacement can be added, but it has a character essentially different from a space displacement.*

For present purposes, the E-operators by which Eddington arrived at this conclusion are not essential. The same categorical distinction between space and time is implicit in the formula for interval, indeed, in the commonsense distinction between time and space.

Common sense, then, recognizes the distinction. How does mathematics recognize it? To answer this we have only to note how the imaginary coefficient was created. In fact, it would be better to say we should note how the imaginary coefficient *forced itself* on the attention of mathematicians. For it only emerged when mathematical formalism took on a life of its own and dictated the necessity of entities to which no physical correlates could at first be found.

For those not familiar with the subject, a brief historical

---

*Sir Arthur Eddington, *New Pathways in Science* (1935), Ann Arbor Paperback ed. (Ann Arbor: University of Michigan Press, 1959), p. 276.

account is in order. In order to solve problems (of an algebraic sort), it was found possible to set up an equation such as: $x^2 = b$ (x times itself = B What is b?).

The solution, $x = \pm\sqrt{b}$ is straightforward (for example, $x^2 = 4$, hence, $x = \pm 2$), but there arose situations in which x^2 was equal to a negative quantity: $x^2 = -1$.

The solution, $\sqrt{-1}$, had no physical correlate, and was for some time set aside as *imaginary,* by which word it was judged to be "false," or "illusory."

However, Gauss and others found a representation for the imaginary as a "direction." Thus, if plus and minus are represented as opposite extremes of a line:

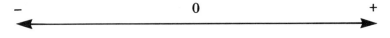

Fig. 35

we can think of the square root of minus as a 90° angle perpendicular to this line.

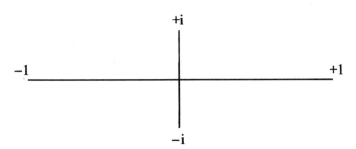

Fig. 36

This new idea emphasizes that the oppositeness of plus and minus, like the oppositeness of North and South, is but a partial account of a more extensive domain of *direction,* and that what had been dubbed imaginary is actually the mathematical counterpart of spatial direction, which is not just North and South, but East and West as well.

This expansion of mathematical usage has been tremendously fruitful, and the imaginaries now have a usefulness equal to that of their counterpart, the real numbers.

The lesson this teaches can be applied to the question of the origin of dimension. As we said before, the usual concept of dimension is additive. First we have the point with no dimension, then the line with one dimension, the plane with two, and the volume with three. But where have these added dimensions come from? They seem, like rabbits out of a hat, to be materializing out of nowhere. Surely, if we are sincere ontologists, we must give a better account than this. (Even if we abandon the deductive account of dimension and draw on the physical universe, then we must take the position outlined in the beginning of our essay and view dimensions as existing in the first place as freedoms, or blanks to be filled in. Such a course limits us to three dimensions and no more.)

However, if we recall that *the operation of taking the square root of negation produced an independent dimension* perpendicular to the other "diametrical" opposition of plus and minus, we are alerted to the fact that, at least in this case, a dimension is created "out of nothing" by the operation of taking the square root.

This leads to a further step. How did the dimension of oppositeness (plus and minus) emerge? *By taking the square root of unity!* In other words, if we find the square root of 1 we obtain two opposite values, $+1$ and -1, for either of these multiplied by itself produces unity.

In other words, we obtain one dimension (the range of values between plus and minus) by extracting from unity its square roots. The extraction of roots unfolds the dimension. We obtain the two dimensions of a plane by a second application of extracting a square root, which is the fourth root of the original unity. Note that the first extraction of the square roots of unity is disguised by the *convention of plus and minus*, which does not so much create a new kind of *number* as it does a *dimensionality of existence* for the operation of number.

The second application similarly creates a two-dimensionality, but here the convention is to refer to the $+i$ and $-i$ as square roots, whereas (+) and (−) do not have this other name. We can therefore trace the origin of dimension to successive extractions of roots of unity, the two applications thus far producing two dimensions whose typical equation is the circle.

A third application of a square root generates a quartic. A typical equation is:

$$\frac{x^2 + y^2}{y^2} = 1$$

This equation can represent a circle rotated 180° about a point on its circumference. By continued rotation such a circle produces a torus with an infinitely small hole, and is presumably a more general figure than the more commonly recognized sphere or cylinder, either of which would be created by moving the circle perpendicular to itself.

Here I would refer my reader to a very interesting paper by Dr. Charles Musès, "Time, Dimensionality and Experience."* Using algebraic methods, Musès shows that dimensionalities can be coded as the degrees of certain characteristic equations. In carrying out his program by successive increase in the degree of his equations, he performs the equivalent of what we have shown is accomplished by extracting successive roots of unity. Both methods produce the same results, but for the present ontological survey, derivation from unity by the extraction of roots has an advantage, because it stresses the dimensionless origin of dimension. It is interesting that p has the property $p^2 = 0$, $p \neq 0$, which means that the origin is on the surface.

Musès shows the generation of dimensionalities of a higher order, and gives characteristic equations of sextic and octic degrees. What is most interesting, in the present context, is that as the degree of these equations (referring to higher kinds

*Annals of the New York Academy of Sciences 138 (1967): 646–60. The above equation is Musès' p-field and represents two circles of half-unit radius tangent at the origin. The torus with an infinitely small hole is Musès' circular *umbilicoid,* given by $(x^2 + y^2 + z^2)^2/y^2 = 1$.

of number or dimensionalities) increases, the successive laws of algebra break down. First the commutative law, then the associative and distributive laws, until finally the field forms become multiform. In other words, in sufficiently higher dimensionalities, results are increasingly indeterminate. A given set of conditions will not produce a predictable result. This I would interpret as equivalent to saying that dimensionalities beyond the third see the reintroduction of freedom.

With this encouragement from Musès, let us pursue the implications of the scheme that we have so far laid down. This involves an extrapolation as to what must be beyond the three constraints which apply to physical objects. To perform this extrapolation, we have need of another principle not heretofore noted.

THE CUMULATIVENESS OF DIMENSION

This principle is that, as we move from one dimension to the next, we retain what has gone before. Each dimension builds on the one before it. This is easy to see in the case of three dimensions, for in making measurements of a real object, we have access both to the two-dimensional concept, or shape, and to one-dimensional extension, or scale. We say the box is $10'' \times 4'' \times 2''$, which describes its proportions and *assumes* the inch to be known. Only when we have to deal with unfamiliar units such as the cubits which measured Noah's ark are we made aware that this normally taken for granted aspect of measure requires a separate determination. So the dimensionality which describes form has dimensionality which describes scale to back it up. The latter, in turn, draws on the point of origin for its meaningfulness, not only in that we must start the extension from a point, but in that we must be able to swing this extension about in whatever direction is required. This is the angular freedom with which a point is endowed, if it is to be the origin for observations. A telescope that could not be swung about would have a very limited usefulness. So dimensions are cumulative. This is clear from an inspection of the expression for distance in spherical coordinates. Note that:

$$ds^2 = dR^2 + R^2(d\theta)^2 + R^2\sin^2\theta d\phi^2 \qquad \begin{aligned} R &= \text{radius} \\ \theta &= \text{longitude} \\ \phi &= \text{latitude} \end{aligned}$$

The first term on the right of the equal sign includes R only, the second includes R and θ, and the third includes R, θ, and ϕ. Each successive term includes the ones before. So we can expect of higher dimensionalities that they must include three-dimensionality.

This inclusiveness has philosophical significance by reason of the integrity with which it endows higher dimensions. Were higher dimensions just added on, they would not contribute the "organic" properties we will now consider.

HIGHER DIMENSIONAL OBJECTS

Coming now to the higher dimensions, we may expect, in accordance with the cumulative principle above, that higher dimensions must include lower dimensions. We have correlated the simpler or lower dimensionalities to the orders of entities recognized by physics:

zero-dimensional	photons	3° of freedom
one-dimensional	nuclear particles	2° of freedom
two-dimensional	atoms	1° of freedom
three-dimensional	molecules and molar matter	0° of freedom

Can we extend this correlation to entities of a more complex nature than molecules? A hint comes from the fact that Musès finds that for the next higher dimensionality after p, entropy is negative.

Now it so happens that there are physical objects for which entropy is negative, for example, seeds. If we regard a seed, or, indeed, any living cell as a higher-dimensional object, we can suppose that the extra dimensionality it involves is somehow interior to it. Externally, it resembles any three-dimensional object; but unlike an *inert* object, it contains an internal organization which can store or expend energy.

How may we designate such an attribute in its simplest terms? The answer is to equate this capacity of the seed with *power,* for power is the technical term for the capacity to convert energy. We may thus view a seed, or, more correctly, the plant cell, as endowed with an extra *internal* dimension, a *power* not possessed by the lifeless object.

Biologists have long sought for criteria to distinguish the living from the non-living. Is it growth? Reproduction? Sensitivity? Photosynthesis? If growth, then we would have to include polymer chemicals. If self-reproduction, we should include chemicals—notably the virus (which by other criteria is a kind of molecule rather than a life form). If sensitivity, then we must include metals, whose sensitivity to fatigue and poisons was specifically studied years ago by Bose,* and is now generally recognized. If photosynthesis, then the fungi, which are classed as plants, would not qualify . . . and so it goes. It is difficult to find a criterion to separate the living from the non-living.

The true distinction between life and inert matter is that life—and we here mean vegetable life—moves against the current of entropy; it locally violates the second law of thermodynamics. This law states that any given distribution of states, such as hot and cold, ordered and disordered, tends to average out. A glass of cold water, left in a room of usual temperature, gets warmer; a glass of hot water gets colder. Stones roll down mountains and fill valleys. The sheets of a manuscript, when scattered by the wind, lose their ordered position; cards, when shuffled, attain more random distribution. All non-uniformity eventually evens out, becomes more average.

Plants, on the other hand, manage to store energy within their own confines. This, of course, is a local phenomenon. The plant does not create this energy, it draws it from sunlight and stores it as starch and sugar. So the plant's activity does not affect the general situation; it is still true that the outpouring energy of the sun is wasting away; the second law of thermodynamics still applies in the larger framework. But

---

*Sir Jagadis Chandra Bose, *Response in the Living and Non-Living* (London: Longmans, Green, 1910).

locally, there is *negative* entropy; for as the plant grows, it stores more energy than it expends.

The plant uses this energy for its own metabolism and growth; without it there could be no metabolism, no growth. We can also view this storage of energy as storage of order. In fact, it is because the energy is stored as order that it is available as fuel to propel the life process, the metabolism which we could call manufacture.

Thus, the more correct statement of the law of entropy is that order tends to revert to disorder. Energy itself neither decreases nor increases, but the *availability* of energy decreases. In other words, order is available energy.

What the plant does, then, is to make energy available, to draw on the energy of sunlight and store it as order in starch molecules. It also *expends* this energy. This is important because, without this ability, the plant would merely store energy; this would replace one kind of determinism with another. It would have no "freedom." The plant has the power to store *or* expend, and thus has a kind of freedom which exempts it to some extent from determinism. It enables it to grow and to reproduce.

Reverting now to the dimensionalities we discussed at the beginning, it will be recalled that the first constraint was to give a fixed value to R, which is extension. The capacity of the plant for unlimited growth, which is not just its own growth, but its self-multiplication into seeds which themselves grow, may therefore be interpreted as a kind of inversion of this original constraint. The plant has the power to increase its extension: it inverts the initial constraint and makes it a power. This is not the same as dropping the constraint, as would occur if a nuclear particle reverted to a photon, or a forest burned up and converted to radiant heat. The constraints of the earlier levels are retained, the plant or the cell accepts the limitations of matter and operates with them to grow into something millions of times its original size. In short, it *escapes the limitations of determinism.*

Of course, the concept of the plant as a higher-dimensional entity does not completely account for how the plant is able to move against entropy, but it does give formal recognition to negative entropy as a principle, and thus provides scientific status for what ordinarily has to be regarded as an exception to scientific law, an accidental or fortuitous emergence.

The problem of plant growth has always been a mystery, and, since science has no place for mystery, there has resulted a sort of internecine war between the mechanists and the vitalists about the explanation of life. The mechanists, clinging to their faith in mechanical law and insisting that life is "merely" chemistry, shove the issue under the carpet and do a disservice to science by failing to recognize that the existence of a problem that transcends available techniques should inspire better technique. The vitalists are equally at fault for obvious reasons, for to hypothecate a mystique which cannot be solved is to retreat from the challenge that motivates scientific progress.

This struggle between vitalism and mechanism is futile, because both parties fall short of the mark. Their adherence to "policies" is a failure to adapt to a new challenge. (The word "adapt" has the same shortcoming here that it has in evolution—what is required is a creative thrust which penetrates more deeply into meaning.)

It is here that Musès' inquiry has such value. For to discover in mathematics itself, or, rather, in implications which spring from the same foundations that have produced calculus, quadratic equations, and tensor analysis, a new and different mathematics, which deductively anticipates what has heretofore had to be classed as a mystery, is precisely what is needed for the growth of science.

It would be very gratifying if we could go further and anticipate more of the character of life from other implications of the mathematics of higher dimensionalities. What does the breakdown of the basic laws of algebra—the commutative law, the associative law, and the distributive law—mean in terms of life or other possible manifestations of nature?

Recognizing that there are several dimensionalities beyond that in which negative entropy first emerges, and that there are animal, and possibly higher orders beyond vegetation (or organization at the cellular level), is it possible that the higher dimensionalities correlate to the higher orders of entity?

Such speculations, could they be confirmed, would indeed close the gap between the two worlds—the ideal world of mathematical relationship, and the real world of actual objects (to which we can now add living entities).

SUPPLEMENTARY NOTE

I believe that the real function of science today is the exploration of the living spirit of man. We are witnessing a revival of interest in psychic phenomena, ESP, dreams, telepathy, clairvoyance, altered states of consciousness, and even immortality and other manifestations not within the established disciplines of science.

What is lacking is obviously not proof of psychic phenomena. Rather, it is a theory capable of structuring and integrating the facts. Here we must recognize that, as Kuhn points out in *The Structure of Scientific Revolutions,* a scientific theory is never displaced by facts, it can only be displaced by another theory.

It has been assumed that these phenomena are outside the realm of physical theory. Sporadic efforts to invoke scientific procedures for investigating them (e.g., screening for telepathy) have been without issue. I believe it is time for some very careful rethinking, however, for it appears that the discoveries of modern science actually confirm mental aspects at the most basic levels of the *physical* universe. To show what I mean, I should first clarify the distinction between the physical and mental aspects of a situation.

When the message "come home" is sent by telegram, what actually happens is the transmission of electrical impulses along a wire. The electrical impulses, per se, have nothing to do with meaning. They must be translated into words by the telegraph operator, and finally into meaning by the recipient of the message. While the electrical impulses obey the laws of matter and occur in space and time, the *recognition of the meaning* of the message "come home" is not the kind of space-time occurrence that science deals with. The faculty of recognition is quite different from the means used in the transmission of messages. Mind and matter are separate and distinct aspects.

While science purports to deal only with matter, I believe that quantum physics has been affirming what is at least an equivalent of mind in the concept of the *quantum of action*. In science, action is the product of energy and time. It occurs in discrete units (quanta) which cannot be further subdivided, and are the fundamental building blocks of the physical universe. One of its forms is the photon, or pulse of light.

I believe that events at this fundamental level of the physical universe can provide us with a language for dealing with the mental aspects of a situation, such as the fact that recognition occurs outside of time. While its antecedents take place in time, recognition itself is instantaneous. Similarly, time does not exist for the photon, since, according to relativity, clocks stop at the speed of light. The effect of the photon (i.e., altering the state of an atom) and the act of recognition are both quantized—the former by definition, and the latter in that you cannot recognize somebody 1½ times. But, most significantly to this discussion, both are events in which some *non-material* entity alters the state of the recipient. (The photon is non-material because it has no rest mass.)

From the point of view of developing a science of ESP—that is, a science in which something immaterial has primary status—these discoveries of quantum physics have great significance, in that they establish a scientific precedent for the existence of non-material influence, opening the door for the acceptance of a more extended range of phenomena.

However, the successes of quantum physics have depended upon mathematics to reach what is not available to the physical senses. I believe that the science of the mind can be developed only by the use of more advanced mathematical equipment. This is what first attracted my interest to Dr. Charles Musès' hypernumber theory.

For many years I have speculated that the imaginary numbers have some correspondence with the non-physical. Consequently, I was intrigued when I read Dr. Musès' paper on higher numbers,* which extends the imaginary to hypernumber, and shows that there are a number of dimensionalities beyond those by which we measure physical objects.

My conviction of the importance of Musès' mathematical discoveries stems in part from the correspondence I have found between his dimensionalities and what I call the levels of process. My theory maintains that process must, by ontological necessity, involve sequential stages, each one building on the one before. The first step is aim or purpose. The succeeding steps see the initial aim, which is potential only, investing first in material, then in form, then in both together to produce what we call the physical world (particular objects or events). This is the world of determinism or law which classical science has long recognized.

The physical world constitutes a basis or foundation which is then "used" by process to create organisms (plants), living entities which employ physical law to grow and reproduce. This state of process is in my system the fifth, and I was struck by the fact that Musès' fifth dimensionality was described as having negative entropy. Since negative entropy is the distinguishing characteristic of plant life, I was moved to look for further correspondences.

I found several—more than enough to convince me that, in quite different ways, we were both uncovering a basic structure in the universe. While these correspondences cannot all be described here, I will mention one.

This is the correspondence between what I call "freedoms" and what he describes as breakdowns of the three laws of ordinary arithmetic. Musès states that first the law of commutation breaks down, and indicates that in yet higher number realms the laws of association and of distribution break down, too.

For my own part, I have shown that my higher states "win back" first two, and then three "degrees of freedom." It is

---

Annals of the New York Academy of Sciences (1967). See also the same author's "Hypernumbers II: Further Concepts and Computational Applications," *Applied Mathematics and Computations* 4 (1978): 45, and "Explorations in Mathematics," *Impact* 27 (1977): 66.

significant to show a correlation between these stages and the successive breakdowns of the laws of ordinary arithmetic.

This I have recently succeeded in doing. The sixth state in my system is characterized by mobility, typical of animals, which exhibit voluntary motion (contrasting with the relatively "one-dimensional" freedom of plant growth). How to correlate this with the breakdown of the law of association?

To do this, find an interpretation of *a, b,* and *c* for which *(ab)c ≠ a(bc)*.

Let *a* = brother
Let *b* = John
Let *c* = mother

Then *(ab)c* = the (brother of John)'s mother = John's mother, whereas *a(bc)* = the brother of (John's mother) = John's uncle. What is it about this example that is responsible for the breakdown of the law *(ab)c = a(bc)*?

It is obvious that we are dealing with two kinds of relationship, parental and fraternal. These are two dimensions of relationship, the vertical relationship of parentage and the horizontal relationship of siblings. Hence, where two-dimensional choice (minimally necessary for true mobility) is involved, the law of association breaks down.

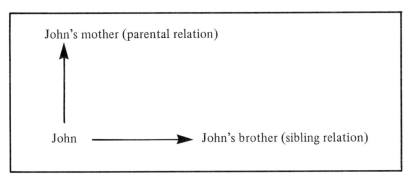

Fig. 37

The breakdown of the law of distribution correlates with my seventh state of process. In my system, this stage is subject to but one law, the law of hierarchy. How does this correlate with the breakdown of the law of distribution?

Choose an interpretation of a, b, and c which will cause the law of distribution to break down.

$$\begin{aligned} \text{Let } a &= \text{commander of} \\ b &= \text{company b} \\ c &= \text{company c} \end{aligned}$$

Then $ab + ac$ means the commander of company b plus the commander of company c. But $a(b + c)$, the commander of companies b and c together, is an officer superior to the captains of b and c individually—a colonel, at least—and two captains do not equal a colonel. This is the very essence of hierarchy.

I admit these are analogies. They are not what is customarily sufficient in mathematical proof. They do, however, have the merit of revealing the correlation between two distinct approaches—my theory of process and Musès' theory of dimensionality. I offer them in the hope that the non-mathematical reader may be aided in grasping sometimes unfamiliar conceptual levels in hypernumber theory.

I see in the breakdown of the laws of ordinary arithmetic in Musès' higher number spaces or dimensionalities a correspondence with the breakdown of physical laws in the realm of parapsychological phenomena, and I believe that his ideas may help in formulating theories covering these phenomena.

The Brain Scale of Dr. Brunler

The essay which follows is one which some readers may find unacceptable—not only because Brunler's* findings are based on the use of the pendulum, but because they evaluate persons on a scale, a concept which runs counter to the whole concept of democracy. There is also the objection that, unlike the other essays in this volume, which are about my own ideas, this one is about the ideas of someone else—hearsay evidence, as it were, and thus, by the criterion of legal procedure, not acceptable in a court of law.

It can also be objected that the list of examples at the end of the essay, all of whom are prominent figures, implies that only prominent people have high brain radiation—or perhaps even that Brunler was influenced by their prominence to include them.

To answer the last charge first, let me explain that Dr. Brunler took the radiations of some 25,000 people, and that the list I was able to obtain from him consisted of those to whom he made reference in his lectures and in my meetings with him. They were individuals in the public eye and were selected because to mention unknown persons would not serve the purpose of illustration. Further, as I note at the

---

*Dr. Oscar Brunler was a London physician who received the Bessemer prize in physics for the invention of the Brunler flame (used in underwater demolition work, especially during World War II). He is also credited with the development of the detergent marketed under the brand name Surf. Dr. Brunler died in Santa Barbara, California, in 1952, at the age of fifty-eight.—ED.

end of the essay, my reason for getting even those names that I did get was to make a study of their astrological charts to discover if a correlation could be found. If so, I would have a confirmation of Brunler's findings, albeit from an unorthodox source. To this end, I have had to confine my study to people for whom dates of birth are available, and hence for the most part to people of prominence. I do have the brain radiation of people not on the list, friends who have given me their time of birth and on whom I get Brunler's readings, but the inclusion of their names would serve no purpose, inasmuch as they would not be known to the reader.

Michael Polanyi, a scientist and philosopher, has made a plea for what he calls an ultra-biology—a biology which would look not just to man's animal heritage, to the chain of being leading to man, but to what lies "beyond" man, to the stages of evolution beyond that at which we currently find ourselves. I don't see how such an ultra-biology could have data to work on if it does not include not only obvious supermen, like Leonardo, but other intervening geniuses, be they great artists, writers, or political leaders.

As to the relevance of this essay to the others in the book, my principal theme, developed also in The Reflexive Universe,* is the evolution of the monad. This involves certain key issues:

> 1. The existence of a free entity, life spark, or spiritual monad which pushes evolution along. This we have correlated to the quantum of action, not only because it is not subject to the laws of determinism (it is indeterminacy), but because it is more fundamental than any particle, indivisible, and non-physical.
>
> 2. The slow evolution of this life spark, or quantum, through levels of organization—i.e., atoms, molecules, cells, organisms, animals, and human beings—implies that there are stages beyond man (though still in the same grand division).

If, then, we are to make a comprehensive study of man, we need to know something of man's evolution, and this includes, as it does with animals, a survey of the whole range of the human potential. As observed in The Reflexive Universe, modern man is at a stage corresponding to the clam in the animal kingdom. This introduces perhaps the most important consideration of all, and one totally neglected by current science.

---

*Arthur M. Young, *The Reflexive Universe: Evolution of Consciousness* (New York: Delacorte Press, 1976).

This is that the type of evolution pertinent to man is not that of the cellular organism, nor that of animal instinct, which is pertinent to the group soul of animals, but that of the individual *soul*. To this Dr. Brunler's evidence has great pertinence for, even if it is not correct, it gives us something to work on, a plan—and a plan, even if the wrong one, is better than no plan.

While some of the other essays in this volume do not pertain directly to man's evolution, almost all deal in one way or another with establishing the priority of what I have called the projective levels of being. "A Formalism for Philosophy" establishes these levels on philosophical grounds, "Constraint and Freedom" by the study of dimensions, "The Mind-Body Problem" by analysis into the four aspects common to both mind and body; "Rotation, the Neglected Invariant" and "Toward a Life Science" deal with the monad itself. An earlier version of "Time to Go Home" was the first published essay in which I stressed the immortality of the soul. This concept is essential to Dr. Brunler's thesis, which would make no sense otherwise.

It can, therefore, be seen that since the main thrust of this volume is to establish a cosmological basis for evolution, man's evolution in particular, Brunler's work is important. There is no point in defending the use of the pendulum—as in dowsing, it is a gimmick to draw on the unconscious and only to be trusted when it can be confirmed. In dowsing the confirmation is obtained when water is found at the point indicated by the dowser. There is no equivalent confirmation possible in the case of brain radiation.

On the other hand, the theory that Brunler deduced from his results was to me evidence that here at last was the first explanation of human evolution. Brunler's concept of sequential development over many lifetimes, first of the physical organism (sensation), then of feeling and intuition, followed by logic with the suppression of feeling, followed in turn by will, made much more sense to me than Jung's theory that these functions were developed all in one lifetime. Furthermore, I was impressed by Brunler's finding that before the self can manifest the creative genius of the great painters, composers, and political leaders, it has to go through lifetimes of heightened sensitivity, exemplifed in clairvoyance or other functioning as a passive or sensitive "medium"; such were the great performers, especially in music, where the creative role is contributed by the composer and its rendition by the performer. This made such good sense that how the readings were obtained was not a critical issue.

In other words, the correctness of the readings themselves is a *separate issue. It could be checked by meeting and getting to know people in the various ranges. This I made every effort to do, and found the readings a remarkably accurate guide to people despite their bewildering variety in other ways. Even this, however, did not satisfy me. I wanted to confirm Brunler's findings in a way that could be confirmed, and therefore began the arduous task of analyzing over 500 astrological charts. It is my hope that this will complete what was left undone in* The Reflexive Universe, *and provide a scheme or map of* human *evolution.*

As I have noted, human evolution is evolution of the individual person through many lifetimes. It is a subject that has been totally overlooked in the fracas about evolution of species. *Since all of humanity is but one species, all the fuss about survival, DNA, gene pools, and so on, is beside the point as far as you and I are concerned. In our next lifetime we will have a body that need have no biological link whatever with our present one. We need not be the same sex, or the same race, and our body may not be an improvement over the one we have now—it may be crippled at birth, blind, even Mongoloid, because we choose to take on some new challenge that we feel will help in our overall evolution. But one thing we won't lose is the overall competence which we have earned through this and perhaps hundreds of previous lifetimes.*

It was in 1951 that my wife and I first met Oscar Brunler, certainly one of the most interesting men I have known—not only because of his unique theory of "brain radiation" but for his vast knowledge of psychic phenomena, in pursuit of which subject we had come to California from our home, then in New York. We were introduced to Dr. Brunler and his wife (a British psychoanalyst) by Harwood White, brother of Stewart Edward White, the author of *The Betty Book.**

The immediate purpose of our visit was to learn our own brain radiation, but what proved of even greater interest to me was the significance of the measurement as an *evolutionary*

***The Betty Book: Excursions into the World of Other-Consciousness Made by Betty *[pseud.] between 1919 and 1936, Now Recorded by Stewart Edward White* (New York: E. P. Dutton, 1937); see, too, Stewart Edward White, *The Unobstructed Universe* (New York: E. P. Dutton, 1940).

THE BRAIN SCALE OF DR. BRUNLER 187

typology (Dr. Charles Musès' term), a way of evaluating what might be called the "wattage" of a given person—especially as applied to the great artists and creative geniuses of history, and which even more significantly provided an account of the evolution of consciousness.

The "brain radiation," which was rather a sort of psychic interplay factor, was registered with a pendulum (the type of process used in so-called "radiesthesia") and measured with the assistance of a long mahogany box about five inches square and four feet long—on which was a metal scale graduated in biometric units, five to a centimeter. Dr. Brunler stood at one end of the instrument swinging the pendulum over a diamond-shaped insert at the end of the scale. The person whose radiation was to be measured wore a headpiece from which a silk cord led to a slider adapted for motion along the scale. Mrs. Brunler operated the slider, pushing it along the scale until Dr. Brunler said "Stop!" The position of the slider gave the reading, which for people could be anywhere from 200 to 500 biometric units (50–100cm), or more for very exceptional invididuals.

The measurement, of course, was not of a kind recognized by ordinary science; it was rather of the nature of the emanations picked up by dowsers, and proof of its existence by scientific instrumentation is not likely to be possible. While both my wife and I were able after a series of lessons to make "brain radiation" measurements which involved long practice on fruits and vegetables, it was agonizingly difficult—and after serveral years we gave it up, mainly because we did not want to take the responsibility. (I should add that in making a reading we always did so independently, and only when we both got the same reading would we consider it valid.)

In any case, due to the highly subjective component in its detection, I would prefer to pass over this aspect and get to what really interested me—the fact that even if Brunler's theory cannot be confirmed (and I would scarcely trust enthusiasts of the pendulum any more than skeptics), the readings which he himself obtained for historical persons, when joined to his deductions as to their significance, make it a contribution of the greatest importance to the study of consciousness. To explain further, I should briefly review how he came to discover the theory.

THE BIRTH OF BRUNLER'S THEORY

Dr. Brunler was a practicing physician in London, using various unorthodox methods of treatment such as colored light and homeopathy, when he heard of the work of a Dr. Bovis in France, who used a "biometer" to detect the falsification of wine and to measure the health of various organs of the body—which he was able to do from a radiation from the finger tips, each finger corresponding to a different organ system. The radiation of a healthy organ was 100 biometric units (20 cm), which he was able to measure by having patients slide their fingers along a scale, at the other end of which he swung a pendulum. When the swing of the pendulum was displaced from a straight back-and-forth to a diagonal pattern, the scale reading was taken. Brunler found this worked and used it for diagnosis. Although the thumb was not mentioned by Bovis, he found that it, too, had a radiation, considerably higher than the other fingers. Not knowing how to interpret this, he nonetheless kept a record of it. Only when an idiot girl was brought to him as a patient, and he found the radiation from her thumb lower than any reading he had previously made, did it occur to him that the radiation from the thumb might have to do with consciousness.

The next person to come to his office was the newsboy. Brunler seized the opportunity, and confirmed that the radiation from the newsboy's head and from his thumb had the same value. What he'd been getting from the thumbs of his patients was a measure of consciousness!

THE BRUNLER SCALE

Brunler turned to his records and noted that a number of big executives had brain radiation readings ranging from 420 to 440. In this range, too, were actors who could hold an audience for a full performance—as well as successful novelists. Details began to fill in. The great majority of people, 95 percent, were below 300, and 90 percent were below 240. A critical point was 310. Below this, physical skills reached a high point of development; above it, the mind started to open out (320–30: mind readers and hypnotists). Intuition and feeling underwent a steady development above 330 to 370, at which point there was a reversal; from here on intuition was repressed and pure reason was stressed, until eventually at 390 intuition returned and was integrated with reason.

From 390 to 420, which Brunler called the fear range, the self is acutely aware of its deficiencies, until at 420, in the ego range, will emerges, and the whole nature changes again to a self-assertive independence or leadership of others.

In other words, between the reason range, where are found the top orthodox professors, defenders of the faith as it were, and the ego range, the self evolves from a receiver to an originator.

The ego range begins to fade at 440, and at 450 (T. E. Lawrence was an example) the ego is thrust aside and the self rejoices in anonymity. Here we find a number of cases of musical interpreters (Toscanini, 474) and political leaders (e.g., Wendell Wilkie, 475) characterized by sensitivity beyond that of the ego. At 480 we find a number of outstanding clairvoyants—Eileen Garrett, 482; Edgar Cayce, 482; Rudi Schneider, 485—as well as outstanding musical performers, who exhibit sensitivity by making themselves a vehicle for the composer: Artur Schnabel, 484; Chaliapin, 482; Myra Hess, 474.

It will now be evident what is going on. The self, which we must assume passes through these stages in succession, requiring many lifetimes to do so (Brunler insisted that with rare exceptions the radiation changed by only a few points during a lifetime), is gradually unfolding new potentialities and powers. Having mastered the normal faculties of sensation, intuition, and reason (note that these are developed in sequence, and not all at once as Jung implied), the self then moves on to develop the will in the ego range.

After will and independent action are mastered, a new sensitivity opens up, one which responds to wider horizons than are visible within any given age (for example, the Edgar Cayce readings).

This preparation is necessary for the next major development, what Brunler called personality or genius, but which he defined as capable of creations which would survive the death of their creators. Here in the 500's occur the great political leaders: George Washington, 512; Bismarck, 511; Queen Elizabeth I, 510. (I should explain that the radiation can be taken from a manuscript or from a painting, most effectively from signatures.) Writers: Emerson, 500; Whitman, 510; playwrights: Shaw, 518; Coward, 520; Yeats, 529; composers: Haydn, 535; Wagner, 538; Liszt, 538; and impressionist painters: Manet, 545; Monet, 538; Cézanne, 570.

Brunler went about the galleries of Europe and was able to persuade them to let him take down paintings so he could put salt on the signatures, from which he subsequently derived their radiation.

To me the placement of high clairvoyance in the 500's immediately below the great artists is a most impressive contribution, enabling us to see how things fit together, how genius is possible. The artist or leader whose work survives his death must first develop the sensitivity to see beyond what is currently accepted as good taste. That the acquisition of this sensitivity itself has to be acquired in earlier lifetimes is understandable when we realize that sensitivity and creativity are essentially antithetical, like inbreath and outbreath: one cannot do *both at once,* though each is necessary to the other.

But to go on with Brunler's Scale: 538, he said, was critical; he called it "the fork." Here the self must begin to give up the personality, which at this stage reaches a culmination (Paganini and Greta Garbo are examples). More powers await, but they are no longer centered in glorifying the person. At 575 the personality is given up altogether. Sri Aurobindo (at 585) wrote a book on giving up the personality.

At 590 we find Macaulay, who could memorize a page in a single reading. Also in this group are Napoleon, 598; Goethe at 608; and Francis Bacon at 640. In the high 600's we find the great renaissance painters, Michelangelo, 688; Titian, 660;

Raphael, 649; Veronese, 646, as well as Rembrandt, 638, culminating in Leonardo at 720, the highest reading that Brunler got.

But I've neglected the lower ranges, which while less glamorous are also of interest. The lowest readings Brunler found were among the hill tribes in India—around 180. (Indian holy men doing various feats were mostly in the ego range, developing their wills.) Below 210 the capacity to read and write is being developed. Once he found a man practicing driving a nail, which he just couldn't do; his level was 210. From here on physical skill develops rapidly, becoming pronounced at 240. He found many bridge players here, as well as stockbrokers (the transactions in both cases deal with quite concrete symbols). At 240 an interest in ESP and occult subjects, especially of the more sensational type, is observed, and these people wander from one guru to another. In the high 200's exceptional mechanical and external skills are observed: toolmakers, chefs, and hotel managers were found here.

Perhaps the major crisis point of the whole range (with the possible exception of 538) is 310, which Brunler called the battle with the physical. There he found drug addicts, alcoholics, sex maniacs and, interestingly enough, test pilots. Perhaps the latter especially exemplify the crisis here, where the will has to conquer the physical urges and instincts. The test pilot is in continual need of overcoming his fear; the alcoholics, drug addicts, and sex maniacs, to control physical urges which at this range reach their peak. It is also a characteristic of this group that they have visions far beyond their capacity to realize. The only *public* figure I have in this range is King Farouk I of Egypt, at 307.

In closing I can only add that Brunler's concept of the self's evolution made a deep impression on me. It makes sense, and it accounts for many of the basic differences between people, their different interests, competences, and scope of influence. Douglas Dean, the parapsychologist, has found, like Brunler, that business executives have a high degree of ESP.*

*See Douglas Dean et al., *Executive ESP* (Englewood Cliffs, N.J.: Prentice Hall, 1974).

Yet parapsychology is still, even after one hundred years of probing and verifying, beginning with the remarkable work of Reichenbach, a top scientist of the nineteenth century, on up to Rhine of recent fame, denied by most scientists, and accepted only with the kid gloves of "instrumentation" by parapsychologists. Why? Because the academic world which constitutes the stronghold of orthodox science is dominated by the professor type in the logic range—who, rather compulsively, suppresses the intuitive faculty in order to develop reason.

This is not true of the range in which top executives are found—between 420 and 440, where reason and intuition are interpreted to the service of will. These persons, according to Brunler, admit freely to ESP and are, in fact, natural healers. Douglas Dean found to his surprise that his executive subjects not only used ESP but were quite ready to come out into the open about it.

Needless to say, I have made every effort I could think of to check his theory. Not only did we go to California three times in 1951 to learn as much as we could about it (fortunate since Brunler died by 1952), but my wife and I went to England in 1952 to look up many of the people whose brain radiation he had found to see if what he said was borne out. We found that it was, and that knowing the person's brain radiation made it possible to adjust to them in a way we might not normally have done. Those in the ego range especially—who are not always what one would normally call cultivated persons—are often difficult to relate to.

Moreover, and I say this with some hesitation, since it will do little to convince the skeptic, I spent a great deal of time studying the astrological correlations to brain radiation. I cannot go into this subject here—it would require a full-length book to cover the 500 charts and biographies involved, but I will say that out of some eight examples at the fork (538), I found that three had the opposition of Mars and Jupiter with Mercury in T-square to both. (The opposition of Mars and Jupiter was considered by the late psychological astrologer C. E. O. Carter to be the most difficult of aspects.) With

Mercury in square to both, the probability of its single occurrence by chance (using an orb of 10°) is one in 600. However, three times out of eight reduces this to one in a million.*

BRUNLER'S "BRAIN RADIATION" LIST

720 Leonardo da Vinci
700 Giorgione
688 Michelangelo
675 Cheiro (palmist)
600 Madame Blavatsky
660 Titian
657 Frederick the Great
649 Raphael
646 Veronese
640 Francis Bacon
638 Rembrandt
633 Rubens
613 Goya
608 Goethe
598 Napoleon
590 Macaulay
590 Turner
585 Sri Aurobindo
585 Botticelli
580 Blake
577 Whistler
577 Domenico Scarlatti
575 Frederick Marion (psychometrist)
575 Walter Russell (sculptor)
570 Cézanne
568 Annie Besant
567 Tchaikowsky
562 Sir Walter Scott
561 Johann Strauss
555 Rossini
550 Renoir

550 Chopin
550 El Greco
545 Manet
542 Degas
540 Dickens
538 Greta Garbo
538 Wagner
538 Liszt
538 Monet
538 Alice Bailey
535 Haydn
529 Yeats
529 Elizabeth Browning
527 Kipling
526 Rasputin
525 Lawrence Olivier
525 Tennyson
522 Byron
520 Noel Coward
520 Mary Pickford
520 Nehru
520 Stalin
520 Galileo
519 Brahms
518 Bernard Shaw
515 Picasso
514 Rommel
512 Swift
512 George Washington
511 Bismarck
511 Longfellow
510 Whitman

*Of equal interest is the fact that this configuration of planets is referred to in alchemical tracts as "the Way."

510 Queen Elizabeth I
508 Eddington
507 James Clerk Maxwell
500 Lenin
500 Emerson
495 Salvador Dali
495 Joe E. Brown
493 Garibaldi
492 Franklin Roosevelt
492 Madame Curie
492 The Duke of Wellington
490 Tesla
490 Franco
489 Poe
485 Rudi Schneider
485 Woodrow Wilson
484 Artur Schnabel
482 Chaliapin
482 Edgar Cayce
482 Eileen Garrett
480 Marconi
477 Herbert Hoover
477 Gandhi
475 Wendell Wilkie
474 Toscanini
474 Myra Hess
473 Tagore
470 Mussolini
470 Edison
469 Einstein
462 Lincoln
462 David Hume
458 Queen Victoria
456 Charles Darwin
452 T. E. Lawrence
448 H. G. Wells
448 Henry Ford

447 Lewis Wallace
442 General Montgomery
439 G. K. Chesterton
439 Cardinal Richelieu
434 Ernst Haeckel
429 Samuel Insull
427 Winthrop Rockefeller
427 Savonarola
426 Sir Arthur Conan Doyle
426 Maria de Medici
423 Krishnamurti
422 The Duke of Windsor
421 John F. Kennedy
420 Bertrand Russell
420 Sigmund Freud
418 General de Gaulle
417 Catherine de Medici
410 Clive Bell
407 Walter Winchell
406 Billy Graham
405 Poincaré
397 Thomas Lipton
394 Anne Boleyn
391 Jackie Kennedy Onassis
390 Czar Nicholas
385 Gurdjieff
385 Carl Jung
382 Harry Hopkins
378 Gorki
378 Count Keyserling
375 General Chiang Kai Shek
360 Kinsey
355 Frank Buchman
    (Oxford Movement)
338 George Carpentier
308 Carol II of Romania
307 Farouk I of Egypt

A Letter to Dean Robert J. Jahn

Foundation for the Study of Consciousness,
1812 Delancey Place,
Philadelphia, Pennsylvania 19103

22 December 1972

Dean Robert G. Jahn
School of Engineering
Princeton University
Princeton, New Jersey

Dear Dean Jahn:

Your letter of November 1972, asking me to fill out a questionnaire [about the value of the courses in Engineering at Princeton], is acknowledged.

While I'm not sure I can be of help to you, I would like to try. Please forgive this rather long letter, but your inquiry touches on my interests and your statement "open a channel of communication whereby the school can benefit in the long term from alumni experience and opinion" indicates possibilities I have always hoped for.

To answer your first question. At Princeton, I was in the

class of '27. Eisenhart, Fine, Magee, Wedderburn, Trowbridge, Compton, and Alexander were teaching the courses I took. My first choice for my major was astronomy but when I was put to work counting craters on the moon, I applied for transfer to math. Years later Stewart recalled my argument with Henry Norris Russell about my preference for theory.

In mathematics I had the unique privilege of a two-year course with Veblen in relativity—a subject which had aroused my curiosity. I was also stimulated by Bertrand Russell's "Introduction to Mathematics," but both Russell and Einstein left me dissatisfied, principally because of their neglect of the importance of time and the essential difference of time from space, and before I left college I started to draw up what I called a Theory of Process which would answer the paradoxes discussed by Russell.

But, launched into the world, I found myself running out of ideas in philosophy. I began to try to do something "practical." Knowing nothing about aerodynamics, I thought I'd invented a method of lifting an aircraft for which the drag would not increase as the velocity. This would enable speeds to be reached sufficient to reach the escape velocity and thus travel to Mars.

After a year of preliminary experiments to verify the principle, mostly consumed in the difficulty of making things (such as a crude wind tunnel), I went to the New York Library and discovered the fallacy in my lift device. I also found that a vast amount of aerodynamic research had been done which I hadn't known about.

Recognizing the impossibility of my Martian project I struck about for something more feasible. I had not given up my project for the Theory of Process, I held this sacred, but I needed, as I put it then, to first do problems in which I could, as in the physics course, "look up the answers in the book." In this case the book was nature—so I wanted to apprentice myself to philosophy by developing some mechanical device.

I went to Washington and spent weeks in the Patent Office looking up tentative ideas—sound on film, television, stereo, and color movies, etc. I settled on the helicopter, which had not been successful up to that time, although there had been numerous attempts.

I then set about to experiment with models. This time I proved in model form the feasibility of the idea—which was the type of helicopter in which a large rotor is driven by propellers on the rotor blades. Isacco and Bleeker were both working on this type, but I believed I had an improvement in that I put the engine in the fuselage and conducted the power by shafting to the propellers (located on the rotor tips). My first discovery was a gear which would convey power from the fuselage to the rotor system without torque, which was a good idea, but blinded me to the other disadvantages of the auxiliary propeller drive.

I spent years trying to build a machine that would hold together under the tremendous gyroscopic and centrifugal forces involved in this design. In the twenty-horsepower model this involved propellers turning 4000 rpm having to make a complete reversal in "absolute" space 800 times per minute. (I had good reason to be dissatisfied with relativity, which did not even give consideration to the fact that rotation is *not* relative.)

This was a great waste of time as far as helicopter design was concerned. I spent years building and rebuilding. But I learned about stress and lightweight design the hard way. At last, with forged magnesium propellers, the blades made exactly alike on my own profiling machine, I got it to hold up under overspeed tests long enough to satisfy myself the job could be done.

But by this time, I began to realize I was using the wrong design. Sikorsky was already making headlines with a single rotor with tail propeller to offset the torque, and I could feel that the time had come to give up the idea I'd been pursuing for eight years and start over. This is perhaps the most excruciating decision for an inventor to make. Most of the time, 99 percent of the time, he must doggedly pursue with a dedication that amounts to madness, his original inspiration—he must fly in the face of authority, he must refuse to be turned aside by failure, etc., but still on top of this, know when it is time to abandon everything and start over.

This one issue—the most important not only for inventors but for any enterprise—is reason in itself that what is important *cannot be taught.* It cannot even be learned. It is purely existential.

In any case, I dropped the propeller-driven lift rotor and went back to small models. Here my long training in making things, in the use of gears, ball bearings, fabrication from dural, and lightweight construction paid off. By use of light models propelled by electric motors overloaded 200 percent, I could make model flights under controlled conditions. To my utmost surprise, all conventional configurations were highly unstable in flight. (My original flights had bypassed the question of instability because I'd used a rigid rotor.) So I had a new problem—stability.

I tried to devise a stabilizing device consisting of a pendulum which would hold the swash plate (which controls the rotor tip-path plane) in a horizontal position. This was a failure. I could see that the pendulum could not distinguish between the acceleration due to gravity and flight acceleration, and would swing with the fuselage as the pendular motion developed and not act as a corrective. I tried various combinations of what I called mast dependent and mast independent control systems with but minor success, until I introduced a stabilizer bar or flywheel which controlled the rotor and made it independent of the mast position in space. This device, while independent of the mast for short time intervals, was over a long period maintained perpendicular to the mast and hence horizontal. Hence it acted as an artificial horizon. It was highly satisfactory. The model could be flown indefinitely in smooth or rough air and remain perfectly stable, even when the mast was forcibly rocked.

It was now no trick to introduce remote control (electrical via solenoids on the swash plate) and I was able to maneuver the model about the interior of my barn or make it go out the door and return.

A LETTER TO DEAN ROBERT G. JAHN

This was the model I demonstrated at Wright Field in 1941. (Colonel [H. F.] Gregory rescued me from the Draft Board with promise of an Army contract.) Soon after, in September 1941, I demonstrated this model at Bell Aircraft, then in Buffalo, New York. My choice of Bell was based on the hope that their experience with a geared propeller (whose long shaft led from a rear-mounted engine through the cockpit) would stand in good stead for the challenge of helicopter design. As it happened, Larry Bell, a man of real vision, was looking for a helicopter much as I was looking for a backer—so a deal was closed. I assigned my patents to Bell, they agreed to build two helicopters to my specifications. Bart Kelley, a boyhood friend, who had helped me wind up the model experiments came with me to Bell at the modest salary of $35.00 per week.

But my troubles had just begun. I had expected I would have all-knowing engineers to make sophisticated full-scale designs from my experimental models. Instead, nothing happened. Such spasmodic suggestions as were forthcoming were ludicrous. When I asked the stress department to establish a wall thickness for the all-important mast, whose outside diameter other considerations established as $2^{5}/_{16}''$, the answer came back $^{1}/_{16}''$ wall, one fifth of what I thought necessary (they designed for ultimate, I for fatigue).

Other efforts were equally wide of the mark. They insisted on designing the hub very stiff to resist the bending load imposed by the lift of the blades, ignoring the fact that if the hub is permitted to bend centrifugal force reduces the load.

Besides which, the project, which came under the drafting department, was conceived as a design project in which manufacturing blueprints were to be the outcome. I wanted to make helicopters and test designs before making drawings. I found that their cost schedule would consume the funds allotted ($250,000) entirely in drawings. In desperation, I went to Russ Creighton, in charge of manufacture, and told him what I wanted to do. He sympathized and underwrote my alternative budget to *build* two helicopters for the same amount. He added a proviso "that the engineering department should have nothing to do with it."

Larry Bell OK'd my budget. I got an old garage some six miles from the main plant, drew on manufacture for a few top mechanics—lathes, milling machine, shaper, etc.—and we went to work. In six months after our establishment at "Gardenville," as the project came to be known, we had built a helicopter which we very gingerly began to try to fly in January 1943. I had never flown an airplane, so the experience was doubly uncertain.

But to make a long story short, after considerable difficulties, we emerged in June 1943 with a machine that flew beautifully (by this time a Bell pilot had been assigned). From then on, the problem became one of restraining the enthusiasm of management, who wanted publicity and demonstrations, whereas we could see we had a long battle ahead of us. We found steel chips in the gear oil, engines had to be replaced in 20 hours—there were still many bugs. It was not until two-and-a-half years later, after we had built three machines searching out the definitive design, that on December 8th, 1945, we launched the production prototype Model 47. Even then, our troubles were not over, but the design must have been basically sound, as this same Model 47, with a more powerful engine, is still in production.

Two years later, after most of the production bugs were ironed out, and the main engineering department indoctrinated into the helicopter, I left Bell to return to my original interest in philosophy.

In a way, those last two years were the most difficult part of my helicopter career. In any case, they are most pertinent to your inquiry on the subject of training engineers.

It was in this period that my opinion was formed. I cannot honestly feel that an academic engineering training contributes to the kind of development work involved in helicopter design. Let me explain why. In a large aircraft company, the engineering department tends to become a sort of bureaucracy. In any case, it was at Bell, and others have told me that other aircraft companies are similar. When I first arrived, Bell was busy with war contracts and the men assigned to me were whatever I could get from a department which needed all the good men it could find. An auto mechanic—the son of a banker friend of the boss—some very young apprentices—not the cream certainly—but they were willing to learn and caught on fast; they pitched right in with a will.

Later, after our success, we had the entire engineering department to indoctrinate. And let me add, without Bart Kelley, and the banker's son who turned out a genius, it could not have been done. A majority of the engineering department were trained specialists and they were completely at sea with the helicopter, it was hard for them to learn, and they were conditioned to avoid contact either with flight tests or with the shop—they simply didn't know what the act of child bearing and child rearing was about. They wouldn't dream of going into the shop to see if a part was made properly or to flight test to see if it worked. They were so housebroken they couldn't pee.

So we had to invent a decontamination course. We would put a man on a particular part: he had to design it—see it through the shop—take it to flight test—get the pilot's report, or better, fly it in the shop and see how it worked—then go back to the drafting board and fix it. This ordeal, which we called the "taste of blood," had a wonderful effect. The person became a new man, he found he could make things happen. Instead of beefing about the system, and prostrating himself before his own delusions, he learned something and enjoyed it.

Naturally there was a lot of grief for the huge aircraft company to be taken over by a few helicopter nuts. Three vice presidents had to hang themselves before Kelley was put in charge—but the miracle happened. Bell Aircraft is now Bell Helicopter. Kelley is vice-president in charge of engineering; Jack Byers—the junior draftsman, is project engineer for the Huey—perhaps the most successful helicopter to date, certainly the most used; and Bell has been cited by Congress for efficient engineering and economical manufacture.

The rest of my story wouldn't interest you. While I kept in touch with my old friends at Bell, and even took time off to make a variable diameter rotor intended for convertiplanes, which was later taken over by Bell but has since died on the vine—I have for the last twenty-two years been at work on my Theory of Process.

I doubt that I can make any positive suggestions that would be applicable to your problems, except perhaps to say that I think what benefit I did derive from Princeton was a consequence of the relative leisure then possible, which gave me time to think for myself and be self-motivated. I recall my enjoyment of Taylor's lectures in physical chemistry, in which I was not enrolled but attended for the fun of it.

As for teaching helicopter theory or practice, that is nonsense—this can be learned much better in industry where the action is. The vogue for hiring Ph.D.'s has fortunately met its deserved decease—it was only put in motion by government extravagances which by now have transferred to greener pastures. Besides the possibility of new helicopters radical enough to require ideas not currently being tested is as unlikely as it is for new automobiles.

What would be truly valuable, provided it could be taught, would be a basic *understanding* of physics. I mean by this, the grasp of principles which would have saved me from thinking I could devise a method of lifting which was independent of velocity. Most inventions fail because of the confusion between force and power. Even trained aerodynamics engineers have their heads so full of coefficients, induced drag, profile drag, etc., that they cannot recognize basic relations of similitude. One of my great advantages was that I instinctively avoided German aerodynamics and stuck to the French, which was less caught up in complicated formulations which missed the point.

I would like to briefly mention something of my career since 1948 when I retired from helicopters, not because it has been my most important life work, but because in some vague way it might relate to Princeton, or at least to the hope mentioned in your letter which I've already quoted.

What I've been doing in the last twenty years, in continuation of the Theory of Process that I started in 1927, has been to take my studies of the sciences much further than I did in college—largely because science has progressed very rapidly since then. Quantum physics, biochemistry have broken new ground. In addition, I did not get into biology at all in college and this has been important to my theory, which is now just about finished (though such things never end).

In this endeavor, I have tried to get help from Princeton on many occasions. In fact, in retrospect I have spent more time going back to Princeton than most of my classmates, and never to football games. In the thirties I tried to persuade Eisenhardt to have a library on aerodynamics because I had spent so much time going to libraries to get the information I needed for helicopters.

In the fifties, I was asked by Nickolsky to take charge of aerodynamic research, but decided against it because I felt my approach would be too radical. (I wanted to build something that had not been done before and realized the students wouldn't have time.)

I was a frequent visitor to Veblen, who finally insisted I take courses in logic (so that I could talk to logicians) before introducing me to Gödel. Veblen unfortunately died a few weeks after I'd sent him my paper on logic and I never met Gödel. I also visited Hempel, beseeching him to notice what had happened in quantum physics, which I'd thought was of tremendous importance to philosophy. And more besides—including President Goheen. My quest with President Goheen was to get assistance from various departments on the task of including many sciences in a comprehensive new paradigm which I believed would assist in the then proposed integration course. President Goheen, while interested in my efforts, realistically pointed out he could not invade the departmental sovereignty.

So I've had to go it alone—even though after all these years I find myself headed in the academic direction, for I'm scheduled to give courses in the integration of science in two small colleges in 1973.

I would rather not guess where Princeton finds itself in the current crisis. Indeed, from what I can gather from the Alumni Weekly, there is none. But I think there is and the academic stronghold is receiving impact because of its responsibility for setting the pattern for our contemporary culture. It has created the mores, the forms that are currently found wanting.

While I consider myself a scientist insofar as I have made use of science in invention, I think that science has been overrated in two ways. In the first place the advance of science has been possible largely because of inventions which have always had to make their way in the face of prevailing opinion, often as dogmatic as were the views of the scholars who refused to look through Galileo's telescope. Again because the credo of science is often at variance with its true findings. This is especially the case today, when the findings of quantum physics are still not appreciated or understood. The philosophies of science, presuming to interpret the laws of nature, fall victim to a superficial rationalism, and idolize determinism even so far as to deny spontaneous intercession, human or divine, in their billiard ball universe.

I therefore claim that the superior conceptual framework attributed to science, and in whose name religion and spiritual insights are set aside, is the malady of science and not its important contribution. By this I am in no sense condemning science, I am exonerating it. This conceptual framework, attributed to science, has eroded our moral sense, the dignity and even the existence of man. (What science gives to man any existence as such?)

But I can hardly expect to persuade anyone of this by argument. My evidence is that the intellegentsia, the enlightenment which has discredited our instinctive heritage and replaced it with the philosophy of science, is in great trouble today. The new generations are groping for some deeper understanding of themselves than they receive from the academic departments—have discovered Zen philosophy—astrology—Tarot—the very things that the enlightenment thrust aside. In short, I think the kids have something, they are right in rejecting the devitalized pablum that the philosophy of science is turning out.

But you've heard all this before and I'm sure you are concerned. Everyone is. I am and I think there is a possibility I can help, for it so happens that the younger generation has broken into the same feed bin I was exploring over twenty years ago and finding real nourishment and food for thought. The young mechanic at the garage converses with me on the book I happen to be reading by the Dalai Lama—he has already read it. It was these subjects, rejected by science, that prodded me into a long-term effort to make a deeper study of science to see if they could be incorporated in a unified cosmology or synthesis. In 1952 I set up the foundation whose letterhead I am using, and began my investigation. I discovered my first clue in quantum physics in the principle of action. A second was the recognition that control, which had proved so important in helicopter work, could be identified with the third derivative (rate of change of acceleration) and hence given status equivalent to that of other scientific measurements.

The first of these factors, action, provided the thrust of process, and the second, control, its guidance. The next clue was that process has an internal structure (strictly topology) such that there are a precise number of stages, with relations between them which define their role and nature.

These stages had their principal exemplification in nature, in the kingdom or levels of organization, nuclear particles, atoms, molecules, etc.

It now became possible to develop my theory of process beyond anything I had originally imagined. It became apparent that not only was what was once called the Great Chain of Being a process, but that each stage was itself a process. In atomic physics, the buildup of shells in the atom; in chemistry (with the help of Dr. Charles Price of the University of Pennsylvania), in the grades of complexity in molecules, from simple salts to proteins and DNA; in biology, in the levels of organization of both plants and animals. Through each of these grand divisions, a definite process of development could be recognized.

The theory achieves the goal I set in my college days, it accounts for much that is currently not understood, it integrates a number of sciences, and, by the judgment of young people who have attended my seminars, it serves as a mnemonic device which clarifies and simplifies the study of science generally, not unlike the Mendeleev table does for chemistry. I might add that it suggests many possible directions for research.

This brings me to present time. Science needs new challenges, the young people are interested in new frontiers. I would like to see the academic fraternity join in what I think can be the Great Instauration and Bacon's vision of a true science enacted in the present age. Mundus sensibilis made enormous contributions, let mundus intellectus now come forth.

Sincerely,

Arthur M. Young

By Arthur M. Young

The Reflexive Universe: Evolution of Consciousness (1976)
The Geometry of Meaning (1976)
The Bell Notes: A Journey from Physics to Metaphysics (1979)

available from Delacorte Press, 245 East 47th Street,
New York City, 10017.

Which Way Out? And Other Essays (1980)

available from Robert Briggs Associates,
2924 Benvenue Avenue, Berkeley, California 94705.

The Jews in Late Ancient Rome

Evidence of Cultural Interaction
in the Roman Diaspora

Leonard Victor Rutgers

BRILL
LEIDEN • BOSTON • KÖLN

This journal is printed on acid-free paper.

Design: TopicA (Antoinette Hanekuyk), Leiden

Library of Congress Cataloging-in-Publication Data

The Library of Congress Cataloging-in-Publication Data
are also available.

ISBN 90 04 11928 0

© Copyright by Koninklijke Brill NV,
P.O. Box 9000, NL-2300-PA Leiden,
The Netherlands

All rights reserved. No part of this publication may be reproduced,
translated, stored in a retrieval system, or transmitted in any form
or by any means, electronic, mechanical, photocopying, recording
or otherwise, without prior written permission of the publisher.
Authorization to photocopy items for internal or personal use is
granted by Brill provided that the appropriate fees are paid directly
to Copyright Clearance Center, 222 Rosewood Drive,
Suite 910, Danvers, MA 01923, U.S.A.
Fees are subject to change.

PRINTED IN THE NETHERLANDS

Nichts befürchtet der Mensch mehr
als die Berührung durch Unbekanntes

Elias Canetti, *Masse und Macht*

CONTENTS

Acknowledgements xi

Abbreviations xiii

Introduction xv

Chapter One: The Study of Jewish History and Archaeology in Historical Perspective: The Example of the Jewish Catacombs of Rome 1
    The Vanished Past: 500-1550 1
    The Unassimilated Rediscovery of Jewish Rome:1550-1632   5
    The Age of the Epigones and the Collectors:1632-1750 14
    The Unappealing Catacombs: 1750-1840 25
    The Excavation of Jewish Rome: 1840-1940 30
    Recent Developments: 1945-Present 43

Chapter Two: The Archaeology of Jewish Rome: A Case-Study in the Interaction Between Jews and Non-Jews in Late Antiquity 50
    Jewish Funerary Architecture in Late Ancient Rome 50
    Artistic Production in Late Antiquity: General Trends 67
    The Wall Paintings in the Jewish Catacombs of Rome 73
    Jewish Sarcophagi from Rome 77
    The Jewish Gold Glasses 81
    Lamps From Jewish Rome 85
    Miscellaneous Finds from the Jewish Catacombs of Rome   88
    Artistic Production in Roman Palestine: Some Parallels 89
    Evaluation of the Phenomenon of Workshop-Identity 92
    Conclusion 96

Chapter Three: References to Age at Death in the Jewish Funerary Inscriptions from Rome: Problems and Perspectives 100
    Introduction 100
    References to Age at Death: a Greco-Roman Custom 101
    An Analysis of Jewish Inscriptional References to Age at Death: Introductory Remarks 107
    Approach 1: The Traditional Approach 109

viii   CONTENTS

    Approach 2: A Critique of the Traditional Approach 112
    Approach 3.a: A More Sophisticated Approach 114
    Approach 3.b: Medians 116
    Approach 3.c: The Survival Rate 118
    Approach 4: The Problem of Age-Rounding 119
    Approach 5.a: The Pattern of Mortality 124
    Approach 5.b: The Hypothetical Life Table and the
    Archaeological Remains 129
    Sociological Inferences: Jewish Women in Antiquity 131
    Implications 136

Chapter Four: The Onomasticon of the Jewish Community of
Rome: Jewish vis-à-vis non-Jewish Onomastic Practices in
Late Antiquity 139
    Introduction 139
    A Critique of Leon's Interpretation of Jewish Onomastic
    Evidence From Rome 139
    Aspects of Jewish Onomastic Practices in Late Ancient
    Rome: Looking for a Pattern 143
    Onomastic Preferences Among Jews in Other Parts of
    the Roman Empire 151
    The Influence of Roman Name-Giving Practices on the Jew-
    ish Onomasticon in Late Ancient Rome: The Question
    of the Duo and Tria Nomina 158
    The Semitic Names in the Jewish Epitaphs from Rome 163
    The Onomastic Practices of Jewish Women 165
    Names Borne by Roman Jews as Indicators of Social Status? 166
    Onomastic Practices as an Indication of Interaction 170

Chapter Five: The Jewish Funerary Inscriptions from Rome:
Linguistic Features and Content 176
    Introduction 176
    The Languages of the Jewish Funerary Inscriptions from
    Rome 176
    The Linguistic Features of the Jewish Epitaphs from Rome   184
    The Content of the Jewish Funerary Inscriptions 191
    Implications 201

Chapter Six: The Literary Production of the Jewish Community
of Rome in Late Antiquity 210
    Introduction 210

The Collatio: General Characteristics	213
The Collatio: A Christian or a Jewish Work?	218
The Pentateuch in early Christian Thought	219
Why is the Collatio not a Christian Work?	233
The Pentateuch in Jewish Thought	240
The Collatio as a Late Ancient Jewish Treatise	247
The Letter of Annas to Seneca	253
Implications	256
Chapter Seven: Conclusions	260
Appendix: *Dis Manibus* in Jewish Inscriptions from Rome	269
Bibliography	273
Index	281

ACKNOWLEDGEMENTS

This book represents the revised version of a dissertation that was submitted to the Department of Religion of Duke University, Durham, North Carolina, U.S.A., in May of 1993. I should particularly like to thank Prof. dr. Eric M. Meyers of Duke University, my dissertation supervisor, for providing an ideal intellectual environment, and for his support and friendship. Furthermore I should like to express my appreciation to the members of my disseration committee including Prof. dr. Kalman Bland, Prof. dr. Elizabeth A. Clark, Prof. dr. E. P. Sanders, and Prof. dr. Annabel J. Wharton, all of Duke University, for providing essential advice.

Over the years, several other people including Prof. dr. Johannes S. Boersma (Free University, Amsterdam), Prof. dr. Günter Stemberger (Institut für Judaistik der Universität Wien), and dr. Gideon Foerster (Hebrew University, Jerusalem) have supported my work on the Jews of ancient Rome when the project was still in an early phase. Here I gladly take the opportunity to thank them for their support.

Still other people have given freely of their time to discuss ideas expressed in this book and have provided me with useful information. I am grateful to Prof. dr. Cesare Colafemmina (University of Bari, Italy), Prof. dr. Harold A. Drake (University of California, Santa Barbara), the late Prof. dr. Jonas Greenfield (Hebrew University, Jerusalem), dr. Amos Kloner (Israel Antiquities Authority, Jerusalem), Prof. dr. Frank H. Stewart (Jacob Blaustein Institute for Desert Research, Sede Boqer, Israel), Prof. dr. Guy G. Stroumsa (Hebrew University, Jerusalem), and Prof. dr. A. Watson (University of Georgia, Atlanta). I am particularly grateful to dr. Boudewijn Sirks (University of Amsterdam) for detailedly reviewing Chapter 6 and to Prof. dr. Pieter W. van der Horst (University of Utrecht) for reviewing the entire manuscript of this book. Finally I should also like to thank dr. Steven Goranson of Durham, North Carolina, who corrected my English and made various useful suggestions to improve the text.

While at Duke I was awarded a Nathan Perilman fellowship in Judaic Studies (1990-1991) and an Andrew W. Mellon Foundation Fellowship (1992-1993). In 1991-1992 the Annenberg Research Institute (now Center for Judaic Studies of the University of Pennsylvania) supported me to do research for what has now become chapters 3, 4, and 6 of this book. In 1993-1994 the Netherlands Organization for Scientific Research (NWO) and the Dr. M. Aylwin Cotton Foundation of Guernsey enabled me to revise the dissertation and to gather essential information in Rome, Italy. It is a pleasant duty to thank these organizations for their generous financial support.

Acknowledgements are also due to the Dutch Institute in Rome for enabling me to use its facilities in 1993-1994 and to the editors of the series Religions in the Graeco-Roman World for accepting this book into their series.

Utrecht, October 1994.

ABBREVIATIONS

CCL	*Corpus christianorum, series latina* (Turnhout: Brepols, 1957f.).
CIJ	J. B. Frey, *Corpus inscriptionum Judaicarum* I (Rome Pontifio Istituto di Archeologia Cristiana, 1936), II (Rome: Pontificio Istituto di Archeologia Cristiana, 1952); B. Lifshitz, *Addenda* (New York: Ktav, 1975).
CIL	*Corpus inscriptionum latinarum* (Berlin: Akademie der Wissenschaften, 1862f.).
CIS	*Corpus inscriptionum semiticarum* (Paris: Reipublicae, 1881f.).
CPJ	A. Tcherikover *et al.* (eds.), *Corpus Papyrorum Judaicarum* (Cambridge Mass.: Harvard University Press, 1957-64).
CSCO	*Corpus scriptorum christianorum orientalium* (Paris and Leipzig: Reipublicae and Harrassowitz, 1903f.).
CSEL	*Corpus scriptorum christianorum latinorum* (Vienna: Geroldi, 1866f.).
DACL	F. Cabrol and H. Leclercq, *Dictionnaire d'archéologie chrétienne et de liturgie* (Paris: Letouzey et Ané, 1907-1953).
GCS	*Die griechischen christlichen Schriftsteller* (Berlin: Akademie, 1897f.).
ICVR	*Inscriptiones christianae urbis Romae. Nova Series* (Rome: Befani and Pontificio Istituto di Archeologia Cristiana, 1922f.).
IGLS	*Inscriptions grecques et latines de la Syrie* (Paris: Geuthner, 1929f.).
Noy	D. Noy, *Jewish Inscriptions of Western Europe.* Vol. 1. *Italy (excluding the City of Rome), Spain and Gaul* (Cambridge: Cambridge U. P., 1993).
PL	J. P. Migne, *Patrologiae cursus completus, series latina* (Paris: Garnier, 1844-64).
PG	J. P. Migne, *Patrologiae cursus completus, series graeca* (Paris: Garnier, 1857-1886).
PS	I. Ortiz de Urbina, *Patrologia Syriaca* (Rome: Pontificio Istituto di Studi Orientali, 1965).
RE	G. Wissowa and W. Kroll (eds.), *Paulys Realencyclopädie der classischen Altertumswissenschaften* (Stuttgart and Munich: Metler and Druckenmüller, 1893-1978).
SCh	*Sources chrétiennes* (Paris: Cerf, 1947f.).

For abbreviations of classical authors, see the list of abbreviations in N. G. L. Hammond and H. H. Scullard (eds.), *The Oxford Classical Dictionary* (Oxford: Clarendon, 1970).

INTRODUCTION

"Every history is contemporary history." Thus Benedetto Croce (1866-1952) summed up what he and his historicist predecessors viewed as one of the main methodological problems confronting the professional historian. Today few historians will disagree with Croce. Some, however, will argue that Croce identified only one out of many problems. Hayden White, for example, maintains that an element of subjectivity or contemporaneity enters the work of the historian not at the moment he or she starts interpreting the evidence he or she has collected, but at a much earlier stage. White believes that "facts" do not exist independently as immutable and objective entities, but rather that they are shaped by the account into which they have been fitted.[1]

The present book focusses on the Jewish community of Late Ancient Rome. It is not a book about historical method. It is not necessary, therefore, to discuss the methodological problems raised in the following pages with the same level of abstraction as does White in his *Metahistory*. Following in the footsteps of Croce, however, a review of earlier scholarship on the Jews of ancient Rome is in order. Such a review will help to set the stage for the chapters that follow. Furthermore, it will allow us to see more clearly the extent to which preconceived ideas have shaped and continue to shape studies on the Jewish community of Rome. A review of the secondary literature on the Jews of Rome is above all useful, however, because it enables us to study how, over a period of no less than four hundred years, scholars have treated Jewish archaeological and epigraphical materials.

Before we can embark on an analysis of all these materials, and on the way in which they have been studied in the past, it is necessary to clarify several issues. Studies such as Edward Said's *Orientalism* or Martin Bernal's *Black Athena* have recently illustrated the truism of Croce's adage referred to above.[2] While Said has argued that nineteenth-century imperialist concerns have affected negatively (and continue to influence negatively) much scholarly work on the cultures of the Near East, Bernal has maintained that classicists have categorically refused to explore the Afro-Asiatic roots of Classical civilization. Said and Bernal have shown that to study the ideas that underlie the

---

[1] H. White, *Metahistory: The Historical Imagination in Nineteenth Century Europe* (Baltimore: John Hopkins U.P., 1973); D. D. Roberts, *Benedetto Croce and the Uses of Historicism* (Berkeley: University of California Press, 1987); F. R. Ankersmit, "Historiography and Postmodernism," *History and Theory* 28 (1989) 137-53.

[2] E. W. Said, *Orientalism. Western Conceptions of the Orient* (London: Penguin, 1978); M. Bernal, *Black Athena. The Afroasiatic Roots of Classical Civilization.* Vol. 1. *The Fabrication of Ancient Greece 1785-1985* (London: Vintage, 1991).

work of previous generations of scholars leads to a recognition of the mechanisms that influence scholarly discourse today. Yet, what is much less clear in the work of both Said and Bernal is this: if history is contemporary history, then this also holds for the interpretations Said and Bernal have given to the works of previous scholars. Thus, one may argue that by criticizing the work of such scholars, Said and Bernal have documented in the very first place what they themselves as products of and participants to the late twentieth century regard as problematic. In the case of Said's work one could argue that his "anti-Zionist" and "pro-Arab/Palestinian" perspective explains not only why he has selected certain orientalists as typical representatives of the European orientalist establishment of the nineteenth century, but also why he has failed to include others. One might in fact wonder whether Said has succeeded in identifying how past European orientalists distorted our view of the Near East. There are good arguments to maintain that Said himself has painted a distorted picture of which issues were central to much of nineteenth-century European orientalist scholarship. In the case of Bernal, one need not read more than a few pages to reach the conclusion that his tome lacks much of the methodological rigor that characterizes the work of those he criticizes. More often than not, Bernal simply fails to back up his claims.

If one accepts, as I do in this study, that every history is contemporary history, then one also has to admit that it is impossible to avoid what Said and Bernal have done, namely to read one's own concerns into the material one is studying. It is possible, however, to state one's concerns clearly at the outset of a study, and to explain why one has chosen a particular approach.

In analyzing scholarly works of the last four hundred years on the Jews of ancient Rome (Chapter 1), as well as in the chapters that follow, I have paid attention to one question in particular: How do Jewish archaeological, epigraphical, and literary materials relate to their contemporary non-Jewish archaeological, epigraphical, and literary materials? This question arose in the light of the evidence that has been preserved in the volcanic soil of the Roman Campagna. There, several Jewish catacombs and hypogea have come to light. At first sight, no differences seem to exist between these catacombs and the much better-known early Christian catacombs of Rome. Consisting of long underground galleries, Jewish and early Christian catacombs look remarkably alike. Moreover, they contain approximately the same type of graves, and their walls have been decorated with the same kinds of wall paintings. Finally, both Jewish and early Christian catacombs are located in the same general areas, that is along the consular roads that connected Rome with its Empire.

I have written this book in an attempt to answer a question that seems inevitable once one takes into account these formal similarities between Jewish and non-Jewish archaeological materials: What is Jewish about the Jewish catacombs of Rome? Or, put differently: If the Jewish and the early Christian catacombs look so much alike, how then did the Jewish community of ancient Rome view itself and how did they relate to their non-Jewish contemporaries?

The question of interaction between Jews and non-Jews in the city of Rome during the third and fourth centuries C. E. has never been studied in any comprehensive fashion. This is surprising for at least two reasons. First, there is an abundance of primary sources that makes the study of this question possible. In fact, there is no other Jewish Diaspora community of Antiquity for which so much material has been preserved as there is in the case of Rome. Second, the study of the interaction between Jewish and non-Jewish cultures has long been a central concern in studies on the Jews of Hellenistic and Roman Palestine. Names of scholars such as Lieberman, Bickermann, Hengel, and Feldman readily come to mind. Why this issue has become such a central concern need not detain us at this point. Suffice it to say here that it has its roots in the nineteenth century. It must be seen as resulting from the enormous influence exercised by exponents of *klassische Altertumswissenschaft* on the first two or three generations of modern historians of the Jewish people. But it must also be understood in the broader context of the Jewish political and social emancipation of the nineteenth century. In short, to study the interaction between Jews and non-Jews in Late Ancient Rome requires entering into a scholarly discourse that is at least a century old, and that is not likely to be concluded in the foreseeable future.[3]

In order to compare Jewish with contemporary non-Jewish materials, it is necessary to establish first the exact chronology of these Jewish materials. In an earlier essay, I had argued that all archaeological and epigraphical materials from the Jewish catacombs of Rome should be dated to the late second through early fifth centuries C. E.[4] The analysis of onomastic and linguistic evidence contained in Chap-

---

[3] S. Lieberman, *Greek in Jewish Palestine* (New York: Jewish Theological Seminary, 1942); *id.*, *Hellenism in Jewish Palestine* (New York: Jewish Theological Seminary, 1950); M. Hengel, *Judaism and Hellenism. Studies in Their Encounter in Palestine During the Early Hellenistic Period* (Philadelphia: Fortress Press, 1974); E. J. Bickermann, *The Jews in the Greek Age* (Cambridge: Harvard U.P., 1988); L. H. Feldman, "Hengel's *Judaism and Hellenism* in Retrospect," *Journal of Biblical Literature* 96 (1977) 371-82; *id.*, *Jew and Gentile in the Ancient World. Attitudes and Interactions from Alexander to Justinian* (Princeton: Princeton U.P., 1993); L. V. Rutgers, "Attitudes to Judaism in the Greco-Roman Period: Reflections on Feldman's *Jew and Gentile in the Ancient World*," forthcoming in *Jewish Quarterly Review*. S. Krauss, *Griechische und lateinische Lehnwörter im Talmud, Midrasch und Targum* (Berlin: Calvary, 1899), 2 vols. can be considered as starting point for this discussion.

[4] L. V. Rutgers, "Überlegungen zu den jüdischen Katakomben Roms," *Jahrbuch für Antike und Christentum* 33 (1990) 140-57.

ters 4 and 5 of this book now confirms that a dating of the Jewish catacombs to the Late Ancient period is indeed correct. Having established the chronology of Jewish archaeological and epigraphical remains from Rome, it then becomes possible to select the evidence with which to compare these Jewish remains. Such evidence includes (1) the early Christian catacombs of Rome, which date to the same general period as the Jewish catacombs and (2) Late Ancient pagan and early Christian inscriptional remains.

To date, only one monograph has been written about Rome's Jewish community, H. J. Leon's *The Jews of Ancient Rome*. It was published in 1960.[5] In the pages that follow I will disagree with Leon on many occasions. The reason for our differences in opinion is not only to be sought in the fact that Leon never succeeded in dating the archaeological and epigraphical materials he studied. The main difference between Leon's *Jews* and the present study is one of approach. While Leon's study is thematic, the present study is problem-oriented. Leon addressed issues such as community organization, the language situation, and name-giving. The present study addresses the same issues, but in doing so, it always asks the question: In what respect are these Jewish materials different from non-Jewish materials? In other words, where Leon opted for a description of only the Jewish community itself, without reference to the non-Jewish world that surrounded it, the present study seeks to understand the Jewish community of Rome on the basis of a comparative and interdisciplinary perspective. It may thus be clear that Leon and I have written very different books. Despite the many criticisms that I aim at his book, there can be little doubt that Leon's study on the Jews of ancient Rome remains fundamental. Nor do I want to suggest that a comparative approach is the only approach that is valuable or methodologically justifiable. I do believe (and hope to show in this study), however, that a comparative and interdisciplinary perspective provides us with the means to determine what ideas and ideals concerned the Jews in third and fourth century Rome.

After an introductory chapter in which I describe the history of research into the Jewish catacombs (Chapter 1), I will study different groups of material in a comparative fashion. In Chapter 2 I will describe the formal appearance of the Jewish catacombs and of the archaeological finds that have been discovered there against the background of Late Ancient—that is non-Jewish—funerary architecture.

---

[5] H. J. Leon, *The Jews of Ancient Rome* (Philadelphia: The Jewish Publication Society of America, 1960). A recent study that also deals extensively with the Jews of ancient Rome is H. Solin, "Juden und Syrer im westlichen Teil der römischen Welt. Eine ethnisch-demographische Studie mit besonderer Berücksichtigung der sprachlichen Zustände," *Aufstieg und Niedergang der römischen Welt* II, 29.2 (1983) 587-789.

Thus, I take up and expand on issues I have addressed previously in another context.[6] I have studied most of the archaeological materials discussed in this chapter "*in situ*," that is, in the Jewish catacombs of Rome and in the various museums where they have been preserved. Although new excavations in the Jewish catacombs certainly remain a desideratum, it was not possible to carry out such excavations within the framework of the present study.[7] Chapter 3 deals with references to age at death on Jewish and non-Jewish funerary inscriptions and with their potential for studying interaction between Jews and non-Jews. In Chapter 4 the onomastic data that can be found in the Jewish epitaphs from Rome are compared with non-Jewish evidence from Rome to show that the majority of names used were typically Late Ancient names rather than specifically Jewish ones. Chapter 5 likewise focusses on Jewish epigraphical remains from Rome. There, I analyze the linguistic features of the Jewish inscriptions in Greek and Latin within the context of the changes that generally affected the use of these languages in Late Antiquity. The content of the Jewish funerary inscriptions, and in particular the important question of how these inscriptions differ from contemporary non-Jewish inscriptions, is also addressed in this chapter. In Chapters 3 through 5 I have made ample use of statistics. These chapters are a good example of what Braudel has once called "simplifier pour mieux connaître."[8] In Chapter 6 we then turn to the literary remains. In it, I propose a new approach to answer the as-yet unresolved question of the authorship of the *Collatio legum Mosaicarum et Romanarum*. I will further argue that the *Collatio* is a prime example of the adoption and adaptation of non-Jewish literary forms by Jews and that it illustrates the changing social and intellectual climate of the fourth century. Chapter 6 will also include a brief presentation of the evidence relating to the *Letter of Annas to Seneca*. Chapter 7 finally draws together the evidence presented in the previous chapters and offers some more general reflections on Diaspora Judaism in Late Antiquity.

We can know much about the Jewish community of Late Ancient Rome. There is, however, also much we do not know. Practically all Jewish archaeological and epigraphical evidence from Rome known today belongs to the funerary sphere. We hardly know, therefore, which part(s) of the Late Ancient city Jews considered home. We

---

[6] L. V. Rutgers, "Archaeological Evidence for the Interaction of Jews and non-Jews in Late Antiquity," *American Journal of Archaeology* 96 (1992) 101-18.

[7] For that reason I have decided not to include pictures in this study. Pictures of many of the finds can easily be found in Leon 1960, and in E. R. Goodenough, *Jewish Symbols in the Graeco-Roman Period* (New York: Pantheon Books, 1953-68), vol. 3, to which should be added U. M. Fasola, "Le due catacombe ebraiche di Villa Torlonia," *Rivista di Archeologia Cristiana* 52 (1976) 7-62.

[8] F. Braudel, *Écrits sur l'histoire* (Paris: Flammarion, 1969) 81.

know even less about how Jews organized their daily lives, and how these activities brought them, or did not bring them, into contact with their non-Jewish contemporaries. One would like to write a comprehensive social history of the Jews of Rome and to raise questions similar to those raised by Toaff in his work on the Jews of medieval Italy, or by Geremek in his study on the "marginaux parisiens" of the fourteenth and fifteenth centuries. One would also like to compare the funerary remains with archaeological evidence from the world of the living. As Morris has shown well in his recent book on mortuary practices in Classical Greece, such a comparison is essential to determine the extent to which funerary evidence reflects the social and religious values held by people before they die.[9] Unfortunately, none of this is possible here. New archaeological discoveries in Rome will perhaps one day enable us to complement our knowledge. At present, we have to be content with the archaeological materials from the Jewish catacombs. Paradoxically enough, even there much work remains to be done.

---

[9] A. Toaff, *Il vino e la carne. Una comunità ebraica nel Medioevo* (Bologna: Mulino, 1989); B. Geremek, *Les marginaux parisiens aux XIVe et XVe siècles* (Paris: Flammarion, 1976); I. Morris, *Death-Ritual and Social Structure in Classical Antiquity* (Cambridge: Cambridge U.P., 1992) 103-99.

CHAPTER ONE

THE STUDY OF JEWISH HISTORY AND ARCHAEOLOGY IN HISTORICAL PERSPECTIVE: THE EXAMPLE OF THE JEWISH CATACOMBS OF ROME

The Vanished Past: 500-1550

By the sixth century C.E. Rome's urban landscape had changed dramatically. Already in the 330s C.E. Constantine shifted the seat of government from Rome to a capital he had newly constructed in the East. Not until several centuries later did it become apparent, however, how profoundly Constantine's decision and the events following it would affect the daily life of Rome's urban population. It has been suggested that in Rome a population of around 800,000 in the year 400 C.E. had been reduced to a mere 100,000 only a hundred years later. The population was to decline een further. Towards the middle of the sixth century, over a period of fifteen years, Rome suffered attacks by hostile armies no less than five times. The city was captured in 546 C.E. by Gothic forces after it had already withstood a twelve-month siege by the Ostrogoths several years earlier (537-538 C.E.). Then, in 549-50 C.E, there followed a renewed and successful siege by the Goths, but Byzantine armies succeeded in re capturing the city two years later, in 552 C.E.[1]

There is no literary or archaeological evidence that indicates if, and if so, how Rome's Jewish population was affected by the rapid sociopolitical and demographic changes that generally affected the inhabitants of Rome in the early Middle Ages. We cannot even be sure whether Jews continued to live in Rome. Perhaps the Jewish community suffered a sharp decline in numbers in the same manner as did Rome's non-Jewish inhabitants.

In the sixth century, burial of Rome's non-Jewish population shifted from the large catacombs outside the Aurelian city wall to a number of sites inside the walls, where the relationship between *abitatio* and *disabitatio* had changed sensibly in Late Antiquity. The latest dated Christian burial in a catacomb, which dates to 535 C.E. (*ICVR* 5, no. 13123), gives a rough indication when burial in these subterranean complexes was discontinued. Such evidence cannot be

---

[1] In general, see R. Krautheimer, *Rom. Schicksal einer Stadt 312-1308* (Leipzig: Koehler and Amelang, 1987) esp. 75-78. Note that the figures suggested by Krautheimer must remain speculative.

taken to mean, however, that from this time onwards the use of catacombs ceased entirely. In the early Middle Ages, Christian catacombs became the focus of intense artistic and devotional activity. Pilgrims came to Rome in search of the graves of those who had died for the greater glory of the Church. *Itineraria*—guides in which the location of the graves of the most famous martyrs were described in a concise fashion—helped them to find their way among the tens of thousands of underground burials excavated under Roman soil. Numerous graffiti scratched on the walls next to the tombs of early Christian martyrs indicate that in the early Middle Ages pilgrims must have flocked to Rome's suburban shrines regularly, and in considerable numbers. Pilgrimage to Rome increased further after Palestine had been conquered by Arab armies in 638-40 C.E., and access to Christianity's holy places there had become more difficult. It continued into the eighth century and beyond.[2]

While the memory of Christian catacombs was thus preserved long after large-scale burials had ceased to take place in them, by the early Middle Ages, knowledge concerning the existence of Jewish catacombs seems to have disappeared entitrely. Not surprisingly, early Medieval Christian *itineraria* abstain from referring to Jewish suburban cemeteries. Only one piece of evidence suggested to Medieval visitors that Rome had once, in a distant past, accommodated Jews: on one of the walls of the church of St. Basilio an inscription had been preserved that commemorated a treaty of friendship Rome had concluded with the Maccabees around the middle of the second century B.C.E. References to this inscription are not very common, however. They first appear in *itineraria* of a relatively late date, as for example the *Mirabilia*, which dates to the middle of the twelfth century, and in the slightly later vulgarization of the same work, known as *Le miracole di Roma*.[3]

Lacking the literary and archaeological evidence, we simply cannot tell whether, in the early Middle Ages, Jews also stopped using catacombs for burial, or whether the early Medieval Jewish cemeter-

---

[2] P. Ariès, *Geschichte des Todes* (Munich: DTV, 1982) 45 and 51; M. Dulaey, "L'entretien des cimetières romains du 5e au 7e siècle," *Cahiers Archéologiques* 26 (1977) 7-18; J. Osborne, "The Roman Catacombs in the Middle Ages," *Papers of the British School at Rome* 53 (1985) 278-328; M. Greenhalgh, *The Survival of Roman Antiquities in the Middle Ages* (London: Duckworth, 1989) 183-201. See also P. Brown, *The Cult of the Saints. Its Rise and Function in Latin Christianity* (Chicago: University of Chicago Press, 1981).

[3] The treaty was historical, see J. D. Gauger, *Beiträge zur jüdischen Apologetik* (Cologne: Hansten 1977) 188f.; R. Valentini and G. Zucchetti, *Codice topografico della città di Roma* (Roma: Tipografia del Senato: 1946) 54, § 24 and 121 § 7 respectively. See also the discussion in *ICVR* 2, 302-3 (de Rossi). The inscription seems to have disappeared.

ies were located inside or outside the city wall. N. Müller, the excavator of a Jewish catacomb that was discovered in the city-quarter of Monteverde Nuovo, south of Trastevere, once suggested that the epitaphs he had spotted on the surface of the catacomb possibly postdated the ones that had been found inside the catacomb. But he then concluded that only further research could clarify the precise chronological relationship between these two groups of funerary inscriptions.[4] Müller passed away unexpectedly before he could carry out such research himself. Even more unfortunately, Müller's excavation notes, which were still accessible in the years directly following his untimely death, have since disappeared. Therefore, it is no longer possible to determine to which inscriptions Müller was referring. The only concrete evidence attesting to the existence of a Jewish above-ground cemetery in the area comes from a Medieval Jewish cemetery that was located near the church of S. Francesco a Ripa. It is first mentioned in sources dating to the thirteenth century. Inasmuch as it was situated between the old Roman period Porta Portese and the new one, constructed in 1643 under Urban VIII, it was located approximately 800 meters to the North of the Monteverde catacomb.[5] Thus the evidence from this Medieval Jewish cemetery is not to be confused with the remains from the Jewish catacomb at Monteverde.[6] These cemeteries developed separately.

It is not until the High Middle Ages that we find isolated evidence suggesting that Jews had again become (or, perhaps, were still) aware that in times long past some of their ancestors had found a final resting place in Roman subterranean cemeteries. In the account of his travels to Southern Europe and to the Eastern part of the Mediterranean in the years 1165-73, Benjamin of Tudela, a Spanish rabbi, reports how he passed through Rome. Among the things he claims to have visited there, he refers to "a cave in a hill on one bank of the river Tiber where are the graves of the ten martyrs."[7] Because of its location near the Tiber, some scholars have suggested that Benjamin visited the Jewish catacomb at Monteverde Nuovo. As

---

[4] N. Müller, *Die jüdische Katakombe am Monteverde zu Rom* (Leipzig, Fock: 1912) 22. For a similar observation (but probably copying Müller), see E. Loevison, *Roma Israelitica* (Frankfurt am Main, 1927) 225.

[5] A. P. Frutaz, *Le piante di Roma* (Rome, 1962) pl. CL, table 333; pl. CLXIX a, 10 table 406. See also N. Pavoncello, "L'antico cimitero ebraico di Trastevere," *Rassegna Mensile di Israel* 32 (1966) 207-16. I would like to thank Dr. G. Brands and Dr. A. Effenberger of Berlin for their efforts, alas unsuccessful, to locate Müller's excavations notes.

6 As does A. Berliner, *Geschichte der Juden in Rom* (Frankfurt am Main: Kaufmann, 1893) 46.

[7] M. N. Adler, "The Itinerary of Benjamin of Tudela," *Jewish Quarterly Review* 16 (1903-04) 453-73; 715-33; *ibid.* 17 (1904-05) 123-41; 286-306; 514-30; 762-81 and *ibid.* 18 (1905-06) 84-101; 664-91.

early as the seventeenth century, others have, however, disputed such a claim.[8] The truth of the matter is that the evidence provided by Benjamin is too brief to determine whether he visited the Monteverde catacomb. What we do know is that the excavations that were carried out in the Monteverde catacomb in 1904-6 and in 1909 have not brought to light any evidence (such as graffiti) that would hint at intense building activity or visitation at this site during any period postdating the fourth century C.E.[9]

In its predilection to include miracle-stories, Benjamin of Tudela's *Itinerary* belongs to the same tradition as contemporary non-Jewish guidebooks such as the *Mirabilia* referred to earlier. Commonly, in such accounts literary motifs, legends, and historical "facts" are hard to separate. With his Jewish contemporaries, nonetheless, Benjamin shares the common trait that he viewed recent events by associating them with important events from the Jewish past. In that sense, Benjamin and other Jews of the Middle Ages were heirs to traditions that had their direct origin in the rabbinic writings of Late Antiquity.[10]

The "Story of the Ten Martyrs," also known as *Midrash Elleh Ezkerah*, is a particularly appropriate example to document that in Jewish writings of the Middle Ages fact and fiction are often hard to separate. Relating the martyrdom of ten famous *Tannaim* and having been transmitted in many manuscripts, this midrash contains indeed little evidence that is historically verifiable. In 1915, P. Bloch attempted to locate the "Story of the Ten Martyrs" in the Jewish milieu of sixth or seventh century C.E. Rome, but his suggestion has not met with universal approval. Most scholars rather prefer to underline the essentially literary character of this story.[11] However this may be, it is clear that Benjamin and his Roman hosts revelled in such stories. As is evident from the other sights they brought to Benjamin's attention, some Roman Jews were quite eager to associate the city's monuments with memorable events taken from Jewish history. Thus Benjamin's informants assured him that two bronze columns in the basilica of S. Giovanni in Laterano were in reality nothing but "the handiwork of King Solomon." Pointing out that they had originally been taken from the First Temple in Jerusalem, they

---

[8] P. Aringhi, *Roma subterranea novissima* (Roma, 1651) 238; J. Basnage, *Histoire de l'église depuis Jésus-Christ jusqu'à présent* (Rotterdam: Leers, 1699) vol. 2, 1033.

[9] Müller 1912.

[10] Y. H. Yerushalmi, *Zakhor. Jewish History and Jewish Memory* (Seattle and London: University of Washington Press, 1982) 31-52; and G. Stemberger, *Die römische Herrschaft im Urteil der Juden* (Darmstadt: Wissenschaftliche Buchgesellschaft, 1983).

[11] I. Gruenwald, *Apocalyptic and Merkavah Mysticism* (Leiden: Brill, 1980) 157-58 and G. Reeg, *Die Geschichte der zehn Märtyrer* (Tübingen: Mohr, 1985) 1-2. The latter study also contains an edition of the different recensions of this midrash.

told him that these highly remarkable pieces of metalwork were known to exude moisture on the ninth of Ab (the date traditionally associated with the destruction of the Temple in Jerusalem). They also took Benjamin to yet another cave, the location of which remains unknown. There, they maintained, Titus had once stored the Temple vessels upon his return from Jerusalem.

It hardly needs stressing that the evidence provided by Benjamin cannot be used to determine whether in twelfth century Rome the existence of Jewish catacombs was still widely known. No less unfortunate, sources other than Benjamin are not very informative on this matter either. In 1190, Pope Clement III prohibited, among other things, the spoliation of Jewish cemeteries. On the basis of Clement's ordinance it has been argued that "jüdischen Katakomben von diebischen Händen durchwühlt wurden."[12] Such a suggestion, however, is pure speculation. We simply do not know which Jewish cemeteries Clement wanted to protect.

It is true that knowledge concerning the existence of these catacombs may not have dissipated entirely during the period stretching from Late Antiquity through the Renaissance. Upon his rediscovery of a Jewish catacomb in 1602, A. Bosio noted, for example, that "curiosi e avidi cavatori" had already preceded him at some undetermined time.[13] Yet whatever the character of such visits may have been, there can be little doubt that the virtual absence of references to the Jewish catacombs of Rome implies that as early as the fifth century C.E. public interest in underground Jewish cemeteries had decreased dramatically. Such interest was not to revive until the early seventeenth century. Only then, as Counterreformation scholars were searching intensively for Christian catacombs and early Christian antiquities, was the Jewish Monteverde catacomb accidentally rediscovered. It was a discovery that was as momentous as it was unexpected. The archaeology of underground Rome would never be the same again.

The Unassimilated Rediscovery of Jewish Rome: 1550-1632[14]

In addition to encouraging a new sense of spirituality (in the founda-

---

[12] Contra Berliner 1893, vol. 2, 10-11; see also N. Güdemann, *Geschichte des Erziehungswesens und der Cultus der Juden in Italien während des Mittelalters* (Wien: Hoelder, 1884) vol. 2, 87.

[13] A. Bosio, *Roma sotterranea* (Rome: Facciotti, 1632) 143.

[14] For a checklist of publications on this topic, see H. J. Leon, "The Jewish Catacombs and Inscriptions of Rome: an Account of Their Discovery and Subsequent History," *Hebrew Union College Annual* 5 (1928) 299-314.

tion of new orders and Spanish mysticism), revising existent doctrine (at the Council of Trent, 1545-63), and carrying through other comparable reforms (such as the Gregorian calendar and the Clementine Vulgate), the Catholic movement of renewal known as the Counterreformation is especially notable in that it gave rise to and actively supported a new kind of history writing and archaeology: sacred, or ecclesiastical historiography and "early Christian archaeology."[15]

In the second half of the sixteenth century, Protestant critics attacked the Holy See in writing with ever increasing force. Such critics based their arguments not only on theological convictions, but also on considerations of a more historical nature. In Protestant circles, a new interest in Church history manifested itself particularly strongly from 1559 to 1574. During this fifteen year period, a group of scholars gathered around Matthias Flacius Illyricus and produced a history of the Church known as the *Magdeburg Centuries*. In particular, these scholars systematically studied the question of how Christianity had come into being and how it had developed through the ages. Discovering the study of history to be a rewarding activity for settling disputes on Christian doctrine, the students associated with the *Magdeburg Centuries* concluded that Christianity as represented by the Roman Catholic Church was nothing but a thoroughly degenerated phenomenon.[16]

Such criticisms having been aimed at it, the Holy See could not remain silent. Protestant claims were to be countered by availing oneself of the weapon that the *Magdeburg Centuries* had used in their attack on the Catholic Church: historiography. The task of writing a comprehensive, authoritative history of the Church of Rome fell to Cesare Baronio (1518-1607). He was a priest who also headed the

---

[15] For a short discussion of the terminology (Counter-Reformation as opposed to Catholic Reform), see *Theologische Realenzyklopädie* 18 (1983) 45-47 s.v. "Katholische Reform und Gegenreformation." In general, see G. Ferretto, *Note storico-bibliografiche di archeologia cristiana* (Vatican City: Tipografia Poliglotta Vaticana, 1942); G. Spini, "Historiography: The Art of History in the Italian Counterreformation," in E. Cochrane (ed.), *The Late Italian Renaissance, 1525-1630* (London: Macmillan, 1970) 91-133; S. Bertelli, *Ribelli, libertini e ortodossi nella storiografia Barocca* (Florence: La Nuova Italia, 1973) 61-91; W. Wischmeyer, "Die Entstehung der christlichen Archäologie im Rom der Gegenreformation," *Zeitschrift für Kirchengeschichte* 89 (1978) 136-49; E. Cochrane, *Historians and Historiography in the Italian Renaissance* (Chicago and London: University of Chigago Press, 1981) 446f.; G. C. Wataghin, "Archeologia e 'archeologie.' Il rapporto con l'antico fra mito, arte e ricerca," in S. Settis (ed.), *Memoria dell'antico nell'arte italiana* (Turin: Einaudi, 1984) 171-217; V. Aiello, "Alle origini della storiografia moderna sulla tarda antichità: Costantino fra rinnovamento umanistico e riforma Cattolica," *Studi Tardoantichi* 4 (1987) 281f.

[16] P. Meinhold, *Geschichte der kirchlichen Historiographie* (Freiburg and Munich: Alber, 1967) vol. 1, 268-95; *Theologische Realenzyklopädie* 11 (1983) 206-14 s.v. "Flacius Illyricus" (O.K. Olson).

Oratorio, one of the Counterreformation's central religious organs that had been founded in 1575. Baronio's *Annales Ecclesiastici,* published from 1598-1607, were the result of many years of painstaking work. They served several purposes. They formed an alternative to humanist historiography, which had always favored profane, and especially political, history over sacred history. They also provided devout Catholics with morally uplifting stories from the past. Having been written, however, as a response to the counterhistory documented in the *Magdeburg Centuries,* Baronio's *Annales* functioned above all as a tool for combatting Protestant heresies.[17]

In the same years that Baronio spent long hours pouring over the writings of the early Church, additional evidence became known that seemed to favor the Catholic, that is Counterreformist view of history. In 1578, workers in the Vigna Sanchez on the Via Salaria accidentally exposed the entrance to a subterranean complex. It turned out to be an early Christian catacomb (known today as the *coemeterium Jordanorum*). Saving the catacombs from oblivion, this remarkable discovery marks the beginning of a period characterized by intense research into the archaeological remains of early Christian Rome. That catacombs were an outstanding feature of the Roman campagna had, of course, never been forgotten entirely. Pomponio Leto (1428-1497/98) and his friends of the *Accademia romana degli antiquari,* for example, seem to have visited catacombs in the same years that a completely preserved (and purportedly ancient) body was discovered on the Via Appia.[18] Yet, before the late sixteenth century, visits and discoveries such as these were never followed up by serious investigations. Rather, it was the spiritually charged climate of the Counterreformation that first led to a preoccupation with early Christian antiquities that was of more than a transitory nature. As a contemporary report on the discovery in the Vigna Sanchez, now preserved in Trier, has it, "there [that is, in the catacomb] was to be seen the *religio,* care, and diligence of those friends of God vis-à-vis the burial of bodies—a confirmation to our indubitable and most certain Catholic faith and Catholic rites."

The catacomb of the Vigna Sanchez and other early Christian catacombs that were now discovered in rapid succession contained the remains of thousands of people. Counterreformation scholars identi-

---

[17] *Dizionario Biografico degli Italiani* 6 (1964) 470-78 s.v. "Baronio" (A. Pincherle); Cochrane 1981, 448, 454, and 457f. On counterhistory, see A. Funkenstein, *Perceptions of Jewish History* (Berkeley: University of California Press, 1993) 36-44.

[18] G. Mercati, "Paolo Pompilio e la scoperta del cadavere intatto sull'Appia nel 1485," *Rendiconti della Pontificia Accademia Romana di Archeologia* 3 (1925) 25-40; Ferretto 1942, 45-86.

fied them as holy martyrs who had all given their lives to glorify the Roman Church. It is not difficult to see how such materials could be appropriated easily to argue that the Holy See was not merely the spiritual, but also the physical descendant of one of the oldest and most respectable communities in the Christian world. Nor does it require much imagination to realize how, after the discovery of early Christian wall painting, Protestant iconoclasm could be depicted as deeply antithetical to traditional Christian practice. At the Council of Trent (1545-1563) it had been decided that contemporary art should serve the Church. Now art from the past, too, could be appropriated for such purposes.[19]

Despite the Counterreformation's fascination with catacombs, the first comprehensive study of Christian catacombs was not published until fifty years after the discovery in the Vigna Sanchez had first stirred the imagination of large segments of Rome's population. With his *Roma sotterranea*, published posthumously in 1632, Antonio Bosio, "antiquario ecclesiastico"(1575-1629), rendered an account of a life spent devoted entirely to elucidating the material history of the early Church. Just as Baronio had benefitted from the particular type of religiosity that permeated the Roman Oratory, it was, again, the intellectual-religious climate of the Oratory that provided the setting for Bosio's groundbreaking work on the catacombs. It was the people from the circle of the deeply religious founder of the Oratory, Filippo Neri (1515-1595), who had first aroused Bosio's interest in early Christian antiquities. They then provided Bosio with an intellectual climate in which his work could flourish, and this even though Bosio never officially became a member of the Oratorio. Finally, after Bosio's death, it was again the clerics of the Oratorio who published his work, first in an Italian edition (1632) and then in a Latin translation (1651).[20]

Given the spirit of the times, Bosio's *Roma sotterranea* is remarkably non-apologetic in tone. Yet in every other aspect, it is, unmistakably, a product of the Counterreformation. According to its subtitle the work was to give an account of "li veri, et inestimabili

---

[19] Ferretto 1942, 108 (citation); E. Josi, "Le pitture rinvenute nel cimitero dei Giordani," *Rivista di Archeologia Cristiana* 5 (1928) 167-227 and *id.*, "Le iscrizioni rinvenute nel cimitero dei Giordani," *ibid.* 8 (1931) 183-284; G. Toscano, "Baronio e le imagine," in R. de Maio (ed.), *Baronio e l'arte. Atti del convegno internazionale di studi. Sora, 10-13.X. 1984* (Sora: Centro di studi Sorani "V. Patriarca," 1985) 411-23 and M. G. Ronca, "La devozione e le arti," *ibid.*, 425-42.

[20] C. Cecchelli, *Il cenacolo filippino e l'archeologia cristiana*. Quaderni di Studi Romani (Spoleto: Paretto e Petrelli, 1938); L. Spigno, "Considerazioni sul manoscritto Vallicelliano G. 31 e la Roma sotterranea di Antonio Bosio," *Rivista di Archeologia Cristiana* 51 (1975) 281-311 and *id.*, "Della Roma sotterranea del Bosio e della su biografia," *ibid.* 52 (1976) 277-301.

tesori, che Roma tiene rinchiusi sotto le sue campagne, i sacri cimiterii di Roma, e dei martiri in essi risposti." Its frontispiece carried an engraving showing, among other things, the sufferings of early Christian martyrs. The same engraving also showed how the burials of such martyrs in ill-lit, rather lurid catacombs might originally have looked. Addressing the "benign" reader in a short preface, Severano, Bosio's editor, stated that the book contained "arsenals from which to take the weapons to combat the heretics, and in particular the iconoclasts, impugners of sacred images, of which the cemeteries are plenty." Such a frontispiece and such an introduction could leave readers with little doubt about the character of the book they held in their hands. This was a work that had been designed specifically to "help illuminate the intellect, and to inflame affection and good will."[21]

As an explorer of catacombs, Bosio was indefatigable. Visiting many Christian catacombs, he soon became an authority on everything relating to underground Rome. For that reason, the owners of a vineyard in the city quarter of Monteverde called upon him after they had discovered some galleries belonging to a catacomb that extended under their property. Little did they know that they were about to discover a Jewish catacomb. Entering the site in December of 1602, Bosio and his hosts first came upon a large painted menorah (the painting has been lost subsequently). Then they found an oil lamp whose disc was decorated with the same motif. Finally they also deciphered a fragmentary inscription, which read ΣΥΝΑΓΟΓ. It was thus clear that the Monteverde catacomb had once be used by Rome's Jewish rather than by the city's early Christian community. Bosio realized that this was so. Nor was he afraid to state his observations publicly. In fact, the few pages dedicated to the Monteverde catacomb in his *Roma sotteranea* are a good example of scholarly integrity. But, at the same time, they are also more than that.[22]

Given the apologetic character of much historical scholarship of the Counterreformation, it is remarkable that Bosio included a discussion of a Jewish catacomb in a work on early Christian cemeteries at all, and Bosio was well aware of this. In his description of the Jewish Monteverde catacomb here therefore hastened to add that "it should not appear strange that in this work on sacred cemeteries we include the cemetery of the Jews." Why? "Not in order to mix them up, but to separate all these materials, it appeared necessary to us to mention it [that is the Jewish Monteverde catacomb]: so that it will be

---

[21] For Bosio's biography, see *Dizionario Biografico degli Italiani* 13 (1971) 257-59 s.v. Bosio (N. Parise). For a survey of his work, see Ferretto 1942, 132-61. Citation: Severano in Bosio 1632, 602.

[22] Bosio 1632, 140-43.

10   CHAPTER ONE

known that our cemeteries have never been profaned nor contaminated by the bodies of either Hebrews or Gentiles."[23] Short and to the point as this justification may appear at first sight, it illustrates well the spirit that permeates the description of the Monteverde catacomb that Bosio appended to his introductory remarks. Two elements in this description merit further attention.

First, in his description of the Monteverde catacomb, Bosio never abandoned the non-apologetic tone that characterizes the *Roma sotterranea* as a whole. That his account is essentially descriptive and void of verbal aggression against Jews or Judaism is surprising. For the Jews of Rome, conditions were far from favorable in the later sixteenth century. In 1553 the Talmud had been burned publicly. Then, under pope Paul IV Caraffa (1555-59) the Roman Jewish community had been locked up behind the walls of a ghetto (1555). It separated them physically, although not intellectually, from the larger outside world. Attempts to convert Jews were carried through with renewed intensity in the same general period. A *Casa dei neofiti* had been instituted in 1543. In 1577 it was followed by a *Collegio dei neofiti*. In the same year obligatory sermons were also instituted.[24] That such a policy went hand in hand with verbal abuse, even in publications of a scholarly nature, might be expected. P. Aringhi's discussion of the Jewish Monteverde catacomb, to which we will turn shortly, is a prime example of how profoundly the religious climate of the Counterreformation could affect the vocabulary of scholarly works written during this period. Bosio, by contrast, never used depreciative terms to describe the Jews. As a scholar he was interested, above all, in catacombs and the physical remains of early Christianity. Contemporary theological debates on Jews could not capture his imagination. But to suppose that Bosio thus had not been at all influenced by the theological controversies of his time, is, however, incorrect. This brings us to our second point.

Several years before Bosio discovered the Jewish catacomb in Monteverde, Baronio had studied the burial customs of the ancient world. Using as his source Genesis 23, which describes the burial of

---

[23] Bosio 1632, 141.

[24] R. Bonfil, "Changes in the Cultural Patterns of a Jewish Society in Crisis. Italian Jewry at the Close of the Sixteenth Century," in D. B. Ruderman, *Essential Papers on Jewish Culture in Renaissance and Baroque Italy* (New York and London: New York U. P., 1992) 401-25; F. Parente, "Il confronto ideologico tra ebraismo e la chiesa in Italia," *Italia Judaica. Atti del I convegno internazionale, Bari 18-22.V.1981* (Rome: Ministero per i beni culturali e ambientali, 1983) 303-73, esp. 310f.; R. Segre, "Il mondo ebraico nei cardinali della Controriforma," *Italia Judaica. "Gli Ebrei in Italia tra Rinascimento e età Barocca." Atti del II convegno internazionale. Genova 10-15.VI.1984* (Roma: Istituto Poligrafico, 1986) 119-38; Funkenstein 1993, 215-19.

Sarah in the cave of Machpelah, Baronio had observed that "it seems that the practice of burying the dead in crypts and in loculi that have been hollowed out in these same crypts has been derived from the Hebrews." He then pointed out that the tomb of "our Lord"(Matthew 27) was likely to stand in the same Jewish tradition. Elsewhere in his *Annales* Baronio referred to a passage in Dio Cassius that relates how Jewish insurgents under Bar Kokbah had dug underground passages in order to wage their war against the Romans. Such underground hideaways could be used in many ways. That they had never been used for sepulchral purposes, however, did not matter to Baronio. He merely wanted to make the point that the "underground roads" in Palestine (which he himself had never seen!) displayed the same formal characteristics as the Christian catacombs of Rome.[25] Baronio thus wanted to explain the use of catacombs by Rome's early Christian community by tracing back its origins to the funerary architecture of the Holy Land. To be sure, there was nothing new to postulating a connection between the burial customs of the earliest Christian communities and those of the Patriarchs.[26] What was new to Baronio's hypothesis, however, was that it was based on a complex of unproven assumptions, namely that (1) from Abraham's days onwards, Jews always and invariably buried in underground cemeteries; that (2) in funerary matters, Christians in faraway Rome copied Jewish-Palestinian customs; and that (3) there are no conceptual differences between burial in small underground cemeteries (hypogea, such as the one in which Jesus had been buried) and burial in large catacombs.

That Baronio's explanation of the origin of the Christian catacombs was to enjoy enormous popularity well into the twentieth century need not detain us at this point. Rather, the following fact requires our attention: Bosio, the first great explorer of the catacombs and author of a study on catacombs that would not be surpassed until the second half of the nineteenth century, never addressed the question of the origin of the Christian catacombs in connection with the Jewish catacomb he had discovered. Bosio rid himself of the question in a single phrase. He noted in passing that the underground tombs of the early Christian community had been "ad imitatione de' Patriarchi antichi"—at least generally speaking. He then observed that Christians had constructed catacombs primarily because "they could not fabricate monuments and conspicuous sepul-

---

[25] *Annales*, vol. 2, 81, and, 346-47; Dio 69.12.3; Ferretto 1942, 127; A. Kloner, "The Subterranean Hideaways of the Judean Foothills and the Bar Kokba Revolt," *Jerusalem Kathedra* 3 (1983) 114-35 and *id.* and Y. Tepper, *The Hiding Complexes in the Judaean Shephelah* (Tel Aviv: Hakibbutz Hame'uchad, 1987)(Hebrew).

[26] See, e.g., Athanasius, *Vita s. Antonii* 90 (*PL* 26, 968).

cres in public places."²⁷

How was it possible that Bosio never theorized about the relationship between Rome's early Christian catacombs and the Jewish Monteverde catacomb? Did he not live in a society that made archaeological discoveries in the catacombs into one of the cornerstones to justify its own legitimacy? And had not Baronio, whose work so deeply influenced Bosio, hinted at a possible Jewish origin for the practice of burial in Christian catacombs?

To answer this question, we must return to the discovery of the Monteverde catacomb and to the observations to which it led Bosio. After the discovery of 1602, Bosio could do one of three things: he could leave out any reference to Jewish materials from his *Roma sotterranea*; he could include a brief description of these materials; or he could explore the implications of his discovery in an attempt to settle the question of the origin of burial in catacombs once and for all. Upon closer reflection, the first and third option offered no real alternatives. For a scholar who had spent the greater part of his life meticulously researching any catacomb he could find, it would have been inconceivable to omit a reference to his discovery of an ancient Jewish underground cemetery. As an intellectual heir to the ideals of the Counterreformation, on the other hand, Bosio had to avoid probing into the deeper implications of his exceptional discovery. In a society in which catacombs and Christian antiquities served primarily apologetic purposes—a society which, moreover, forcefully tried to convert the Jews, shut them up in ghettos, and even expell entire Jewish communities from large territories—queries that were likely to lead to alternative interpretations of the past would have met with very little sympathy. In choosing the second option, that is, providing the reader with a purely descriptive analysis of the Jewish Monteverde catacomb, Bosio ingeniously found the middle road.

It may thus seem that Bosio was not unlike his contemporary Joseph Scaliger (1540-1609) in that he solved incongruities between Christian doctrine and non-Christian primary sources by including a description of the latter without using it to criticize the tenets of the former.²⁸ When seen against the larger background of Christian theological thinking on the Jews, Bosio's treatment of the Monteverde catacomb was, however, more than just a solution in the style of Scaliger to a tricky problem. It was the application of a well-tried formula that had it roots in Late Antiquity. Then, in an attempt to define orthodoxy, both Church and state had come to agree on a

---

²⁷ Bosio 1632, 2.
²⁸ On Scaliger, see A. T. Grafton, "Joseph Scaliger and Historical Chronology: The Rise and Fall of a Discipline," *History and Theory* 14 (1975) 156-85.

division of the world in two groups: orthodox Christians on the one hand, and all the "others" (including pagans, heretics, schismatics, and Jews) on the other.[29] This is exactly the division we also find in Bosio.[30] Like the lawgivers and theologians of Late Antiquity, Bosio accepted the presence of the Jews as a given. Yet, Jews never constituted a real presence that could be taken seriously. Reconstructing the material history of the early Church through study of the Roman catacombs, it was simply unthinkable for Bosio that Jewish remains could ever illuminate the origins of orthodox Christianity.

Bosio's contribution to the study of underground Rome would affect scholarship on catacombs profoundly, and for centuries to come. Where earlier humanist antiquarians had failed to draw a sharp line of distinction between Christian and pagan Rome and had cared only little about early Christian antiquities, Bosio had rescued such finds from oblivion, giving them a fixed place in the archaeological study of the past. Yet, as a result of the Counterreformation spirit that permeated his *Roma sotterranea,* Bosio did not succeed in winning the humanists for his cause. Quite the contrary. Just as a line of division between ecclesiastical and profane historiography became ever sharper as a result of the Counterreformation, antiquarians and students of early Christian material remains were soon to go separate ways. Antiquarians continued to associate the study of antiquities with research into Greek and especially Roman realia.[31] Not before the twentieth century would the interests of Classical and early Christian archaeologists finally start to converge. As a result, the Jewish catacombs remained the exclusive territory of early Christian archaeologists.

As for the Jewish community in Counterreformation Rome, they apparently never studied the remains from the Monteverde catacomb for themselves. Earlier, during the Renaissance, some Jews in Italy had started, like their non-Jewish contemporaries, to collect antiquities, but they seem to have been unaware of the existence of Jewish antiquities. Comparably, in the hundred years following the expulsion of the Jews from Spain in 1492, a variety of historical studies had been published by Jewish authors, but the archaeology of the Jews of ancient Rome plays no role in such works: in the sixteenth century and beyond, study of Jewish history meant research of the literary sources (as opposed to other kinds of sources). Translations of Josephus' *Jewish War* and *Antiquities* into the vernacular were especially popular during the period 1550-1599. Moreover, by the

---

[29] A. Berger, "La concezione di eretico nelle fonti Giustinianee," *Rendiconti dell' Accademia dei Lincei* 8.10 (1956) 353-68.

[30] E.g. Bosio 1632, 593; 596.

[31] Cochrane 1981, 441, 478.

time Jewish antiquities started to come to light in Rome itself, the political and theological climate was rapidly changing for the Jews of Rome. Not only did ghetto regulations make access to Jewish antiquities more difficult, the intellectual interests of the Jewish community itself were changing too. They now shifted away from study of history towards private devotion and a preoccupation with Lurianic Kabbala.[32]

The discovery of a Jewish catacomb in Monteverde in 1602 by Antonio Bosio can thus be said, in conclusion, to have evoked only little interest at first. Bosio, the most serious scholar of Late Ancient material remains of his age, did not reflect, at least not publicly, upon the deeper meaning of his discovery. Not surprisingly, his Christian colleagues did so even less. Within the Roman Jewish community, the discovery of an old Jewish catacomb made little impact. In the early seventeenth century, their intellectual and religious horizon was too far removed from the antiquarian concerns of their non-Jewish contemporaries to be able to appreciate fully the importance of this discovery.

The Age of the Epigones and the Collectors: 1632-1750

By the seventeenth century most historians, both humanist and ecclesiastical, agreed that the main aim of their endeavors was to establish the truth. They also agreed that only the study of authentic documents could help them in achieving such an aim. Yet, they disagreed as to what constituted an authentic document. The question of the relation between sacred and profane history which had increasingly dominated the scholarly debate of the preceding century, continued to interest not only Giambattista Vico (1668-1744), but other scholars as well. During this period, many scholars became increasingly pessimistic regarding the feasibility of ever attaining real, ob-

---

[32] M. A. Shulvass, "Knowledge of Antiquity among the Italian Jews of the Renaissance," *Proceedings of the American Academy of Jewish Research* 18 (1948-49) 298; A. Milano, *Il Ghetto di Roma. Illustrazioni storiche* (Rome: Staderni, 1964) 385-96; P. Burke, "A Survey of the Popularity of Ancient Historians, 1450-1700," *History and Theory* 5 (1966) 133-52; Yerushalmi 1982, 57-75; R. Bonfil, "Riflessioni sulla storiografia ebraica in Italia nel Cinquecento," *Italia Judaica. "Gli Ebrei in Italia tra Rinascimento e età barocca." Atti del II convegno internazionale. Genova 10.15.VI.1984* (Roma: Istituto Poligrafico, 1986) 54-66; A. Melamed, "The Perception of Jewish History in Italian Jewish Thought of the Sixteenth and Seventeenth Centuries. A Reexamination," *ibid.* 139-70; L. Segal, *Historical Consciousness and Religious Tradition in Azariah de' Rossi's Me'or 'Einayim* (Philadelphia: The Jewish Publication Society of America, 1989); R. Bonfil, "How Golden Was the Age of the Renaissance in Jewish Historiography?," in D. B. Ruderman, *Essential Papers on Jewish Culture in Renaissance and Baroque Italy* (New York and London: New York U. P., 1992) 219-49.

jective historical knowledge. The widespread existence of such pessimism explains why a shift occurred that led away from the exclusive study of literary sources to a preoccupation with archaeological and epigraphical remains. The study of the physical remains of the past increased to such an extent that J. Spon felt it necessary to introduce a special term to denote such an activity: archaeology. The importance attached to inscriptions originated in the awareness that they had never been subject to corruption by the hands of later editors or Medieval scribes. The fascination with the tangible remains of the ancients led to an increase in the collecting of such remains. Paradoxically, it did not result in new, large-scale excavations. F. Bianchini's excavations on the Palatine in the years following 1705, for example, were to remain an isolated phenomenon. No less paradoxically, real efforts to protect sites from spoliation or destruction never materialized during this period either. The "ufficio dei custodi delle sante reliquie," instituted by Clement X in 1672, seems to have accomplished little or nothing in terms of rescuing sites from destruction. Not without reason Montesquieu observed that: "Rome nouvelle vend pièce à pièce l'ancienne."[33]

In the later seventeenth century as well as in the eighteenth century, Catholic scholarship was to draw its strength from literary studies such as the *Acta Sanctorum*, begun by the Bollandists in 1685, and from the scholarship of the Maurists. The more critical Catholic minds of the later seventeenth century now agreed with Protestant critics in showing little sympathy for dubious relics from the catacombs.[34] In the specific case of the catacombs, no new research comparable to that of Bosio was undertaken. His historical-archaeological *magnum opus* would become the standard reference work on catacombs, yet this did not happen until after it had been translated into Latin. Severano, who had edited Bosio's hand-written manuscript and who had seen it through the press in the three years following the latter's death, set out to translate the *Roma sotterranea* into Latin.

---

[33] Ferretto 1942, 201-6; A. Momigliano, "Ancient History and the Antiquarian," *Journal of the Warburg and Courtauld Institute* 13 (1950) 285-315; C. Mitchell, "Archaeology and Romance in Renaissance Italy," in E. F. Jacob (ed.), *Italian Renaissance Studies* (London: Faber and Faber, 1960) 455-83; A. Momigliano, "Vico's *Scienza Nuova*: Roman 'Bestioni' and Roman 'Eroi'," *History and Theory* 5 (1966) 3-23; *Dizionario Biografico degli Italiani* 10 (1968) 187-94 s.v. "Bianchini" (S. Rotta); Cochrane 1981, 446-67; A. Momigliano, "Storia antica e antiquaria," in *id.*, *Sui fondamenti della storia antica* (Turin: Einaudi, 1984) 3-45; H. Gross, *Rome in the Age of Enlightenment. The Post-Tridentine Syndrome and the Ancient Regime* (Cambridge: Cambridge U.P., 1990) esp. 310f. (citation Montesquieu).

[34] On Mabillon, see A. Momigliano, *Essays in Ancient and Modern Historiography* (Middletown: Wesleyan University, 1977) 277-93. See also *Dizionario Biografico degli Italiani* 11 (1969) 248 s.v. "Boldetti." And see Gross (previous note) 264-68. Among Catholic scholars, Arlinghi's literary approach to studying Jewish burial customs remained popular, see e.g. J. Ciampini, *Vetera monumenta* (Rome: Komarock, 1690) 161-66.

Severano's translation, however, was never published.³⁵ In 1650, on the occasion of the Holy Year, a quarto edition of Bosio's *Roma sotterranea* was published. Known as "Bosietto" or "little Bosio," it was identical to the *editio princeps* of 1632 in every respect except for its format.³⁶ A Latin translation of the *Roma sotterranea* finally appeared in 1651 as *Roma subterranea novissima*. It had been prepared by P. Aringhi (1600-1676), who was a cleric at the Roman Oratory.

Having been written in Latin, Aringhi's translation soon gained a measure of widespread popularity that the Italian original had never enjoyed. In 1659, eight years after publication, a reprint of it was published in both Paris and Cologne. Further reprints followed soon, first in 1668, and then in 1671 (both were published in Arnhem in the Netherlands, the latter was a pocket edition). Thus, within a few years scholars in every part of Europe learned about a subject that had remained, for almost half a century, the exclusive territory of Catholic scholars who lived in Rome. Yet little could such scholars realize that Aringhi was not a very faithful translator. In fact, Aringhi's translation of Bosio's *Roma sotterranea* is not a translation at all, but a free rendering into which Aringhi inserted many observations of his own.³⁷ It is characterized by a militant, typically Counterreformation tone—a tone that had been largely absent from the original. The preface Aringhi composed on the occasion may serve to illustrate his style. Greeting the "pious and studious reader," Aringhi observed that what he or she was about to read did not deal with "rubbish or mere remains." It dealt instead "with gold, nay gems and precious stones." The real purpose of the work to which Aringhi so proudly appended his signature was to help those "who aspire to God in their orthodox faith."³⁸

Whatever his real intentions might have been, Aringhi thought he helped his readers when he proclaimed his anti-Judaic sentiments. He assured his readers, for example, that the seven-branched candelabrum, sculpted on Titus' arch on the Forum "had been robbed of its light by the brightness of the Gospel." Comparably, Aringhi ascertained that the Jewish martyrs referred to by Benjamin of Tudela could really not be anything but "pseudo-martyres." Given his dislike for Judaism, it is hardly surprising to note that the same Aringhi al-

---

[35] It has been preserved in the Bibliotheca Vallicelliana in Rome as *Cod. Vall.* 32.

[36] A. Bosio, *Roma sotterranea* (Rome: Grignani, 1650).

[37] Note, for example, that Aringhi wrote as many as twelve pages on the Jewish Monteverde catacomb were Bosio had dedicated only three pages to a discussion of this site.

[38] Aringhi 1651, preface.

ways distinguished carefully between the "Hebraei" of the Old Testament—or forefathers of the Church in its capacity as *Verus Israel*—and the "Judaei" of Roman and later times.[39] Verbal aggression reaches its peak in Aringhi's translation of the chapter in which Bosio related his discovery of the Jewish Monteverde catacomb. Where Bosio had merely written about "Hebrews," Aringhi now wrote about "perfidious Jews." Like Bosio, Aringhi stressed that Jews and Christians had never been buried together. Yet where Bosio did not explain why this had never happened, Aringhi ventured to suggest that the lack of common burial ground had to be ascribed to timeless Christian principles such as the practice to keep the "sacred and religious" separated from the "profane."[40]

Having little or no independent scholarly value, the polemicist Aringhi nevertheless merits our attention, if just for one reason only: Aringhi makes explicit what had always remained implicit in the work of Bosio. Much as he might have disliked Jews, Aringhi too had to deal with the problem that had confronted Bosio twenty years earlier: to explain the presence of a Jewish underground cemetery amidst Rome's early Christian catacombs.

It may be recalled that Bosio had found an easy way out of this problem. He simply accepted the existence of the Jewish Monteverde catacomb as a given. For Aringhi such a solution was not satisfactory. Instead, Aringhi revived the discussion of ancient burial customs that had been published by Baronio. Aringhi thus argued that the earliest Christian communities refrained from cremating their dead in imitation of the patriarchs Abraham, Isaac, and Jacob. He also maintained that the first Christians derived the practice of burial in catacombs from the *antiquorum Patriarcharum sepulchra*. He then argued that this "patriarchal" tradition of underground burial had been transmitted directly to the Christians, suggesting that the graves of the Kings of Judah and the tomb of Jesus were the missing links that connected early Christian to patriarchal burial. Aringhi finally maintained that the Jewish cemetery of Monteverde was not part of this tradition, but he then failed to explain why he believed this to be so.[41]

---

[39] Aringhi 1651, 372 (Arch of Titus); 238 (martyrs); and *passim* (Hebraei and Judaei). The *locus classicus* for the Christian distinction between "Jews" and "Hebrews" is in Eusebius, *Praeparatio evangelica* 6 (*GCS* 43.1, 368-69; also contrast Eusebius' considerations with the explanation proposed later by Ambrosiaster, *Quaestiones* 44.7, 80.1 and 108.2 in *CSEL* 50, 78, 138, and 252). Note also that the distinction between "Jews" and "Hebrews" surfaced in early Christian literature well before Eusebius, e.g. Tertullian, *Apologeticum* 18.6 (*CCL* 1, 119) and Aristides' Apology 2.5 (E. Hennecke, *Die Apologie des Aristides* [Leipzig: Hinrichs'sche Buchhandlung, 1983] 8).
[40] Aringhi 1651, 231-39.
[41] Aringhi 1651, 5-8, 238-39.

While Bosio had never specified why the Jewish catacomb and its early Christian counterparts had (in his view) nothing in common, Aringhi argued that Jewish and early Christian underground cemeteries *could not* have anything in common. Suggesting that Christians had constructed catacombs in direct imitation of practices described in the Hebrew Bible, Aringhi presented Rome's early Christian community in exactly the same manner as early Christian apologists had presented Christianity more than a millennium and a half ago, namely as heir to the traditions of the Hebrew Bible. Yet although Aringhi's argumentation was identical to that found in early Christian apologetic literature, it served a very different purpose. Aringhi no longer had to justify Christianity as such. He used the Hebrew Bible only because it helped him to free himself of evidence (that is the Jewish Monteverde catacomb) that could not be fitted into his history of the early Church. Like Bosio before him, Aringhi operated on the basis of the *a priori* assumption that the Jewish burial ground at Monteverde and the Christian catacombs did not have anything in common. But while Bosio had merely defined Christian catacombs as places for burial, Aringhi managed to establish a link between archaeological remains and Christian theological reasoning. It resulted in a history of the catacombs in which there was little or no place for Jewish evidence.

Although Bosio and Aringhi thus refused to take seriously the study of Jewish archaeological remains, they nonetheless contributed enormously to what was then called "early Christian archaeology." As a result of Bosio's and Aringhi's publications, scholars who had never been to Rome could now acquaint themselves with catacombs and the archaeological materials they contained. They could also familiarize themselves with the interpretation the Church gave to these materials. It was an interpretation that was not to convince everyone. In fact, from the start Protestant scholars had doubted that the catacombs contained evidence to show that Catholic doctrine and religious practice had remained unchanged through the ages. Toward the end of the seventeenth century, Protestant criticism grew ever louder—as is evident, for example, from the writings of the French Huguenot J. Basnage (1653-1723).

In the second volume of his *Histoire de l'église depuis Jésus-Christ jusqu'à présent,* Basnage briefly discussed the phenomenon of "catacombs." Among other things, Basnage wondered about the reasonability of the claim that catacombs contained the earthly remains of Christian martyrs. He wanted to know how one could distinguish the archaeological remains of orthodox martyrs from those of heretics and schismatics. In the end, he believed that it was impossible to make such a distinction. Using Aringhi's translation and referring to the illustrations contained therein, Basnage further argued that coins, lamps, and other archaeological materials suggested that burial in catacombs was an old pagan practice rather than a Christian innovation. Referring to the consular data contained in ancient inscriptions, he finally also asserted that major constructional activity in the Christian catacombs did not predate the fourth century C.E. Such theses were unheard of in Basnage's days, at least in the camp of Catholic scholars.[42]

It may be clear that in a scheme that ascribed the origin of burial in catacombs to Rome's pagan population, the Jewish catacombs had no role to play. In fact, Basnage briefly disputed Bosio's identification of the Monteverde catacomb as Jewish. He argued that the Monteverde catacomb could never have been used by Jews because no Hebrew inscriptions were ever found there. He furthermore also denied the existence of any connection between the Monteverde catacomb and the graves visited by Benjamin of Tudela.[43]

Seventeen years later, in his enormously influential *Histoire des Juifs*—the first comprehensive history of the Jews since Josephus—Basnage was to return to the question of Jews and burial customs in ancient Rome. This time Basnage was eager to show that before 70 C.E. the Jews of Rome had lived near the Vatican rather than in Trastevere. To support this view, Basnage marshalled all the evidence he could find. Like his contemporaries, he attempted to distinguish carefully between authentic and non-authentic sources. In his *Histoire des Juifs* there was no place for legends, at least not where it concerned the Jewish community of Rome. Such methodic rigor notwithstanding, Basnage was much less critical, however, when it came to the interpretation of authentic documents. He listed seven arguments indicating that before 70 C.E. Jews had never lived in Trastevere. None of these reasons is likely to strike the modern reader as convincing. Where only a few years before Basnage had argued, for example, that Benjamin's description was too general to determine the location of the "cave of the Ten Martyrs," he did not hesitate now to locate this cave in the Vatican area! Similarly, having discovered, in a source dating to the year 1220, that the Ponte S. Angelo (near the mausoleum of Hadrian or Castel Sant'Angelo) was also known as the *Pons Judaeorum*, Basnage argued that such evidence

---

[42] J. Basnage, *Histoire de l'église depuis Jésus-Christ jusqu'à présent* (Rotterdam: Leers, 1699) vol. 2, 1027f.; 1030; 1034-37. Views similar to those held by Basnage were also expressed by his contemporary, G. Burnet, a protestant and bishop of Salisbury, see R.W. Gaston, "British Travellers and Scholars in the Roman Catacombs," *Journal of the Warburg and Courtauld Institutes* 46 (1983) 150-51; Ferretto 1942, 248-68.

[43] Basnage (previous note) 1033.

showed that Jews had always lived in the Vatican area.[44] How was it possible for Basnage to make such unreasonable claims? And why did Basnage use all his efforts to establish a connection between the Jews of ancient Rome and the Vatican hill in the first place?

Basnage himself hints at the answer to these questions. Towards the end of his study on which city quarters accommodated Jews he wrote: "The location which we assign to the Jews must embarrass the Catholics somewhat; for it is on the Vatican and in Cafarello that the relics of so many Christian martyrs are found." And he then concluded his discussion with a rhetorical question: "But if the Jews buried their dead in those places [namely catacombs], how can their bones be distinguished at the present day from those of Christians?"[45]

What Basnage inserted in his *Histoire des Juifs*, therefore, was a discussion of exactly the same question that had bothered him several years earlier while writing his *Histoire de l'église*: is there any evidence to argue that catacombs are the tombs of early Christianity's martyrs? Although this time Basnage's argumentation differed from the line of reasoning he had followed in 1699, his views had remained unaltered. Basnage still believed that it was impossible to distinguish the bones of Christian martyrs from those of ordinary dead. But, rather than criticizing Aringhi's interpretation of early Christian materials, this time Basnage used—not to say twisted—evidence concerning the Jewish community in Rome to prove his case. So, despite different arguments, it was the same old conclusion all over again.

It is characteristic of the concerns underlying Basnage's work that even in a history *of the Jews*, Jewish evidence was never elevated to a status where it would be evaluated on its own terms. Thus the difference between Bosio's and Aringhi's *a priori* of Jewish archaeological materials and the specific ways in which Basnage studied such materials is a difference in degree only and not in kind. All three scholars have in common that they viewed Jews and Judaism, whether of the Roman period or contemporary times, through distinctly Christian eyes. Just as Bosio and Aringhi did not speculate about the Jewish Monteverde catacomb because they believed such considerations would help them little in explaining the origin of the *Christian* catacombs and *Christian* burial practices, so Basnage used literary and archaeological evidence concerning the Jews in ancient Rome only because it served to disprove Catholic claims. Ultimately, Basnage's

---

[44] I have used the following English translation of the *Histoire des Juifs*: "Continuation of the History of the Jews to the Reign of Hadrian" in C. E. Stowe, *History of the Hebrew Commonwealth* (London: Ward, 1828) 261-65.

[45] Basnage in Stowe (previous note) 266.

study of the Jews of ancient Rome served no other purpose than to strengthen his own religious beliefs. Basnage differed from contemporary scholars in believing that Jewish history had not come to an end in post-biblical times. Regarding everything else, he remained unmistakably a child of his times. Strongly disliking Talmudic reasoning, in publishing the *Histoire des Juifs,* Basnage did not want to write about the Jews *per se.* He merely wanted to prove the veracity of Christianity over against Judaism.[46]

With Basnage's work, discussions of Jewish evidence from Rome came to a premature end. Serious discussions of Jewish archaeological remains from Rome would not be resumed until the 1860s. In the years following the publication of Basnage's *Histoire des Juifs,* scholars such as G. Bianchini claimed to have visited the Monteverde catacomb. Whatever the character of such visits might have been, and whatever the answer to the question of whether such visits actually took place, the remarks of Bianchini have no independent scholarly value. In reports such as these, Jewish antiquities are merely mentioned to add an exotic flavor to accounts that list the many sights that can be visited in Rome.[47]

Insofar as epigraphical materials were concerned, antiquarians continued to be active in the late seventeenth and early eighteenth centuries. They were especially interested in the collection into corpora of the many ancient inscriptions in Greek and Latin that had come to light over the preceding centuries. The continued popularity of the collecting of antiquities much facilitated access to such inscriptions.[48] In their attempts to survey extant materials as comprehensively as possible, antiquarians not infrequently included Jewish inscriptions in their publications too. This was the case with Spon's *Miscellanea eruditae antiquitatis* of 1685, Fabretti's collection of ancient inscriptions of 1699, Muratori's *Thesaurus* of 1739-1743, Lupi's *Dissertatio* of the same year, as well as with several other works of lesser importance.[49] Inscriptions were published with little

---

[46] See in general, L. A. Segal, "Jacques Basnage de Beauval's *Histoire des Juifs*: Christian Historiographical Perception of Jewry and Judaism on the Eve of the Enlightenment," *Hebrew Union College Annual* 54 (1983) 303-324 and G. Cerny, *Theology, Politics and Letters at the Croassraods of European Civilization* (Dordrecht: Nijhoff, 1987) esp. 182-202.

[47] G. Bianchini, *Delle magnificenze di Roma antica e moderna* (Rome: Chracas, 1747) LXX. Leon 1928, 303 suggests that Bianchini never visited the Monteverde catacomb, but that he copied Bosio.

[48] Ferretto 1942, 163f.; W. Larfeld, *Griechische Epigraphik* (Munich: Beck, 1914) 7-105; *DACL* 7 (1926) 830-1090 s.v. "Inscriptions (Histoire des recueils d'inscriptions)" (H. Leclercq).

[49] J. Spon, *Miscellanea eruditae antiquitatis* (Lyon: Huguetan, 1685) 371, nos. 118-20; R. Fabretti, *Inscriptionum antiquarum quae in aedibus paternis asservantur explicatio etc.* (Rome: Herculis, 1699) 389, nos. 246-48; and 465, no. 101; L. A. Muratori, *Novus*

or no commentary. Epigraphers did not usually pause to reflect on the possible significance of the Jewish epigraphical materials they included in their surveys. More often than not, we do not even know where the Jewish inscriptions retrieved during this period were found.

No archaeological research worthy of that name was carried out in either the Christian or the Jewish catacombs of Rome during the two centuries following Bosio's death in 1629. During this period, the catacombs did not disappear entirely from the public eye, as had been the case in the high Middle Ages. Instead, they now became an almost inexhaustible treasure-trove for all those "apassionionati" who wanted to enrich their collections of antiquities with new materials. F. Buonarotti, who was the first to publish, in 1716, three Jewish so-called gold glasses, described this renewed fascination with antiquities well. He and his friends would enter the catacombs, and empty one underground gallery after the other. For them it was pure "divertimento."[50] The methods employed by his contemporary M. A. Boldetti (1663-1749) who, in 1720, published a Jewish lamp supposedly found at Rome, were no less rigorous. Boldetti's research method was in fact so crude that it has been described as follows: "Boldetti partit de ce principe qui'il faillait déménager tout le mobilier funéraire des premiers fidèles. L'exploration se faisait méthodiquement en ce sens qu'une gallerie visitée était une gallerie vidée." The nature of scholarly endeavours during this period is perhaps best illustrated, however, by the fact that the illustrations of Bosio's *Roma sotterranea* were republished in 1737 by Bottari as *Sculture e pitture sacre estratte dai cimiteri di Roma*, but that the accompanying text was left out altogether.[51]

The general fascination with antiquities of the late seventeenth and early eighteenth century resulted in the discovery of new Jewish inscriptions, Jewish lamps, and Jewish gold glasses. Like early Christian antiquities, Jewish antiquities were avidly sought after by collectors. Generally speaking, such collectors were aware of the Jewish

---

thesaurus veterum inscriptionum (Milan, 1739-43) vol. 1, 152, no. 4; vol. 2, 708, no. 3; 1129, no. 6; vol. 3, 1674, no. 3 and vol. 4, 2045, no. 7; A. Lupi, *Dissertatio et animadversiones nuper inventum Severae martyris epitaphium* (Palermo, 1734) 51, 177 and table VII, 1-2. Of lesser importance are the publications of T. S. Baier, R. Venuti, and P. Wesseling.

[50] F. Buonarotti, *Osservazioni sopra alcuni frammenti di vasi antichi di vetro ornati di figure trovati ne' cimiteri di Roma* (Rome, 1716) XII. Gold glasses usually consist of the decorated base of a larger glass vessel. The decoration consists of a piece of goldleaf between two thin layers of translucent glass. For more details, see Chapter 2.

[51] M. Boldetti, *Osservazioni sopra i cimiteri dei SS. martiri ed antichi cristiani di Roma* (Rome: Salvioni, 1720) 526; Citation: *DACL* 2 (1925) 974-76 s.v. "Boldetti." See also *Dizionario Biografico degli Italiani* 11 (1969) 247-49 s.v. "Boldetti" (N. Parise).

character of the Jewish antiquities they could add to their collections on rare occasions. They were, however, not interested in the provenance of these Jewish materials. Nor were they interested in the possible relationship between Jewish and early Christian artifacts. At the same time, collectors and scholars did not feel uncomfortable about including Jewish archaeological materials in their surveys of ancient art. Such a procedure reflects the general tendency among the antiquarians of the time to collect antiquities rather than to think systematically about the larger historical issues these materials were raising. This was a world which tried to capture the "visible" in order to classify and then describe it in verifiable terms. It was an approach that enjoyed widespread popularity well into the nineteenth century.[52]

Even though the publications of this period generally provide little or no information on how their authors viewed Jews and Judaism, it is nevertheless possible to show that traditional views on this subject continued to be popular. Agitating against the fact that the portico of Santa Maria in Trastevere had been adorned, in 1742, by M. A. Boldetti with reused pagan inscriptions, an unnamed cleric declared this church to have been profaned. It was a complaint Church authorities had to take seriously. As a contemporary observed, the incident not only led to much commotion within Catholic circles, but also carried with it the danger that it could provide critics such as J. Spencer with additional evidence to argue that the Church of Rome had derived its most important rites from pagan religious practice.[53]

To justify the reuse of "cose gentilesche" in a Christian context, G. Marangoni (1673-1753) devoted an entire treatise to the question in 1744. His defense of the "eruditissimo signor canonico" Boldetti consisted of two parts. First Marangoni analyzed several passages taken from the Hebrew Bible to show that the reuse of pagan spoils had been permitted to the Jews on many occasions. Then, Marangoni established a link between the practices described in the Hebrew Bible and the behavior of Boldetti. This was not difficult to accomplish. Like Aringhi and early Christian apologists before him, Marangoni adopted the typically Christian view which held that the Catholic Church was the legitimate heir to "the precepts, the priesthood, the principate of the synagogue" and, by implication, to the Hebrew Bible. The idea that the Church is *Verus Israel* thus helped

---

[52] For an analytic description of Jewish materials, see, e.g., F. Buonarotti, *Osservazioni sopra alcuni frammenti di vasi antichi di vetro* (Florence: S.A.R., 1716) 19-21. In general, see M. Foucault, *Les mots et les choses* (Paris: Gallimard, 1966) 143, 157-58; G. B. de Rossi, "Prefazione" in *Nuovo Bulletino di Archeologia Cristiana* 1863, s.p.

[53] G. Marangoni, *Delle cose gentilesche e profane trasportate ad uso, e adornamento delle Chiese* (Rome: Pagliarini, 1744) *passim*.

Marangoni in defending the reuse of pagan materials by the Church. At the same time it also helped him to argue that many of the Church rites were of biblical rather than of pagan origin: Marangoni discovered that early Christian theology could serve the double purpose of countering criticisms of Catholic extremists and of warding off attacks by Protestant skeptics.[54]

The importance attached to early Christian apologetic literature, and to the idea of *Verus Israel* in particular, also explains why Jews fulfill identical roles in the works of Aringhi and Marangoni. In these works, Jews were never seen as existing in their own right. The Jews of the Hebrew Bible, for example, were believed to have been created only to help prepare the world for the arrival of Christianity. Jews in seventeenth and early eighteenth century Italy were defined in comparably negative terms. Such Jews were either seen, in Aringhi's words, as *testes nostrae redemptionis*, or they were regarded people who had *not yet* been converted (note that these notions are not compatible!).[55] Along similar lines, Jewish writings were considered as having no other purpose than to legitimize the Church, its doctrine, and its practices. Such writings could be (and were) appropriated to defend the Church against its critics, or they were misused to help effect the conversion of the Jews.

Understandably, in the climate described above, finds from the Jewish catacombs of Rome could not attract the scholarly attention they deserved. In the seventeenth century and beyond, the new fascination with archaeology went hand in hand with a view of Jews and Judaism that was inherently Christian and that had changed little since the days of the Counterreformation. Put differently, the treatment of Jewish archaeological materials documents the extent to which, during the eighteenth century, the study of the catacombs remained within the parameters that had been set long ago by Bosio and, more explicitly, by Aringhi. Given the state of catacomb studies in this period, it does not come as a surprise to note that in even a work of the scholarly status of Diderot's *Encyclopédie* of 1751, Jewish archaeological materials play no role. In 1751 catacombs were still defined as places where one goes to worship martyrs, and as places where bones were found that were sent to every Catholic country as relics, that is, after the pope had established a saint's name

---

[54] Marangoni (previous note) 21 and 77f.

[55] Aringhi 1651, 397. Similar views had already been expressed by Augustine. For a good example of how "a history of the Jews" was often nothing but a history of their conversion to Catholicism, see G. Moroni, *Dizionario di erudizione storico-ecclesiastica da S. Pietro ai nostri giorni* (Venice: Emiliana, 1843) vol. 21, 5-43, esp. 23f. Although his dictionary postdates the period discussed here, Moroni sums up a view of Jewish history that has its roots in this period.

to go with them.⁵⁶

The Unappealing Catacombs: 1750-1840

In the second half of the eighteenth century, after they had been studied uninterruptedly for two hundred years, the catacombs of Rome ceased to attract the attention of scholars. By this time, Classical and early Christian archaeology had established themselves as separate fields of study. This development was much to the detriment of the latter. Often organizing themselves into learned societies, students of both history and art history preferred to write about Greece, or about Republican and Imperial Rome, rather than about Late Antiquity. They treated Late Antiquity as "a dunghill strewn with diamonds, crying out to be pillaged and badly needing to be cleaned out."⁵⁷

As a result of the Enlightenment's preference for secular history, and for the kind of rationality associated with Greek civilization in particular, Christian views on history came under attack. Conventional ideas about the history of the Jews in biblical times were also reexamined. In fact, by criticizing such biblical notions as providence and revelation, the *philosophes* found a way to disguise attacks that were aimed at the Church rather than at the Jews. Many *philosophes* hoped that the socially inferior position of the Jews of their own time would finally come to an end. This hope was based on ideas that had already been current in Europe for some time, even if in a somewhat different form. Yet, even though, in Enlightenment thought, anti-Judaic sentiment served to a large extent as a means to an end, more traditional views on Jews and Judaism did not disappear entirely. Basnage's influence remained especially palpable.⁵⁸

In the second half of the eighteenth century, Italy not only remained the artistic center it had always been; it also developed into a country where Classical scholarship flourished and in which intellectual pursuits were highly valued. Inasmuch as, during this period, the scions of Europe's wealthiest families concluded their education by

---

⁵⁶ *Encyclopédie ou dictionnaire raisonné des sciences, des arts et des métiers* (Stuttgart: Frommann, 1988 [1751-80]) vol. 1, 758 s.v. "catacombe."

⁵⁷ P. Gay, *The Enlightenment. An Interpretation. The Rise of Modern Paganism* (New York: Norton, 1977) 323.

⁵⁸ Momigliano 1950, 307f.; S. Ettinger, "The Beginnings of the Change in the Attitude of European Society Towards the Jews," *Scripta Hierosolymitana* 7 (1961) 193-219; H. Liebeschütz, *Das Judentum im deutschen Geschichtsbild von Hegel bis Max Weber* (Tübingen: Mohr, 1967), 2f.; Gay (previous note) 72f., 93; C. Hoffmann, *Juden und Judentum im Werk deutscher Althistoriker des 19. und 20. Jahrhunderts* (Leiden: Brill, 1988) 12-14.

traveling extensively through Europe, visiting Italy became especially fashionable. The accounts written on the occasion of the "Grand Tours" are interesting in many respects, not least because they show the extent to which early Christian art and the catacombs had disappeared from view. During the two years J. W. von Goethe spent in Italy (1786-88), for example, he visited the catacombs only once, and only when he was about to leave Rome for good. He found these "dumpfige Räume" thoroughly repellent and preferred to read Bosio rather than to visit such places himself. Some of Goethe's compatriots, such as J. Führich, shared his feelings completely. So did the Reverend Richard Burgess. Publishing a study on *The Topography and Antiquities of Rome* in 1821, he reports that he spent "more than half an hour" in the catacombs. The Frenchman Stendhal thought such trips superfluous to begin with. In his *Promenades dans Rome* he compiled long lists of Roman monuments he considered worthwhile visiting, but included not even a single reference to the catacombs. The *Histoire de la peinture* by the same author starts with a description of the invasions of the "guerriers du Nord, sauvages dégradés" around 400 C.E., yet, paradoxically, it deals with Late Ancient and Early Medieval painting from Rome, for which the most important evidence has been preserved in the catacombs, in a most superficial manner. Even the handful of authors who wrote about catacombs during this period had to admit that "on est porté à croire que ces souterrains vont exciter la curiosité des voyageurs: cependant il faut convenir qu'ils attirent peu leur regards."[59]

The lack of interest in catacombs must be seen in the light of the fact that the most important archaeological discoveries of this period occured neither in catacombs nor in Rome. Since the 1720s, new finds drew attention to Italy's Etruscan past and to the study of other pre-Roman civilizations. Then, in the period from 1738 to 1766, Herculaneum was discovered and partly excavated. Excavations in nearby Pompeii started in 1748.[60] Yet the main reason why public

---

[59] J. W. von Goethe, *Italienische Reise* (Frankfurt am Main: Insel, 1976) vol. 2, 725; for the account of J. von Führich, see E. Haife, *Deutsche Briefe aus Italien. Von Winckelmann bis Gregorovius* (Munich: Beck, 1987) 277; Burgess remarks are cited in Gaston 1983, 157; Stendhal in V. de Vitto and E. Abravanel (eds.), *Oeuvres complètes* (Genève and Paris: Champion, 1969 [1829]) vol. 1, 7-8; 19-22; vol. 3, 273-78; vol. 26, 3f., 61f. Artauds de Montor, *Voyage dans les catacombes de Rome. Par un membre de l'Académie de Cortone* (Paris: Schoell, 1810) 2. In general, see L. Schudt, *Italienreisen im 17. und 18. Jahrhundert* (Vienna-Munich: Schroll, 1959); J. Burke, "The Grand Tour and the Rule of Taste," in R. F. Brissenden, *Studies in the Eighteenth Century. Papers Presented at the D. N. Smith Memorial Seminar, Canberra 1966* (Canberra: Australian National U. P., 1968) 231-50.

[60] Momigliano 1950, 304f. and Appendix II, pp. 314-15; F. Gigante, Le scoperte di Ercolano e Pompei nella cultura europea del XVIII secolo," *La Parola del Passato* 34

interest shifted away from the catacombs so dramatically in the later eighteenth and early nineteenth century is not such archaeological discoveries alone. The reason is to be sought in a general change in taste and in the revolutionary development, or, more precisely, the birth of Classical Archaeology.

As the result of the antiquarianism that had characterized the study of archaeological remains during the later seventeenth and early eighteenth century, scholars now had in their possession endless catalogues, but they had no idea of how Greek and Roman art had developed chronologically. In retrospect, and despite claims of contemporaries to the contrary, the antiquarians had not really been successful in "concilier les monuments avec l'histoire."[61] This situation changed completely in 1764, upon the publication of the *Geschichte der Kunst des Altertums* by J. J. Winckelmann (1717-1768). Not only did this history of the rise and decline of ancient art fill a lacuna that had been felt for decades. Nor did Winckelmann merely define the parameters for archaeological, or, more precisely, for art historical research for decades to come. His *Geschichte* was the first systematic history of Greek and Roman art ever to be written.

Despite its importance and scope, Winckelmann's *Geschichte* was, to be sure, a history in which there was no place for Late Antiquity. Having been disappointed by the paintings that had been discovered at Herculaneum and Pompeii, there was in fact little room in this history of ancient art for anything but Greek sculpture of the Classical period (fourth century B.C.E.) Studying such sculpture, ironically, through mediocre, Roman-period marble copies of the bronze originals, and having been influenced by the views on art of writers such as Pliny the Younger, Winckelmann maintained that the arts had started their decline as early as the time of Augustus. That being the case, he wondered why one of the less successful portraits of the emperor Claudius could not be used as a counterweight to the bells of the Escorial. Winckelmann believed that after the reign of Marcus Aurelius (161-180 C.E.) the imminent decline had truly become unstoppable. Material remains postdating the second century C.E. were, in his view, really "beyond the history of art." The decline of form could best be compared to "a maiden standing on the shore of the ocean, following with tearful eyes her departing lover with no hope of ever seeing him again, fancying that in the distant sail she sees the image of her beloved."[62] Seroux d'Agincourt, who wrote an impor-

---

(1979) 377-404.

[61] Citation: Comte de Caylus (1692-1765) as cited in M. Käfer, *Winckelmann's hermeneutische Prinzipien* (Heidelberg: Winter, 1986) 23.

[62] J. J. Winckelmann, *Geschichte der Kunst des Altertums* (Vienna: Phaidon, 1934 [1764]) 363; 379-83; 390. In general, see H. L. van Dolen and E. M. Moormann, *Johann*

tant six volume history of Medieval art, which he began publishing in 1823, agreed heartily and concluded that already in the second century C.E. "most people were better Christians than good painters." So did Gibbon, to judge from the books in his library, among which the publications of Winckelmann are as noticeably present as the works of Bosio and Aringhi are absent. [63]

Arranging his material around the central theme of decline—just as Gibbon would do several years later when writing his *History of the Decline and Fall of the Roman Empire*—Winckelmann thus wrote a history of ancient art in which his personal views on aesthetics are as unmistakable as are his ideas on the spiritual value and moral potential of art.[64] Yet Winckelmann recognized that in order to evaluate them properly, the remains from the past needed to be studied in a comprehensive fashion, and not just collected (as the antiquarians had done). Fundamental as such a recognition was for the subsequent development of Classical Archaeology, from the point of view of the study of catacombs, Winckelmann's influence was decidedly negative. As a result of his views on ancient art, archaeological materials from the catacombs were now largely neglected. For all his appreciation of the historic importance of Medieval art, even Seroux d'Agincourt had to admit that artistically his study represented nothing but a tedious catalog of "objets défigurés." Popes such as Clement XIV (1769-1774) and Pius VI (1775-1799) agreed. Rearranging and expanding the Vatican's collections, their passion was for Greek and Roman art, and not for early Christianity's physical remains. The time of the Counterreformation was definitively over.[65]

Not surprisingly, in Winckelmann's days, Jewish antiquities attracted no attention. The two magnificent reliefs commemorating the Roman victory of 70 C.E. over the Jews that were mounted on the inside of Titus' arch on the Forum are the only Jewish antiquities re-

Joachim Winckelmann. Een portret in brieven (Baarn: Ambo, 1993) 9-73. And see H. Wrede, "Die Opera de' Pili von 1542 und das Berliner Sarkophagcorpus. Zur Geschichte von Sarkophagforschung, Hermeneutik und Klassischer Archäologie," *Jahrbuch des Deutschen Archäologischen Instituts* 104 (1989) 373-414.

[63] J. B. L. G. Seroux d'Agincourt, *Histoire de l'art par les monuments dépuis sa décadence au IVe siècle jusqu'à son renouvellement au XVIe* (Paris: Trentel and Würtz, 1823), vol. 2, 22; G. Keynes, *The Library of Edward Gibbon. A Catalogue of his Books* (London: Cape, 1940) s.v.

[64] P. Burke "Tradition and Experience: The Idea of Decline from Bruni to Gibbon," in *Edward Gibbon and the Decline and Fall of the Roman Empire* (Cambridge Mass. and London: Harvard U. P., 1977) 87-102; P. R. Gosh, "Gibbon's Dark Ages: Some Remarks on the Genesis of the *Decline and Fall*," *Journal of Roman Studies* 73 (1983) 1-23.

[65] Seroux d'Agincourt (previous note) vol. 1, IIIf., 4-5; vol. 2, 10f.; 36f.; H. Beck *et al.* (eds.), *Antikensammlungen im 18. Jahrhundert* (Berlin: Mann, 1981) 149-65.

ferred to with some regularity, yet they evoke no more than stereotypical comments in travelers' accounts of the period.[66] One of the few people who wrote about catacombs during this period was Artaud de Montor, yet his work has little independent value. His remarks on the Jewish Monteverde catacomb, for example, are not based on personal observations, but were copied from Bosio.[67]

On a methodological level, the study of Jewish history was not affected by the developments that shaped and transformed Classical studies in these years. It has been suggested that there exists a parallelism between the beginnings of the "klassische Altertumswissenschaft" as reflected in the work of C. G. Heyne (1729-1812) and the interest of Göttingen's orientalist J. D. Michaelis (1717-1791) in the physical remains collected by the Orient expedition of C. Niebuhr in the years 1761-1767.[68] Such a suggestion, however, is not correct; the suggested parallelism is more apparent than real. Heyne used his philological experience and knowledge of Classical literature to come to a better understanding of the works of art themselves. Michaelis, by contrast, used archaeological finds because they helped him explain the literary sources. More specifically, they served him as independent evidence in support of the apologetic thesis that Mosaic law was superior to all other law. Throughout the eighteenth century, the term *antiquitates iudaicae* or *antiquitates hebraicae* continued to be used in the traditional sense, that is, to indicate discussions of Jewish religion which were based exclusively on an analysis of the literary sources. As Simonis explained in the introduction to his lengthy comments on H. Reland's *Antiquitates sacrae veterum hebraeorum* (1769), the notable lack of reference to physical remains in such works was lamentable, but it was equally impossible to remedy (archaeological excavations had not yet taken place in the Holy Land). Yet, even if archaeological finds had been available in any quantity, one wonders whether this would have much altered the purpose such *antiquitates* ultimately served: to illuminate the Holy Scriptures as flawlessly as possible. In the late eighteenth century, people who were interested in Jewish or biblical archaeological remains had simply not yet endorsed the same

---

[66] F. Haskell, *Past and Present in Art and Taste* (New Haven and London: Yale U. P., 1987) 16-31. For references to the Arch of Titus in reports of visitors to Rome, see E. and R. Chevallier, *Iter Italicum. Les voyageurs français à la découverte de l'Italie ancienne* (Genève: Slatkine, 1984) 274-91, esp. 278f.

[67] F. Artaud de Montor, *Voyage dans les catacombes de Rome. Par un membre de l'Académie de Cortone* (Paris: Schoell, 1810) 123-24. Note also his peculiar discussion of the scholarly value of the works of Bosio and Aringhi on pp. 13-16.

[68] Hoffmann 1988, 27-30; On Michaelis, see R. Smend, "Johann David Michaelis und Johann Gottfried Eichhorn-zwei Orientalisten am Rande der Theologie," in B. Moeller, *Theologie in Göttingen. Eine Vorlesungsreihe* (Göttingen: Vandenhoeck & Ruprecht, 1987) 58-81, esp. 59-71.

methodological principles as practiced by contemporary specialists of Greek and Roman art.[69]

As a result of changes in taste and emphasis in scholarly research, Jewish archaeological materials from Rome, along with the Christian catacombs, had thus ceased to interest the leading eighteenth century students of the past. By 1805, the general lack of interest in the catacombs and the accompanying lack of efforts to protect the materials they contained was complete. The resulting "devastamento" encompassed all of underground Rome.[70]

The Excavation of Jewish Rome: 1840-1940

As had been the case in the preceding centuries, also in the nineteenth century the study of the Jewish catacombs of Rome was intimately linked with research on the Christian catacombs in that city. In contrast to earlier work on the catacombs, however, scholars were now prepared to take Jewish realia more seriously. Such a readiness was due to several factors. From the second half of the nineteenth century onwards, several new Jewish catacombs and hypogea were discovered in and around Rome. Thus, it became difficult for scholars to gloss over such Jewish materials in the same way as Bosio and Aringhi had once done. Even more important, the revival of interest in the catacombs—which was due to the activities of G. Marchi (1795-1860) and of G. B. de Rossi (1822-1894) in particular—went hand in hand with a new sensitivity to the methodological problems that surrounded (and still surround) the study of the Roman catacombs. In the words of a French contemporary, scholars agreed that the study of catacombs should no longer be seen as an exercise in theology: "c'est l'histoire pure et simple qu'il faut faire." Such claims notwithstanding, progress was only relative during this period, certainly insofar as the Jewish catacombs were concerned. Traditional ways of looking at Jewish materials did not disappear altogether from the publications of Christian scholars.[71]

---

[69] For a short appreciation of Heyne's work, see U. Hausman, *Allgemeine Grundlagen der Archäologie* (Munich: Beck, 1969) 33-34; J. Simonis, *Vorlesungen über die jüdischen Alterthümer nach Anleitung H. Relands Antiquitatum Sacrae Veterum Hebraeorum* (Halle: Curt, 1769) 1-2. H. Reland, *Palaestina ex monumentis veteribus illustrata* (Leiden: Broedelet, 1714) 2 vols. is another example of the literary approach the characterizes Christian scholarship on Jews and Judaism during this period.

[70] See Ferretto 1942, 298.

[71] Ferrretto 1942, 296f.; *Dizionario Biogragrafico degli Italiani* 39 (1991) 201-205 s.v. de Rossi (N. Parise); T. Roller, *Les catacombes de Rome. Histoire de l'art et des croyances religieuses pendant les premiers siècles du Christianisme* (Paris: Morel, 1881) vol. 1, II (citation).

The nineteenth century also saw the emergence of historical scholarship authored by Jews (*Wissenschaft des Judentums*). Yet, generally speaking, Jewish historians showed little interest in archaeological remains. They rather preferred to study the past on the basis of the literary sources. Thus, insofar as the Jewish catacombs of Rome are concerned, the nineteenth century was as much an age of rediscovery as it was an age of stagnation.[72]

The first serious study of the nineteenth century that dealt with Jewish archaeological materials can be found in a work on "the early Christian arts" published in 1844. In it, its author, Marchi, reported about an unsuccessful effort to locate the Jewish Monteverde catacomb. He also elaborated on how he viewed the relationship between the Jewish and Christian catacombs.[73]

The fact that Marchi tried to establish the location of the Monteverde catacomb at all, is highly significant, for it illustrates the new, more dynamic approach that was to henceforth characterize the study of Rome's catacombs. Yet, a brief analysis of Marchi's account also shows the extent to which, around the middle of the nineteenth century, religiously inspired sentiments still determined the course of an argument that presented itself as rational and straightforward.

The idea that stood at the basis of Marchi's observations on the Jewish and early Christian catacombs of Rome was his conviction that the catacombs of Rome were a local phenomenon. Contrary to Aringhi, Marchi did not believe that the cave of Machpelah could explain the origin of catacomb burial in Rome. Such a starting point left Marchi with three possible alternatives: (1) the Christian community of ancient Rome had itself invented the practice of constructing and burying in catacombs; (2) Roman Christians had derived the practice from the Jewish community; or (3) Christians had adopted burial customs that were of pagan derivation.

To Marchi, a cleric and a pious Catholic, the last solution was least attractive of all. Thus he was left with only two alternatives: either the Christians got it from the Jews or vice versa. Upon further reflection, Marchi concluded that there really was only one alternative: the Jewish community of ancient Rome had taught the Christians to dispose of their dead in this way. Why? Because, as Marchi maintained, the Jews had always been "far too tenacious in their rites and proper customs...to have wanted, in this completely religious prac-

---

[72] For further literature on the *Wissenschaft des Judentums*, see below, this section.

[73] For this and for the discussion that follows, see G. Marchi, *Monumenti delle arti cristiane primitive nella metropoli del cristianesimo* (Roma: Puccinelli, 1844) 19-21. J. G. H. Greppo, *Notices sur des inscriptions antiques tirées de quelques tombeaux juifs à Rome* (Lyon: Barret, 1835) has no independent value. It copies Bosio and Aringhi and is written in the same spirit (e.g. p. 25: "ce peuple, qui fut autrefois le peuple de Dieu" etc.).

tice...to imitate the Christians."

Although based on prejudice rather than on verifiable observations, Marchi found this a most satisfactory solution for one reason in particular: it opened the way to construing catacomb burial as an illustration of the typically Christian view that Jesus had come not to abrogate the Law, but rather to perfect it. Put differently, just as Christianity fulfilled the promises that had once been given to the Jewish people, so the architects of the Christian catacombs brought to perfection a mode of burial that had remained imperfect among the people who had invented it.

Even though Marchi's and Aringhi's starting-points were different, it is remarkable to see, upon closer reflection, how little the former had moved away from the latter's ideas on Jews. The lack of decoration in the Jewish Monteverde catacomb that had struck Bosio as "unnoble," had, in Marchi's view, to be explained as resulting from the "raffreddemento della pietà in questi Giudei...e dalla ingenita loro sordidezza"—an explanation that would have appealed to Aringhi.

It may be clear that in 1844 little conceptual progress had been made in the study of the catacombs. Like his predecessors, Marchi, self-professed "amico del vero," was interested in the Jewish connection. Like his predecessors, his interest was heavily charged theologically. Again, it resulted from the wish to understand better the world of early *Christianity*. To be sure, it was a world in which the existence of the Jews could be justified only because they fulfilled the same role they had fulfilled in early Christian apologetic literature, and in the works of Aringhi and Marangoni: to be Christianity's midwife.

Several years after Marchi tried to locate the Jewish Monteverde catacomb, a Jewish catacomb that had not been known previously was discovered along the Via Appia. It was the beginning of a sixty year period (1859-1919) during which a total of four Jewish catacombs (large underground cemeteries) and two Jewish hypogea (small underground cemeteries) were found and explored. R. Garrucci, who replaced E. Herzog in this capacity, excavated a sizable Jewish catacomb, the so-called Vigna Randanini catacomb (1859-1860). Located between the Via Appia and Via Appia Pignatelli, this catacomb is still accessible today.[74] In 1866 a small

---

[74] E. Herzog, "Le catacombe degli Ebrei in Vigna Randanini," *Bulletino dell' Instituto di Corrispondenza Archeologica* 1861, 91-104; R. Garrucci, *Cimitero degli antichi Ebrei scoperto recentemente in Vigna Randanini* (Roma, 1862) and *id.*, *Dissertazioni archeologiche di vario argomento* (Roma, 1864-65); O. Marucchi, *Breve guida del cimitero giudaico di Vigna Randanini* (Roma, 1884).

Jewish hypogeum came to light in the Vigna Cimarra, near the catacomb of S. Sebastiano. It was never properly investigated, and it has disappeared since.[75] In 1882 there followed the discovery of a similar Jewish hypogeum along the Via Labicana. It was briefly studied by O. Marucchi, until the collapse of underground galleries made further access impossible. It has never been relocated.[76] Several years later, in 1885, N. Müller found another hypogeum along the Via Appia Pignatelli, not far away from the Vigna Randanini catacomb. He tried to identify this site as yet another Jewish underground cemetery, but his attemps have not met with universal approval. The precise location of this hypogeum is presently unknown.[77] In 1904-1906 and in 1909 the Jewish Monteverde catacomb which Bosio had first entered in 1602, was excavated by N. Müller. The catacomb was destroyed in the early 1920s.[78] In 1919, finally, a large complex of underground galleries was accidentally discovered under the Villa Torlonia, on the Via Nomentana. Subsequent research carried out in 1973-1974 has shown that it actually consisted of two separate Jewish catacombs that were connected to each other only in a later stage. This catacomb complex is still accessible today.[79]

Although it is possible that some Jews in ancient Rome were buried in places other than catacombs or hypogea, the excavation reports concerning Jewish extra-catacombal burials are so unspecific that it is impossible to determine today, what such burials might originally have looked like.[80] Finally, there is no evidence, either in-

---

[75] G. B. de Rossi, "Scoperta d'un cimitero giudaico presso l'Appia," *Bulletino di Archeologia Cristiana* 5 (1867) 16 and Berliner 1893, 48 and 90-92.

[76] O. Marucchi, "Di un nuovo cimitero giudaico sulla Via Labicana," *Dissertazioni della Pontificia Accademia Romana di Archeologia* 2 (1884) 499-532; *id.*, *Di un nuovo cimitero giudaico sulla Via Labicana* (Rome, 1887).

[77] N. Müller, "Le catacombe degli Ebrei presso la Via Appia Pignatelli," *Römische Mitteilungen* 1 (1886) 49-56; on the supposed Jewish character of this site, see O. Marucchi in *Bulletino di Archeologia Cristiana* 3 (1884-85) 140 and C. Vismarra, "I cimiteri ebraici di Roma," in A. Giardina (ed.), *Società romana e impero tardoantico* (Bari: Laterza, 1986) vol. 2, 389-92.

[78] Müller 1912; *id.*, "Cimitero degli antichi Ebrei posto nella Via Portuense," *Dissertazioni della Pontificia Accademia Romana di Archeologia* 12 (1915) 205-318; R. Kanzler, "Scoperta di una nuova regione del cimitero giudaico della Via Portuense," *Nuovo Bulletino di Archeologia Cristiana* 21 (1915) 152-57; G. Schneider-Graziosi, "La nuova Sala Giudaica nel Museo Cristiano Lateranense," *Nuovo Bulletino di Archeologia Cristiana* 21 (1915) 13-56; R. Paribeni, "Iscrizioni del cimitero giudaico di Monteverde," *Notizie degli Scavi di Antichità* 46 (1919) 60-70; N. Müller and N. Bees, *Die Inschriften der jüdischen Katakombe am Monteverde zu Rom* (Leipzig: Fock, 1919).

[79] R. Paribeni, "Catacomba giudaica sulla Via Nomentana," *Notizie degli Scavi di Antichità* 46 (1920) 143-55; H. W. Beyer and H. Lietzmann, *Die jüdische Katakombe der Villa Torlonia in Rom* (Berlin: de Gruyter, 1930); Fasola 1976

[80] See e.g. *CIJ*, LXI, and nos. 3, 4, and 286; A. Konikoff, *Sarcophagi from the Jewish Catacombs of Ancient Rome. A Catalogue Raisonné* (Stuttgart: Steiner, 1986) nos. 1, 3, 12-15, 16-20; *Notizie degli Scavi di Antichità* 1911, 139.

scriptional or iconographic, in support of Gressmann's and Goodenough's thesis that the so-called Vibia catacomb contains the graves of the adherents of a Jewish mystical group.[81]

A look at the professional careers and backgrounds of the excavators of the Jewish catacombs of Rome shows that by the end of the nineteenth century, the field of catacomb studies was more diversified in terms of "practitioners" than it had ever been. While some of them wrote about early Christian art, early Christian catacombs, and early Christian epigraphy (Garrucci, de Rossi, Marucchi), others devoted their professional careers to studying the world of the New Testament and the history of the early Church (Müller, Beyer, Lietzmann). And, also in terms of religious affiliation, there was more diversity than before. While some of these scholars had a Catholic background (Garrucci, de Rossi, Marucchi), others were Protestants (Müller, Lietzmann). The excavators of the Jewish catacombs had in common, however, that none of them was Jewish or a student of Jewish history.

Where Bosio and Aringhi had effectively succeeded in removing the Jewish evidence from scholarly discussions on catacombs, the archaeological discoveries in the years following 1859 made unavoidable the question that had never received the attention it deserved: What is the relationship between the Jewish and Christian catacombs? As it turned out, it was a question to which there were many possible answers. Let us turn to an investigation of work of scholars who discussed this question. They include Herzog, Garrucci, de Rossi, Krauss, and Müller.

Herzog, who briefly worked in the Jewish Vigna Randanini catacomb, believed that Jewish and early Christian underground burial had both sprung from the wish to provide co-religionists with a proper burial. He was convinced, however, that the network of underground galleries in the Vigna Randanini catacomb had been constructed in imitation of the Christian catacombs. This was evident, in the words of Herzog, from "tutto il sistema del nostro ipogeo giudaico." Failing to specify what he intended with the expression "the whole system," it is not clear what Herzog meant. Supposing, however, that Herzog wanted to say that the underground galleries in the Jewish and the Christian catacombs look similar, one is still left to wonder how Herzog arrived at the conclusion that the Jewish catacomb imitated its Christian counterpart rather than vice versa. One

---

[81] H. Gressmann, "Jewish Life in Ancient Rome," in G. A. Kohut (ed.), *Jewish Studies in Memory of Israel Abrahams* (New York: Jewish Institute of Religion, 1927) 170-91, esp. 173-75; E. R. Goodenough, *Jewish Symbols in the Graeco-Roman Period* (New York: Pantheon Books, 1953) vol. 2.

also wonders on the basis of what evidence did he conclude that the Jewish catacomb was constructed in the time of Alexander Severus (222-235 C.E.) and that it postdated the Christian catacombs by about a century.[82]

Continuing the work Herzog had begun in the Vigna Randanini catacomb, Garrucci discussed the issues Herzog had raised, but he interpreted them very differently. Garrucci's main concern was to prove that the Jewish catacombs did not take precedence over their Christian counterparts. He argued that the quintessential Jewish grave form was the hypogeum, or small underground family grave. The use of other types of graves by Jews was inconceivable because, according to Garrucci, "the synagogue and synhedria that ruled the Jews" would never have permitted them to deviate from the practices of their forefathers. Adding a discussion of Jewish, mostly postbiblical, sources that deal with Jewish burial customs, Garrucci then argued that the Jewish Vigna Randanini catacomb in Rome belonged to the class of small underground tombs that are described in Genesis. He even went so far as to maintain that the Vigna Randanini catacomb was not a catacomb at all, but rather a collection of crypts and caverns "according to their national custom." Garrucci further observed that the use of inscriptions and sarcophagi by the Jews of Rome did not originate in "their fatherland," but that "the Jews, capital enemies of the Church" had derived it from the Church; "We know, after all, that idolaters and heretics tempted in this deceitful way all those who turned to the Church, the only restorer of corrupted humankind." The fact that the loculi or burial slots look identical in both Jewish and Christian catacombs was explained by suggesting that such loculi had a common pagan origin. Yet, the two most important arguments to document the supposed originality of the Christian catacombs were the following. First, Garrucci maintained that inscriptions from Christian catacombs dated to the first century C.E., and that none of the epigraphic materials from the Vigna Randanini catacomb predated the third century C.E. Second, Garrucci argued that the early Church had been the only institution in Antiquity that had held itself responsible for the burial of its members. Such an assertion permitted him to suggest that burial in catacombs was a Christian phenomenon *per definition*.[83]

Writing about Garrucci's scholarly qualities, Theodor Mommsen once observed that *omnibus inimicus et maxime sibi ipsi* (everyone's enemy, but to himself the worst enemy).[84] Reading Garrucci's re-

---

[82] Herzog 1861, 95-6.

[83] Garrucci 1862, 11-25.

[84] See C. Perone, "Per lo studio della figura e dell'opera di R. Garrucci, 1812-1885," *Miscellanea greca e romana* 13 (1988) 17-50. Note that Perone's evaluation of Garrucci's

flections on the Vigna Randanini catacomb, it becomes evident how his work could upset the more critical minds of the nineteenth century. In fact, Garrucci's account is peculiar mixture of prejudice, unproven assumptions, and inconsistent reasoning. The idea that Jews always and invariably bury in hypogea is a claim that we have already encountered in the works of Baronio and Aringhi. It is based on the Christian belief that petrification characterizes Jewish religion and Jewish practices. That Garrucci upholds the traditional Christian view of Jews and Judaism is particularly evident from his vocabulary (note, for example, the idea that Jews are heretics). What is more important, some of Garrucci's observations are simply incorrect. The assertion that the Vigna Randanini catacomb consists of a collection of hypogea, and that it is not a catacomb at all, is highly misleading. The Vigna Randanini catacomb originated in a collection of hypogea that were originally unconnected. Yet there can be no doubt that the complex under Vigna Randanini soon developed into a true catacomb (for details, see Chapter 2). Comparably, there are no inscriptions from first century Rome that can be identified as Christian. Yet, even if one allows for the fact that Garrucci did not—and perhaps could not—know this, he should in the very least have explained why he believed the Jewish inscriptions to postdate the Christian ones by at least two centuries.

How inconsistent Garrucci's reasoning can become at times is particularly evident from his discussion of inscriptions, sarcophagi, and loculi. How did Garrucci know, for example, that in using sarcophagi Jews imitated Christians rather than pagans? Have not many sarcophagi with pagan iconography been preserved in Rome? Conversely, why should the Jews have derived their loculi from pagan examples rather than from Christian catacombs? Or, why should not Christians and pagans have derived their loculi from the Jews? It is against the background of such inconsistent reasoning that we must also see Garrucci's detailed discussion of Jewish literary sources concerning grave forms. This discussion did not serve to explain the finds in the Vigna Randanini catacomb, but was meant to convince the reader that Jewish and early Christian burial customs have little in common. It is obvious that a discussion of Jewish literary sources that do not speak of catacombs at all and that, moreover, mostly postdate the ancient period, can only draw the reader's attention away from what Garrucci in his capacity as excavator should have done: to give a precise description of the formal appearance of the catacomb itself.

---

scholarly qualities is far too positive.

It is thus fair to conclude that the main aim of Garrucci's discussion was to argue what Aringhi had argued some two hundred years earlier, namely that the Jewish catacombs *could not* have anything in common with the early Christian catacombs of Rome. Defining Jewish culture in strictly Christian theological terms, that is, as a religion that had been supplanted by Christianity long ago, Aringhi and Garrucci simply preempted the possibility of ever evaluating Jewish materials properly. Garrucci's later work, such as his comprehensive survey of early Christian art, would serve, among other things, to advance similar views.[85]

G. B. de Rossi elaborated on the relationship between Jewish and Christian underground tombs in his *La Roma sotterranea cristiana*. The publication of this book marks the beginning of a new period of catacomb research. Criticizing the work of Aringhi, de Rossi substituted a theological by an archaeological-historical approach. De Rossi's work set the tone for research on catacombs until far into the twentieth century. At the same time, however, even de Rossi could not free himself entirely from the typically Christian view of history. In fact, his discussion of the Jewish catacombs shows particularly well how a combination of sound reasoning and deep respect for the Church characterizes his work on the archaeology of underground Rome.

Like his predecessors, de Rossi underlined the formal differences between the hypogea of Roman Palestine and the Roman catacombs. Unlike his predecessors, however, de Rossi reflected on the possibility that the "genesis" of the Christian catacombs was perhaps to be found in the Jewish Monteverde catacomb. In the end, however, de Rossi believed that in constructing catacombs Christians did not imitate Jews, nor Jews Christians. He rather preferred to view the construction of catacombs by Jews and Christians as having occurred separately, without one community having influenced the other. Thus, in his view, both Jewish and Christian catacombs were strictly *sui generis*. Finally, de Rossi also believed that while the Christian catacombs had come into existence towards the end of the first century C.E., Jewish underground cemeteries such as the Vigna Randanini catacomb did not predate the third, or, in any event, the second century C.E.[86]

It does de Rossi credit to have taken seriously the formal similarities between the Jewish and early Christian catacombs of Rome: he

---

[85] See R. Garrucci, *Storia dell'arte cristiana nei primi secoli della Chiesa* (Prato, 1873-1880) vol. 1, 9-16; vol. 6, commentary on table 489f.

[86] G. B. de Rossi, *La Roma sotterranea cristiana* (Rome: Salvucci, 1864) vol. 1, 87-92, 97, 184-85. And see *id.*, "Delle nuove scoperte nel cimitero di Domitilla," *Bulletino di Archeologia Cristiana* 3 (1865) 33-40.

was the first scholar to do so in the two hundred fifty years that the existence of Jewish catacombs in Rome had been known. No scholar before de Rossi had ever thought of presenting evidence from the Christian catacombs of Callisto and Ciriaca next to evidence from the Jewish catacomb under the Vigna Randanini. In fact, such a presentation of the evidence would have been inconceivable even to de Rossi's contemporaries such as Garrucci. Yet it is remarkable that, at the same time, even de Rossi never really explored these similarities. Instead he made the *a priori* statement that Jewish and early Christian catacombs developed as separate entities. Nor did he explain why archaeological materials that look identical should date to the first century in one case (the Christian catacombs) but to the second or third in another (the Jewish catacombs).

Why did de Rossi not remove such inconsistencies? Because even though times had changed, de Rossi was facing the same dilemma his great predecessor, Bosio, had once faced: to write about Jewish catacombs in an intellectual climate in which the study of catacombs still served, in de Rossi's own words, to illuminate the "fede e carità" displayed by the early Church and its members. Like Bosio, de Rossi found a middle way: refraining from verbal aggression against the Jews, he acknowledged the formal similarities between Jewish and Christian catacombs, but he then refrained from exploring the implications of his observations. Ultimately, for de Rossi, research of the "ampia e cattolica necropoli cristiana" still was what it had been for Bosio, namely "archeologia sacra," that is the archeological investigation of holy things.[87]

In the nineteenth and early twentieth century, several Protestant scholars also turned their attention to the catacombs. Some of them, such as V. Schultze, criticized their Catholic contemporaries. He argued that nineteenth century Catholic scholarship on catacombs could still not be distinguished in any way from the study of Christian dogma.[88] A brief look at discussions of the Jewish catacombs by Protestant scholars shows, however, that on a methodological level the differences between them and their Catholic counterparts were not as large as their polemics would suggest. Schultze himself believed that the Jewish Vigna Randanini catacomb had served as an example for the Christian catacombs, but only in a most general way.

---

[87] See De Rossi (previous note) 91, 110-11; and see the matter-of-fact approach to Jewish archaeological materials in *id.*, "Verre représentant le Temple de Jérusalem," *Archives de l'Orient Latin* 11 (1883) 439-55.

[88] V. Schultze, *Der theologische Ertrag der Katakombenforschung* (Leipzig: Dresden, 1882) 3f. See also F. Piper, *Einleitung in die monumentale Theologie* (Gotha: Besser, 1867).

He argued that once Rome's Christian community had copied the basic idea from the Jews, Christian catacomb burial had developed independently. Such ideas were remarkably similar to Marchi's view on the origin of the catacombs. Similarly, Schultze criticized Garrucci because the latter had not realized that the Jewish catacombs predated the Christian ones. Yet Shultze's treatment was not so very different from Garrucci's in that he too never indicated on the basis of what evidence he concluded that the Jewish catacombs were older than the Christian catacombs. Along similar lines, preconceived notions about Jews and Judaism permeate Schultze's work in very much the same way as they permeate the work of his Catholic contemporaries. Writing about the inscriptions on Jewish gold glasses, for example, Schultze observed that the defective orthography indicated, how "imperfect the Jews were in acquiring the language of the country in which they lived." That many early Christian gold glasses carry exactly the same inscriptions was a fact that never bothered or even occurred to Schultze.[89]

N. Müller, another Protestant scholar and excavator of the Monteverde catacomb, disagreed with Schultze and, like de Rossi, maintained that the Jewish and the Christian catacombs of Rome developed completely independently from one another. Like de Rossi, Müller never explained how he reached this conclusion. Furthermore, Müller ascribed the origins of Jewish catacomb burial in Rome to oriental influences in general, and to old "Palestinian" practices in particular. Most likely, such a thesis was not a revival of the old idea that the cave of Machpelah served as a prototype for all subsequent Jewish underground burials. It rather betrays the influence of the works of J. Strzygowski and of certain Roman legal historians who viewed the Orient as the ancient world's cultural center and most important artistic source.[90]

In short, it has become clear that from a methodological point of view, similarities far outweigh the differences in the works of Herzog, Garrucci, de Rossi, Schultze, and Müller. Today it seems no less than obvious that in order to determine the relationship between Jewish and Christian catacombs, one must first date these catacombs properly. It is true that chronological considerations play a role in the works of above mentioned scholars, yet none of them ever addresses this issue in an analytical and systematic fashion. Instead, they all operate on the basis on unproven assumptions.

---

[89] V. Schultze, *Die Katakomben. Die altchristlichen Grabstätten. Ihre Geschichte und ihre Monumente* (Leipzig: von Veit, 1882) 19-22; 193-94.

[90] Müller 1912, 15-16; 26-27; E. Weigand, "Die Orient-oder-Rom-Frage in der frühchristlichen Kunst," *Zeitschrift für die Neutestamentliche Wissenschaft* 22 (1923) 233-56; L. Mitteis, *Reichsrecht und Volksrecht* (Leipzig, 1891) 31

The work of the above mentioned scholars has in common, furthermore, that they considered pagan, Jewish, and early Christian art as separate entities. Such a consideration resulted from the compartmentalization of Classical studies, the study of early Christian literature, and Jewish historiography. As has been pointed out, early Christian and Classical Archaeology had gone completely separate ways since the publication of Bosio's *Roma sotterranea*. Jewish antiquities were hardly studied at all. The Viennese school of Art History tried to break down this compartmentalization, but their efforts do not predate the very end of the nineteenth century. Besides, their efforts were slow to bear fruit. Because early Christian archaeologists did not bother to study pagan archaeological remains and because their notion of the chronology of early Christian art was flawed, students of Christianity's physical remains did not realize that in the first two to three centuries of the Common Era, a specific Christian iconography had yet come into being. Nor were they willing to admit that when artifacts with a Christian iconography finally started to appear in Rome in the course of the third century C.E., such products were manufactured in the same workshops that catered to pagans and Jews.

The compartmentalization of scholarship on the ancient world explains why above mentioned authors always try to define the relationship between the Jewish and Christian catacombs in terms of influence: the Jewish catacombs served as an example for the Christian catacombs (Marchi, Schulze), or vice versa (Garrucci), or they were independent from one another (de Rossi, Müller). Not even one author considered the possibility that will be explored in Chapter 2 of this book, namely that the Jewish and Christian catacombs are contemporary. Thus they did not consider the possibility that the usage of catacombs is a phenomenon that is neither exclusively Jewish nor typically Christian, but rather generally Late Ancient.[91]

---

[91] F. Wickhoff, *Die Wiener Genesis* (Wien: Tempsky, 1895); A. Riegel, *Stilfragen. Grundlegungen zu einer Geschichte der Ornamentik* (Berlin: Schmidt, 1923); id., *Spätrömische Kunstindustrie* (Vienna: Österreichische Staatsdruckerei, 1927) esp. 1-22; and M. Dvorák, *Kunstgeschichte als Geistesgeschichte. Studien zur abendländischen Kunstentwicklung* (Munich: Pier, 1928) esp. 3-40; J. Wilpert, *Erlebnisse und Ergebnisse im Dienste der Christlichen Archäologie* (Freiburg i.B.: Herder, 1930) 140; G. P. Kirsch, "L'archeologia cristiana. Suo carattere proprio e suo metodo scientifico," *Rivista di Archeologia Cristiana* 4 (1927) 49-57, esp. 50, 54-6. On workshops, see J. Wilpert, *Principienfragen der christlichen Archäologie mit besonderer Berücksichtigung der "Forschungen" von Schultze, Hasenclever und Achelis* (Freiburg: Herder, 1889) 3-5; id., *Die Malereien der Katakomben Roms* (Freiburg: Herder, 1903) 16-17; J. Engemann, "Altes und Neues zu Beispielen heidnischer und christlicher Katakombenbildern im spätantiken Rom," *Jahrbuch für Antike und Christentum* 26 (1983) 128-51; id., "Christianization of Late Antique Art," *The 17th International Byzantine Congress. Major Papers*, 3-8.VIII.1986 (New Rochelle: Caratzac, 1986) 83-105.

Before 1940 there was only one scholar that considered this possibility, namely P. Styger. Being, in his time, the most important student of Roman catacomb topography and construction, Styger observed that the supposed formal differences between the Jewish and the Christian catacombs did, in reality, not exist. Yet, unlike scholars that had previously discussed this issue but had refrained from elaborating on its implications, Styger drew a conclusion that is not merely consistent but that seems inevitable: if Jewish and Christian catacombs look perfectly alike in so many respects, then they must have developed simultaneously. Styger suggested that Jews and Christians had synchronically transformed existing excavation techniques and had adapted traditional-Roman funerary architecture in response to local needs.

Such observations were revolutionary, even for Styger himself. Only a few years before, he himself had written a comprehensive history of the Roman catacombs in which the Jewish catacombs were mentioned not even once. In these same years, Styger had also written a booklet on Jews and Christians in Rome, in which he expounded views that can hardly be called philosemitic.[92]

The nineteenth century also witnessed, finally, the publication of the first modern studies on Jews and Judaism whose authors were Jews. Although it lacked official institutional backing and was practiced by scholars with widely differing interests, the movement known as *Wissenschaft des Judentums* led to fundamental conceptual and methodological changes in the study of Jewish history. The first two major monographs on the Jews in Rome, Berliner's *Geschichte der Juden in Rom* of 1893 and H. Vogelstein and P. Rieger's joint study of 1896 which carries an identical title, stand in this *Wissenschaft* tradition. This is evident from their overall conception of Jewish history, from the comprehensive collection of primary source materials, from the way in which such materials were evaluated, as well as from the language in which they were written. What was largely absent in these studies, however, was the concern with more general questions of Jewish emancipation that had characterized the work of an earlier generation of *Wissenschaft* scholars. These then were the kind of monographs that E. Gans had so emphatically asked for many years before. For the first time, discussions of the Jewish catacombs of Rome were now inserted into

---

[92] P. Styger, "Heidnische und Christliche Katakomben" in T. Klauser and A. Rücker (eds.), *Pisciculi. Studien zur Religion und Kultur des Altertums (FS F. J. Dölger)* (Münster: Aschendorf, 1939) 266-75; *id., Die römischen Katakomben. Archäologische Forschung über den Ursprung und die Bedeutung der altchristlichen Grabstätten* (Berlin: Verlag für Kunstwissenschaft, 1933); *id., Juden und Christen im alten Rom* (Berlin: Verlag für Kunstwissenschaft, 1934) 7, 24, 37, 41f., 57.

accounts that were written specifically to illuminate the history *of the Jews*. A. Berliner in particular is known to have taken a lively interest in Rome's Jewish archaeological heritage. He was also the first to publish several Jewish inscriptions from the Jewish hypogeum in the Vigna Cimarra. Yet for all such interest, Berliner as well as Vogelstein and Rieger approached the history of the Jews in ancient Rome on the basis of the literary sources. They discussed the Jewish catacombs primarily because the inscriptions that were found there could be fitted easily into a general history of the Jews of Rome. They refrained, however, from studying the catacombs, or any of the other archaeological materials that had come to light in them.[93]

It is thus fair to conclude that the rediscovery of four Jewish catacombs and two Jewish hypogea in Rome between 1859 and 1919 did not, paradoxically enough, result in major changes in the ways scholars thought about these catacombs and hypogea. The archaeologists responsible for excavating the Jewish underground cemeteries could not free themselves of preconceived notions that had shaped discussions of Jewish catacombs for centuries. It is certainly significant that in a two-page review of Müller's discoveries in the Jewish Monteverde catacomb, A. de Waal, a Catholic scholar, kept lamenting about the fact that no Jewish inscriptions had been discovered that carried the names of people mentioned in the New Testament.[94]

Jewish scholars, on the other hand, were interested in the Jewish catacombs from Rome. Yet, because their interest in Jewish history was essentially philological, their work did little to encourage innovative thinking on Jewish archaeological materials. Jewish archaeologists such as T. Reinach cared only about the inscriptions from the Jewish catacombs.[95]

---

[93] In general, see G. Herlitz, "Three Jewish Historians. Isaak Markus Jost, H. Graetz, E. Täubler," *Yearbook of the Leo Baeck Institute* 9 (1964) 69-90; Liebeschütz 1967, 65-66; I. Schorsch, "From Wolfenbüttel to Wissenschaft. The Divergent Paths of Isaac Markus Jost and Leopold Zunz," *Yearbook of the Leo Baeck Institute* 22 (1977) 109-28; M. A. Meyer, "The Emergence of Jewish Historiography: Motives and Motifs," *History and Theory Beiheft* 27 (1988) 160-75. Monographs on Rome: Berliner 1893, esp. 50-51; 90-91; H. Vogelstein and P. Rieger, *Geschichte der Juden in Rom* (Berlin: Mayer and Müller, 1896) esp. 49f.; see also A. Berliner, *Gesammelte Schriften* (Frankfort on the Main, 1913)188f.; E. Gans, "Gesetzgebung über Juden in Rom," *Monatsschrift für die Wissenschaft des Judentums* 1 (1823) 25-67, esp. 25.

[94] A. de Waal, "Die jüdische Katakombe an der Via Portuensis," *Römische Quartalschrift* 19 (1905) 140-42.

[95] T. Reinach, "Le cimetière juif de Monteverde," *Revue des Études Juives* 71 (1920) 113-26 and *id.*, "Une nouvelle nécropole judéo-romaine," *ibid.* 72 (1921) 24-28.

Recent Developments: 1945-Present

In the last fifty years, the field of Jewish studies has developed more rapidly and changed more fundamentally than ever before. Several factors have contributed to this process. New archaeological discoveries in the Holy Land and throughout the Mediterranean have brought to light a wealth of archaeological, literary, and epigraphical evidence. Such discoveries have enriched our knowledge immeasurably. At the same time they have also led students of Jewish history to study the past in a more interdisciplinary fashion than was the case previously. Then, the institutionalization of Jewish studies, especially in Israel and in the United States, has contributed to the professionalization of a field of academic inquiry that had remained the domain of Christian theologians for centuries. Put differently, the ideals of the *Wissenschaft* scholars of the nineteenth century have been realized to the full extent only after 1945: the recognition of Jewish studies as an independent academic discipline; and the study of Jewish history *qua* Jewish history.

In the postwar period, the Jewish community of ancient Rome has attracted a fair amount of scholarship. The Jewish catacombs have been studied by E. R. Goodenough. H. J. Leon and H. Solin have reviewed the literary and epigraphical evidence bearing on the Roman Jewish community. U. M. Fasola has re-excavated the Jewish catacomb under the Villa Torlonia. Following the example set by the monographs of Berliner and of Rieger and Vogelstein, these recent studies have in common that they evaluate Jewish evidence on its own terms (except for the work of Fasola, whose excavation report is purely descriptive). These studies are particularly important, because in surveys of Rome's early Christian catacombs the Jewish catacombs continue to play the same insignificant role they have always played in studies of this type.[96]

At the same time, however, the rupture between earlier and postwar scholarship on the Jewish catacombs is not as complete as one would perhaps imagine. Just as previous scholars had advanced views that were not supported by the archaeological evidence (such as the idea that the Jewish catacombs postdate the Christian ones by several centuries), Goodenough and Leon too can sometimes be found to ar-

---

[96] Goodenough 1953-68, vol. 2 and 3; Leon 1960; Fasola 1976; Solin 1983. Other studies such as Vismara 1986 and Konikoff 1986 have no independent scholarly value; on the latter, see L. H. Kant and L. V. Rutgers in *Gnomon* 62 (1988) 376-78. D. Poliakoff, *The Acculturation of Jews in the Roman Empire: Evidence from Burial Places* (Ph.D. Boston University 1989) is not a serious scholarly work. References to Jewish catacombs in surveys of Christian catacombs, see P. Testini, *Archeologia cristiana* (Bari: Laterza, 1980 [1958]) 317 and 523f ; F. W. Deichmann, *Einführung in die christliche Archäologie* (Darmstadt: Wissenschaftliche Buchgesellschaft, 1983) 52.

gue on the basis of *a priori* assumptions. The use of such *a priori* assumptions in their work must be understood against the background of previous scholarship on the Jews of Rome. Because Christian scholars had systematically and continuously denied the importance of Jewish materials, Goodenough and Leon not merely had to describe such materials, but were automatically forced to enter into long-standing arguments over the interpretation of individual finds. One example will suffice to illustrate how old polemics determined the course of an argument in the work of Goodenough and Leon.[97]

In the Vigna Randanini catacomb two frescoed rooms had been discovered by Garrucci (the so-called Painted Rooms I and II). Inasmuch as both rooms contained (and still contain) wall paintings that are blatantly paintings of iconography, these rooms had long baffled scholars. Before the 1930s there was little or no archaeological evidence to suggest that the Jews of Antiquity had ever employed artists on a large scale. Observing that the commandment forbidding graven images (Exodus 20, 3-4) had been in full force during the Roman period, scholars such as R. Garrucci and J. B. Frey, the editor of the *Corpus Inscriptionum Judaicarum*, therefore concluded that these rooms had been pagan in origin. By the 1950s, however, evidence from excavations indicated that in Late Antiquity the second commandment had not at all prevented Jewish iconic artistic production. Thus Goodenough argued that the two painted rooms had to be considered Jewish after all, their non-Jewish iconography notwithstanding. Leon agreed. Yet rather than basing their conclusions on a study of the evidence preserved in the catacomb itself, Goodenough and Leon preferred simply to turn the old argument around. A visit to the catacomb would have enabled them to observe that the archaeological evidence there does indeed not support the contention of Garrucci and Frey. Such a visit would also have permitted them to see, however, that it does not support their own contention either. Painted Rooms I and II originally formed a separate area that was not connected to the rest of the Jewish catacomb and that was accessible through a separate entrance (phase I). The epigraphic remains do not permit us to determine whether these rooms were used by Jews or by non-Jews during this first phase. Only at a later stage, a small gallery was excavated that connected Painted Rooms I and II to the Jewish catacomb (phase II). It was only then that these rooms were reused by Jews for burial.[98]

---

[97] That Goodenough's work is problematic for other reasons as well is a well known point that is not relevant here, see, in general, M. Smith, "Goodenough's *Jewish Symbols* in Retrospect," *Journal of Biblical Literature* 86 (1967) 53-68.

[98] See the discussion of the archaeological evidence in Rutgers 1990, 147-48; for a

Insofar as the influence of traditional views on Jews and Judaism in postwar scholarship on the Jews of ancient Rome is concerned, there is still another point which deserves our attention. In an important review of Leon's monograph of 1960, Momigliano pointed out that self-containedness and lack of contact with the larger, non-Jewish outside world struck him as the most outstanding characteristic of the Jewish community of Rome. More recently, other scholars have reached very similar conclusions. They too have concluded that the Jews of ancient Rome spent much of their lives in isolation from and in antithesis to contemporary society.[99]

What is noteworthy about such observations is that they are remarkably similar to the Christian view that had always regarded Jewish and non-Jewish culture as irreconcilable entities. As early as 416 C.E. Augustine remarked that Jews were the only people conquered by Rome never to become true Romans and attributed this "failure" to integrate to the Jews' continued faithfulness to their old religious practices. Agreeing with Augustine, other Christian scholars including Bosio, Aringhi, Basnage, Diderot, Tillemont, Gibbon, Herder, Renan, Gregorovius, and Seeck further elaborated on the idea that "Judaism" and "isolation" are nothing but interchangeable categories.[100]

How was it possible for scholars of the postwar period to agree so strongly with previous scholarship on the question of the social status of Rome"s Jewish community? Did the Jews of ancient Rome really live in splendid isolation? Or can we identify other factors that have

---

picture, see Goodenough 1953-68, vol. 3, fig. 743f.; Garrucci 1865, 165 (but nuanced), Frey in *CIJ*, CXXIV; Goodenough 1953-68, vol. 2, 21; Leon 1960, 60; and more recently D. Mazzoleni, "Les sépultures souterraines des Juifs d'Italie," *Les Dossiers de l'Archéologie* 19 (1976) 82-98, esp. 92; P. Prigent, *Le Judaïsme et l'image* (Tübingen: Mohr, 1990)153 and T. Rajak, "Inscription and Context: Reading the Jewish Catacombs of Rome," in J. W. van Henten and P. W. van der Horst (eds.), *Studies in Early Jewish Epigraphy* (Leiden: Brill, 1994) 237. On the second commandment, see E. E. Urbach, "The Rabbinical Laws of Idolatry in the Second and Third Centuries in the Light of Archaeological and Historical Facts," *Israel Exploration Journal* 9 (1959) 149-65 and 229-45.

[99] A. Momigliano in *Gnomon* 34 (1960) 178-82; Solin 1983, 602, 617, 717, 720; but see 695 n. 238a; C. Vismara, "Orientali a Roma. Nota sull'origine geografica degli Ebrei nelle testimonianze di età imperiale," *Dialoghi di Archeologia* 5 (1987) 119-21, esp. 119.

[100] *Enarrationes in Psalmos* LVIII s.1.21 (*CCL* 39, 744); Bosio 1632, 141; Aringhi 1651, 393-94 (citing Augustine); Diderot in *Encyclopédie*, vol. 1, 24, s.v. Juif (copying Basnage); L. S. L. de Tillemont, *Mémoires pour servir à l'histoire écclesiastique des six premiers siècles* (Venice: Pitteri, 1732) vol. 1, 146-47 and vol. 2, 169 and 341; E. Gibbon, *History of Christianity* (New York: Eckler, 1883; reprint Arno Press and New York Times, 1972) 108, 113, 208; Herder: Liebeschütz 1967, 23f.; Renan: S. Almog, "The Racial Motif in Renan's Attitude to Jews and Judaism," in *id.* (ed.), *Antisemitism Through the Ages* (Oxford: Pergamon Press, 1988) 260; F. Gregorovius, *Wanderjahre in Italien* (Munich: Beck, 1967) 205; O. Seeck, "Die Ausrottung der Besten," in K. Christ (ed.) *Der Untergang des römischen Reiches* (Darmstadt: Wissenschaftliche Buchgesellschaft, 1970) 40.

influenced this view?

To answer this question, we have to return, once again, to the beginnings of modern Jewish historiography in general, and to the work of H. Graetz in particular. In 1846 Graetz published a programmatic study, entitled *Die Konstruktion der jüdischen Geschichte* in which he proposed a theory of Jewish history. According to it both Judaism and pagan culture could best be defined by identifying their essence ("Grundidee," or "Grundwesen"). Having defined their essence, Graetz then pointed out that Judaism and paganism were mutually exclusive. He furthermore also believed that contacts between these two cultures would carry the character of a "Kampf auf Leben und Tod." Adopting a Hegelian view of history, Graetz proposed that such battles would consist of a "three step" scenario. The existence of non-Jewish cultures such as Hellenistic Greek culture (thesis) *by definition* threatened its antithesis, namely Judaism. Judaism would, of course, defend itself against such threats. Having defended itself successfully, Judaism would then succeed in integrating Hellenism's less incompatible elements without giving up its own essence (synthesis). Battles could repeat themselves endlessly, but the scenario would always remain the same (thesis, antithesis, synthesis). In short, Graetz proposed a very rigid view of Jewish history according to which Jewish culture could relate to other cultures only by fighting and conquering them.[101]

That Graetz conceived of Jewish history in this way reflects the influence Hegel's thinking exerted on nineteenth century historians.[102] In the particular case of Graetz there was, however, still another factor that shaped his view of Jewish history as a history of persecution and intellectual isolation. Inasmuch as the nineteenth century was a time in which the struggle for social and political emancipation of the Jews had only just begun, it was practicaly inevitable for historians to present the Jewish past in the same way as they saw the Jewish present, namely as a "Leidens- und Gelehrtengeschichte." The extent to which the sufferings of the Jews informed discussions on the Jewish communities of Antiquity is particularly evident in the case of Rome. There, anachronistically, a Jewish ghetto existed until the last days of the Risorgimento (1870). It was common knowledge that in it, Roman Jews had to confront a life that could hardly be called alluring. The often squalid situation

---

[101] H. Graetz, *Die Konstruktion der jüdischen Geschichte* (Berlin: Schocken, 1936 [1846]) 7, 41 and 43.

[102] Droysen viewed Paganism as thesis, Judaism as antithesis, and Christianity as synthesis; see A. Momigliano, "Droysen between Greeks and Jews," *History and Theory* 9 (1970) 139-53; Hoffmann 1988, 78-79.

in which they had to live made such an impact on the minds of those who lived outside the ghetto and explains that it was only natural for R. Lanciani to suppose that in ancient Rome, from the days of Cicero onwards, the Jews had likewise lived in a ghetto. Vogelstein and Rieger could not agree more. Others were satisfied with a few isolated sources to argue that the Jews of Rome had always been greedy beggars, or, better still, pariahs. Jewish women in Rome were believed to be especially impervious to change. By speaking of the "judéogrec" and the "judéolatin du Trastevere," T. Reinach went so far as to suggest that Jews in ancient Rome had, like their Medieval descendants, even developed their own language. Nor is it possible to miss the parallelism between some descriptions of ancient Trastevere as wetlands infested with disease and later descriptions of the ghetto as a site prone to frequent inundation by the Tiber. Given the prevalence of such views, it should come as no surprise to note, finally, that Ascoli, author of the first corpus of Jewish inscriptions from Italy, did not discuss all inscriptions that had been discovered in the 1850s in a Jewish catacomb in Venosa (Basilicata), but only those that had been written in Hebrew.[103]

To return to recent studies on the Jews of ancient Rome, it is not possible to demonstrate that nineteenth-century views of Jewish history have directly influenced postwar scholarship on the Jewish community of Rome. Only in one case have nineteenth-century concepts of Jewish history been revived, namely by reintroducing the infelicitous phrase that, hundred years ago, led to the so-called "Berliner Antisemitismusstreit."[104] Yet, despite the absence of a direct link between nineteenth-century scholarship on the Jews and recent studies of Rome's Jewish community, nineteenth-century views of Jews and Judaism nevertheless continue to shape modern scholarship on the Jews of ancient Rome in one important respect, namely in a methodological one: nineteenth-century beliefs make a comparative or interdisciplinary perspective seem futile. If nineteenth-century

---

[103] Gregorovius 1967, *passim* and 215; R. Lanciani, *New Tales of Old Rome* (London: MacMillan, 1901) 216-17, 229, 241-42; Vogelstein and Rieger 1896, 36. Berliner 1893, 104, however, disagreed strongly: "Das Ghetto ist eine Erfindung des 16. Jhs.;" J. Carlebach (ed.), *Von Duldern und Kämpfern. Aus dem Nachlasse von H. Hildesheimer* (Frankfurt on the Main, 1911) 8-9; G. Bardy, *La question des langues dans l'église ancienne* (Paris, 1948) 82; A. Momigliano, "A Note on Max Weber's Definition of Judaism as a Pariah-Religion," *History and Theory* 19 (1980) 313-18; T. Reinach, "Le cimetière juif de Monteverde," *Revue des Études Juives* 71 (1920) 125; G. I. Ascoli, *Iscrizioni inedite e mal note, greche, latine e ebraiche di anitichi sepolcri giudaici del Napoletano* (Turin and Rome: Loescher, 1880) 47 (279). See also K. Hoheisel, *Das antike Judentum in christlicher Sicht* (Wiesbaden: Harassowitz, 1978) 13f.

[104] Solin 1983, 686 and 690 n. 224. The phrase that Jews were a "ständiges Ferment der Unruhe" derives from T. Mommsen, *Römische Geschichte* (Berlin: Weidmannsche Buchhandlung, 1889) vol 3, 550. See, in general, Hoffmann 1988, 87-132. Conceptually, Solin is heavily indebted to the work of T. Mommsen and A. von Harnack.

scholars before him had not already argued so consistently that Jews had always lived detached from the larger non-Jewish world surrounding them, how was it possible for Leon to study, among other things, the names used by the Jews in ancient Rome without discussing contemporary non-Jewish onomastic practices? Did Leon not himself admit that non-Jewish names were prevalent among the Jews of Rome?

It is true that Leon's methodology is one of many possible methodologies. Yet what matters here is that Leon's approach has far-reaching implications for the question of "isolation." If one studies Jewish inscriptions and Jewish catacombs without taking into account contemporary non-Jewish materials, a picture will emerge that must *by definition* be one of isolation. Put differently, the fact that Leon has isolated Jewish evidence from non-Jewish evidence cannot be taken to indicate that the Jews of ancient Rome were themselves actually isolated. This is, however, the conclusion that Momigliano and Solin have drawn.[105]

In addition, one may also wonder what these scholars mean by the rather vague term "isolation." Were Jews isolated because of their religion, because of their social status, because of the languages they spoke or did not speak, or because of a combination of these factors? Solin has recently argued that the "isolation" of the Jews is evident from the fact that while Syrian immigrants disappear from the epigraphic record within one or two generations after immigration, Jews always remain recognizable. Such a line of reasoning is, however, unconvincing for two reasons. First, it presupposes, wrongly, that recognizability of identity equals social isolation.[106] Second, it compares two groups who defined their identities in very different ways. On inscriptions Syrians describe themselves by including the name of the village from which they came. That such additions would disappear after Syrian immigrants had been in Rome for several decades was inevitable. After all, the children of immigrants could no longer claim to have come from the village of the parents for the simple reason that they were born in Rome. With Jews, on the other hand, the situation was very different. Roman Jews were not merely people that had once been inhabitants of Judaea. Jews, not only in Rome but throughout the Roman Empire, formed a religious

---

[105] Why Momigliano would have done this is a complex question that cannot be addressed here. To understand the background of Momigliano's reasoning, see A. Momigliano, *Pagine Ebraiche* (Turin: Einaudi, 1987) 129-51, 237-39, 242; M. Gigante, "Precisazioni sul rapporto Croce-Momigliano," *Annali Pisa* III.17.4 (1987) 1045-60. And contrast Momigliano's experiences with the observations of H. Arendt, *Ebraismo e modernità* (Milan: Feltrinelli, 1986) e.g. 117-37.

[106] See further the discussion in Funkenstein 1993, 220-56.

community, one which derived its internal coherence not merely from the fact that they upheld the same religious beliefs and practices, but also from the knowledge that they shared in the same history. In other words, in ancient Rome Syrian immigrants never had nor could have the sense of belonging that many Jews in this city must have felt. The Jews of Rome were a *people* in the sense that their Syrian neighbors were not.[107]

In the chapters that follow I will explore the thesis that the Jews of ancient Rome lived in isolation. I will try to answer the question that has never been raised in previous studies on the subject, namely: If Jews lived in isolation, were they isolated in every possible way, or just in some ways and not in others? Were the Jews of ancient Rome isolated because they wanted to be isolated, or because they viewed themselves as isolated, or both? Or were these Jews isolated because non-Jews tried to isolate them? Or did these non-Jews present the Jewish community as "isolated" because they wanted the Jews to be isolated even though in reality Roman Jews were not isolated at all?

Because Christian scholars and, more recently, students of Jewish history have always approached the history of the Roman Jewish community by studying some sources of evidence at the expense of others, I have decided to study this question on the basis of a comparative approach. In Chapter 2 I will study Jewish and non-Jewish archaeological remains; in Chapters 3 through 5 Jewish and non-Jewish inscriptions; and in Chapter 6 Jewish and non-Jewish literary evidence. In Chapter 7 I will then take up the question with which we finish this chapter: In what respects were the Jews of ancient Rome isolated?

---

[107] *Contra* Solin 1983, 720. On the identity of groups in the eastern part of the Mediterranean, see the important discussion in F. Millar, "Empire, Community and Culture in the Roman Near East: Greeks, Syrians, Jews and Arabs," *Journal of Jewish Studies* 38 (1987) 143-64. For a general discussion of peoplehood, see Y. Leibowitz, *Het geweten van Israel* (Amsterdam: Arena, 1993) 17f. (Dutch translation of *Am, Eretz, Medina*).

CHAPTER TWO

THE ARCHAEOLOGY OF JEWISH ROME: A CASE-STUDY IN
THE INTERACTION BETWEEN JEWS AND NON-JEWS IN
LATE ANTIQUITY

Jewish Funerary Architecture in Late Ancient Rome

It is generally agreed that among all cultural phenomena burial customs are among those least susceptible to sudden change. Examples which illustrate this assertion abound. Studying the Romanization of Etruria, Kaimio observed that around the turn of the Common Era, Etruscan burial customs were still very similar to those practiced several centuries before—even though parts of Etruria had been subjugated by Rome as early as the fourth century B.C.E. and Latin had completely replaced local languages in the late Republic.[1] That there was noting unusual about these developments in Etruria becomes evident when we turn to Roman Britain. There, as late as the early third century C.E., burial practices were still very similar to those documented for pre-Roman times. During this period, Romanization was confined to the inclusion a new range of grave-gifts deposited in the graves.[2]

Yet, for all the conservatism surrounding death, burial, and mourning, from time to time burial customs do change. Sometimes, as in early fifth-century B.C.E. Athens, legislation triggered or speeded up such changes. At other times, as in second century C.E. Rome, changing practices accompanied changing attitudes towards the disposal of the dead. In antiquity Jewish funerary customs formed no exception to such patterns of continuity and change. The following survey of Jewish tombs shows that Hecataeus of Abdera was correct when he observed that Jewish burial customs were transformed after the Jews had come under foreign rule. In fact, Hecataeus' observation

---

[1] J. Kaimio, *Studies in the Romanization of Etruria = Acta Instituti Romani Finlandiae* 5 (1975) 226.

[2] C. J. S. Green, "The Significance of Plaster Burials for the Recognition of Christian Cemeteries," in R. Reece (ed.), *Burial in the Roman World* (London: Council for British Archaeology, 1977) 48; R. Philpott, *Burial Practices in Roman Britain. A Survey of Grave Treatment and Furnishing A.D. 43-410* (Oxford: Tempus Reparatum, 1991) 221-22.

holds equally for Late Antiquity as it does for the early Hellenistic period.[3]

One of the most consequential changes to transform Roman burial practices under the Empire was a general shift from cremation to inhumation. In the city of Rome itself, this occurred in the second century C.E. and led to profound changes in traditional tomb construction. Ultimately, as the Christianization of Rome's city populace intensified in the course of the fourth century C.E., the change resulted in the construction of large underground cemeteries commonly known as catacombs. These were situated in the suburban areas that circled the city on all sides.

As a result of the intensive archaeological investigations over the centuries (see Chapter 1) the development that culminated in the excavation of large underground cemeteries can be reconstructed with a fair amount of detail. The practice started in and around Rome in the late first century C.E. Then the large *columbaria*—the most characteristic tomb type during the early Imperial period—were increasingly replaced with freestanding, masonry family graves. The size of such masonry structures suggests that they were primarily used by extended families or by *collegia funeraticia* of moderate size. From the second century C.E. onwards, when inhumation increasingly gained popularity, many such constructions of modest dimension were enlarged, sometimes repeatedly, to accommodate extra burials through the construction of additional burial chambers. Such chambers were normally excavated underground. Usually they accommodated graves that were of a less representative nature than the ones located in the original tomb above ground. Discoveries in the necropolis of the Isola Sacra show that such underground extensions could hold up to 150 burials.

Not all graves constructed during this period, however, consist of a mausoleum above ground and galleries underground. Some tombs lack the architectural arrangements *sub divo* and consist exclusively of subterranean galleries and/or rooms. The walls of these subterranean burial complexes have been hollowed out with graves, the majority of which are rather unostentatious. Only occasionally—as in case of the so-called Villa Piccola in the catacomb of Sebastiano or the hypogeum

---

[3] A. D. Nock, "Cremation and Burial in the Roman Empire," *Harvard Theological Review* 25 (1932) 321-59; H. Brandenburg, "Der Beginn der stadtrömischen Sarkophagproduktion der Kaiserzeit," *Jahrbuch des Deutschen Archäologischen Instituts* 93 (1978) 277-32; H. Wrede, "Die Ausstattung stadtrömischer Grabtempel und der Übergang zur Körperbestattung," *Römische Mitteilungen* 85 (1978) 411-33; S. C. Humphreys, "Family Tombs and Tomb Cult in Ancient Athens. Tradition or Traditionalism?" *Journal of Hellenic Studies* 100 (1986) 96-126, esp. 102. For Hecataeus, see M. Stern, *Greek and Latin Authors on Jews and Judaism* (Jerusalem: Israel Academy of Sciences and Humanities, 1974-84) vol.1, no. 11 with commentary *ad.loc.*

of the Flavi at Domitilla—do such subterranean tombs contain graves of a more ostentatious nature.[4]

As late as the fourth century C.E., small underground tombs of the type just described continued to be produced, not only for pagan customers, who appear to have always preferred burial in tombs of smaller dimension over inhumation in large catacombs, but also for Christians. Since such tombs were located mostly in the same general areas, it was not unusual that they would be connected with one another upon excavation for further expansion. Coalescing in the course of time, collections of private hypogea thus turned into large underground complexes, or catacombs. The presence of pagan materials in what at first sight appear to be Christian catacombs should be understood in this light, namely that many catacombs evolved into their present shape only gradually. Of course, such an explanation does not exclude the additional possibility of Christian use of pagan materials, by choice.[5]

Not all catacombs originated in such independent underground tombs that slowly "grew" into extensive funerary complexes. Christian catacombs such as Callisto, Pretestato, and Priscilla display a pattern that varies from the "typical" second and third century C.E. Roman tomb just described. The formal characteristics in the earliest galleries in these catacombs indicate that they were excavated from the start with an eye to future, systematic expansion. Put differently, the formal characteristics suggest that these catacombs were designed

---

[4] For this and for what follows, see U. M. Fasola and P. Testini, "I cimiteri cristiani," *Atti del IX congresso internazionale di archeologia cristiana* (Vatican City: Pontificio Isitituto di Archeologia Cristiana, 1978) 103-39 and 189-210; L. Reekmans, *Die Situation der Katakombenforschung in Rom* (Opladen: Westdeutscher Verlag, 1979); H. Brandenburg, "Überlegungen zu Ursprung und Entstehung der Katakomben Roms," *Jahrbuch für Antike und Christentum Erg. Band* 11 (1984) 11-49 (important); L. Reekmans, "Spätrömische Hypogea," in O. Feld and U. Peschlow (eds.), *Studien zur Spätantiken und Byzantinischen Kunst FS Deichmann 2* (Bonn: Habelt, 1986) 11-37; H. von Hessberg, "Planung und Ausgestaltung der Nekropolen Roms im 2. Jh. n.Chr." in *id.* and Zanker 1987, 43-60; *id.* and M. Pfanner, "Ein augusteisches Columbarium im Park der Villa Borghese," *Jahrbuch des Deutschen Archäologischen Instituts* 103 (1988) 46f.; U. M. Fasola and V. Fiocchi Nicolai, "Le necropoli durante la formazione della città cristiana," *Actes du XIe congrès international d'archéologie chrétienne. Lyon, Vienne, Grenoble, Genève et Aoste (21-28IX.1986)*(Vatican City and Rome: Pontificio Istituto di Archeologia Cristiana and École Française de Rome, 1989) vol. 2, 1153f.

[5] Engemann 1983; A. Ferrua, "La catacomba di Vibia," *Rivista di Archeologia Cristiana* 47 (1971) 7-62 and 49 (1973) 131-61; *id.*, "Iscrizioni pagane della catacomba di Sant'Ippolito," *Rendiconti della Pontificia Accademia Romana di Archeologia* 50 (1977-78) 107-14; *id.*, "Iscrizioni pagane della catacomba di S. Domitilla," *ibid.* 53-54 (1980-82) 387-402; *id.*, "Iscrizioni pagane della via Salaria," *ibid.* 60 (1987-88) 221-36; *id.*, "Iscrizioni pagane della catacomba di S. Agnese," *Rendiconti della Accademia Nazionale dei Lincei* 35 (1980) 85-96 and *id.*, "Iscrizioni pagane della catacomba di Priscilla," *Archivio della Società Romana di Storia Patria* 110 (1987) 5-19.

specifically as cemeteries that served to bury large groups of people, as opposed to graves destined for the direct kin alone.

Due to the destruction of some Jewish catacombs and hypogea and as a result of the absence of reliable plans for some sites, it is no longer possible to reconstruct the topographical history of every Late Ancient Jewish grave site in Rome. Müller's preliminary description of the Monteverde catacomb, for example, as well as the plans he and his successors published are so unspecific that they simply do not permit us to reconstruct in any detail the genesis and subsequent development of the site. Despite such drawbacks it is nevertheless possible to reconstruct the general development of Jewish funerary architecture in Late Ancient Rome. We possess good plans in the cases of three Jewish catacombs, namely the Jewish Vigna Randanini and the two catacombs under the Villa Torlonia. Even more important, it is still possible to enter these catacombs and study the formal appearance of their galleries *in situ*. As we will now see, a detailed study of the general layout, of the types of volcanic soil in which they were excavated, and of the individual grave types attested in these underground Jewish cemeteries reveals in a stunning way the extent to which Jewish funerary architecture evolved as part of the larger developments that shaped tomb construction in and around the Rome in Late Antiquity.[6]

The only published map of the Vigna Randanini catacomb—Frey's sketch published in the 1930s—is neither as detailed nor as reliable as one would wish. Yet, it is useful in that it reveals one important feature of the Vigna Randanini catacomb, namely that this catacomb was never constructed systematically as a coherent complex. The rather irregular shape of the galleries indicates that the Vigna Randanini catacomb originally consisted of various underground hypogea that were separate and independent. Such hypogea coalesced only in the course of time when they underwent underground extensions in the same way as occured in many contemporary non-Jewish tombs in and around Rome. The existence of at least three separate entrances—one situated near its northern edge, on the Via Appia

---

[6] For the Monteverde catacomb, see Müller 1912; *id.*, "Cimitero degli antichi Ebrei posto nella Via Portuense," *Dissertazioni della Pontificia Accademia Romana di Archeologia* 12 (1915) 205-318, Table 9-10; R. Kanzler, "Scoperta di una nuova regione del cimitero giudaico della Via Portuense," *Nuovo Bulletino di Archeologia Cristiana* 21 (1915) 153. For the Vigna Randanini catacomb, see J. B. Frey, "Nouvelles inscripitons inédites de la catacombe juive de la Via Appia," *Rivista di Archeologia Cristiana* 10 (1933) 386, reproduced in Goodenough (1953-67) vol. 3, fig. 734. I have used a much more detailed but unpublished map that was made under auspices of the Pontificia Commissione di Archeologia Sacra. For the Villa Torlonia catacomb, see Fasola 1976, map facing p. 62. The maps of this and the other Jewish catacombs and hypogea are most easily available in Vismara 1986.

Pignatelli and two others located in the southern part of the complex, near the Via Appia Antica—further confirms that the Vigna Randanini catacomb was never a systematically planned catacomb.

It is impossible to determine whether these hypogea in the Vigna Randanini catacomb correspond to one or more mausoleums above ground. The area on top of the catacomb was never excavated and access to it is practically impossible today. Likewise it is impossible to determine when the Vigna Randanini catacomb turned from a collection of hypogea into a catacomb. The construction of underground galleries (as opposed to hypogea) may have started in the course of the third century C.E. There can be little doubt, however, that the construction of such galleries was a relatively late phenomenon in the history of this catacomb. Study of the so-called Painted Rooms I and II can illustrate that the constructional history of the Vigna Randanini catacomb is very similar to that of several Christian catacombs located in the same general area.

Painted Rooms I and II were decorated with wall paintings in a red and green linear style that is attested frequently in early third century C.E. Rome, as well as in Ostia.[7] Such decorative lines divide a white surface into asymmetrical fields. In turn, these fields are decorated with animals, and sometimes they also contain pagan mythological figures. In Chapter 1 we have already seen that it is impossible to determine today whether these rooms were originally used by Jews or by others. Whatever the correct answer to this question may be, however, in the present context it suffices to note that the formal characteristics of the galleries directly outside these two rooms suggest that Painted Rooms I and II must be considered as an underground hypogeum that was initially independent. The marks on the walls of the gallery that gives access to these painted rooms indicate that the original excavators of this area came from an eastern direction. They had started to work their way underground at a staircase that lies to the East of Painted Rooms I and II. Although this staircase is now covered with dirt it is still discernible on Frey's map. The diggers-marks left on the walls of the narrow gallery that connects the area of Painted Rooms I and II with the rest of the Jewish catacomb suggest, on the other hand, the existence of another digging crew. This group came the opposite, westerly, direction. They had begun excavating at a second staircase that was also situated in the same general area, but to the West of Painted Rooms I and II. That Painted Rooms I and II did not form part of the rest of the Jewish catacomb from the outset is furthermore confirmed by the narrowness

---

[7] For parallels and references, see Rutgers 1990, 146-47.

of the connecting gallery. It contrasts markedly with the spaciousness of the area directly in front of Painted Rooms I and II. Last but not least, it can also be observed that the rest of the Jewish catacomb is at a slightly lower level than the area in and around Painted Rooms I and II. Along with the mentioned formal characteristics, such a difference in height shows indisputably that Painted Rooms I and II and the rest of the Jewish catacomb were not constructed as a coherent whole at one and the same time.

In its entirety, the evidence relating to Painted Rooms I and II compares very well with the finds in the so-called Villa Piccola in the catacomb of Sebastiano, directly across the street from the Vigna Randanini catacomb, and with the so-called hypogeum of the Flavi in the catacomb of Domitilla, somewhat further to the West. Both the Villa Piccola and the hypogeum dei Flavi are underground burial complexes of rather limited dimension. Their walls were decorated in the same red and green linear style one encounters in the Painted Rooms I and II of the Vigna Randanini catacomb. The Villa Piccola and the hypogeum dei Flavi were originally constructed to accommodate pagan burials. Only in a later phase were they connected to the large underground galleries of the Christian catacombs of Sebastiano and Domitilla, respectively. In the case of the latter hypogeum it was only after this connection that a decoration was added that was iconographically explicitly Christian.[8]

When seen against the background of developments in contemporary funerary architecture, the building history of Painted Rooms I and II in the Jewish Vigna Randanini need no longer baffle us, but can be reconstructed as follows. Painted Rooms I and II were excavated with an eye to constituting a family grave for people whose religious preferences can no longer be reconstructed. Later on, when workers in the Jewish catacomb were digging a gallery going in an easterly direction, the Painted Rooms were (accidentally?) integrated into the larger network of underground galleries of the Vigna Randanini catacomb. The paintings, with their rather explicit pagan iconography, were not destroyed or in any sense "Judaized." Rather, the two rooms were reused for Jewish burials, as *kokhim* in the back wall of Painted Room II and in a gallery branching off the entrance to this area indicate.[9]

---

[8] For the Villa Piccola, see Brandenburg 1984, 21f. On Domitilla, see L. Pani Ermini, "L'ipogeo detto dei Flavi a Domitilla. I. Osservazioni sulla sua origine e sul carattere della decorazione," *Rivista di Archeologia Cristiana* 45 (1969) 119-73 and *ead.*, "L'ipogeo detto dei Flavi a Domitilla. II. Gli ambienti esterni," *ibid.* 48 (1972) 235-69.

[9] *Contra* Vismara 1986, 377 who argues this part of the catacomb was not reused by Jews. On *kokhim*, see *infra*.

In contrast to the Vigna Randanini catacomb, the building history of the Villa Torlonia catacombs was radically different. The most striking feature about both the upper and lower Jewish catacombs under the Villa Torlonia catacomb is the regularity of their galleries. The entire complex consists of long, more or less straight, high and carefully hewn galleries, from which other galleries, equally straight, branch off. Thus there can be little doubt that these galleries were excavated systematically to accommodate as many burials as possible. Some cubicula and arcosolia in the upper catacomb housed the graves of people who appear to have been quite well off financially. The great majority of the graves, however, consists of simple slots carved high into the walls of the seemingly endless galleries. The simplicity of most burials in both Jewish catacombs under the Villa Torlonia is also borne out by the omnipresence of graffiti and *dipinti*. According to my estimates, they outnumber the more expensive funerary inscriptions on marble by a relationship of 3:1. Clearly, many of the people who were buried here did not have much money to spend, but their aspirations were the same as those of people who could afford marble inscriptions and pictorial decoration. Cumulatively, such evidence suggests that—not unlike the early Christian catacombs of Callisto, Pretestato, and Priscilla already referred to earlier—the two Jewish catacombs under the Villa Torlonia were constructed specifically as catacombs to provide large portions of the Jewish community in third and fourth century C.E. Rome with a proper burial.

Two smaller Jewish burial sites that not yet been discussed so far, the hypogea on the Villa Labicana and in the Vigna Cimarra. They display characteristics that are different from those of the Vigna Randanini and Villa Torlonia catacombs. The value of the map of the Jewish hypogeum on the Via Labicana (now Casilina), is of only limited value.[10] For the Jewish hypogeum in the Vigna Cimarra, no maps exist at all. Despite such drawbacks, the meager evidence suffices to assert that in terms of their formal appearance these two underground complexes look quite similar to the pagan and early Christian hypogea mentioned earlier: they constituted small burying facilities that were constructed underground. Not destined for the entire community, and never developping into real catacombs their use was confined to individual families.

In summary, then, as concerns their formal appearance, the Jewish underground cemeteries of Late Ancient Rome can be said to mirror exactly the developments that characterize Late Ancient tomb con-

---

[10] For this map, see Marucchi 1887, opposite of p. 36. The schematization of the map suggests that it was drawn from memory.

struction in general. Like Late Ancient Christian and especially pagan hypogea, some Jewish hypogea remained essentially family tombs (Via Labicana, Vigna Cimarra); other hypogea (Vigna Randanini) evolved over time into catacombs, as was also the case with several Christian catacombs; still others (Villa Torlonia) were planned systematically and, like some Christian catacombs, may have served from the outset as large burial grounds for significant segments of the Roman Jewish community.

In view of such striking similarities, it should cause no surprise that the correspondence between the Jewish catacombs and hypogea of Rome with contemporary underground cemeteries used by pagans and Christians goes further than the layout of their galleries or their building history alone. From a technical perspective too, Jewish and non-Jewish underground burial complexes have much in common. For example, the diggers of pagan and early Christian catacombs and hypogea are known to have adapted preexisting underground cavities for funerary purposes. This can be observed in one of the earliest regions of the (Christian) Priscilla catacomb where a deserted network of subterranean cisterns and water channels was subsequently transformed into a burial ground. Also, other underground tombs such as those now part of the catacomb complex of Sebastiano had been adapted from preexisting structures. There, quarries that had originally served for the extraction of *pozzolana* (volcanic soil used by the Romans for the production of concrete) were later adapted for funerary purposes.[11]

Evidence from the Jewish catacombs and hypogea indicates that also in these catacombs the use of preexisting cavities was quite customary. It can be shown, for example, that several of the galleries in the upper catacomb under the Villa Torlonia were nothing but an adaptation of water channels that had been dug before this area came into use as a catacomb. Although further rescarch is necessary, some areas in the Vigna Randanini catacomb may similarly have been *pozzolana* quarries before they were transformed into a subterranean burial ground. Along the same lines, the cross-section of one of the underground galleries of the Jewish hypogeum along the Via Labicana shows a profile that indicates unmistakably that this gallery had originally functioned as a water channel and that its walls had been arranged with graves only during a later stage. Finally, the large width of some galleries in the Monteverde catacomb—often seen as one of the most outstanding features of the Jewish catacombs of Rome—is in

---

[11] F. Tolotti, "Influenza delle opere idrauliche sull'origine delle catacombe," *Rivista di Archeologia Cristiana* 56 (1980) 7-48 with further examples. On the Sebastiano catacomb, see Brandenburg 1984, 28-9.

reality an indication that parts of this catacomb had been constructed in an old quarry. Both in Roman and in modern times, the Monteverde area was famous for the quality of the *pozzolana* that could be found there. In fact, the destruction of the Jewish Monteverde catacomb in the early 1920s was due to collapse of quarries situated on a lower level, that is, directly under the catacomb.[12] Moreover, the constructional similarities between Jewish and non-Jewish tombs are due to the fact that the diggers of Jewish and of the Christian catacombs faced identical problems. The diggers of the Vigna Randanini catacomb and the hypogeum in the Vigna Cimarra, for example, had to construct galleries out of the same "tufo granulare" that *fossores* encountered while excavating some of the major Christian catacombs in that area including Domitilla, Callisto, Pretestato, and Sebastiano.[13]

The formal similarities between Jewish and early Christian funerary architecture do not stop here. Not only are there similarities between the Jewish and Christian catacombs in terms of building history, layout, and formal appearance of the galleries, but also in terms of the tomb types the Jewish catacombs are identical to the Christian catacombs.

Müller once suggested that as many as eleven different kinds of grave forms may be encountered in the Jewish Monteverde catacomb. Such a suggestion is, however, slightly misleading. In reality, most of the tombs listed by Müller are but variations on a few basic types of tombs.[14] The following tomb types can be encountered in the Jewish catacombs of Rome: the *loculus*, the *arcosolium*, the *forma*, the sarcophagus, and the amphora. As is the case in the Christian catacombs, the *loculus*, or burial slot excavated parallel to the gallery, is the grave form that is most common in the Jewish catacombs. It was sealed with bricks or a marble plate that sometimes carry an inscription. Somewhat less common, but still attested quite frequently in both Jewish and Christian catacombs, are the *arcosolia*, or graves consisting of a semicircular arch under which a rectangular container has been cut out for burial. Graves of the arcosolia type are usually found in *cubicula* (rectangular rooms). Sometimes, however, arcosolia have been arranged parallel to the galleries containing loculi, as in the upper catacomb under the Villa Torlonia. The *forma* or grave cut into the floor is the least representative of all rockcut graves. What matters here is that such *formae* can be found in Jewish and Christian catacombs alike.

---

[12] Personal observations. And see Fasola 1976, 32-34: galleries C1 and B 2-3; Marucchi 1887, opposite of p. 36; Müller's 1912, 25-6 (description).

[13] G. de Angelis d'Ossat, *La geologia delle catacombe romane* (Vatican City: Pontificio Istituto di Archeologia Cristiana, 1943) 23, 26, 177-78, 203, and 252.

[14] Müller 1912, 30-46.

Inhumation in sculpted marble sarcophagi is much less common than burial in loculi, arcosolia or *formae*. Yet, although the corpus of Jewish sarcophagi is small and above all fragmentary, the evidence is nevertheless sufficient to maintain that also in this case Jews made arrangements quite similar to those documented for the Christian catacombs.[15]

The extent to which Jewish and non-Jewish archaeological materials run parallel becomes evident when we turn to an inspection of the more unusual tomb types preserved in the Jewish catacombs of Rome. A few sarcophagi made of clay rather than of marble can still be found in the courtyard of the Jewish Vigna Randanini catacomb. Similar sarcophagi apparently have been discovered in the Monteverde catacomb. Their limited number suggests that terracotta sarcophagi did not constitute the most common form of burial during any period of Jewish burial in Rome.[16] Be this as it may, recent archaeological discoveries indicate that in Late Antiquity burial in terracotta sarcophagi constituted an option that was considered attractive by people other than Jews. Literary evidence suggests that burial in such sarcophagi may have been common as early as the first century B.C.E. Archaeological remains further complement this picture. One tomb, discovered off the Via Padre Semeria near the Porta Ardeatina, not far from the Via Appia, contained, among other things, a completely preserved terracotta sarcophagus dating to the period of the Severi (late second, early third century C.E.). Another discovery along the Via Portuensis (the same road along which the Jewish Monteverde catacomb was located) now documents that the general use of such sarcophagi was still comparably common in the third and fourth century C.E.[17]

Likewise rather unusual is the arrangement of a sarcophagus that has been preserved in the so-called Painted Room IV of the Vigna Randanini catacomb. The sarcophagus had been installed in a hollowed out space that was deep enough to contain the sarcophagus including its lid. Thus, the richly sculpted front of the sarcophagus was hidden from view completely. An identical kind of arrangement can also be found in one of the still unexplored side-galleries in the northern part of the Vigna Randanini catacomb. The rationale behind this rather unusual arrangement need not detain us. Suffice it to say here that sarcophagi hidden away from view can also be found else-

---

[15] For details, see the section entitled "Jewish Sarcophagi from Rome" below.

[16] Personal observation. And see Garrucci 1862, 12; Müller 1912, 37-38 and *CIJ* 475.

[17] Pliny, *NH* 35.160 with the discussion by Brandenburg 1978, 324; E. L. Caronna, "Roma. Via Padre Semeria. Muri in *opus testaceum* e tombe," *Notizie degli Scavi di Antichità* 36 (1982) 421-24; L. Cianfriglia *et al.*, "Roma. Via Portuense, angolo via G. Belluzzo. Indagine su alcuni resti di monumenti sepolcrali," *Notizie degli Scavi di Antichità* 40-41 (1986-87) 37-174, esp. 95 and 125.

where in underground Rome. In those cases as well, such an arrangement constitute exceptions rather than the rule. One such example comes from the Christian catacomb of Callisto. Another, the so-called mausoleum IV, can be found in the catacomb of Peter and Marcellinus. Finds in Antioch (Syria) show that the "concealment" of sarcophagi occurred elsewhere in the Roman world as well.[18]

Inhumation in amphoras or *enchytrismos* is another kind of burial that was practiced in many parts of the Roman world. It was less common than most other types of burial described earlier. It has been suggested that burial in amphoras somehow derives from similar kinds of burials found at many sites on Sardinia dating to the second century B.C.E and later, but such a suggestion is perhaps not entirely correct. In Italy itself the custom was known during the last few centuries before the beginning of the common era, as discoveries at Ostia show. Nor should it be forgotten that in the Greek world, burial in large vessels of pottery started much earlier—not to mention examples from Iron Age Syria dating to the seventh century B.C.E. Children in particular often found their last resting place in graves of this sort.[19]

For the present purpose, the supposed Sardinian or Punic origin of this type of burial is immaterial. What matters here is that amphoras were used for burial by Jews and non-Jews alike. To cite only a few examples that are roughly contemporary to the amphora burials that were discovered in the Monteverde catacomb: they can be found in Rome in a hypogeum on the Via Appia Pignatelli, in pagan tombs on top of the Christian catacomb of Thecla, in the necropolis of Isola Sacra near Rome, in a large necropolis at Portorecanati (Macerata), as well as in several Late Ancient necropoleis in Ravenna, where it is impossible to distinguish between the pagan and the Christian dead.[20]

---

[18] L. V. Rutgers, "Ein *in situ* erhaltenes Sarkophagfragment in der jüdischen Katakombe an der Via Appia," *Jewish Art* 14 (1988) 16-27; A. Ferrua, "Ultime scoperte a S. Callisto," *Rivista di Archeologia Cristiana* 52 (1976) 212-216; J. Guyon, *Le cimetière aux deux lauriers. Recherches sur les catacombes romaines* (Rome: Pontificio Isitituto di Archeologia Cristiana and École Française de Rome, 1987) 279; J. Lassus, *Sanctuaires chrétiennes de Syrie* (Paris: Geuthner, 1947) 228.

[19] G. Maetzke, "Florinas (Sassari). Necropoli ad enkytrismos in località Cantaru Ena," *Notizie degli Scavi di Antichità* 89 (1964) 280-314; Vismara 1986, 366; C. Vismara, "Un particolare tipo di sepoltura della Sardegna romana: Le tombe 'ad enchytrismos,'" in C.d'Angela *et al.* (eds.), *Le sepolture in Sardegna dal IV al VII secolo* (Vico Aquila: S'Alvure, 1990) 33-35; M. Floriani Squarciapino (ed.), *Scavi di Ostia III. Le necropoli I. Le tombe di età repubblicana e augustea* (Rome: Istituto Poligrafico, 1958) 17; J. P. Thalman, "Tell 'Arqa (Liban Nord). Campagnes I-III (1972-1974)," *Syria* 55 (1978) 88-93. Literary references: R. Garland, *The Greek Way of Death* (Ithaca: Cornell U. P., 1985) 144. Children: E. Breccia, *La necropoli di Sciatbi* (Cairo: Institut Français d'Archéologie Orientale, 1912) XXIII; Garland (this note) 160.

[20] Müller 1912, 42-43; *Notizie degli Scavi di Antichità* 1885, 158; V. Santa Maria Scrinari, "Il complesso cimiteriale di Santa Tecla. I. La necropoli pagana," *Rendiconti della Ponitificia Accademia Romana di Archeologia* 55-56 (1982-84) 394; G. Calza, *Le*

In most other parts of the Roman world a very similar pattern is documented.[21] Literary evidence finally likewise attests to the practice of using pottery for funerary purposes.[22]

While even the less common grave types in the Jewish catacombs and hypogea of Rome can thus be said to have little exclusively Jewish about them, there is one kind of grave that forms an exception to the rule that in Rome Jewish and non-Jewish grave forms look identical. It is to be found in one region of the Vigna Randanini catacomb only. Having been cut at right angles rather than parallel to the underground galleries, these graves go straight into the wall. They are known by their Hebrew name as *kokhim*.[23]

Although *kokhim* have been observed in number of Christian catacombs including Comodilla, a grave in the catacomb on the Via Anapo, several graves in a hypogeum near the Circus of Maxentius, some graves in the anonymous catacomb at the Villa Doria Pamphili, two graves in the catacomb of Ermete ai Pairioli, a few graves in the hypogeum of the Flavi in the catacomb of Domitilla, and finally, a few examples in the Calepodio catacomb, these Christian examples should nevertheless not be confused with their Jewish counterparts.[24] The formal characteristics of the *kokhim* in the Jewish Vigna Randanini catacomb clearly set the Jewish examples apart from the Christian

---

necropoli del Porto di Roma nell'Isola Sacra (Rome: Libreria dello Stato, 1940) 44, 54, and 60; M. Mercando, et al., "Portorecanati (Macerata) La necropoli romana di Portorecanati," *Notizie degli Scavi di Antichità* 28 (1974) 149, 365, and 411, M. G. Maioli, "Caratteristiche e problematiche delle necropoli di epoca tarda a Ravenna e in Romagna," XXXV *Corso di Cultura sull'Arte Ravennate e Bizantina* (Ravenna 1988) 323-25, 329-35, 340, 342, and 343-45.

[21] E.g. *Yugoslavia*: N. Cambi, "Salona und seine Nekropolen," in von Hesberg and Zanker 1987, 254 and 272; *Africa*: R. Guéry, *La nécropole orientale de Sitifis (Sétif, Algérie). Fouilles de 1966-67* (Paris: Centre National de la Recherche Scientifique, 1985) 106 n. 80 and 282 n. 317; *Britain*, see Philpott 1991, 22-25; *Egypt*, see A. Adriani, *Repertorio dell'arte d'Egitto greco-romano* (Palermo: Ignazio Mormino, 1966) 63 and Table 35, 126; *Israel*: N. Avigad, *Beth She'arim. Report on the Excavations During 1953-1958.* Volume III. *Catacombs 12-23* (Jerusalem: Israel Exploration Society, 1976) 55 and 183. Further examples in H. P. Kuhnen, *Nordwest-Palästina in hellenistisch-römischer Zeit* (Weinheim: VCH Acta Humaniora,1987) 59, 60, 67, and 77; N. Feig, "Notes and News," *Israel Exploration Journal* 38 (1988) 76; M. Peleg, "Persian, Hellenistic, and Roman Burials at Lohamei Hageta'ot,"'*Atiqot* 20 (1991) 131-52.

[22] W. Hilgers, *Lateinische Gefässnamen* (Düsseldorf: Rheinland-Verlag, 1969) 98 s.v. *alveus*, 101 s.v. *amphora*, 115 s.v. *olla*, 197 s.v. *hydria*, 199 s.v. *labellum*, 200 s.v. *labrum*, 235 s.v. *operculum*, 284 s.v. *solium*, 287 s.v. *testa*, 299 s.v. *unceus* and 303 s.v. *urna*.

[23] For a picture, see Leon 1960, Table 8, fig. 10. The observation that there is also a *kokh* at the end of gallery D 5 in the lower Villa Torlonia catacomb (e.g. Beyer and Lietzmann 1930, 6; Vismara 1986, 370) is incorrect (communication by the late U.M. Fasola).

[24] For references to the various excavation reports, see Rutgers 1990, 155 ns.100-105, to which should be added *Rivista di Archeologia Cristiana* 61 (1985) 252 (the Calepodio catacomb).

tombs listed here. The Jewish graves are always cut at ground level. They all contain a ledge on the inside that enabled one to bury at least two people separately in one *kokh* at one and the same time (although we have no way of telling if this actually happened, because these *kokhim* were never excavated properly). In addition, whereas the Christian *kokhim* occur in limited number and in the oldest parts of these catacombs only, the Jewish *kokhim* constitute a very common grave type, at least in large parts of the Vigna Randanini catacomb.

In Rome, the *kokh* can therefore be considered as a typical Jewish grave form. It is true that elsewhere in the Mediterranean, *kokhim* are far from being an exclusively Jewish grave form. Evidence of *kokhim* exists in places that are as far apart as Cyrene, Alexandria in upper Egypt, Cyprus, Petra, several sites in the Decapolis in Jordan, Palmyra, Dura-Europos, various sites in Greece, and, finally, many sites in Israel itself.[25] Yet the widespread existence of *kokhim* in other parts of the Roman world does not alter the fact that the occurence of numerous *kokhim* in the Vigna Randanini catacomb—and in this Jewish catacomb only—is exceptional. It deserves an explanation.

In the past it has been argued on both linguistic and archaeological grounds that the use of *kokhim* was intimately linked with the custom of secondary burial. Recent discoveries, among other places, at Jericho have shown that such a point of view is not entirely correct. *Kokhim* could also be used for primary burial. Previous scholarship also held that secondary burial ceased after 70 C.E. Recent archaeological discoveries now indicate, however, that the practice of secondary burial (as opposed to reinterment in the Land of Israel) may very well have continued in Palestine after 70 C.E.—albeit on a smaller scale than before. [26]

---

[25] Jewish examples at Carthage, Gamarth Hill, see Goodenough 1953-68, vol. 2, 63-68 and vol. 3, 865f. and Egypt, see E. Naville, *The Mound of the Jews and the City of Onias* (London: Seventh Memoir of the Egypt Exploration Fund, 1890) 13 and at numerous sites in Israel itself. For non-Jewish examples, see J. P. Peters, H. Thiersch, and S. A. Cook, *Painted Tombs in the Necropolis of Marissa (Mareshah)* (London: Palestine Exploration Fund, 1905) 81f.; O. Vesberg and A. Westholm, *The Swedish Expedition 4.3. The Hellenistic and Roman Periods in Cyprus* (Stockholm: The Swedish Cyprus Expedition, 1956) figs. 16-18 and p. 150; A. Adriani in *Enciclopedia dell'Arte Antica ed Orientale* 1 (1958) 210-14 s.v. Alessandria; J. M. C. Toynbee, *Death and Burial in the Roman World* (Ithaca: Cornell U. P., 1971) 219-34; D. C. Kurtz and J. Boardman, *Greek Burial Customs* (Ithaca: Cornell U. P., 1971) figs. 65 and 71; E. D. Oren and U. Rappaport, "The Necropolis of Maresha-Beth Govrin," *Israel Exploration Journal* 34 (1984) 114-53, esp. 150; J. Dent, "Burial Practices in Cyrenaica," in G. Barker *et al.* (eds.), *Cyrenaica in Antiquity* (Oxford: BAR International Series 236, 1985) 327-36, esp. 329-31.

[26] E. Y. Kutscher, "Kokh and Its Cognates," *Eretz Israel* 8 (1967) 273-79 (Hebrew); E. M. Meyers, *Jewish Ossuaries. Reburial and Rebirth* (Rome: Biblical Institute Press, 1971) 63-69; I. Gafni, "Reinterment in the Land of Israel: Notes on the Origin and Development of the Custom," *The Jerusalem Cathedra* 1 (1971) 96-104; Avigad 1976, 259; A. Kloner, *The Necropolis of Jerusalem in the Second Temple Period* (Unpublished Ph.D.

It is not possible to determine whether the *kokhim* in the Vigna Randanini catacomb in Rome were also used for primary or secondary burial, or for both. While none of the these *kokhim* was found intact, independent evidence hinting at the possible existence of secondary burial among the Jews of Rome is also ambiguous. One inscription (*CIJ* 483) from the Monteverde catacomb, now in the Museo Nazionale delle Terme, displays a eight-petalled rosette as decoration. It looks surprisingly similar to the rosettes frequently encountered on ossuaries from the Jerusalem-area (ossuaries are small containers that were used for secondary burial). The Roman piece is in too fragmentary a condition, however, to establish if this really was the fragment of an ossuary rather than a marble plate, used for a regular Jewish burial in Rome. The decoration alone does not solve the issue, for on Christian inscriptions too, petalled rosettes were sometimes carved.[27] No less ambiguous is the evidence provided by several Jewish funerary inscriptions from Rome. The inscriptions in question commemorate two, or more rather than one, person. Because most graves are rather small, it could be argued that such inscriptions were used to seal off graves in which secondary burial was practiced. Upon closer reflection, such a conclusion is, however, unwarranted. In the Vigna Randanini catacomb some inscriptions were not put in front of the actual graves, but on the wall, in between the graves. The same can be observed in the catacombs under the Villa Torlonia. Thus, it is quite conceivable that inscriptions placed in such spots contained information meant to refer not to one grave alone, but rather to several graves in the same general area.[28] Finally, it is not clear whether a Jewish sarcophagus that contained the skeletons of several people when it was discovered can be taken as an indicator of secondary burial. We

---

dissertation: Hebrew University of Jerusalem, 1980) (Hebrew) 223f., 247f.; R. Hachlili and A. Killebrew, "Jewish Funerary Customs During the Second Temple Period in the Light of the Excavations at the Jericho Necropolis," *Palestine Exploration Quarterly* 115 (1983) 105-39, esp. 110. Reburial after 70 C.E.: R. A. S. Macalister, *The Excavation of Gezer 1902-1905 and 1907-1909* (London: Palestine Exploration Fund, 1912) vol. 1, 338; Y. Aharoni *et al.*, *Excavations at Ramat Rahel. Seasons 1961 and 1962* (Rome: Università degli Studi, 1964) 74-78, 80-81; B. Mazar, *Beth She`arim. Report on the Excavations During 1936-1940*. Vol. I. *Catacombs 1-4* (Jerusalem: Massada Press, 1973) 22, 56-57, 74, 94, 134-35, 183; Avigad 1976, 73, 94, 102, esp. 114-15; E. Meyers *et al.*, *Ancient Synagogue Excavations at Khirbet Shema,' Upper Galilee, Israel 1970-72* (Durham: Duke U. P., 1976) 140-41; Oren and Rappaport (previous note) 121, 125, 127-28; A. Kloner, "The Cemetery at Horvat Thala," *Eretz Israel* 17 (1984) 325-32; G. Avni *et al.*, "Notes and News," *Israel Exploration Journal* 37 (1987) 73. And see B. R. McCane, *Jews, Christians, and Burial in Roman Palestine* (Ph.D. dissertation, Duke University, 1992) esp. 87-92.

[27] E.g. *ICVR* 3 (1956) Table XII, second row.

[28] Inscriptions, e.g.*CIJ* 305, 347, 361, 385, 391, 417, 418, 470, 497, 502, 511 (a sarcophagus = Konikoff 1986, no. 20), 535, 543, 732; Fasola 1976, 57 and Solin 1983, 655 nos. 2, 6 (?), 7, 10, 68. On the Torlonia catacomb, see Fasola 1976, 12.

simply do not know if these bones are the remains of the people for whom the sarcophagus was originally destined. Besides, inasmuch as sarcophagi are large enough to accommodate the bodies of several people and collective burial was not unusual in the ancient world, it is usually very difficult to determine whether sarcophagi containing the bodies of several people point to primary or to secondary burials.[29] In short, then, the evidence that could possibly document the existence of the custom of secondary burial among the Jews of Late Ancient Rome is extremely scarce.

The *kokhim* in the Vigna Randanini catacomb as such are nevertheless unusual enough to propose that this grave form was brought to Rome by Jews from the eastern part of the Mediterranean as a tomb type with which they had long been familiar. In fact, the *kokhim* in the so-called "Tomb of the Prophets," located on the Mount of Olives in Jerusalem look so similar to the *kokhim* of the Vigna Randanini catacomb that it almost seems as if they were made by the same crew of diggers.[30] It is impossible to tell what considerations, if any, played a role for the Jews who in Rome favored *kokhim* over the many other types of graves available. But it is clear that one of the conclusions Leon reached in his monograph on the Jews of ancient Rome can no longer be upheld entirely. Leon argued that the Jews buried in the Vigna Randanini catacomb belonged to the most Romanized segment of the Roman Jewish population. The widespread occurrence of *kokhim* suggests, however, that some of the people buried here may have valued traditional practices more than Leon was willing to admit. Further evidence adduced by Leon to support his thesis is likewise questionable. We have already seen that there is no valid reason to attribute the paintings in Painted Room I and II, with their pagan iconography, to Jewish patrons. Consequently, one can no longer argue on the basis of this pictorial evidence that the Jews of the Vigna Randanini had somehow become thoroughly Roman in their appreciation of contemporary, non-Jewish art. As will be shown in Chapter 4, the onomastic evidence similarly cannot be used to characterize the

---

[29] The sarcophagus is *CIJ* 511 (=Konikoff 1986, 20). And see G. Koch and H. Sichtermann, *Römische Sarkophage* (Munich: Beck, 1982) 21. For collective burial (non-Jewish) see Peters, Thiersch, and Cook 1905, 63; E. Brandenburg, *Die Felsarchitektur bei Jerusalem* (Kirchhain, 1926) 262; G. Tchalenko, *Villages antiques de la Syrie du Nord* (Paris: Geuthner, 1953) vol. 1, 68; A. Negev, "The Nabataean Necropolis of Mampsis (Kurnub)," *Israel Exploration Journal* 21 (1971) 121-24; D. Barag, "Hanita, Tomb XV. A Tomb of the Third and Early Fourth Century," *'Atiqot* 13 (1978) 7; J. J. Davis, "The Third Campaign at Abila of the Decapolis (1984)," *Near East Archaeology Society Bulletin* 24 (1985) 74; Vismara 1990, 34.

[30] Personal observation. And see H. Vincent, "Le tombeau des Prophètes," *Revue Biblique* 10 (1901) 72-88, esp. the cross-sections on p. 74. Note that Vincent's dating of this complex to the fourth and fifth centuries C.E. must remain hypothetical.

Jews buried in the Vigna Randanini catacomb as more Romanized than their co-religionists who had been entombed in other Jewish catacombs around the city. Thus, contrary to Leon, there is little evidence to argue that the Vigna Randanini catacomb represents the most Romanized segment of the Roman Jewish community.[31]

Concluding this discussion of Jewish funerary architecture, it should be noted that the existence of formal similarities between the Jewish and Christian catacombs is by no means an extraordinary phenomenon that was confined to the city of Rome alone. Elsewhere in the Diaspora it can also be observed that in their funerary architecture, Jews participated in what was common locally. In Naples, Late Ancient Jewish tombs have been discovered whose formal characteristics approach the type of tomb that was widespread locally and that is known as "a(lla) cappuccina" (a shallow grave covered with roof tiles). Inscriptions show that some of them were used for burial by Jews who may have been recent immigrants from Palestine and North Africa; others, however, served the non-Jewish population in that same locality.[32] In the southeastern part of Sicily, at Noto Antica, one Jewish tomb can be found in the middle of a row of Christian hypogea. All these tombs date back to the fifth century B.C.E. and were slightly recut in Late Antiquity. Most importantly, in their Late Ancient phase, they all look essentially alike from a formal point of view.[33] On Sardinia, at Sulcis, a Jewish hypogeum dating to the fourth century C.E. or later exhibits little to distinguish it from contemporary Christian hypogea in the area.[34] Elsewhere on Sardinia, at Porto Torres, in Late Antiquity Jews were buried alongside Christians in the deserted thermae of the town in tombs "alla cappuccina" and in amphoras.[35] On Malta in Rabat (Mdina) Jewish hypogea are dispersed among Christian ones and can be distinguished only on the basis of inscriptions and graffiti, and not on the basis of formal

---

[31] *Contra* Leon 1960, 258.

[32] G. A. Galante, "Un sepolcreto giudaico recentemente scoperto in Napoli," *Società Reale di Napoli. Memorie della Reale Accademia di Archeologia, Lettere e Belle Arti* 2 (1913) 233-45 and E. Serrao, "Nuove iscrizioni da un sepolcreto giudaico di Napoli,"*Puteoli* 12/13 (1988-89) 103-17.

[33] Personal observation. For pictures of these graves, see Rutgers 1992, 112-13. I would like to thank V. La Rosa who enabled me to visit this site.

[34] A. Taramelli, "Scavi e scoperte di antichità puniche e romane nell'area dell'antica Sulcis," *Notizie degli Scavi di Antichità* 1908, 150. See, in general, G. Zilliu, "Antichità paleocristiane del Sulcis," *Nuovo Bulletino di Archeologia Sardo* 1 (1984) 283-300.

[35] G. Maetzke, "Scavi e scoperte nel campo dell'archeologia cristiana negli ultimi dieci anni in Toscana ed in Sardegna," *Atti del secondo congresso nazionale di archeologia cristiana 25-31.V.1969* (Rome: L'Erma di Brettschneider, 1971).

characteristics.³⁶ At Doclea (Titograd, ex-Yugoslavia), one Jewish *fossa*-grave (a rectangular grave dug into the soil) lies among a collection of identical graves used by non-Jews.³⁷ In Tripoli, a Jewish hypogeum was discovered (and has now been lost) which looked exactly the same as a Christian hypogeum in Sirte.³⁸ In Teucheira, amphoras were used for Jewish burial in the same way as non-Jews used amphoras for funerary purposes at various sites in Roman North Africa.³⁹ In Korykos, in Asia Minor, Jewish inscriptions are found on several sarcophagi. Other sarcophagi of the exact same kind contain inscriptions referring to the non-Jewish population in the area.⁴⁰ In Alexandria, Jewish and non-Jewish graves look identical.⁴¹ Even as far away as Egra (in the Hejaz), a tomb used by a Jew can only be identified as such by the inscription it carries. In every other aspect the tomb looks explicitly Nabatean. The inscription follows local epigraphic traditions notably closely.⁴² Evidence recently discovered at Celarevo along the Danube in what formerly was Yugoslavia suggests that the Jewish community practice of using grave forms that had been developed locally continued at some sites into the early and even into the high Middle Ages.⁴³

Several preliminary conclusions can be drawn on the basis of the evidence presented in this section. We have seen that the formal characteristics of the Jewish catacombs are very similar to those displayed

---

[36] Personal observation. See also, with maps, M. Buhagiar, *Late Roman and Byzantine Catacombs and Related Burial Places in the Maltese Islands.* BAR International Series 302 (1986) 29 and E. Becker, *Malta Sotterranea. Studien zur altchristlichen und jüdischen Sepulkralkunst* (Strassburg, 1913) 71.

[37] A. Cermanovic-Kucmanovic and D. Srejovic, "Jeversjka grobnica u Duklji," *Jeverjski Almanak* 1963-64, 56-62. I am indebted to J. Schaeken for the translation of this article.

[38] P. Romanelli, "Una piccola catacomba giudaica di Tripoli," *Quaderni di Archeologia della Libia* 9 (1977) 116-17.

[39] For the Jewish examples, see G. Lüderitz, *Corpus jüdischer Zeugnisse aus der Cyrenaika* (Wiesbaden: Dr. Ludwig Reichert, 1983) 124. For non-Jewish examples, see A. Campus, "L'uso delle anfore nelle tombe della Sardegna Imperiale," in A. Mastino (ed.), *L'Africa romana. Atti del'VIII convegno di studio Cagliari, 14-16.XII.1990* (Sassari: Gallizzi, 1991) 931.

[40] J. Keil and A. Wilhelm, *Monumenta Asiae Minoris Antiquae. III. Denkmäler aus dem Rauhen Kilikien* (Manchester: Manchester U. P., 1931) esp. 121. Conceivably, the same holds true for the Jewish graves at Seleucia on the Calycadnos, *ibid.* 18.

[41] A. Adriani, *Repertorio dell'arte d'Egitto greco-romano* (Palermo: Ignazio Mormino, 1966) 60, 110, and Table 34.123.

[42] A. Jaussen and R. Savignac, *Mission archéologique en Arabie (mars-mai 1907). De Jérusalem au Hedjaz Médain-Salez* 1 (Paris: Leroux, 1909) 339, 344 and nos. 4 and 172; A. Negev, "The Nabataean Necropolis at Egra," *Revue Biblique* 83 (1976) 216.

[43] The interpretation of the finds is particularly difficult. See, in general, R. Bunardzic, *Celarevo. Risultati delle ricerche nelle necropoli dell'alto medioevo* (Roma, 1985: catalog of an exhibit that never took place); C. Bálint, *Die Archäologie der Steppe* (Vienna and Cologne: Böhlau, 1989) 174 and esp. 186 with C14- dating.

by the Christian catacombs. Together with the fact that the archaeological materials from the Jewish catacombs date to the same period as the archaeological materials from the Christian catacombs, this must be taken to mean that the Jewish and Christian catacombs developed synchronically. Neither did the Jewish catacombs serve as example for the Christian catacombs nor vice versa. Both the Jewish and Christian catacombs are part of a development that is generally Late Ancient and that was not confined to the adherents of one religion only.

Such a conclusion should perhaps not surprise us. From the third century C.E. onwards a general trend manifested itself towards standardization or homogenization in funerary architecture. It is attested in places that are as far away as Pannonia and Britain.[44] Such developments cannot be taken as assuring that from then on grave forms were henceforth entirely the same in all parts of the Roman Empire. They do imply, however, that in large areas of the Roman world people with different cultural and religious backgrounds all relied on the same crews or workshops for the construction of their graves. Insofar as the Jews were concerned, the development of funerary architecture in Late Antiquity suggests that they did not form an isolated group, one impenetrable to outside influences—at least insofar as their funerary architecture was concerned. Rather, the opposite was the case. We have seen that in tomb architecture almost exactly the same kind of the formal developments took place in the Jewish as in the non-Jewish sphere. Insofar as this aspect of their material culture was concerned, Jews did not behave differently than did other ethnic groups.

The presence of *kokhim* in the Vigna Randanini catacomb, on the other hand, draws our attention to the fact that the adoption of local building traditions could go hand in hand with a preference for more traditionally Jewish grave forms. This suggests that interaction between Jews and non-Jews cannot be conceived of as a one-way process only. I will deal with this question in greater detail in the conclusion to this chapter. Let us first turn to an inspection of archaeological remains from the Jewish catacombs in general, and to the investigation of the mechanisms that stand at the basis of artistic production in antiquity in particular.

Artistic Production in Late Antiquity: General Trends

Just as with funerary architecture, a study of archaeological materials from the Jewish catacombs of Rome shows that the influence of the

---

[44] Philpott 1991, 226; Morris 1992, 33 and 68.

non-Jewish world on the artistic products used by the Jewish community of ancient Rome was strong. Such a study also shows, however, that in making use of what was generally available, Jews sometimes had preferences that differed from the preferences of non-Jews. As we will now see, investigation of the technical and iconographical features of archaeological remains from the Jewish catacombs of Rome provides us with a unique means to define precisely one particular aspect of interaction between Jews and non-Jews in antiquity. To appreciate fully the mechanisms that determined the production of Jewish art in Late Ancient Rome, it is necessary, however, to sketch briefly some general trends chraracterizing artistic production in Late Antiquity.

Down through Hellenistic times, craftsmen who formed workshops of moderate size had always refrained from specialization. Athenian sculptors, for example, are known to have produced different types of monuments and to have worked with different kinds of material including marble and bronze. Similarly, in Hellenistic Rhodes, sculpture workshops remained family affairs and stayed small. As in Athens, Rhodian sculptors accepted different kinds of commissions and worked a variety of materials in their professional careers.[45]

It was not until the end of the Republic that such traditional ways of artistic production changed fundamentally. As the Roman Empire expanded, so did the city of Rome. With building construction booming, demand increased. To meet such demands workshops were forced to reorganize and make more efficient the production process. Such production changes could be achieved most effectively in two ways, namely through a lowering of the quality of the product or through specialization of the work force employed.

Evidence provided by marble capitals and by sarcophagi shows that both types of manufacturing change were put into practice. Around the middle of the first century B.C.E., a type of marble capital decorated with acanthus leaves was created in Rome that—in contrast to the capitals produced by Athenian craftsmen—could be carved quickly and without much technical skill. Soon such capitals were to enjoy enormous popularity, not only in Rome, but in other parts of the Empire as well. Further attempts at streamlining and economizing the production of capitals were made in the second half of the first century C.E. Because this was a time-consuming activity that required a well-trained sculptor, the actual carving of the stone was now reduced as much as possible. Inasmuch as this activity demanded a much lower

---

[45] See V. C. Goodlett, "Rhodian Sculpture Workshops," *American Journal of Archaeology* 95 (1991) 669-81.

level of specialization on the part of its practitioner, parts of the stone were henceforth worked with drills.[46]

Study of the large-scale production of sarcophagi as it came of age in Rome in the course of the second century C.E. sheds further light on the process of systematizing that determined the course of artistic production in the later Roman world. Perhaps more than any other form of ancient craftsmanship, the production of sarcophagi was completely routinized. From the moment such stone containers were first quarried until the moment they were finally sold, everything had been organized to run as smoothly as possible. The technical features of these sarcophagi show remarkably well the extent to which in Late Antiquity the actual carving of these pieces in massive numbers was determined by continuing systematization, especially in the use of the running drill. The physical appearance of early Christian sarcophagi in particular can serve as a prime example of a continuing shift from quality to quantity.

To be sure, the trend towards an ever increasing coordination of production affected not just work in sculpture, but in all fields of ancient artistic production. In the case of the production of pottery, and that of *terra sigillata* in particular, the process of production was increasingly aimed at the manufacturing of huge quantities of pottery (and lamps). Technical developments in the building industry also illustrate the general tendency towards a more pragmatic use of the available resources. The change from *opus incertum* to *opus reticulatum,* for example, reflects such pragmatism just as do developments in brick manufacture. Interestingly, even the Latin language of late Republican times reflects the changes in the process of production quite closely. H. von Petrikovits has shown that the increase in terms to denote various sorts of specialized craftsmanship is nothing short of explosive during this period.[47]

---

[46] M. Pfanner, "Über das Herstellen von Porträts. Ein Beitrag zu Rationalisierungsmaßnahmen und Produktionsmechanismen von Massenware im späten Hellenismus und in der römischen Kaiserzeit," *Jahrbuch des Deutschen Archäologischen Instituts* 104 (1989) 157-257. On the drill, see K. Eichner, "Die Produktionsmethoden der stadtrömischen Sarkophagfabrik in der Blütezeit unter Konstantin,"*Jahrbuch für Antike und Christentum* 24 (1981) 85-113, esp. 104f.; M. Pfanner, "Vom 'Laufenden Bohrer' bis zum bohrlosen Stil. Überlegungen zur Bohrtechnik in der Antike," *Archäologischer Anzeiger* 1988, 667-76, esp. 675-76. And see K. S. Freyberger, *Stadtrömische Kapitelle aus der Zeit von Domitian bis Alexander Severus. Zur Arbeitsweise und Organisation stadtrömischer Werkstätten der Kaiserzeit* (Mainz: von Zabern, 1990) esp. 133-35.

[47] Pfanner 1989, 171-75, and 224f.; D. P. S. Peacock, *Pottery in the Roman World; an Ethnoarchaeological Approach* (London and New York: Longman, 1982) esp.114-28 and W. V. Harris, "Roman Terracotta Lamps: The Organization of an Industry," *Journal of Roman Studies* 70 (1980) 126-45; M. Steinby, "Ziegelstempel von Rom und Umgebung," *RE Supplementband* XV (1978) 1489-1531; T. L. Heres, *Paries. A Proposal for a Dating System of Late-Antique Masonry Structures in Rome and Ostia* (Ph.D. dissertation Am-

Having streamlined their production, one of the most outstanding features of artistic production in Late Antiquity is that individual workshops satisfied the demands of customers with widely diverging tastes. There were workshops, for example, that produced altars decorated with traditional Roman iconographical themes. At the same time, however, such workshops also produced altars carrying much less traditional, or even non-Roman iconographical motifs. A good example of such "all-inclusive" production is provided by an altar, dedicated in Rome in the second or third century C.E. by a Tiberius Claudius Felix, possibly a Palmyrean, and his family to Sol Sanctissimus-Malachbelus and the Palmyrene gods. It carries an inscription in Latin on its front and another inscription, this time in Palmyrean, on its left side. It also contains various representations in relief. Although the iconography expresses a concern for Palmyrean theology, the sculptural decoration of the piece is thoroughly Roman from a stylistic point of view. Little or nothing of the schematic frontality that characterizes much of the artistic production at Palmyra itself is to be found here. Rather, the piece can best be compared stylistically with Roman altars and dating to the same general period dedicated to other non-Palmyrean gods. A second altar that displays very similar characteristics suggests that there was nothing unusual about Roman workshops satisfying the artistic (and religious) demands of Palmyrean immigrants. This second piece is dated to the year 236 C.E. It was dedicated through a bilingual inscription in Greek and Palmyrean by a family which had likewise originated from Palmyra. This second piece takes the form of an aedicula in which the gods Aglibol and Malakbel have been carved. From a stylistic perspective, everything about it is again Roman: the shape of the columns including their capitals, as well as the decoration of pediment with its superficially-carved wreath and fanciful acroteria. Even Aglibol himself is dressed as an actual Roman general.[48]

It is possible to expand almost interminably the list of non-Roman religious figures and themes that were represented in a thoroughly Roman fashion. That the figural representations manufactured for the devotees of other gods of Oriental origin were often rooted technically

---

sterdam: Free University, 1982); H. von Petrikovits, "Die Spezialisierung des römischen Handwerks II," *Zeitschrift für Papyrologie und Epigraphik* 43 (1981) 285-306, esp. 289 and 293.

[48] E. E. Schneider, "Il santuario di Bel e delle divinità di Palmira. Communità e tradizioni religiose dei Palmireni a Roma," *Dialoghi di Archeologia* 5 (1987) 69-85, esp. 73-77. On Palmyrean art in general, see K. Parlasca, "Das Verhältnis der palmyrenischen Grabplastik zur römischen Porträtkunst," *Römische Mitteilungen* 92 (1985) 343-56. And see S. R. Tufi, "Le due statue funerarie 'palmirene' di Arbeia (South Shields) sul vallo di Adriano," *Dialoghi di Archeologia* 5 (1987) 97-105 and L. Bianchi, "Uno scultore partico a Gorsium?," *ibid.*, 101-5.

and stylistically in local Roman artistic traditions is evident, for example, in the case of depictions of Mithras. They include a fourth century C.E. relief from Rome found in the Circus Maximus and a painting on the back wall of a Mithraeum at Marino (on the Lago di Albano). More to the North, along the Rhine in Germania Superior, cult reliefs displaying the tauroctone Mithras bear characteristics that suggest strongly that this particular type of relief was created and developed locally rather than in Asia Minor, as has sometimes been intimated. This being the case, it should come as no surprise that also for the representation of Mithras' assistants, the torchbearers *Cautes* and *Cautopates*, Greco-Roman prototypes have been postulated.[49] Comparably, Isis and Serapis can usually be recognized as such only on the basis of their attributes. Depictions of Sabazius are likewise so variable that it has recently been observed that "there exists no generally accepted standard for the recognition of this god in art." Certain syncretistic tendencies (for example the identification of Sabazius with Zeus) certainly contributed further to the fact that representations of so many different gods look stylistically so much alike.[50]

When the number of Christian customers started to increase appreciably in Late Antiquity, it was only natural that the old custom of individual workshops catering to clients with diverging preferences continued. Studying the arch of Constantine, L'Orange and von Gerkan were the first to clarify, in 1939, the extent to which early Christian art had originally been rooted in non Christian artistic traditions. On the basis of a stylistic analysis of the arch's sculptural decoration, they were able to show that the "non-Christian" reliefs on this arch and some Christian frieze sarcophagi of the Constantinian era had originated in one and the same workshop.[51]

In recent years, much additional evidence has been discovered to suggest that the phenomenon of "Werkstattgleichheit" (common workshop-identity) was once widespread. Publishing a corpus of gold

---

[49] G. M. A. Hanfmann, *The Season Sarcophagus in Dumbarton Oaks* (Cambridge Mass.: Harvard U. P.,1951) no. 72; M. J. Vermaseren, *Mithriaca III. The Mithraeum at Marino* (Leiden: Brill, 1982) 8 and 24; E. Schwertheim, *Die Denkmäler orientalischer Götter im römischen Deutschland* (Leiden: Brill, 1974) 280-90; L. A. Campbell, *Mithraic Iconography and Ideology* (Leiden: Brill, 1968) 29.

[50] V. Tram Tan Tinh, "État des études iconographiques relatives à Isis, Sérapis et Sunnaoi Theoi," *Aufstieg und Niedergang der RömischenWelt* II.17.3 (1984) 1710-38; E. N. Lane, "Towards a Definition of the Iconography of Sabazius," *Numen* 27 (1980) 9-33 (citation p. 9). Note also that Simon Magus had himself represented as Jupiter and his wife Helena as Minerva, see Ireneus, *Adversus Haereses* 1.23 (*PL* 7, 571). And see G. Wilpert, "La statuta di Simon Mago sull'Isola Tiberina," *Rivista di Archeologia Cristiana* 15 (1938) 334-339 (unconvincing).

[51] H. P. L'Orange and A. von Gerkan, *Der spätantike Bildschmuck des Konstantinsbogen* (Berlin: de Gruyter, 1939) 219, 222-25.

glasses, Morey managed to distinguish several workshops on the basis of the designs and decorative motifs used. He found that one and the same workshop could cater to pagan as well as to Christian customers. Although recent scholarship has questioned some of Morey's attributions, it has also shown that his view of common workshop-identity continues to be persuasive. Similarly, assignment to one and the same workshop helps to explain why a collection of twenty-five glass bowls that were decorated with incised pagan mythological and early Christian themes look identical in everything except their iconography. Common workshop-identity must finally also be assumed for certain pottery lamps from Athens as well as from North Africa. The formal characteristics of these lamps suggest that in the fourth century C.E., individual workshops produced such lamps for pagan and for Christian customers synchronically.[52]

That early Christian art came into existence in pagan workshops and that it was at first heavily indebted to the specific figural types and modes of representation developed by these workshops for a pagan audience, needs no further demonstration here. Klauser, among others, has described the process in detail in a series of well known articles on the earliest history of Christian art.[53] In the present context, it is important merely to keep in mind the significance of the phenomenon of common workshop-identity for the development of artistic production in Late Antiquity. In the following survey it will be argued that Jewish art from the Jewish catacombs and hypogea of Rome originated in the same workshops that also produced for pagan and

---

[52] In general, J. B. Ward-Perkins, "The Role of Craftmanship in the Formation of Early Christian Art," *Atti del IX congresso Internazionale di archeologia cristiana* (Vatican City: Pontificio Istituto di Archeologia Cristiana, 1978) 637-52 and Engemann 1986. Sarcophagi: Eichner 1981, 88 and ns. 22 and 23. Glass: C. R. Morey, *The Gold-Glass Collection of the Vatican Library. With Additional Catalogues of Other Gold-Glass Collections* (Vatican City: Bibliotheca Apostolica Vaticana, 1959) e.g. nos. 21, 27, 28 and 32; D. B. Harden, "The Wint Hill Hunting Bowl and Related Glasses," *Journal of Glass Studies* 2 (1960) 45-81; R. Noll, "An Instance of Motif Identity in Two Gold Glasses," *Journal of Glass Studies* 15 (1973) 31-34; L. Faedo, "Per una classificazione preliminare dei vetri dorati tardoromani," *Annali della Scuola Normale Superiore di Pisa*, Ser. 3, vol. 8.3 (1978) 1025-70, esp. 1030f. Pottery: F. Bejaoui, "Les dioscures, les apôtres et Lazare sur des plats en céramique africaine," *Antiquités Africaines* 21 (1985) 173-77.

[53] T. Klauser, "Studien zur Entstehungsgeschichte der christlichen Kunst I-IX," *Jahrbuch für Antike und Christentum* 1 (1958) - 10 (1967). On the use of traditional figural types and formulae on a mosaic floor in Cyprus, see the discussion by J. G. Deckers, "Dionysos der Erlöser? Bemerkungen zur Deutung der Bodenmosaiken im "Haus des Aion" in Nea-Paphos auf Cypern durch W.A. Daszewski," *Römische Quartalschrift* 81 (1986) 145-72. Especially in the minor arts, pagan motifs were slow to disappear on objects commissioned by Christians, see in general, with further references, the survey in Deichmann 1983, 366f. and see J. Huskinson, "Some Pagan Mythological Figures and Their Significance in Early Christian Art," *Papers of the British School in Rome* 42 (1974) 68-97.

early Christian customers. I thus hope to show that, from a technical perspective, Jewish material culture in Late Ancient Rome was indissolubly linked to Late Ancient artistic production in general.

The Wall Paintings in the Jewish Catacombs of Rome

The corpus of wall paintings from the Jewish catacombs and hypogea of Rome is small. Bosio noted that he saw a large painted menorah upon visiting the Monteverde catacomb in 1602, but the painting must have been lost subsequently, for it is never referred to in any later literature on the site.[54] Only in the Vigna Randanini catacomb and in the upper Villa Torlonia have wall paintings been preserved that can still be studied today.

In several of the still not fully-explored galleries in the Vigna Randanini catacomb, cubicula may be found, whose walls have been decorated with white stucco, but that lack pictorial decoration. Only four cubicula have been furnished with wall paintings. Since the publication of the second volume of Goodenough's *Jewish Symbols* they are usually referred to as Painted Rooms I-IV.[55]

Painted Rooms I and II have been painted in a red and green linear style that was quite common elsewhere in Rome and Ostia towards the end of the second and in the first half of the third century C.E. in both pagan and early Christian contexts.[56] As has been observed in this chapter and in Chapter 1, it is, however, impossible to determine if these two rooms were originally commissioned by Jews. For that reason, they will be excluded from this discussion of Jewish wall paintings from Rome.

Painted Room III is decorated with various motifs, but loculi take up most of the wall space. The basic pattern of decoration consists of red and green lines, floral bands, and, in the four corners, depictions of palm trees. The walls flanking the entrance carry skillfully-painted marble incrustation. Garrucci maintained that the ceiling of this room once contained a decoration with birds, which was subsequently destroyed. Other scholars have followed Garrucci, but such assertions are not correct. The ceiling of Painted Room III was removed in antiquity. That this is so follows from the fact that in antiquity the additional loculi were cut on top of those already in existence. In other words, the original ceiling of Painted Room III was not destroyed re-

---

[54] Bosio 1632, 143 and Aringhi 1651, 396.
[55] Goodenough 1953-68, vol. 3, figs. 737f.
[56] For parallels, see Rutgers 1990, 146-47.

cently, but must have been removed in antiquity. Leon's suggestion that the red and green lines depicted on both sides of the entrance to Painted Room III are *mezuzot* is likewise misleading. The color and general appearance of these lines, and the fact that they appear on all three sides of the entrance, show that these lines are part of the general decoration in red and green lines. Just as the rest of the decoration of Painted Room III, these lines decorating the entrance have nothing specifically Jewish about them.[57]

Painted Room IV displays a rather crude decoration consisting of red bands on a white stucco underground. It is the only painted room in the Vigna Randanini catacomb with an iconography that is explicitly Jewish. While a menorah was painted above the arcosolium facing the entrance, the white ceiling was decorated in yellowish-brown with what appear to be *ethrogim*.[58]

In the upper catacomb under the Villa Torlonia, by contrast, Jewish iconographic motifs assume a much more prominent position in the wall paintings preserved. There, one cubiculum ("Cubiculum II") and two arcosolia were decorated with a variety of Jewish themes, including the menorah and several Torah shrines. More neutral themes, displaying various sorts of animals, also occur.[59]

For the present purpose, it is not necessary to describe these paintings in greater detail at this point. Here it suffices to place them into the broader context of Late Ancient wall painting to determine in what respect they are typical or a-typical.

A comparison of the wall paintings from the Jewish catacombs with those found in non-Jewish contexts shows that many motifs used on the walls of the Jewish catacombs were not isolated occurrences. They rather belong to the stock repertoire of Late Ancient wall decoration as it can be found elsewhere in Rome as well. The marble incrustation in Painted Room III in the Vigna Randanini catacomb, for example, reflects a particular style of wall decoration that attained empire-wide diffusion in Late Antiquity. Examples range from the Via Latina catacomb in Rome to the so-called "Hanghäuser" in Ephesos and a room in Luxor embellished for the worship of the Roman emperor—to cite only a few.[60] Individual motifs depicted in Painted Room III, such

---

[57] R. Garrucci in *La Civiltà Cattolica* 6 (1863) 104; *CIJ* CXXI-CXXII, Goodenough 1953-68, vol. 2, 20; vol. 3, 757-58; Leon 1960, 206.

[58] Goodenough 1953-68, vol. 3, 759-61 and Leon 1960, Table 9, fig. 12.

[59] Goodenough 1953-68, vol. 3, 806-817.

[60] Deichmann 1983, 325f.; W. Tronzo, *The Via Latina Catacomb. Imitation and Discontinuity in Fourth Century Roman Painting* (University Park: College Art Association, 1986) 36-39 to which other examples could be added.

as the oversized kantharos or the black lozenges, likewise belong to the basic repertoire of Late Ancient painting.[61]

For all its Jewish iconography, even the walls of the painted cubiculum and the painted arcosolia in the upper Villa Torlonia catacomb conform rather strictly to the general pictorial repertoire of Late Antiquity. The general division of space on the arches of the two arcosolia in Cubiculum II into a central tondo, flanked by two semicircular fields, is exactly what one usually encounters on the ceiling of arcosolia in the Christian catacombs. Not surprisingly, the lines dividing the arches of the arcosolia as well as the central ceiling of this cubiculum into various fields have been decorated with ovolos and other motifs for which exact parallels have recently been documented in the Christian catacomb of Marcellinus and Peter.[62] The dolphins depicted on the central ceiling of Cubiculum II have many parallels in the Christian catacombs.[63] Similarly, the painted *lenos* sarcophagus on the front of one of the arcosolia is strongly reminiscent of comparable arrangements in the Via Latina catacomb as are the columns sculpted out of tuffa and decorated with a painted marble veneer.[64] The arches of the two arcosolia have been adorned with cassettes that are common everywhere in underground Rome, as they are in other parts of the later Roman Empire.[65] Additional ornamental motifs such as the garlands can again be paralleled with similar motifs in the Marcellinus and Peter catacomb.[66] Remarkable also are the representations on the lower part of the arches of the arcosolia. They display various kinds of animals in an idyllic-bucolic setting. It is well known that such scenes were widespread in Late Antiquity in both pagan and Christian contexts. A recent study by Provoost has in fact shown that in early Christian art, bucolic themes heavily outrank other iconographic themes not only in wall paintings, but also on gold glasses, inscrip-

---

[61] For a similar kantharos, see A. Ferrua, *Le pitture della nuova catacomba di Via Latina* (Vatican City: Pontificio Istituto di Archeologia Cristiana, 1960) Table XIX; for the lozenge: B. M. Appolonj Ghetti *et al.*, *Esplorazioni sotto la confessione di S. Pietro in Vaticano* (Vatican City: Pontificio Istituto di Archeologia Cristiana, 1951), 38 and Table X. Similar lozenges occur in other parts of the empire as well (e.g. Abila in Jordan, personal observation).

[62] J. G. Deckers *et al.*, *Die Katakombe "Santi Marcellino e Pietro." Repertorium der Malereien* (Vatican City: Pontificio Istituto di Archeologia Cristiana, 1987) 384 no. 14; 385, no. 17.

[63] J. B. Frey, "Il delfino col tridente nella catacomba giudaica di Via Nomentana," *Rivista di Archeologia Cristiana* 8 (1931) 301-14.

[64] Ferrua 1960, Tables XXXII and CXVII; fig. 24 and, more in general, Deichmann 1983, 325.

[65] Ferrua 1960, Table LXV, LXXIII,2; Deckers 1987, Table 66 and 67.

[66] Deckers 1987, 381 no. 1.

tions, and on pottery lamps.⁶⁷ Perhaps most surprising is that the depiction of the sun and moon that flank the Torah shrines found on the back wall of several arcosolia are arranged in a way that is strongly reminiscent of representations of *Sol* and *Luna* on non-Jewish monuments, such as the Mithraeum at Marino, south of Rome.⁶⁸

Given the present state of knowledge concerning wall paintings in the Jewish catacombs of Rome, it is as yet impossible to identify the particular workshop(s) that executed these paintings. There can be little doubt, however, that the workshop(s) responsible for embellishing several cubicula and arcosolia in the Vigna Randanini and upper Villa Torlonia catacombs were identical to the workshops that executed wall paintings in Christian catacombs and pagan hypogea. This is likely not only because it would hardly have been possible to make a living in Rome as painter if one had to depend on Jewish commissions exclusively, for these must have been relatively few and far between. The limited repertoire of decorative motifs indicates that only a few workshops managed to fulfill the wishes of customers with the most divergent religious convictions.

Unfortunately, a chemical analysis of either the stucco or the pigments of the wall paintings in the Jewish catacombs has never been carried out.⁶⁹ On the basis of personal observations, I believe, however, that there is every reason to suppose that the outcome of such a study will provide additional evidence to support the view that no significant differences exist between the Jewish catacombs and other non-Jewish monuments of Late Antiquity.

Despite all these similarities between the wall paintings from Jewish, early Christian, and pagan monuments, it should be stressed that the phenomenon of common workshop identity can only serve to explain the technical aspects of the wall paintings from the Jewish catacombs of Rome. In several, although not in all, cases, the adoption of painting techniques and pictorial motifs is combined with a preference of iconographical themes that are unmistakably Jewish. As we will now see, this combination of artistic techniques that are generally Late Ancient with a predilection for Jewish iconographical themes is characteristic of much Jewish art from Rome. The implications will be

---

⁶⁷ A. Provoost, "Das Zeugnis der Fresken und Grabplatten in der Katakombe S. Pietro e Marcellino im Vergleich mit dem Zeugnis der Lampen und Gläser aus Rom," *Boreas* 6 (1986) 152-72.

⁶⁸ M. J. Vermaseren, *Mithriaca III. The Mithraeum at Marino* (Leiden: Brill, 1982) pl. XIV.

⁶⁹ For this type of analysis, see e.g. J. Riederer, *Archäologie und Chemie. Einblicke in die Vergangenheit* (Berlin: Rathgen-Forschungslabor SMPK, 1987) 202-13; K. Hangst in Deckers *et al.* 1987, 21-22.

discussed towards the end of this chapter, after additional evidence has been presented.

Jewish Sarcophagi from Rome

In ancient Rome, some Jews used sarcophagi in their funerary practices. A small corpus of these sarcophagi has survived, but considerable problems exist insofar as their identification as Jewish sarcophagi is concerned.[70] In Chapter 1 we have seen that over the centuries catacombs remained accessible and that archaeological materials were removed constantly from these subterranean sites. The introduction of intrusive pieces may have started long before the sixteenth century. The fact that according to Roman law, a grave in which someone had been buried was considered *res religiosa* did not necessarily prevent people from violating such graves. The stripping and illegal use of materials originating from tombs concerned lawgivers on various occasions in the fourth, and then again in the fifth century. Still later, in the Middle Ages individuals seeking marble fragments to burn in kilns to prepare plaster roamed these ancient burial grounds, breaking up marble sarcophagi into smaller pieces so that they could be more easily transported. Traces of their destructive activity have been found all over Rome.[71]

Sarcophagi for Jewish burial have been discovered both inside and outside the Jewish catacombs and hypogea. Because the respective excavation reports are too unspecific or because such reports are absent altogether, it is impossible to tell if independent Jewish burial in sarcophagi outside the Jewish catacombs or hypogea ever took place on a large scale. It is certainly conceivable that sarcophagi found outside the Jewish catacombs ended up there only after they had been removed from their original location in the catacombs.

That some sarcophagi discovered inside the Jewish catacombs were used by Jews for burial is beyond doubt. This can be inferred from their iconography or from the epitaphs they carry. In the case of other sarcophagi, Jewish usage is, however, more ambiguous. Several sar-

---

[70] For a recent catalog, see Konikoff 1986. Several pieces should be added to this corpus, see Kant and Rutgers in *Gnomon* 1988, 377 and Rutgers 1992, 104 n. 15.

[71] *CIJ* 350, 9* and 36*; Müller 1912, 34, 39-41; F. de Visscher, *Le droit des tombeaux romains* (Milan: Giuffrè, 1963) 44, and esp. 49f.; Koch and Sichtermann 1982, 23; Rutgers 1990, 153-54; *Codex Theodosianus* 9.17; *Novels of Valentinian (III)* 23.1 of 447 C.E.

cophagi display an iconography that is plainly pagan. They furthermore lack inscriptions that could inform about the person(s) buried in them. Because it has been assumed (correctly, I believe) that the Jewish catacombs and hypogea were used exclusively by Jews, such pieces with a pagan iconography have often been included in the corpus of Jewish sarcophagi from Rome.[72]

Yet such an approach does not solve satisfactorily the problem of attribution. Because Jewish and pagan tombs were located in the same general areas, there exists a possibility that pagan materials from the surface have made their way into the Jewish catacombs.[73] M. Gütschow's study of the sarcophagus collection at the Christian catacomb of Pretestato has shown well how pagan materials that had originally nothing to do with this catacomb were introduced in various underground galleries at an unknown period in time. The pieces in question are often fragmentary in nature and appear to have fallen into the catacomb through the collapse of underground galleries.[74] We have no reason to suppose that in the case of in the Jewish catacombs the situation was radically different. Inasmuch as the Vigna Randanini catacomb is excavated in exactly the same type of collapsible tuffa as the Pretestato catacomb, several of the sarcophagi fragments considered by some researchers as genuine examples of Jewish funerary art might in reality not be Jewish at all, but could very well have originated in the pagan necropoleis in the direct vicinity of this catacomb. The same holds true for the Villa Torlonia and Monteverde catacombs. It should also be noted, finally, that even when sarcophagus fragments are found *in situ*, they may not originally have been destined for Jewish burial. Some funerary plaques that are inscribed on both sides indicate that the reuse of existing materials was widespread.

An identification of an individual fragment as Jewish is only warranted when the following three conditions are met: (1) a sarcophagus is found *in situ;* (2) it carries an inscription that indicates its use by Jews; or (3) the iconography of the piece suggests a Jewish commission. In the following discussion I have only included pieces to which the rule formulated above applies. Other pieces have been excluded. For example, the remains of several sarcophagi, such as those discovered in the Villa Torlonia catacombs, are so fragmentary that one may

---

[72] Most notably by Goodenough 1953-68, vol. 2, 25f. and Konikoff 1986, *passim*.

[73] Personal observations. See also C. L. Visconti, "Scavi di Vigna Randanini," *Bulletino dell'Instituto di Corrispondenza Archeologica* 1861, 16-22; for the Villa Torlonia catacombs, see *CIJ* CXXVI. These pagan tombs await further, systematic investigation. See *CIL* 9.7648-7783. And see Müller 1912, 22.

[74] M. Gütschow, "Das Museum der Prätextat-Katakombe," *Atti della Pontificia Accademia Romana di Archeologia* III.4 (1938) 35-43.

rightly wonder if in their "Jewish phase" they were ever used as full-size sarcophagi.[75]

Most striking about the sarcophagi that were used for Jewish burial, according to the principles enunciated above, is the predominance of decorative frameworks and iconographical motifs that are either neutral or plainly pagan. Only four sarcophagi display an iconography that is explicitly Jewish; yet even on two of these pieces, the Jewish motifs have been fitted into a larger, non-Jewish framework.[76] One of them, the single most famous Jewish sarcophagus from Rome, displays two sculpted Victoriae holding a *clipeus* (tondo) in which an elaborately carved menorah has been rendered in relief. The sarcophagus has been preserved only in part. It is conceivable, therefore, that the piece served as a slab to seal off a loculus rather than a sarcophagus proper. Whatever its precise identification may be, there can be little doubt that this Jewish object belongs iconographically to the larger group of season sarcophagi. This was a type of sarcophagus that enjoyed such popularity in non-Jewish circles during the third and early fourth century C.E. that mass production became necessary. Despite the menorah, which is placed in the most prominent position on the front of the piece, it is evident that the Jewish sarcophagus originated in one of the workshops that specialized in the manufacture of this kind of sarcophagi and that sold its products to Rome's city populace at large. By replacing the portraits traditionally depicted in the central *clipeus* by a menorah, the artisans accommodated the wishes of a Jewish customer. This was an exceptional case. Rather than producing for Jewish customers exclusively, the production of season sarcophagi was geared toward the much larger non-Jewish market.[77]

A brief look at other Jewish sarcophagi from Rome shows that the majority of these sarcophagi must have originated, like the season sar-

---

[75] I am thinking here of pieces such as the ones illustrated by Goodenough 1953-68, vol. 3, nos. 819-26. M. Gütschow provides a description of these pieces in Beyer and Lietzmann 1930, 42-44, but fails to indicate where and how these pieces were originally found. Fasola 1976, 15 n. 10 is more precise as to where the various new sarcophagus fragments he discovered were found. But, again, detailed information is not provided. Other pieces that have been excluded include Goodenough 1953-68, vol. 3, nos. 733, 736, 795-804, 827, 830-34 and Konikoff no. 16. Goodenough 1953-68, vol. 2, 42-43 and vol. 12, 35 attributes sarcophagi to the Jewish community of Rome on the basis of people telling him that the fragments in question were once found in the Jewish catacombs. Inclusion of materials as Jewish on the basis of such premises is, however, unacceptable.

[76] Goodenough 1953-68, vol. 3, nos. 786 and 788 (=Konikoff 1986, 4), 818 (=Konikoff 1986, 13), 789 (=Konikoff 1986, 14) and 787 (=Konikoff 1986, 15).

[77] Goodenough 1953-68, vol. 3, no. 789 (=Konikoff 1986, 14); Hanfmann 1951, Koch and Sichtermann 1982, 21-23; P. Kranz, *Jahreszeitensarkophage* (Berlin: Mann, 1984). On loculus slabs, see J. S. Boersma, "A Roman Funeral Relief in the Allard Pierson Museum, Amsterdam," *Bulletin Antieke Beschaving* 48 (1973) 125-41, esp. 131-32 and Koch and Sichtermann 1982, 82-83.

cophagus, in Roman workshops. One such a sarcophagus was once located in Painted Room IV of the Vigna Randanini catacomb. Although the piece has vanished during the last hundred years, its outward appearance can be reconstructed quite reliably, as I have shown elsewhere.[78] Its sculpted front once showed several figures, one of which is most likely to be identified with the Muse Urania. Thus, the sarcophagus did not represent an isolated example. Rather, as its iconography and its general features show, it belonged to the larger group of Late Ancient sarcophagi that all gave visual expression to the ideal of the *mousikos aner*. In Rome alone, more than two hundred such sarcophagi have been found. They were produced in large numbers and could be ordered from stock. That was exactly what happened with the sarcophagus from the Vigna Randanini catacomb. This time, it was even easier to satisfy the Jewish buyer than in the case of the season sarcophagus discussed above. Without making changes in its decoration, the Urania sarcophagus was installed in Painted Room IV.[79]

A group of nine Jewish sarcophagi all display a front decorated with a *tabula inscriptionis* in the center. Flanking it on both sides are *strigiles* of varying quality.[80] These sarcophagi belong to a group that, from the second century C.E. onwards, enjoyed popularity among the general public in Rome and elsewhere. Again, Jews participated in what was commonly fashionable.[81] So did a certain Faustina who ordered a sarcophagus cover she had decorated with theatrical masks,[82] as did the Jews who ordered the terracotta sarcophagi referred to earlier.

In the entire corpus of Jewish sarcophagi from Rome only two pieces cannot be described on the basis of common workshop-identity. Both sarcophagi are explicitly Jewish in their iconography. One seems to have been employed in the Vigna Randanini, while the other may

---

[78] Rutgers 1988.

[79] Koch and Sichtermann 1982, 197-203 with references to further literature and the standard corpora.

[80] The sarcophagi in question are depicted in Goodenough 1953-68, vol. 3, nos. 828 (= Konikoff 1986, 21), and Konikoff 1986, nos. 1, 2, 6, 17, 18, 19, 20 (in a sense, no. 21 also belongs to this group). And compare the piece described, but not depicted by Müller 1912, 41. Although it displays the same general layout, the sarcophagus of *Iulia Irene Arista* should, however, be excluded from this group. Contrary to what is often asserted, this sarcophagus was never used for Jewish burial, as Ferrua has correctly argued recently, see *ICVR* 8 (1983) no. 21115.

[81] See Koch and Sichtermann 1982, 73-76.

[82] Goodenough 1953-68, vol. 3, no. 787 (=Konikoff 1986, 15). On this group of sarcophagi in general, see T. Brennecke, *Kopf und Maske. Untersuchungen zu den Akroterien an Sarkophagdeckeln* (Berlin, 1970) 166-80.

once have adorned the Villa Torlonia catacomb.[83] Consisting of Jewish iconographical motifs such as the menorah, ethrog, and lulav, the decoration of these two sarcophagi puts them into a unique class. They must have been commissioned and carved separately rather than mass produced. Yet, if we can trust Garrucci's drawing, one side of the Vigna Randanini sarcophagus was carved with a griffin, that is, with a motif that was genuinely common on the sides of non-Jewish Roman sarcophagi. Thus, even on this otherwise outstandingly Jewish piece one finds traces of the traditional Roman style of sarcophagus manufacturing![84]

In terms of production, technique, decorative frameworks employed, and iconography, the sarcophagi used by Jews for burial compare well with the wall paintings in various Jewish catacombs. In both cases, artistic products commissioned by Jews were rooted completely in the decorative practices developed by non-Jewish Roman workshops. Except for one piece, all extant "Jewish" sarcophagi found in Rome appear to have been manufactured locally.[85] There is every reason to suppose that the workshops that fulfilled the wishes of Jewish patrons were identical to those that produced items for pagan and a Christian clientele. It is possible that such workshops also employed Jewish artists (see, for example, *CIJ* 109). Yet, it is not correct to conclude that a sarcophagus with a Jewish iconographic program was necessarily produced by Jewish artists. In fact, in the present context it is really irrelevant to reflect on an artist's possible Jewish background: in Late Ancient Rome Jews worked according to the standards set by the non-Jewish Roman workshops.[86]

The Jewish Gold Glasses

Esthetically, the small collection of Jewish gold glasses from Rome constitute the most appealing group of Jewish realia that have come down from antiquity. A total of thirteen Jewish gold glasses have so

---

[83] Goodenough 1953-68, vol. 3, nos. 786 and 788 (=Konikoff 1986, 4) and 818 (=Konikoff 1986, 13).

[84] Goodenough 1953-68, vol. 3, no. 786. The sarcophagus, now in Berlin, is now missing the side which contained the griffin. I am indebted to T.-M. Schmidt of Berlin for enabling me to study the remains of this sarcophagus.

[85] The corner fragment of a column-sarcophagus (Konikoff 1986, 11) was perhaps imported from Asia Minor, but it remains unclear, whether this piece was ever used for a Jewish burial proper (rather than reused to seal a grave). On imported sarcophagi in general, see G. Koch, "Stadtrömisch oder östlich?," *Bonner Jahrbücher* 180 (1980) 51-104 and *id.*, "Östliche Sarkophage in Rom," *Bonner Jahrbücher* 182 (1982) 167-208.

[86] *Contra* Konikoff 1986, 19 and 65.

far come to light. The exact provenance of these pieces is known only in case of two Jewish gold glasses. They were both found in the late nineteenth century, in Christian catacombs, namely in a cubiculum of the catacomb of Marcellinus and Peter, and, later, in an underground gallery at S. Ermete ai Parioli. Ironically, the only fragments of gold glasses known to have been found in the Jewish catacombs, are, without exception, not Jewish in their iconography.[87]

The rough edges of the gold glasses indicate that most, if not all of them, are fragments of larger glass vessels, rather than of independent medallions. Their decoration consists of representations in gold leaf that have been fitted in between two layers of glass. Cologne and Rome appear to have been the main centers where such gold glasses were manufactured. Archaeological studies have shown that the production of gold glasses is a typically Late Ancient phenomenon, which dates specifically to the fourth century C.E. It is not surprising, therefore, to note that the designation *auricaesor* was introduced in exactly this period.[88]

Most Jewish gold glasses display the same kind of decorative motifs and iconography. They consist of a circular base that is usually subdivided into two semicircular fields. Such fields are adorned with one or more elaborately rendered menorot and often also with a Torah shrine. Lions normally flank either the menorah or Torah shrine. Other Jewish objects such as the ethrog, lulav, and shofar have frequently been included too. Most gold glasses also contain short inscriptions. Only one Jewish gold glass differs from the pattern just described. It represents what appears to be the Temple of Jerusalem and, as has

---

[87] For a catalog, see T.-M. Schmidt, "Ein jüdisches Goldglas in der frühchristlich-byzantinischen Sammlung," *Forschungen und Berichte* 20-21 (1980) 273-80. For a color picture, see *Encyclopedia Judaica* 7 (1971) 620-621. Illustrations are also easily available in Frey, *CIJ* 515-22 and in Goodenough 1953-68, vol. 3, nos. 964-978. Jewish gold glasses in Christian catacombs: G. B. de Rossi, "Insegno vetro rappresentante il Tempio di Gerusalemme," *Bulletino di Archeologia Cristiana* 20 (1882) 121-58 and 21 (1883) 92; G. Bonavenia, "Un cenno sulle recenti scoperte fatte nel cimitero S. Ermete ai Parioli," *Römische Quartalschrift* 8 (1894) 38-44. Non-Jewish gold glasses in Jewish catacombs: Müller 1912, 59; Marucchi 1887, 27-28; R. Paribeni, "Catacomba giudaica sulla Via Nomentana," *Notizie degli Scavi di Antichità* 46 (1920) 153-54; Fasola 1976, 6 and 19.

[88] G. Bovini, *Monumenti figurati paleocristiani conservati a Firenze nelle raccolte pubbliche e negli edifici di culto* (Vatican City: Pontificio Istituto di Archeologia Cristiana, 1950); Faedo 1978, 1046-69; M. Sternini, "A Glass Workshop in Rome (IVth-Vth century A.D.)," *Kölner Jahrbuch für Vor- und Frühgeschichte* 24 (1991) 105-14; D. Whitehouse, "A Glassmaker's Workshop at Rome?" *Journal of Roman Archaeology* 4 (1991) 385-86. Von Petrikovits 1981, 289 (*auricaesor*).

been argued recently, should perhaps be seen as an artistic rendering of the festival of Purim.[89]

Despite such typically Jewish iconographic features, there is good reason to suppose that Jewish gold glasses were manufactured in the very same workshops that also catered to non-Jews. Two gold glasses with Jewish and four gold glasses with non-Jewish motifs all have in common borders which are decorated in an identical fashion and colored in order to highlight a design that was otherwise executed purely in gold leaf. For that reason, twenty-five years ago, J. Engemann attributed these Jewish and non-Jewish gold glasses to the same workshop.[90] A third Jewish gold glass can possibly now be added to this group.[91] Further strengthening Engemann's case, it may be pointed out that other motifs depicted on these gold glasses likewise derive from the larger iconographic repertoire current in Late Antiquity. The two birds on one Jewish gold glass hold a ribbon in their beaks and flank the Torah shrine in a way that is practically identical to heraldic arrangements on Late Ancient sarcophagi. Similarly, most Torah shrines have been depicted in a manner that is strongly reminiscent of earlier and contemporary representations of Roman book shelves. There can be little doubt, then, that the workmen who crafted these pieces used the traditional iconographic models in attempting to render designs that did not belong to their standard repertoire (that is, the Jewish motifs).[92]

In this context, it may also be observed that Jewish and non-Jewish gold glasses have more in common than their decoration alone. Also in terms of the inscription they carry, and even in terms of the function they originally fulfilled, the dividing line between Jewish and non-Jewish gold glasses remains far from clear. Five Jewish gold glasses carry an inscription reading (in one form or another) *pie zeses*, an expression used to express good wishes or serving to convey a toast. This formula is extremely common on gold glasses with a pagan or Christian iconography, as it is on other types of glass or, for that matter, ceramic vessels.[93] The inscriptions on two other Jewish gold

---

[89] Goodenough 1953-68, vol. 3, no. 978; A. St. Clair, "God's House of Peace in Paradise. The Feast of Tabernacles on a Jewish Gold Glass," *Journal of Jewish Art* 11 (1986) 6-15.

[90] J. Engemann, "Bemerkungen zu römischen Gläsern mit Goldfoliendekor," *Jahrbuch für Antike und Christentum* 11-12 (1968-69) 7-25.

[91] Schmidt 1980, 280.

[92] Compare the gold glass in Goodenough 1953-68, vol. 3, no. 967 with, for example, the sarcophagi illustrated in *Jahrbuch für Antike und Christentum* 3 (1960) 128 and Table 7c. A similar arrangement with birds may be found on a funerary altar preserved in the museum of Fermo (Marche).

[93] For Jewish examples, see Goodenough 1953-68, vol. 3, nos. 964, 966, 969, 972 and 975. For pagan examples, see Morey 1959, no. 96. For Christian examples (New Testa-

glasses contain, among other things, the epithet *anima dulcis*, a phrase that is likewise common on non-Jewish gold glasses. On early Christian inscriptions it was used regularly.[94] The expression *vivas cum* ... is far less common, yet again it occurs on Jewish and on non-Jewish gold glasses alike.[95]

It is not entirely clear what purpose gold glasses served. Specimens preserved in a reliable archaeological context are often found cemented in the stucco sealing off loculi in the catacombs. They may have been used to decorate the outside of tombs. They might also have been deposited to facilitate the identification of particular graves. Yet, there are strong indications to suggest that gold glasses were originally manufactured to serve purposes other than those documented by the archaeological record. We have already seen that most gold glasses are probably fragments of larger glass vessels.[96] On the basis of the inscriptions they carry, and because of the iconography of some non-Jewish pieces, scholars usually assume that gold glasses functioned as presents during the celebration of the Roman New Year and comparable festive occasions such as weddings and birthdays. In this connection the evidence provided by Late Ancient consular dyptichs, as well as the contorniates (typically Roman bronze medals used as presents), is often referred to, because they have much in common iconographically with the gold glasses, and because they are known to have served similar purposes.[97]

---

ment scenes), see Morey 1959, no. 366. The examples can be expanded without any problem, see Morey, index s.v. *pie zeses*. On the translation of this term, see L. Vidman, "Inscriptions," in E. Strouhal (ed.), *Wadi Qitna and Kalabsha-South: Late Roman, Early Byzantine Cemeteries in Egyptian Nubia* (Prague: Charles University, 1984) 215-6. See also Harden 1960, *passim* and S. Tau and M. Nicu, "Ein beschrifteter Glasbecher aus der Nekropole von Barcea-Tecuci (4. Jh. u. Z.)," *Dacia* 29 (1985) 165-66.

[94] Goodenough 1953-68, vol. 3, nos. 968 and 975; Morey 1959, index s.v. *anima dulcis*. Early Christian inscriptions, e.g. *ICVR* 4593, 4675, 4682, 9523, 9691, 10108, 10160, 14052, 14242, 14475, 14481, 14523, 14534, 14607, 14645, 14722, 14838, 14845, 14846, 21132, 21437, 21694, 21705, 21877, 22059, 22137d, 22140d, 22216a, 22314, 22325, 22361, 22450, 22456, 22545, and 22570. The closest parallel, on Jewish inscriptions, to the *anima dulcis* of the Christian epitaphs is the *anima bona* is *CIJ* 210 or the *anima innox* in *CIJ* 466.

[95] Jewish: Goodenough 1953-68, vol. 3, 973; for non-Jewish examples, see Morey 1959, index s.v. *vivas cum*.

[96] See Engemann 1968-69, 10; F. Fülep, "Early Christian Gold Glasses in the Hungarian National Museum," *Acta Antiqua Academiae Scientiarum Hungaricae* 16 (1968) 402, n. 5; D. Barag, "A Jewish Gold Glass Medallion from Rome," *Israel Exploration Journal* 20 (1970) 99.

[97] Engemann 1968-69, 12-16; Schmidt 1980, 274; Stuiber in *Reallexikon für Antike und Christentum* 10 (1978) 693-95 s.v. Becher. On the contorniates and especially on their connection with the New Year, see now A. and E. Alföldi, *Die Kontorniat-Medaillons* (Berlin and New York: de Gruyter, 1990) 12-24. And see D. Baudy, "*Strenarum Commercio*. Über Geschenke und Glückwünsche zum römischen Neujahrsfest," *Rheinisches Museum für Philologie* 130 (1987) 1-28.

That the custom of distributing handouts on the occasion of the New Year (*strenae*) cannot have been completely unknown to Jews follows from the fact that on one known occasion such New Year presents were offered to the patriarch Judah II. Should we interpret the Jewish gold glasses from Rome as New Year gifts too? In view of one of the canons accepted during the Council of Laodicea (later fourth century C.E.), which forbade Christians from accepting festive presents from Jews, it should be noted that such a possibility is not as far-fetched as it may seem.[98]

Whatever the answer to this question, the Jewish gold glasses are remarkable in that they show better than any other group of Jewish archaeological materials from Rome the extent to which, on the level of material culture, Jews participated in the larger contemporary non-Jewish world. The Jewish gold glasses were not merely manufactured in workshops that felt equally at home in the production of gold glasses with a Jewish, pagan, or an early Christian iconography. Rather, also as concerns their usage, the differences between Jewish and non-Jewish gold glasses may have been far less substantial than a first look at the iconography of the Jewish gold glasses would suggest. We have seen that, more often than not, Jewish gold glasses carry the same inscriptions as do non-Jewish specimens. Similarly, like early Christian gold glasses, Jewish gold glasses appear to have been used secondarily for funerary purposes only after the larger vessel of which they formed a part had perished (or was broken off intentionally).[99] Even more significantly, some Jews felt no qualms in using gold glasses with pagan iconographical motifs. It is fair to assume that in fourth-century Rome, Jews may have refrained from participating in all the rituals surrounding the celebration of the Roman New Year. Yet, there can be little doubt that Jews took over artistic products that had become fashionable among the general populace in the city of Rome with few, if any, changes.

Lamps From Jewish Rome

Among the grave gifts from the Jewish catacombs and hypogea of Rome, pottery lamps are by far the most common. No systematic

---

[98] G. Blidstein, "A Roman Gift of Strenae to the Patriarch Judah II," *Israel Exploration Journal* 22 (1972) 150-52. In a Jewish context gold glasses could also have functioned as gifts, see *BT Baba Bathra* 146a. On the association of two Jewish gold glasses with Purim, see St. Clair 1985, *passim* and Schmidt 1980, 277-79. J. D. Mansi, *Sacrorum Conciliorum nova et amplissima collectio* (Leipzig: Welter, 1901) vol. 2, 570, canon 37.

[99] For the remains of (Jewish) goldglasses in the catacombs, see Müller 1912, 59; Marucchi 1887, 27-28; Beyer and Lietzmann 1930, 3 and Fasola 1976, 6 and 19.

survey of these lamps has ever been published. At present, information concerning lamps from the Jewish catacombs has to be pieced together from a wide variety of publications and excavation reports. Illustrations are often lacking. Most unfortunately, earlier excavators simply did not care to record smaller finds such as these with any degree of precision. Illustrative of the all-pervasive carelessness in this particular field of research is Leon's offhand observation, unaccompanied by further details, that "in the debris from the Nomentana [Villa Torlonia] catacomb, not long after its discovery, I saw a number of lamps and lamp fragments among which was one bearing a Christian monogram."[100] Müller writes that he discovered hundreds of lamps in the Monteverde catacomb, twenty-six of them in a single grave. Apparently, all that remains of this collection today are the six lamps entered into the list of acquisitions in the Bode Museum of Berlin and a black and white picture of six further specimens in the photo archive of the Vatican Museums.[101] Fasola discovered approximately one hundred lamps during his recent excavations in the upper and lower Villa Torlonia catacombs. Only a few of them have been illustrated. Study of these lamps is impossible because they have been locked away in the catacomb behind an iron fence that cannot be opened.[102] Given the lack of documentation, it need not be stressed that the following remarks on the lamps used by the Jews of Rome are of a preliminary nature.

As a result of carefully controlled excavations, only in recent years have scholars started to understand better the history of pottery production in Late Antiquity in general, and the role played by workshops in North Africa in particular. We now know that in that part of the Roman world an independently operating lamp industry came into being in the second century C.E. By the middle of the fourth century C.E. it had augmented its share in the market to such an extent that North African lamps enjoyed popularity in many parts of the later Roman Empire. This is evident not only from the wide distribution of lamps of the type *Atlante VIII A1a and X A1a*, but also from the fact that African lamps were often imitated locally.

North African potters also produced lamps with a Jewish iconography. Their products and the imitations thereof have been found as far

---

[100] Leon 1960, 225.

[101] Müller 1912, 53-56; photo archive of the Musei Vaticani Negative no. XXXII.85.27.

[102] Garrucci 1862, 8; Goodenough 1953-68, vol. 3, nos. 942-947; Fasola 1976, 59-60.

away as Trier and Augsburg.[103] That such lamps were manufactured together with Christian lamps in the same workshops can be ascertained for a group of lamps found in Carthage that date to the late third century C.E. and that all display identical spouts.[104] There is little evidence to support Le Bohec's contention that these lamps were the products of Jewish craftsmen.[105] Rather, this group of lamps provides yet another example of the phenomenon of common workshop-identity that we have encountered in connection with other groups of archeological materials discussed elsewhere in this chapter.

In the city of Rome, while lamp production remained primarily a local affair, imitation of North African models was widespread.[106] Lamps with Jewish iconographical motifs, mainly the menorah, on their discuses found in Rome fit neatly into this general picture. Some Jewish specimens from Rome are close to the non-Jewish lamps produced in this city in their general shape as well as the crudeness of their design.[107] Other Jewish lamps from Rome, recovered mostly under unclear circumstances, however, display a more refined decoration. Although they may not have been imported from North Africa directly, there can be little doubt that North African lamps provided the example to be imitated.[108]

Further study of the Jewish lamps of Rome is necessary to establish whether they were, like their Carthaginian counterparts, produced by

---

[103] D. Korol, "Juden und Christen in Augsburg und Umgebung in Spätantike und frühem Christentum-Das Zeugnis der Kleinkunst," *Tesserae (FS Engemann)* = *Jahrbuch für Antike und Christentum Erg. Bd.* 18 (1991) 51-55.

[104] J. Deneauve, *Lampes de Carthage* (Paris: Centre National de la Recherche Scientifique, 1969) 85, 220-21.

[105] Y. Le Bohec, "Inscriptions Juives et Judaïsantes de l'Afrique romaine," *Antiquités Africaines* 17 (1981) 165-207, esp. 196. The only evidence he adduces is the name *Sabbatius*. Yet, this name is only generally Semitic and does not automatically indicate the presence of Jews. In addition, the term "Judaïsantes" as used by Le Bohec is misleading: someone carrying a Semitic name need not necessarily have been receptive to Jewish theological ideas or customs, let alone have followed them.

[106] A. Provoost, "Les lampes en terre cuite. Introduction et essai générale avec des détails concernant les lampes trouvées en Italie," *L'Antiquité Classique* 45 (1976) 5-39 and 550-86; D. Whitehouse *et al.*, "The Schola Praeconum I: The Coins, Pottery, Lamps and Fauna," *Papers of the British School at Rome* 50 (1982) 53-101; A. Carandini *et al.*, "Il contesto del Tempio della Magna Mater sul Palatino," in A. Giardina (ed.), *Società romana e impero tardoantico. II. Le merci. Gli insediamenti* (Bari: Laterza, 1986) 27-43; L. Anselmino *et al.*, "Ostia. Terme del Nuotatore," *ibid.* 45-81; L. Anselmino, "Le lucerne tardoantiche: produzione e cronologia," *ibid.* 227-40; C. Pavolini, "La circolazione delle lucerne in terra sigillata africana," *ibid.* 241-50. From a different perspective, but likewise important, is J. M. Schuring, "The Roman, Early Medieval Coarse Kitchen Wares from the San Sisto Vecchio in Rome," *Bulletin Antieke Beschaving* 61 (1986) 158-207.

[107] See, for example, the lamps in Goodenough 1953-68, vol. 3, nos. 942-947.

[108] E.g. Fasola 1976, 59; V. B. Mann (ed.), *Garden and Ghettos. The Art of Jewish Life in Italy* (Berkeley: University of California Press, 1989) 226 no. 29, 227 nos. 30 and 32.

the same Roman workshops as were contemporary lamps with a pagan or a Christian decoration. Such a study will also have to come to grips with the fact that on Late Ancient pottery lamps from Rome and other parts of Italy, a candlestick with five instead of seven branches not infrequently belongs to the standard repertoire of decoration.[109] Overall, many Jews in Late Ancient Rome do not seem to have been excessively particular when it came to the types of lamps used for funerary purposes. Next to the specimen with an explicit Jewish iconography, other lamps with pagan or early Christian decorative themes have been discovered in the Jewish catacombs. More neutral themes can also be observed. Such finds show that Jews bought products that were generally available and that originated in workshops on which also the non-Jewish population of Late Ancient Rome relied.[110]

Miscellaneous Finds from the Jewish Catacombs of Rome

Like the pottery lamps, other minor finds from the catacombs and hypogea of Rome have never been studied systematically. In the present context, it is not necessary to carry out such an investigation. Suffice it to discuss briefly only the two most significant finds that fall into this category.

While excavating the Vigna Randanini catacomb, Garrucci discovered, in what appears to have been an undisturbed grave, a small object made of glass. It represents the head of Medusa and, most likely, is to be interpreted as an amulet. Further amulets that may have been used by Jews in Late Ancient Rome have been discovered, but the original context in which they were found as well as their present whereabouts remain unknown.[111]

The iconography of these amulets seems to suggest that some Roman Jews were receptive to what was customary in non-Jewish "magic" circles. Of course, the Medusa head from the Vigna Randanini catacomb need not necessarily be an amulet, for it could also be interpreted as an item of personal adornment of the type permitted by

---

[109] Rome: Müller 1912, 53-56; Goodenough 1953-68, vol. 3, nos. 942, 945, 946; Musei Vaticani photo archive negative no. XXXII.85.27; Siracusa: P. Orsi, "Nuovi ipogei di sette cristiane e giudaiche ai Cappucini in Siracusa," *Römische Quartalschrift* 14 (1900) 193; Citadella (Sicily): P. Orsi, "Nuove chiese Bizantine nel territorio di Siracusa," *Byzantinische Zeitschrift* 8 (1899) 613-42, esp. 617; Carthage: Goodenough 1953-68, vol. 3, no. 939. See also V. Sussman, *Ornamented Jewish Oil-Lamps* (Warminster: Aris and Philips, 1982) 20.

[110] Pagan: see Goodenough 1953-68, vol. 3, no. 943 (from the Monteverde catacomb); Christian: Müller 1912, 53-56 and Leon 1960, 225; Fasola 1976, 60.

[111] Garrucci 1862, 8 (=Goodenough 1953-68, vol. 3, no. 1044). For references to other amulets, see Rutgers 1992, 109 n. 44.

the rabbis of Roman Palestine for sentimental reasons. Yet independent of how their users viewed these artifacts, and whether they carried the same connotation for Jewish users as they did for non-Jews, the Medusa piece as such documents that in terms of their material culture, the Jews of Rome were part of the larger non-Jewish world of Late Antiquity.[112]

The second object that needs to be mentioned here briefly is a sculptural relief. It is hardly ever discussed in studies of Jewish archaeological evidence from Rome. It represents a pick axe, or, more precisely, an *ascia*. This motif is very common on both pagan and early Christian monuments in various parts of the Empire. Some scholars have interpreted representations of the *ascia* as an indicator of the occupation of the person buried near it, but others have favored a mystical interpretation of the motif. Using epigraphic evidence, de Visscher has maintained that the *ascia* was meant to indicate and somehow guarantee the exclusive ownership (*securitas perpetua*) of the grave by the person over whose tomb the *ascia* had been placed.[113]

It is impossible to determine on the basis of which considerations the person who commissioned the *ascia* relief for the Vigna Randanini catacomb did so. Yet, in and by itself, the occurrence of a relief showing an *ascia* in a Jewish catacomb illustrates once again that in equipping their tombs, Jews in Late Ancient Rome participated fully in many trends that were genuinely Late Ancient rather than uniquely Jewish.

Artistic Production in Roman Palestine: Some Parallels

Only a few of the kinds of artifacts that were found in the Jewish catacombs of Rome were ever exported to Roman Palestine. There, local needs were met by workshops in the region, such as those that produced the massive sarcophagi preserved in the necropolis of Beth She'arim and the series of pagan funerary portraits from the Beth Shean area. Because of differences in style, decorative patterns, and quality of workmanship, it is impossible to mistake the artistic prod-

---

[112] *Semahot* 8.7. In general, see the survey by P. S. Alexander in E. Schürer (G. Vermes *et al.* [ed.]), *The History of the Jewish People in the Age of Jesus Christ (175 B.C.- A.D. 135)* (Edinburgh: Clark, 1986) vol. 3.1, 342-79.

[113] Goodenough 1953-68, vol. 2, 28 and vol. 3, no. 791. Another another such *ascia* that may have been found in the Jewish hypogeum in the Vigna Cimarra, see Goodenough 1953-68, vol. 2, 29 and 33. F. de Visscher, "Ascia," *Jahrbuch für Antike und Christentum* 6 (1963) 187-92 and *id., Le droit des tombeaux romains* (Milan: Giuffrè, 1963) 277-94 (essentially the same, but in French).

ucts manufactured for the Jews of Rome with those originating in Roman Palestine.[114]

Despite such differences in formal appearance, artistic products from Rome and Roman Palestine had much in common in terms of techniques of production. Recent discoveries have clarified the extent to which, in Roman Palestine too, it was customary for one and the same workshop to cater to a variety of customers. We now know that the cancel screens found in churches and in synagogues originated, at least in many cases, in the same workshops. This is shown to be likely by the fact that Jewish and Christian chancel screens share in the overall layout of their decoration.[115] We also know that common workshop-identity is probable in the case of lead sarcophagi found in Beth She`arim and at various other sites throughout Israel. On these sarcophagi, menorot and crosses constitute no more than simple additions (the "finishing touch"), for they were stamped into the sarcophagus only after the general layout had been determined. Thus it was easy to satisfy customers with the most divergent artistic preferences.[116]

Glass workshops could likewise accommodate all customers. The so-called pilgrim vessels (mold-blown vessels of hexagonal or octagonal shape) that were sometimes decorated with Jewish and sometimes with Christian imagery are identical in physical appearance to such an extent that glass specialists have concluded that both Jewish and Christian pieces were manufactured in the same workshops.[117] With regard to mosaic floors, scholars have long noted the strong

---

[114] Avigad 1976; I. Skupinska-Løvset, *Funerary Portraiture of Roman Palestine. An Analysis of the Production in its Culture-Historical Context* (Gotheborg: Åström,1983) esp. 270f. (see also *Gnomon* 56 [1984] 754f.); G. Koch, "Der Import kaiserzeitlicher Sarkophage in den römischen Provinzen Syria, Palästina und Arabia," *Bonner Jahrbücher* 189 (1989) 163-211. And see M. L. Fischer, "Figured Capitals in Roman Palestine. Marble Imports and Local Stones: Some Aspects of 'Imperial' and 'Provincial' Art," *Archäologischer Anzeiger* 1991, 141-44.

[115] See G. Foerster, "Decorated Marble Chancel Sreens in Sixth Century Synagogues in Palestine and Their Relation to Christian Art and Architecture," in *Actes du XIe congrès international d'archéologie chrétienne* (21-28.IX. 1986) (Rome: École Française de Rome and Pontificio Istituto di Archeologia Cristiana, 1989) 1809-20; E. Russo, "La scultura del VI secolo in Palestina. Considerazioni e proposte," *Acta ad Archaeologiam et Artium Historiam Pertinentia* 6 (1987) 123.

[116] Avigad 1976, 173, 178; L. Y. Rahmani, "On Some Recently Discovered Lead Coffins from Israel," *Israel Exploration Journal* 36 (1986) 243; *id.*, "More Lead Coffins from Israel," *ibid.* 37 (1987) 143 and *id.*, "A Christian Lead Coffin from Caesarea," *ibid.* 38 (1988) 247-48; *id.*, "Five Lead Coffins from Israel," *ibid.* 42 (1992) 81-102. On the technical aspects, see now H. Fronig, "Zu Syrischen Bleisarkophagen der Tyrus-Gruppe," *Archäologischer Anzeiger* 1990, 523-35. And see R. Arav, "A Mausoleum Near Kibbutz Mesillot," *Atiqot* 10 (1990) 81-89 (Hebrew).

[117] D. Barag, "Glass Pilgrim Vessels from Jerusalem, Part I," *Journal of Glass Studies* 12 (1970) 62. I owe this reference to E. Marianne Stern.

stylistic and representational similarities that exist between the floors of a synagogue at Beth Alfa and a Samaritan synagogue at Beth Shean.[118] More recently, comparable similarities have been observed and discussed in respect to the mosaic floor of the synagogue at Horvath Sûsiya and the pavement in the chapel of the priest John at Kh. el-Mekhayyat (Nebo). In the case of mosaic floors, identical designs need not necessarily point to the same workshop. They may also have resulted from the use of the same pattern book by different groups of craftsmen. Yet, the execution, well into the sixth century, of very similar arrangements on the floors of Palestinian synagogues and churches in the same general area indicates that even in Roman Palestine itself, Jewish communities were not averse to using artistic techniques and decorative patterns that were current throughout Late Ancient society as a whole.[119] Especially where it concerns the minor arts, the Jews of Roman Palestine, like their co-religionists in Rome, bought artifacts that were generally available and that were also used by non-Jews. Thus we can explain the presence of an a lamp with Chi-Rho found in the Jewish necropolis at Beth She'arim, the discovery of pagan and even erotic lamps in Sepphoris, or the deposition of Samaritan amulets in Jewish graves.[120]

In Roman Palestine, as in Rome, the material culture of the Jews thus reflects quite reliably what artifacts were available to the population of the later Roman Empire at large. It is not correct to see in this phenomenon aspects of assimilation or to view it as a manifestation of syncretism.[121] The person in Rome who ordered the season sarcophagus with a menorah, or the Jews who commissioned the mosaic floors of Roman Palestine, did not aspire to assimilate. Nor were they syncretists. Rather, they opted for art that ingeniously expressed their

---

[118] R. Hachlili, *Ancient Jewish Art and Archaeology in the Land of Israel* (Leiden: Brill, 1988) 390.

[119] G. Foerster, "Allegoric and Symbolic Motifs With Christian Significance from Mosaic Pavements of Six-Century Palestinian Synagogues," in G. C. Bottini *et al.* (eds.) *Christian Archaeology in the Holy Land. New Discoveries* (Jerusalem: Franciscan Printing Press, 1990) 545-52. And see R. Hachlili, "On the Mosaicists of the 'School of Gaza'," *Eretz Israel* 19 (1987) 46-58 (Hebrew).

[120] Avigad 1976, 74 and 188-90; E. C. Lapp, *A Chronological-Typological Study of the Ancient Oil-Lamps of Sepphoris in the Lower Galilee* (M.A. thesis: Duke University, 1991) esp. nos. 121-29; E. M. Meyers, "The Challenge of Hellenism for Early Judaism and Christianity," *Biblical Archaeologist* 55 (1992) 84-91; see also E. M. Stern, "Ancient and Medieval Glass from the Necropolis Church at Anemurium," *Annales du 9e congrès international d'étude historique de verre* (Liège: A.I.H.V., 1985) 47. R. Reich, "A Samaritan Amulet from Nahariya," *Revue Biblique* 92 (1985) 383-88.

[121] M. Smith, "The Image of God. Notes on the Hellenization of Judaism, with Especial Reference to Goodenough's Work on Jewish Symbols," *Bulletin of the John Rylands Library* 40 (1958) 496; U. Schubert, "Assimilationstendenzen in der jüdischen Bildkunst vom 3. bis zum 18. Jahrhundert," *Kairos* 30-31 (1988-89) 165-67.

Jewish identity while at the same time representing what was artistically fashionable in late Roman society in general. The correct term to describe this phenomenon is not assimilation but common workshop-identity.

Evaluation of the Phenomenon of Workshop-Identity

In the previous pages we have seen that no significant formal differences exist between the Jewish and Christian catacombs in layout, excavation techniques, grave forms, decoration, or artifacts used. It is reasonable to conclude, therefore, that Jewish funerary art and archaeology in Rome developed as an integral part of Late Ancient artistic production in general. Such a conclusion has at least two important implications that bear on the scholarly discussions sketched in Chapter 1: (1) the Jewish catacombs as a whole do not significantly predate the Christian ones, nor vice versa and (2) the Jews were not as isolated as previous scholarship has maintained. Let us inspect these points in turn.

In this chapter it has been argued that the Jewish catacombs of Rome came into being only gradually, and that this happened as part of developments that were not limited to the Jewish sphere alone. We saw how the Vigna Randanini catacomb developed into a true catacomb only after several independent hypogea were connected to one another. It has been pointed out that this development mirrors developments that determined the building history of some Christian catacombs that are located in the same area and that date to the same general period. We also saw that other Jewish underground tombs such as the hypogeum on the Via Labicana or the hypogeum in the Vigna Cimarra, never developed into catacombs—just as certain Christian and pagan underground cemeteries also were destined to remain small subterranean burial plots. Finally we saw that there were catacombs that were designed as catacombs from the outset. Again, they include both Jewish and Christian catacombs. The fact that the artifacts found in the Jewish catacombs were made in the same workshops that produced for Christian customers must be taken to mean that the Jewish catacombs cannot predate the Christian ones by two centuries (or vice versa!). Therefore, only one scenario is conceivable: making full use of all available technologies and taking advantage of topographic factors where appropriate, Jews and Christians *simultaneously* started to excavate underground tombs. In constructing catacombs Jews did not copy Christian practices nor did Christians copy Jews. The transformation of these hypogea into large underground burial complexes that could house hundreds, and even thousands, of dead, was a gradual process, which took place *simultaneously* at both Jewish and early Christian sites.

The phenomenon of common workshop-identity offers us a glimpse of an aspect of Jewish daily life in Rome that is as intriguing as it is destined to remain obscure in its details: in Late Antiquity Roman Jews chose artifacts that enjoyed popularity among the city's non-Jewish population; Jews were aware that these artifacts were being produced; Jews had access to and relied on workshops that produced these artifacts; conversely, workshops were willing to satisfy the wishes of Jewish customers. In terms of their material culture, then, Roman Jews can hardly be called "isolated."

Yet, a lack of isolation in these matters cannot be taken to mean that in the process of employing non-Jewish artisans, Roman Jews assimilated or gave up their own identity. The archaeological materials we have studied so far indicate that Jews related to non-Jews in ways that cannot appropriately be described in terms such as isolation, non-isolation, or assimilation. Rather, interaction between Jews and non-Jews was a process where appropriation on one level was paired with affirmation of Jewish identity on another. We have seen that *kokhim* occur in the Vigna Randanini catacomb, next to tomb types that are generally Late Ancient. Such *kokhim* suggest that traditional Jewish funerary practices and beliefs did not disappear entirely.

That appropriation of non-Jewish products did not lead to giving up one's Jewish identity becomes particularly evident when we turn to the iconography of the archaeological remains from the Jewish catacombs. It has already been observed several times that use of artifacts produced in late Roman workshops went hand in hand with a preference for iconographical themes that are often, although not always, markedly Jewish. Jewish iconographic themes occur not only in almost all the categories of archaeological evidence discussed in the previous sections of this chapter; but they almost always appear in the most prominent position.

A few examples will suffice to illustrate the centrality of Jewish themes. This is strikingly evident, for example, on the Jewish season sarcophagus, mentioned earlier, on which the *clipeus* or central medallion is decorated with a menorah. The iconographic importance of such an arrangement becomes particularly clear when it is seen against the larger background of Roman funerary art in general, and the importance attached to the *clipeus* in this art. From the second century C.E. onwards these central medallions normally framed the bust of the deceased or individual couple. Such arrangements are believed to reflect a general trend in Roman funerary art towards the private apotheosis of the dead.[122] The Jewish season sarcophagus was also decorated with a *clipeus*, but this time it did not carry any portraits. Instead, an elaborately carved menorah now served as the focal point in the decoration of the piece.

Similarly, on some wall paintings in the Jewish catacombs, Jewish motifs have been represented in central and thus in the visually most noticeable positions. In the upper Villa Torlonia catacomb Jewish iconographic motifs have been depicted in the central positions of a decorative scheme that is otherwise generally Late Ancient. The Torah shrines with the lighted menorot on the back wall of several arcosolia, together with the menorot on the ceiling of Cubiculum II and on the arches of mentioned arcosolia, are prominent features in the overall layout of the pictorial decoration. In Painted Room IV in the Vigna Randanini catacomb, the pictorial decoration has been reduced to a few crude red lines only, yet, significantly, what is represented is the menorah and *etroghim*.

Important, finally, is the evidence provided by the Jewish epitaphs. They constitute a group of documents that reflects the feelings of the average Roman Jew perhaps more reliably than any other category of archaeological remains discussed so far. Significantly, on funerary inscriptions from Rome, Jewish themes outrank non-Jewish ones. Of all Jewish "symbols," the menorah indisputably outranks all others. In antiquity, the seven branched candelabrum was the Jewish symbol *par excellence*.[123]

To understand better the possible significance of these Jewish themes, it is necessary also to take into account their chronology. Although it is true that Jewish and/or pagan and or neutral iconographic motifs do coexist, chronological differences in their popularity become apparent. The Jewish materials with decoration that is iconographically neutral or pagan date largely, although not exclusively, to the third century C.E. They include Painted Room III in the Vigna Randanini catacomb, as well as a sizable number of sarcophagi used by Jews for burial. The artifacts displaying predominantly Jewish iconographical themes, by contrast, invariably date to the fourth cen-

---

[122] F. Matz, "Stufen der Sepulkralsymbolik der Kaiserzeit," *Archäologischer Anzeiger* 1971, 102-16; J. Engemann, *Untersuchungen zur Sepulkralsymbolik der späteren römischen Kaiserzeit* (Münster: Aschendorff, 1973) 35-39; H. Wrede, *Consecratio in formam deorum. Vergöttlichte Privatpersonen in der römischen Kaiserzeit* (Mainz: von Zabern, 1981).

[123] See *CIJ* s.v. "Emblèmes et représentations." On the menorah, see Rutgers 1992, 110. *Contra* M. Simon, "Un document du syncrétisme religieux dans l'Afrique romaine," *Comptes Rendus de l'Académie des Inscriptions et Belles Lettres* 1978, 500-24 and M. Dulaey, "Le chandelier à sept branches dans le christianisme ancien," *Revue des Études Augustiennnes* 29 (1983) 3-26.

tury C.E. The Jewish season sarcophagus, for example, appears to have been produced in the first decade of the fourth century C.E. The wall paintings in the Villa Torlonia catacomb date approximately to the period 350-370 C.E. The Jewish gold glasses were also manufactured in the fourth century C.E. [124]

When the iconography of the Jewish catacombs is seen against the background of the larger iconographic trends of the fourth century C.E., it is hard to believe that the centrality of Jewish motifs on artifacts dating to the fourth century was purely accidental. While neutral bucolic and Old Testament scenes outnumber other themes in early Christian art in Rome during the third century C.E., the fourth century C.E. witnessed a profound change in this respect. From early on in the fourth century, richly sculpted frieze sarcophagi were produced for Christians. Such sarcophagi contained a more coherent representation of Christian themes than had hitherto been the case. While Old Testament scenes appear to have been kept primarily with an eye to the typological implications inherent (in Christian thinking) in so many events recorded in the Old Testament, scenes taken from the New Testament now quickly gained in popularity. Representations of the figure of Jesus in prominent positions were especially favored.[125] The Christianization of society in general and of the arts in particular can hardly have gone unnoticed among Jews living in Rome. It is perhaps not entirely correct to conceive of the relationship between Jewish and Christian art in fourth century Rome in terms of "cause and effect" in the manner that Dunbabin has described the relationship between pagan and Christian art in Late Antiquity.[126] Yet, the occurrence of motifs that are decidedly Jewish in the most prominent spots may very well have strengthened Jewish identity during a period of increasingly rapid social, religious, and legal change. It is remarkable that not only in Rome, but also in Venosa, Noto Antica, Rabat, Doclea, and the Jewish hypogeum in Tripoli, or for that matter, in Roman Palestine itself, the same Jewish themes, and the menorah in particular, appear over and over again. It is no less remarkable that at all the sites mentioned, events recorded in the Hebrew Bible have left no traces in the funerary art of the Jews of Late Antiquity.

---

[124] Rutgers 1990, *passim*.

[125] F. W. Deichmann *et al.*, *Repertorium der christlich-antiken Sarkophage* (Wiesbaden: Steiner, 1967); Deichmann 1983, 141f.; Engemann 1986, 94f.; E. Struthers Malbon, *The Iconography of the Sarcophagus of Junius Bassus* (Princeton: Princeton U. P., 1990) esp. 42-90.

[126] K. M. D. Dunbabin, *The Mosaics of Roman North Africa. Studies in Iconography and Patronage* (Oxford: Clarendon, 1978) 144.

Conclusion

This survey of Jewish archaeological remains from Rome has clarified how the study of Late Ancient artistic production enables us to determine how, in Late Antiquity, Roman Jews related to non-Jews. Archaeological evidence documents that in Late Ancient Rome, Jewish material culture did not evolve in splendid isolation. Such finds also suggest, however, that the use of what was locally available went hand in hand with a marked preference for representations that were specifically Jewish in their iconography. That Jews continued to affirm their identity is perhaps most clear from the fact that they continued to bury in catacombs in which only Jews were laid to rest. Some time ago I suggested that interaction between Jews and non-Jews in antiquity was particularly intense because they buried in the same general areas as non-Jews.[127] Upon closer investigation, such a line of reasoning is misleading, at least insofar as Rome is concerned. In Rome, Jews used for burial the same general areas as did non-Jews. Yet such a fact does not tell us much about interaction. Inasmuch as the burial grounds in question were owned either by Jews, or Christians, or pagans, and inasmuch these funerary complexes served the inhumation of one's own co-religionists only, they cannot be considered as communal burial grounds. In death, both Jews and Christians wished to preserve their respective communal identities.

The conclusion that the exclusively Jewish catacombs came into use in the late second century C.E. at the earliest raises two questions that are difficult to answer: (1) where and how were Jews buried before the exclusively Jewish catacombs came into use? And (2) why did the Jewish catacombs come into use in the late second century C.E. rather than earlier or later?

A Jewish community is known to have existed in Rome from the first century B.C.E. onward, and some Jews may have settled there even earlier. That we do not know how and where this community disposed of their dead is due to three factors. First, available literary sources never include this kind of information. Second, there are no funerary inscriptions dating to this period that can be identified as Jewish. Last, but not least, a full-fledged Jewish iconography that could help facilitate the identification of Jewish graves dating to the first centuries B.C.E. and C.E. had not yet come into being.

In an attempt to identify these early burials two solutions have been put forward, but neither of them is very satisfactory. The first solution was proposed by Nock many years ago. He argued, on the basis of the

---
[127] Rutgers 1992, 109-17.

occurrence of Oriental/Semitic names on inscriptions in Roman *columbaria*, that immigrants originating from the eastern part of the Mediterranean (including perhaps Jews) quickly conformed to the custom of cremation—despite the prevalence of the practice of inhumation in their respective home-countries. Along similar lines, several years later, Frey maintained that several inscriptions bearing Semitic names pointed to "paganizing Jews" who did not object to having their remains cremated.[128]

Inasmuch as the names in question are only generally Semitic rather than identifiably Jewish, the Nock-Frey solution is based on two presuppositions that cannot be proved, namely that upon immigrating Jews abandoned both their Jewish names and their traditional funerary practice (namely, inhumation). It is possible, at least theoretically, that this happened. As we will see in Chapter 4, in the third and fourth centuries C.E. the nomenclature of Roman Jews largely followed that of their non-Jewish contemporaries, and typically Jewish names were not very popular. One could speculate that the name-giving practices of the Jews who ended up in Rome in the first century B.C.E. and C.E. were similar in that they too abandoned or had to abandon their Jewish names.

The second presumption presents a more serious problem. In view of the fact that there were strong theological reasons for Jews not to cremate their dead, and considering that we lack archaeological evidence suggesting that in antiquity Jews ever did so, there is little reason to suppose that the Jewish community in first century B.C.E. Rome would have changed their traditional funerary practices suddenly and *en masse*. We know that non-Jews originating in the eastern Mediterranean did not give up their traditional burial customs upon immigration. Some scholars have even attributed the origin of the custom, in Rome, of burial in sarcophagi to such immigrants, but others have disputed this claim.[129] Why then would Jews, who always refrained from cremation, have suddenly abandoned a practice that was firmly rooted in their history and traditions? Even if one accepts the hypothesis that some Jews were laid to rest in *columbaria*, it seem rather unlikely that *columbaria* were used for burial by the Jewish community of Rome at large.

---

[128] A. D. Nock, "Cremation and Burial in the Roman Empire," *Harvard Theological Review* 25 (1932) 329; *CIJ* CXXI and 571-74.

[129] H. Gabelmann, *Die Werkstattgruppen der Oberitalischen Sarkophage* (Bonn: Rheinland-Verlag, 1973) 8-9, who develops a thesis first proposed by A. Byvanck; H. Brandenburg, "Der Beginn der stadtrömischen Sarkophagproduktion der Kaiserzeit," *Jahrbuch des Deutschen Archäologischen Instituts* 93 (1978) 278f. Note also that the earliest locally-manufactured sarcophagi in the eastern Mediterranean imitate examples from Asia Minor.

The second solution to the problem of where Roman Jews buried their co-religionists is the suggestion that Roman Jews always buried in catacombs.[130] This argument is based on the assertion that the Jewish catacombs came into use as early as the first century C.E. Proponents of this view all fail to study archaeological materials from the Jewish catacombs systematically and comprehensively. Instead, they base their argumentation on the evidence provided by brickstamps discovered in the Jewish catacombs. As I have shown elsewhere, brickstamps are useless for dating catacombs. In fact, the brickstamps from the Jewish catacombs, and from the Villa Torlonia catacomb in particular, can serve as a prime example to show that in general brickstamps were not produced specifically for catacombs, but that they were merely reused there. Consequently, brickstamps, when found in catacombs, do not provide a *terminus post quem* but only a *terminus ante quem non*.[131] Furthermore, even if brickstamps could serve as reliable chronological indicator, they would still not solve our problem: the custom of stamping bricks did not begin until the reign of Augustus. Thus, even if one were to accept a first century dating for the Jewish catacombs we still would not know where Jews buried during the first hundred or hundred fifty years of their residing in Rome.

Just as it impossible to determine where in Rome Jews disposed of their dead before the late second century C.E., it is not exactly clear what changes or developments triggered the construction of Jewish catacombs in Late Antiquity. It is conceivable that a combination of factors contributed to this process, including the need to provide everyone with proper burial, the need to find space in Rome's overcrowded suburban cemeteries, the need to be buried in the same cemetery as one's co-religionists, and the technical know-how to build these underground cemeteries. Yet none of these factors necessarily explains why the construction of Jewish cemeteries would have occurred in the late second, and especially in the third and fourth centuries C.E. We know that in the case of the Roman Christian community, increase in membership and the wish to ensure the proper burial of the gradually increasing number of converts, and especially the poorer ones, contributed to the construction of Christian catacombs.[132] We also know that in Late Antiquity pagan groups never constructed catacombs, perhaps because they did not feel the same re-

---

[130] This has often been suggested; see now M. H. Williams, "The Organization of Jewish Burials in Ancient Rome in the Light of Evidence from Palestine and the Diaspora," *Zeitschrift für Papyrologie und Epigraphik* 101 (1994) 165-82.

[131] Rutgers 1990, 153-54.

[132] See Brandenburg 1984, 46-49. And see, more generally, P. Brown, *Power and Persuasion in Late Antiquity* (Wisconsin: University of Wisconsin Press, 1992) 77-78.

sponsibility vis-à-vis the dead as did their Christian and Jewish contemporaries. But what considerations led the Jews to construct their catacombs in the same general period as did the Christians? One of the reasons that it is difficult to determine why Jews started to use catacombs is that we cannot answer the other issue just raised, namely the question of where the Jews of ancient Rome buried their dead before the catacombs came into use. Did the Jewish community increase to such an extent that a new mode of burial became necessary? Do changes in grave architecture reflect changes in Jewish self-awareness, sense of identity, their views on death and burial, their role in society at large, or all of these? If so, how?

It is most unfortunate that the archaeological remains in the Jewish catacombs themselves do not provide us with evidence to answer such questions. As is so often the case, archaeology only enables us to see the how, not the why.

CHAPTER THREE

REFERENCES TO AGE AT DEATH IN THE JEWISH FUNERARY INSCRIPTIONS FROM ROME: PROBLEMS AND PERSPECTIVES

Introduction

In the next three chapters, we will turn our attention to the Jewish funerary inscriptions from Rome. One question in particular shall deserve our attention: are the epigraphic formulae in these epitaphs typically and exclusively Jewish, or do the Jewish funerary inscriptions reflect commemorative patterns that are generally prevalent in Late Antiquity? The question of what significance should be attached to those differences and similarities will be discussed briefly at the end of each chapter and then comprehensively in Chapter 7.

This fairly technical chapter concerns the references to age at death that appear in the Jewish epitaphs from Rome. Of the approximately 594 Jewish inscriptions from Rome that are known today, 145, or somewhat less than one third, mention the age someone was believed to have reached when he or she passed away.[1] References of this kind occur in inscriptions commemorating adult males and females, but they are also included in inscriptions referring to children.

Inscriptional references to age at death in Jewish funerary inscriptions have been studied by several scholars.[2] None of the treatments of this question is entirely satisfactory, however, primarily because the conclusions reached and methodology followed in recent studies on references to age at death in non-Jewish inscriptions are not sufficiently taken into account.[3]

---

[1] This figure is based on the evidence published in *CIJ*, *passim*, Leon 1960, esp. 264f., Fasola 1976, *passim* and Solin 1983, esp. 655f.

[2] Leon 1960, 229-30; B. Blumenkranz, "Quelques notations démographiques sur les Juifs de Rome dans les premièrs siècles," *Studia Patristica* 4 (1961) 341-47; P. W. van der Horst, *Ancient Jewish Epitaphs. An Introductory Survey of a Millennium of Jewish Funerary Epigraphy (300 B.C.E.-700 C.E.)* (Kampen: Kok Pharos,1991) 73-84.

[3] Important studies include K. Hopkins, "On the Probable Age Structure of the Roman Population," *Population Studies* 20 (1966-67) 245–64; J. D. Durand, "Mortality Estimates from Roman Tombstone Inscriptions," *American Journal of Sociology* 65 (1959/60) 365–73; I. Kajanto, *On the Problem of the Average Duration of Life in the Roman Empire* (Helsinki: Suomalainen Tiedeakatemia, 1968); M. Clauss, "Probleme der Lebensalterstatistiken aufgrund römischer Grabinschriften," *Chiron* 3 (1973) 395–417; R. Duncan-Jones, *Structure and Scale in the Roman Economy* (Cambridge: Cambridge U.P., 1990) 102. For further references, see *infra*.

Following the method employed in the previous chapter, references to age at death in the Jewish epitaphs from Rome will not be studied in isolation. Along with Jewish inscriptional evidence from Rome, a discussion of Jewish epitaphs from Teucheira (Tocra in Cyrenaica, North Africa), from Tell el-Yehoudieh (Leontopolis, Lower Egypt), and from a variety of non-Jewish sites have been included. Evidence from individual sites will always be evaluated separately and on its own terms, after which it will be compared to materials from other sites. Because epigraphic customs are not the same at all sites, it is not helpful to compound into one undifferentiated category Jewish inscriptions that originate from the entire Mediterranean.[4]

References to Age at Death: a Greco-Roman Custom

Before we can turn to study which patterns underlie the references to age at death in Jewish funerary inscriptions, one introductory observation must be made. The inclusion of references to age at death was not originally a Jewish epigraphic custom; rather, Jews copied it after they had come into contact with the larger, non-Jewish world that surrounded them. Only in this way can we explain the fact that references to age at death are largely absent from Jewish inscriptions found in Roman Palestine itself. Not even a single such reference can be found in the four main corpora of Jewish inscriptions from the area, namely the ossuary inscriptions from Jerusalem, the Jewish epitaphs from Jaffa, the funerary inscriptions from Beth She'arim, and, finally, the Jewish inscriptions from Caesarea.[5]

Some inscriptions from Roman Palestine admittedly do refer to age at death, yet they are few and far between. In addition, they generally postdate by several centuries the period during which Jews had first

---

[4] *Contra* van der Horst 1991, 73, 79, 81.

[5] For Jerusalem, *CIJ*, vol. 2, nos. 1210-1397; other examples in J. Milik and B. Bagatti, *Gli scavi del "Dominus Flevit" (Monte Oliveto, Gerusalemme)* (Jerusalem: Franciscan Printing Press, 1958); E. Puech, "Inscriptions funéraires palestiniennes: Tombeau de Jason et ossuaires," *Revue Biblique* 90 (1983) 480-533; for Jaffa, *CIJ* 891-960; for Beth She'arim, see M. Schwabe and B. Lifshitz, *Beth She'arim*. Volume II. *The Greek Inscriptions* (New Brunswick: Rutgers U. P., 1974); for Caesarea, see see M. Schwabe, "A Jewish Sepulchral Inscription from Caesarea Palestine," *Israel Exploration Journal* 1 (1950) 49-53; *id.*, "Two Jewish Greek Inscriptions Recently Discovered at Caesarea," *Israel Exploration Journal* 3 (1953) 127-30 and 233-38; M. Schwabe in E. Stern (ed.), *Bulletin of the Israel Exploration Society*. Reader A (Jerusalem: Israel Exploration Society,1965) 152 (Hebrew); B. Lifshitz, "Inscriptions grecques de Césarée en Palestine," *Revue Biblique* 68 (1961) 115-26; *id.*, "La nécropole juive de Césarée," *Revue Biblique* 71 (1964) 384-87; *id.*, "Inscriptions de Césarée en Palestine," *Revue Biblique* 72 (1965) 98-107;*id.*, "Inscriptions de Césarée," *Revue Biblique* 74 (1967) 49-59; and add, *Zeitschrift für Papyrologie und Epigraphik* 7 (1971) 162. And see S. Klein, *Sefer ha–Yishuv* (Jerusalem: I. Ben–Zvi, 1977) 150-51.

come into contact with non-Jewish epigraphic practices. That references to age at death constitute the exception rather than the rule in Roman Palestine becomes particularly evident when one lists the inscriptions in question. They include a grave stele from Apollonia, commemorating someone who may not have been Jewish,[6] an inscription that may relate to a Samaritan rather than a Jewish community in the area of present-day Nablus,[7] five inscriptions found near the Sea of Galilee,[8] and an inscription from Zoar dated to the year 389-90 C.E.[9] To this tiny collection, a few other inscriptions from the Negev/Palaestina Tertia can still be added, but they are all Late Ancient, and mostly Christian.[10]

In addition to stressing their numerical insignificance, there is still another way to show that references to age at death do not belong to the traditional repertoire included in Jewish epitaphs, namely by comparing inscriptions composed in Hebrew or Aramaic with those written in Greek or Latin. In a trilingual Jewish epitaph from Tortosa, Spain, which dates, possibly, to the sixth century C.E., and which commemorates a certain Meliosa who died at 24 years of age, the Greek and Latin part of the inscription are identical.[11] The Hebrew portion of the inscription, by contrast, omits any such reference to age at death. Instead, phrases such as "that her memory may be a blessing" and "may her soul live in the world to come" have been substituted. That the Tortosa inscription is not at all exceptional becomes evident when we turn to other Jewish inscriptions in Hebrew. A bilingual Jewish epitaph from Taranto (Calabria, Italy) refers to age at death in its Latin part, but, again, the Hebrew part dispenses with such information.[12] Jewish epitaphs from Zoar near the Dead Sea likewise

---

[6] *CIJ* 2, 891 = Klein 1977, 7.

[7] *CIJ* 2, 1169 = Klein 1977, 35. The names of the three women in the inscription are Semitic, but not necessarily Jewish, *contra* Frey, who maintains that these names are Jewish.

[8] At Heptapegon, Klein 1977, 39; Tiberias: *CIJ* 2, 984-86 and Klein 1977, 63 no. 140.

[9] H. L. Vincent, "Une colonie juive oublié," *Revue Biblique* 36 (1927) 401-7, esp. 402-3. The inscription may also be found in *CIJ* 2, 1209.

[10] A. Negev, *The Greek Inscriptions from the Negev* (Jerusalem: Franciscan Printing Press, 1981) nos. 10, 16, and 17 (from Avdat) 53 (North Church at Sobota [Shivta]); s.a., *Byzantine Inscriptions from Beer-Sheva and the Negev, Negev Museum Beer-Sheva* 2 (1985) nos. 12 and 13 (provenance unclear). M. Schwabe in Stern (above n. 5) 178f. (Ashkelon); A. Alt, *Die griechischen Inschriften der Palaestina Tertia westlich der 'Araba* (Berlin and Leipzig, 1921) nos. 15, 21, 25, 39, 55, 131, 132, and 136; C. A. M. Glucker, *The City of Gaza in the Roman and Byzantine Periods* (Oxford: BAR- IS 325, 1987), 128 no. 14.

[11] *CIJ* 661. For a picture, see F. Cantera and J. M. Millás, *Las inscripciones hebraicas de España* (Madrid: C. Bermejo, 1956) 270-71. See now D. Noy, *Jewish Inscriptions of Western Europe*. Vol. 1. *Italy (excluding the City of Rome), Spain and Gaul* (Cambridge: Cambridge U.P., 1993) no. 183.

[12] *CIJ* 629 = Noy 1993, no. 120.

conform to this pattern. One of these inscriptions that dates to the years 389-90 C.E. was written in Greek and mentions age at death.[13] The other inscriptions were all written in Aramaic and omit this kind of information.[14]

A survey of inscriptions found in several countries on the eastern Mediterranean seaboard, shows, interestingly, that in these inscriptions too, references to age at death were far from common. Among the substantial collection of epitaphs that has been discovered in the Nabatean necropolis at Médain-Sâleh (Egra), not a single one mentions age at death.[15] In Palmyra, four out of 515 inscriptions in Palmyrean include this information,[16] while at Hatra, further to the East, only one out of 341 Aramaic inscriptions refers to age at death.[17] On the Phoenician coast the situation is practically identical; there too, references to age at death were not customarily included in inscriptions.[18] Thus, neither in Palestine nor elsewhere in the ancient Near East, were references to age at death an indigenous epigraphic custom.

Inscriptions from Roman Syria and Arabia are particularly interesting because they indicate who was probably responsible for introducing this custom to the area: the Roman military. In Syria, four inscriptions, written in Latin and including references to age at death, all relate to soldiers who served in the Parthian legion "Severiana."[19] Another inscription in Latin which has been preserved on a mutilated bust from Palmyra, likewise commemorates a soldier, this time an *eques* of the *ala I Ulpia singularium*.[20] Although it is true that not all epitaphs from Syria which mention age at death can be associated convincingly with the Roman military,[21] a certain degree of Romanization can nevertheless be established in a fair number of cases.[22] Similarly, inscriptions from Bostra (Hawran) that detail age at death

---

[13] *CIJ* 1209

[14] J. Naveh, "Another Jewish Aramaic Tombstone from Zoar," *Hebrew Union College Annual* 56 (1985) 103–16 and *id.*, "The Fifth Jewish Aramaic Tombstone from Zoar," *Liber Annuus* 37 (1987) 369–71.

[15] A. Jaussen and R. Savignac, *Mission archéologique en Arabie (mars-mai 1907). De Jérusalem au Hedjaz Médain Saleh*. Volume 1 (Paris: Leroux, 1909) 139f.

[16] *CIS* V.2.3, 4358, 4359, 4562, 4616.

[17] F. Vattioni, *Le iscrizioni di Hatra* (Napoli, 1981) no. 30. Hatra was destroyed by Sassanian forces in the 240s C.E.

[18] *CIS* I.1, *passim*.

[19] W. K. Prentice, *Syria. Publications of the Princeton University Archaeological Expeditions to Syria in 1904–1905 and 1909*. Division III. *Greek and Latin Inscriptions* (Leiden: Brill, 1922) nos. 128, 130, 131, and 134.

[20] H. Seyrig, "Antiquités syriennes," *Syria* 14 (1933) 152-68, esp. 161-62.

[21] E.g. Prentice (above n. 19) nos. 97 and 348.

[22] E.g. Prentice (above n. 19) nos. 112 and 356.

can often be linked with Roman military presence in the region.[23] Finally, in inscriptions from Jerash and Amman (Jordan), references to age at death and Latin onomastics likewise go hand in hand. Besides mentioning age at death, the number of years someone served in the Roman army is a regular feature of the Jerash inscriptions.[24]

Only after Roman epigraphic practices had been known in Roman Arabia for a comparatively long time did references to age at death finally start to occur in more substantial numbers and in contexts other than military. At Umm el-Djimal (northern Jordan), for example, as many as 200 inscriptions carrying mostly indigenous names mention age at death, but none of these inscriptions seems to predate the third century C.E.[25] Similarly, in and around Kerak (in modern day Jordan), age at death is referred to almost without exception. The inscriptions in question date from 375 C.E. to 737 or 785 C.E., but the majority appear to have been carved in the sixth century C.E.[26] To these examples, other inscriptions from Jordan can readily be added.[27] Again, most if not all of these materials date to the Late Ancient period.

In the western part of the Roman empire, on the other hand, the custom of including references to age at death was more widespread. In fact, it is likely that the custom originated in this part of the ancient world, and that this happened at a relatively early stage, namely in Republican times.[28] Its popularity seems to have increased especially in the early Empire, that is, from the time of Tiberius onwards.[29] This increase then continued into the later Roman Empire. By the fourth and fifth centuries C.E., age at death is referred to twenty times as of-

---

[23] *IGLS* 13.1, 9172-73, 9175, 9178, 9180, 9187, 9188, 9191, 9193-95, 9197-98, 9203, 9207, 9210-13, 9217, and 9220.

[24] C. B. Welles in C. H. Kraeling, *Gerasa. City of the Decapolis* (New Haven: American Schools of Oriental Research, 1938), nos. 201, 205, 214, 215, 219, 221, 234, and 235; and *IGLS* 21, 34. And add S. Mittmann, *Beiträge zur Siedlungs- und Territorialgeschichte des nördlichen Ostjordanlandes* (Wiesbaden: Harrassowitz, 1970) 204 no. 48 (an inscription discovered south of Kerak).

[25] E. Patlagean, *Pauvreté économique et pauvreté sociale à Byzance, 4e-7e siècles* (Paris: Mouton, 1977) 96.

[26] R. Canova, *Iscrizioni e monumenti protocristiani del paese di Moab.* (Vatican City: Pontificio Istituto di Archeologia Cristiana, 1954) XLI: 403 out of 415 inscriptions include such information. To this collection can now be added the inscription published by M. Piccirillo, "Un' iscrizione imperiale e alcune stele funerarie di Kerak," *Liber Annuus* 39 (1989) 105-18, esp. 117.

[27] Mittmann 1970 (above n. 24), 167f. nos. 1, 2, 15, 25, 26, 27, 30, 35-42, 45-48; Piccirillo (previous note)114 (from Madaba, dating to 633 C.E.); A. Ovadia, "Greek Inscriptions from the Northern Bashan," *Liber Annuus* 26 (1976) 170-212, esp. 183-85, 187-88, 192, 200-1, and 204 (from northern Bashan) and *IGLS* 21,180-182 (three Christian inscriptions from Dibon). See also *IGLS* 21, 176 (Um-el-Walid).

[28] I. Kajanto, *A Study of the Greek Epitaphs of Rome.* Acta Instituti Romani Finlandiae II:3 (Helsinki: Tilgman, 1963) 12.

[29] Kajanto 1968, 11.

ten as in inscriptions dating to the first two centuries of the common era.[30] According to K. Ery, the custom was especially widespread in inscriptions in Latin. Yet, in inscriptions in Greek it appears to have been fairly frequent too.[31] The fact that such Greek inscriptions not only refer to the number of years, but also include the total amount of months and days someone lived is probably due to Latin influence.[32]

Upon emigrating from the Roman Near East, many groups adopted the custom of referring to age at death in funerary inscriptions. This group included Jews, but it also included many others. A native from the city of Petra, whose bilingual tombstone in Nabataean and Latin was found in the Tiber (in Rome), for example, had a reference to his age at death included in his epitaph—a good example of someone accommodating to an epigraphic practice that differed from traditional Nabataean custom.[33] Similarly, natives from Palmyra, who were not familiar with this habit at home, quickly conformed to the Roman practice of mentioning age at death after they had entered the Roman army and died in places that were as far apart as Rome, Dacia, and Numidia.[34] The influence of Roman commemorative patterns is most evident, however, in epigraphic evidence from North Africa and from Etruria. In Punic inscriptions from Carthage dating to the fourth to second centuries B.C.E.—that is, to a period when Roman influence in Punic epigraphic customs was minimal—not a single reference to age at death is to be found. By contrast, in neo-Punic inscriptions dating to the period after the area had come under Roman control, references to age at death suddenly became fairly common.[35] Similarly, references to age at death are lacking in Etruscan epigraphical materials predating Rome's conquest of Etruria. They appear only after the area had come become part of the Roman Empire, as is evident, for

---

[30] K. Hopkins in F. Hinard (ed.), *La mort, les morts et l'au–delà dans le monde romain. Actes du colloque de Caen 20–22 nov. 1985* (Caen, 1987) 119. See also Kajanto 1968, 12.

[31] K. K. Ery, "Investigations on the Demographic Source Value of Tombstones Originating from the Roman Period," *Alba Regia* 10 (1969) 51–67, esp. 60: 9980 inscriptions in Latin versus 822 inscriptions in Greek mention age at death.

[32] Kajanto 1963, 13.

[33] *CIS* II.1., no. 159.

[34] *CIS* V.2.3. nos. 3905, 3906, and 3908. See also *CIL* 8.515.

[35] H. Benichou-Safar, *Les tombes puniques de Carthage: topographie, structures, inscriptions et rites funéraires* (Paris: CNRS, 1982) 184 and 325-27 who also sees this custom as deriving from the Roman world. For the neo-Punic inscriptions, see H. Donner and W. Röllig, *Kanaanäische und aramäische Inschriften* (Wiesbaden: Harrassowitz, 1971) nos. 128, 133-36, 140, 142, 142-44, 147, 149, 151-52, 157-58, 165, 169, 171, and 180.

example, from the 113 Etruscan funerary inscriptions from Tarquinia and Volterra.[36]

There is no obvious explanation for the absence of references to age at death in inscriptions in the Semitic and Greek world, as opposed to their presence in the Roman world. The rationale behind including or excluding references to age at death has never been explained satisfactorily. Several explanations are plausible. For example, epitaphs commemorating members of the *ordo senatorius* mention age at death almost exclusively when it concerns younger people. Conceivably, such references were included to indicate that the deceased had failed to hold the more prestigious offices in the *cursus honorum* because he had died prematurely and not because of a lack of capability.[37] Concomitantly, among other social groups, age at death may have been added, when other, more interesting information was lacking.[38] With regard to a later period, Duncan-Jones has observed that in twentieth century India people were unwilling to specify their age at the taking of a census because of misgivings of a religious nature.[39] In an entirely different context, Ariès has pointed out that while the custom of erecting inscriptions became again fashionable in Europe in the twelfth century, it was only in the sixteenth century that the inclusion of references to age at death finally became standard practice. Ariès seeks the reason for this phenomenon in the development of a quantifiable conception of human existence that defines life not in terms of its effectiveness, but rather in terms of its length.[40]

It is conceivable that comparable ideas also played a role in Jewish circles. Wisdom of Solomon observes, for example, that "old age is not honored for length of time, nor measured by number of years; but understanding is gray hair for men and a blameless life is ripe old age."[41] Similar ideas are expressed by Philo on several occasions. In his view, old age should not be defined in terms of years but rather by the amount of knowledge gathered by the νοῦς—a concept that had previously been expressed by non-Jewish Hellenistic philosophers in

---

[36] The evidence is presented by J. Heurgon, *Daily Life of the Etruscans* (New York: MacMillan, 1964) 29-31. These materials date to the period 200-50 B.C.E.

[37] W. Eck, "Altersangaben in senatorischen Grabinschriften: Standeserwartungen und ihre Kompensation," *Zeitschrift für Papyrologie und Epigraphik* 43 (1981) 127–34, esp. 132-34. But note that in Late Antiquity, senators had their age at death recorded with painstaking precision, *ibid.*, 128 n. 7.

[38] R. Cagnat, *Cours d'épigraphie latine* (Paris: Fontemoings, 1913) 285.

[39] R. Duncan–Jones, "Age–Rounding, Illiteracy and Social Differentiation in the Roman Empire," *Chiron* 7 (1977) 333–53, esp. 334 n. 3.

[40] P. Ariès, *Geschichte des Todes* (Munich: DTV, 1982) 279-85.

[41] *Wisdom of Solomon* 4:8-9.

terms very similar to those subsequently employed by Philo.⁴² In rabbinic literature, age at death does not appear to have been major concern.⁴³

Whatever the background of the inclusion or exclusion of references to age at death may be, it is fair to conclude that the presence of references to age at death in Jewish epitaphs indicates outside influence. Of course, as a language, Hebrew was capable of detailing age at death as two Jewish inscriptions from Taranto and especially Jewish Medieval tombstones show.⁴⁴ Yet, epigraphic evidence from various parts of the Roman world indicates unmistakably that the custom in question is of Roman and not of Oriental origin. In addition to Rome, Jews in different parts of the Diaspora including Teucheira in North Africa and Tell el-Yehoudieh in Egypt readily adopted the custom—as we will see presently.⁴⁵

An Analysis of Jewish Inscriptional References to Age at Death: Introductory Remarks

The similarities between Late Ancient Jewish and non-Jewish inscriptions go further than the inclusion of references to age at death alone. In Jewish inscriptions from Rome that contain this kind of information, certain patterns recur that are practically identical to the patterns that have been observed for non-Jewish inscriptions found in and around that city. In order to study the similarity between these pat-

---

⁴² C. Gnilka, in *Reallexikon für Antike und Christentum* 12 (1983) 995-1094 s.v. "Greisenalter," esp. 1051-52 and 1072-73.

⁴³ See e.g. L. Feldman, "The Rabbinic Lament," *Jewish Quarterly Review* 63 (1972) 51-75. The very few references to tombstones (*tzlon*) in tannaitic literature concern the question of how to avoid defilement (e.g. *Tosefta, Ohalot* 1:5). Blumenkranz 1961, 342 and, more recently, J. and M. Dupâquier, *Histoire de la démographie: la statistique de la population des origines à 1914* (Paris: Perrin, 1985) 33 have both argued on the basis of 2 Samuel 24 that Jews disliked the taking of a census. It is impossible to determine, however, if such ideas effectively influenced epigraphic customs.

⁴⁴ *CIJ* 622 and 630 = Noy 1993, nos. 126 and 121. And see *CIJ* 595, a Jewish inscription from Venosa: it refers to age at death and was written in Hebrew script but the language employed is Greek. Clearly, in this inscription the Hebrew follows Greek epigraphic patterns. Medieval evidence: B. Wachstein, *Die Inschriften des alten Judenfriedhofes in Wien* (Vienna and Leipzig: Holzhauser, 1912) XXXV-XXXVI; Cantera and Millás 1956, nos. 1-6; I. S. Emmanuel, *Precious Stones of the Jews of Curaçao. Curaçaon Jewry 1656-1957)* (New York: Bloch, 1957) nos. 1171f. and p. 203; A. Scheiber, *Jewish Inscriptions in Hungary from the Third Century to 1686* (Budapest and Leiden: Akadémiai Kiadó Budapest and Brill, 1983); G. Nahon, *Inscriptions hebraïques et juives de France médiévale* (Paris: Corlet 1986).

⁴⁵ For Teucheira, see G. Lüderitz, *Corpus jüdischer Zeugnisse aus der Cyrenaika* (Wiesbaden: Dr. Ludwig Reichert, 1983) 63-145. For Tell el-Yehoudieh, see *CIJ* 1451-1530 and the addenda in *CPJ* 3, 162-63.

terns, it is necessary to compare from a variety of angles the Jewish evidence systematically with non-Jewish data.

The following survey is based on several considerations. First, I have distinguished between inscriptions that commemorate males as opposed to females. It is methodologically incorrect to lump inscriptions relating to males and females together, as some scholars have done. In the ancient world, female mortality patterns differed from those of males. In addition, by studying inscriptions relating to males and females separately, important differences in commemorative practice can more easily be detected.[46]

Furthermore, I have not made any distinctions between inscriptions written in Greek as opposed to those composed in Latin. Scholars have observed that the average age at death according to the Greek inscriptions can be as much as thirty years higher than in inscriptions in Latin (51 years in Greek as opposed to 21 years in Latin inscriptions). This difference is usually accounted for by differences in commemorative practice: Greek inscriptions often refer to age at death when people died at very high ages; Latin inscriptions do the opposite.[47] In Jewish inscriptions from Rome, however, the language used did not have such a distorting effect. According to my estimates, the average age at death that can be inferred from Jewish inscriptions is approximately the same in both languages (Greek: 26.5 years and Latin: 30.4 years). In this particular aspect, Jewish epitaphs from Rome seem to reflect a trend that manifested itself more generally in Late Antiquity: in non-Jewish inscriptions dating to the third and fourth centuries C.E., the differences in the recorded average age at death in Greek as opposed to Latin inscriptions can be shown to have decreased markedly.[48]

Along similar lines, in terms of including references to age at death, no differences can be ascertained between the various Jewish catacombs. 36% of all inscriptions from the Monteverde catacomb refer to age at death, while inscriptions from the Vigna Randanini catacomb include such information in 35%, and inscriptions from the Villa Torlonia catacombs in 31% of all inscriptions. Such percentages show that Leon was wrong when he suggested that Jews buried in one Jewish catacomb were more Romanized than those buried in other Jewish catacombs. The comparable presence of these epigraphic formulae in these inscriptions indicates that the users of the various Jewish cata-

---

[46] For further details, see below, section "Sociological Inferences: Women in Jewish Society."
[47] Claus 1973, 405 and, similarly, Ery 1967, 60. See also Sallust, *Jug.* 17.6.
[48] Ery 1967, 62.

combs behaved quite similarly when it came to borrowing epigraphic formulae from the larger world surrounding them.

Next, it should be noted that no agreement exists in the scholarship on the subject as to the minimum number of inscriptions necessary to constitute a sound database. M. Claus believes 75 inscriptions to be the absolute minimum, while he considers 300 to 400 inscriptions necessary to arrive at truly incontrovertible results.[49]

Finally, it should also be stressed at the outset of this study that Jewish epitaphs reflect the total number of deaths that occurred only very imperfectly. By dividing the total number of years during which such inscriptions were set up in Rome (approximately some 230 years) by the total number of Jewish funerary inscriptions known today (594), it follows that 2.5 inscriptions were erected per annum. Figures derived from the non-Jewish inscriptions from various parts of the Empire are comparably low.[50] Obviously, such figures are not indicative of the total number of deaths per year within the Jewish community of Rome or, for that matter, within any community.

Approach 1: The Traditional Approach

Since the late nineteenth century, scholars have championed a variety of approaches in the study of references to age at death in non-Jewish inscriptions. The question of the reliability of the inscriptional evidence for the reconstruction of ancient mortality and life expectancy rates has been a central concern in much of this research. In this section, several of these approaches will be discussed, with special emphasis on their usefulness for a better understanding of Jewish inscriptional data from Rome, Teucheira, and Tell el-Yehoudieh.

Approach 1 constitutes the least complicated way to establish the average age at death of an "inscriptional population." It has been used ever since researchers first focused their attention on the question of inscriptional references to age at death. According to this approach, the average age at death can be calculated by dividing the total number of years recorded in funerary inscriptions by the total number of funerary inscriptions that contain this kind of information.[51] Table 1 shows the outcome of this calculation for a number of sites.[52]

---

[49] Clauss 1973, 408-9.

[50] Ery 1967, 66 Table 2.

[51] The concept of "average age at death" can be somewhat confusing. Obviously, most people died before or after the rather abstract "average age at death." As Molleson points out in S. C. Humphreys and H. King (eds.), *Mortality and Immortality. The Anthropology and Archaeology of Death* (London: Academic Press, 1981) 22: "When the life expectancy in a static population is twenty years, for example, about half of the deaths oc-

Table 1: Average age at death as derived from epitaphs. The total number of inscriptions is indicated in parentheses.

Jewish	Males	Females
Rome	27.2 (82)	27.6 (63)
Teucheira	26.8 (65)	32.0 (39)
Tell el-Yehoudieh	25.8 (44)	27.1 (25)

Non-Jewish	Males	Females
Rome	23.9	23.0
Italy without Rome	29.3	25.9
Africa and Numidia	49.2	47.1
Mauretania	39.6	35.9
Spain	38.8	34.3
Gaul	30.6	26.9
Dalmatia	29.7	27.0
Moesia	42.3	31.3
Germany	29.0	29.5
Dacia	37.1	31.9
Pannonia	37.1	31.5
Noricum	36.1	35.9

Although this approach seems very straightforward, it is not a reliable method for establishing average age at death. Most problematically, it fails to take into account the distorting effects that result from specific commemorative practices. At first sight, it appears, for example, that Jews in Rome lived longer than the rest of the population. Such an impression is, however, not correct. Jewish inscriptions refer less often to children under the age of ten and more often to people over the age of seventy than do non-Jewish inscriptions from this city (for percentages, see below, Approach 2). Obviously, such a commemorative practice results in a higher average age at death for Jews as opposed to non-Jews. It would be wrong to infer that a life expectancy for the Jews in Rome was generally higher than for everybody else.

Comparably, Jewish females in Rome, Teucheira, and Tell el-Yehoudieh seem to have lived several months or even years longer than their male counterparts. Such evidence contrasts strongly with non-Jewish data, where the average age at death of women is *always* lower than that of men. Yet, again, when other determinants are taken into account, it can be argued convincingly that this difference must

---

cur before five, a quarter after fifty, and only 6.5% occur in the ten-year span centered on the mean age at death."

[52] The non-Jewish data derive from R. Duncan-Jones, "Age–Rounding, Illiteracy and Social Differentiation in the Roman Empire," *Chiron* 7 (1977) 333–53, esp. 349. For Rome, Africa and Numidia and Spain slightly lower figures are given by Hopkins 1966, 247. But the figures calculated by H. Nordberg, *Biometrical Notes*, Acta Instituti Romani Finlandiae II:2 (Helsinki: Tilgman, 1963) 39 are similar to those provided by Duncan-Jones.

be ascribed to sociological and epigraphic rather than to strictly medical factors (see below, section "Sociological Inferences").

Along similar lines, it can also be noted that figures derived from non-Jewish inscriptions, and from those discovered in North Africa in particular, are exceptionally high. In fact, in North Africa, the average age at death for both men and women (fifty years) comes surprisingly close to averages that were attained only at the turn of the twentieth century and then only in countries where proper medical care was available. Concomitantly, it contrasts strongly with recent data on life expectancy in preindustrial societies comparable to the Roman empire.[53] Long ago, R. Étienne sought to account for this difference by suggesting that the African countryside constituted a particularly healthy environment to live in, but this suggestion is not convincing.[54] Although differences in environmental stress existed in various parts of the Empire and are reflected in skeletal remains,[55] the ensuing discrepancies cannot have been so dramatic as to make likely the high average age at death reflected in inscriptions from North Africa. The high average age at death can be explained more convincingly by local epigraphic practice: in North Africa references to age at death were included in the epitaphs of those who died at a high age, while in Rome such references were especially favored in the case of infants.[56] The low average age at death in the Jewish inscriptions from Teucheira can be explained in much the same way: the high percentage of children recorded in the Jewish inscriptions from this site caused the average age at death to drop appreciably (see below, Approach 2).

In short, even a cursory analysis of the method proposed by Approach 1 shows that this approach does not lead to reliable results for establishing the average age at death of either Jews or non-Jews in the Roman world. Rather, it shows that epigraphic customs are a factor that must be taken into account in any analysis of inscriptional references to age at death. It also shows that Jewish funerary inscriptions are not less affected by epigraphic customs than non-Jewish epitaphs.

---

[53] Some of these data will be presented in Table 4, below.
[54] Repeated, with some reservations, by van der Horst 1991, 83.
[55] See e.g. the remarks by P. Smith and E. Arensburg, "The Jewish Population of Jericho," *Palestine Exploration Quarterly* 115 (1983) 136.
[56] Ery 1967, 54-57, esp. figs. 2 and 3.

Approach 2: A Critique of the Traditional Approach

To show that calculations of the average age at death as proposed by Approach 1 are distorted by a disproportional commemoration of the very old and/or the very young, a second approach tries to establish the percentage of inscriptions referring to people who died under ten or over seventy years of age. Table 2 contains these data.[57]

Table 2: Percentage of people who died before reaching the age of ten. In parentheses, the percentage of people who died after age seventy.

Jewish	Males	Females
Rome	31.7 (8.5)	25.0 (11.1)
Teucheira	26.2 (7.7)	15.4 (7.7)
Tell el-Yehoudieh	15.9 (2.4)	8.7 (0.0)
Non-Jewish	Males	Females
Rome	35.4 (4.7)	27.9 (1.6)
Italy, except Rome	22.3 (6.7)	20.3 (4.4)
Spain	5.9 (13.1)	7.1 (6.6)
Africa	7.7 (4.5)	7.7 (4.5)
Kom Abou Billou	15.5 (4.8)	21.3 (6.3)
Tehneh	16.9 (7.0)	10.6 (10.6)

Approach 2 helps explain in greater detail some of the incongruities observed in connection with Approach 1. The figures recorded here for Jewish and non-Jewish inscriptions from Rome indicate that children under the age of ten form an important segment of the total number of the people whose age at death was referred to in inscriptions. In the case of Egypt (Jewish and non-Jewish inscriptions) such percentages drop considerably. In the non-Jewish inscriptions from Spain and North Africa the decrease is even more noticeable. The difference between the various percentages explains why the average age at death in inscriptions from North Africa as calculated on the basis of Approach 1 is excessively high in comparison to inscriptions from Rome. Among the references to age at death in the Roman inscriptions (Jewish and non-Jewish), children under the age of ten are included approximately four times more often than in African inscriptions. Clearly, such a difference results in a much lower overall average age at death in inscriptions from Rome than in those from Africa, when this average is calculated according to the method described here as

---

[57] The non-Jewish data derive from Hopkins 1966, 252-55; the Egyptian data (Kom Abou Billou and Tehneh) from B. Boyaval, "Remarques sur les indications d'ages de l'épigraphie funéraire grecque," *Zeitschrift für Papyrologie und Epigraphik* 21 (1976) 217–43, esp. 219 and 221.

Approach 1.⁵⁸ Similarly, the comparatively low average age at death in Jewish inscriptions from Teucheira can be attributed to the high percentage of children under ten referred to in the inscriptions.⁵⁹

Table 2 also explains why the average age at death of Jewish males in Rome as documented in Table 1 is slightly higher than the figure attested for non-Jewish males in the inscriptions from that city. In the Jewish inscriptions from Rome relating to Jewish males the percentage of males, under the age of ten is somewhat lower than the percentage for non-Jewish males, while the percentage for Jewish males over seventy is slightly higher than for their non-Jewish counterparts. Logically, such differences result in a higher overall average age at death for females when such an average age is calculated in the way described here as Approach 1. Similarly, Jewish females over seventy are more often commemorated than Jewish males falling into that category, while this relationship is reversed in case of Jewish male and female children under ten years of age. Consequently, the average age at death for Jewish females as documented in Table 1 is slanted. Most likely, it does not correspond closely to the average age at death.

In addition to documenting the importance of epigraphic practices in inscriptions that refer to age at death, Approach 2 permits us to compare Jewish and non-Jewish epigraphic practices. It is remarkable, for example, that the percentage of children under ten, both male and female, referred to in Jewish inscriptions from Rome is practically identical to the percentage recorded for the non-Jewish inscriptions for that city (31.7% and 25.0% versus 35.4% and 27.9%, respectively).⁶⁰ The percentage of people over seventy years of age referred to in Jewish as opposed to non-Jewish inscriptions is not entirely identical, but it can nevertheless be compared. In both cases it is lower than the figure for children under ten. Similarly, at Tell el-Yehoudieh, the percentages compare quite well with the percentages recorded for other, non-Jewish sites in Egypt; at least for children under ten the percentages are quite comparable.⁶¹ The Jewish evidence from Teucheira

---

⁵⁸ Thus only in Africa are Jewish children mentioned more often in inscriptions than pagan children. In Rome, Jewish and non-Jewish children are mentioned equally often.

⁵⁹ As in Table 1, the evidence from Teucheira is once again more similar to that from Italy than from North Africa.

⁶⁰ Blumenkranz 1961, 343 is unprecise in this respect. Leon 1960 included also inscriptions carrying the word νήπιος (child) in his calculations. I have excluded such references because it remains unknown until what age one was considered a child.

⁶¹ The evidence from Tell el-Yehoudieh (Jewish) is similar to Tehneh (non-Jewish) in that a higher percentage of males under the age of ten are recorded than females belonging to this group (different is the evidence from Kom Abou Billou). By contrast, in Jewish inscriptions from Egypt relating to people over seventy, more males than females are referred to. On the non-Jewish inscriptions from Egypt this relationship has been reversed.

does *not* compare well with the figures documented for North Africa. It should be stressed, however, that this particular comparison is not entirely satisfactory to begin with. If we would compare the inscriptional evidence from Teucheira with other epitaphs from the region, a different picture would perhaps emerge, since the non-Jewish evidence is based on inscriptions from the whole of Roman North Africa.

The similarities between the percentages recorded for Jewish and non-Jewish inscriptions in Table 2 suggests that in commemorating their dead, Jews and non-Jews in both Rome and Egypt preferred to include references to age at death in the case of children, compared to those over 70.

In conclusion, Approach 2 shows that epigraphic custom seriously affects calculations of the average age at death as arrived at on the basis of Approach 1. Approach 2 also suggests that these epigraphic customs varied regionally. Finally, Approach 2 indicates that with the exception of Teucheira,[62] Jewish epigraphic customs in a given area (and in Rome in particular) are surprisingly similar to non-Jewish inscriptional habits in that same area

Approach 3.a: A More Sophisticated Approach

In the study of average age at death as reflected in non-Jewish inscriptions, three other approaches have been suggested (Approaches 3.a, 3.b, and 3.c). All three approaches have in common that they attempt to reduce distortions as they result from the disproportionate commemoration of young children. The first of these approaches excludes inscriptions referring to children under ten years of age from the calculations. The data resulting from such a calculation are presented in Table 3.[63]

Table 3: Average age at death taking into account only those who survived beyond age 10.

Jewish	*Male/Female*
Rome	37.6
Teucheira	34.1
Tell el-Yehoudieh	33.4

---

[62] Note, however, that the comparison of the data from Teucheira (one single site) with North Africa (an entire region) is not entirely appropriate.

[63] Non-Jewish data derive from Hopkins 1966, 248. For the Theban ostraca, see A. E. Samuel *et al.* (eds.), *Death and Taxes. Ostraka in the Royal Ontario Museum I* (Toronto: Hakkert, 1971) 25 (males only). Because no separate data for non-Jewish males and females were available, this is the only table in which male and female data have been combined. The figures for Rome are 32.6 years (males) and 28.7 years (females), Teucheira 32.1 years (males) and 40.9 years (females) and Tell el-Yehoudieh 32.7 years (males) and 28.9 years (females).

(Table 3 continued)

Non-Jewish	Male/Female
Rome	29.3
Latium	29.6
Cisalpine Gaul	32.1
Brutii, Lucania, Campania, Sicily, Sardinia	33.7
Calabria, Apulia, Samnium	34.8
England	36.5
Asia, Greece	36.8
Aemilia, Etruria, Umbria	37.1
Spain	37.8
Africa	53.3
Theban ostraca	29.4

Approach 3.a offers the advantage of eliminating distortions that result from the high number of commemorations of children under ten in some collections of inscriptions. The advantage offered by Approach 3.a becomes especially clear, when one compares the figures presented in Table 3 to those contained in Table 1. In Table 3 the average age at death in both Jewish inscriptions from Rome, Teucheira and Tell el-Yehoudieh has risen appreciably—as it has in non-Jewish inscriptions from Rome. Now it comes much closer to some of the figures documented for other parts of the empire such as Spain and North Africa.

Yet, even the elimination of references to children does not lead to satisfactory results. Among other things, it does not take into account the distorting effect that may result from a disproportional commemoration of the elderly. The average age at death as calculated for groups of inscriptions that include only a few references to children under the age of ten years to begin with (for example Spain and Africa) will hardly be affected by calculations of the type described here as Approach 3.a. The high average age at death recorded in such groups of inscriptions results precisely from the fact that young children are hardly ever referred to. The fact that in Table 3 the average age at death is higher in the case of the Jewish population in Rome than in the case of the non-Jewish population can also be ascribed to this factor. Jewish males over 70 are twice, and Jewish females seven times, as often referred to in Jewish inscriptions than in non-Jewish inscriptions (compare Table 2 for percentages). The differences between the three Jewish sites as documented in Table 3 result likewise from differences in the commemoration of people over seventy: in Rome, a total of 19.6% of the Jewish inscriptions (both males and female) relate to this particular group, while at Teucheira and Tell el-Yehoudieh the percentages are 15.4% and 2.4%, respectively.

Approach 3.a, like Approach 2, has the advantage of showing that epigraphic custom affects Jewish and non-Jewish inscriptions in very much the same way. However, Approach 3.a resembles Approach 1 in that it has the disadvantage of not offering a truly reliable method for deriving past mortality patterns from inscriptions.

Approach 3.b: Medians

Approach 3.b is methodologically comparable to Approach 3.a. The purpose of this approach is to establish medians, that is the average age by which half of those who survived until their fifteenth birthday would have died. Table 4 contains some figures derived from Jewish and non-Jewish inscriptions.[64]

Table 4: Ages by which one half of those surviving to age 15 would have died

Jewish	Males	Females
Rome	35	29
Teucheira	30	32
Tell el-Yehoudieh	30	27
Non-Jewish	Males	Females
N.W. Africa	48	44
Carthage	38	33
Lambaesis	45	38
N. Africa 4th-6th c.	52	47
Bordeaux, Brindisi, and Merida	44	36
Danube provinces	40	33
England	40	37
N. Italy	52	40
India 1931	48	43
India 1901-1911	44	43
USA 1949-1951	69	76

Comparing the data computed on the basis of Approach 3.b with the figures calculated on the basis of Approach 1, the average age at death in all three collections of Jewish inscriptions has increased (although less than the increase as documented in Approach 3.a). Like Approach 3.a, Approach 3.b results in average ages at death that seem more credible than those calculated on the basis of Approach 1. Yet, Ap-

---

[64] Unfortunately, no figures for non-Jewish inscriptions from Rome are available. The non-Jewish data derive from Hopkins 1966, 249 who follows A. R. Burn, "Hic breve vivitur. A Study of the Expectation of Life in the Roman Empire," *Past and Present* 4 (1953) 2–31. The data for 20th-century India and the U.S.A. derive from J. D. Durand, "Mortality Estimates from Roman Tombstone Inscriptions," *American Journal of Sociology* 65 (1959-60) 365–73.

proaches 3.a and 3.b only correct distortions for collections of inscriptions in which children have been commemorated in disproportionate numbers. Where this is not the case, as for example at various sites in Africa, Approach 3.b, like Approach 3.a, is unable to rule out distortions that result from a disproportionate commemoration of the old.

Most importantly, Table 4 enables one to develop a more precise picture of female mortality. With the exception of the Jewish data from Teucheira, all other figures (Jewish as well as non-Jewish) recorded in Table 4 show that the life expectancy for women who survived their fifteenth birthday was lower than that for males.[65] Given the general lack of sanitary conditions in a preindustrial society like the Roman empire, a lower average age at death for women as opposed to men is what one would expect. Although it has been suggested recently that far fewer women in the Roman world died in childbirth than is usually assumed, death in childbirth, and death resulting from general exhaustion, must nevertheless have been a major factor influencing the life expectancy of Jewish and non-Jewish women in the Roman world.[66] In India, as late as 1931, it was exactly these factors that were responsible for a life expectancy among women that was remarkably low.[67]

Approach 3.b also illustrates the incorrectness of some of the assertions of Blumenkranz. Studying the Jewish epitaphs from Rome according to the traditional approach (Approach 1), he concluded that Jewish women at Rome lived longer than their male counterparts. He then tried to explain the apparent longevity of Jewish women on the basis of epigraphic practices. Blumenkranz correctly observed that children erected epitaphs to their mothers twice as often as to their fathers. He then supposed that children would erect epitaphs to their mothers when their fathers (that is, the mother's husbands) had already passed away. Reasoning along such lines, Blumenkranz's conclusion then became inescapable: the high percentage of children commemorating their mothers indicates that these mothers lived longer than their spouses.[68]

---

[65] I have no explanation for the anomaly regarding the evidence from Teucheira.

[66] Hopkins in Hinard 1987, 118 n. 10. On the changes of death in childbirth in societies in which medical care is in an underdeveloped stage, see also N. Keyfitz, *Population. Facts and Methods of Demography* (San Francisco: Freeman, 1971) 39. The archaeological evidence for women who have died in childbirth is sometimes ambiguous; female skeletons containing the bones of a fetus can also have resulted from deaths that resulted from pregnancy or from disease; see the discussion in W. J. White (ed.), *Skeletal Remains from the Cemetery of St. Nicholas Shambles, City of London* (London, 1988) 71-73 (early medieval). On pregnancy in the ancient world, see also the remarks by Pliny, *NH* 6.6f.

[67] The figures for the U.S.A. 1949-1951 (Table 4) suggest that only in more recent years this pattern was finally reversed.

[68] Blumenkranz 1961, 346-47.

118   CHAPTER THREE

Blumenkranz's inference is logically correct. Nevertheless his argument fails to convince because it does not take into account other factors. In Rome, many Jewish women appear to have married when they were still in their teens. Jewish males, on the other hand, seem to have been several years older when they took such a step.[69] Thus, being older to begin with, more of these males are likely to have predeceased their wives than vice versa. In other words, the evidence discussed by Blumenkranz shows only that a number of women passed away after their husbands did. The evidence does not permit us to infer, however, that such women actually attained higher ages.[70]

In conclusion, Approach 3.b, like Approach 3.a, eliminates one distortion (the distortion resulting from the disproportionate commemoration of young children) while at the same time it allows for another (the distortion resulting from the disproportionate commemoration of old people). Approach 3.b. does indicate, however, that both Jewish and non-Jewish women were more likely to die younger than their male counterparts.

Approach 3.c: The Survival Rate

A third Approach (3.c) tries to exclude distortions resulting from epigraphic practice by establishing the survival rate in percentages of those people who survived until their fortieth birthday. Table 5 contains these percentages for Jewish and non-Jewish inscriptions respectively.[71]

---

[69] For the evidence, see Leon 1960, 230-31.

[70] Hopkins, among others, has systematically argued that inscriptions referring to age at death do not indicate high mortality rates for women, but rather differences in commemorative practice (Hopkins 1966, 246 and 261 and *id.* in Hinard 1987, 125; comparable is P. Salmon, *Population et dépopulation dans l'empire romain* [Bruxelles: Latomus, 1974] 86). Hopkins believes that young wives received an epitaph disproportionally often, because their husbands were still alive to commemorate them. While there is evidence to suggest that this was indeed the case in non-Jewish inscriptions (see Hopkins 1966, 261, esp. Table 8 and see the documentation in Ery 1967, 67 Table 5 and discussion on pp. 61-62), the situation is rather different in Jewish inscriptions from Rome. Here, Jewish women were twice as likely to be commemorated with an inscription by their children than by their husbands (correctly observed by Blumenkranz 1961, 347). Brothers and sisters are likewise known to have commissioned epitaphs for Jewish "mothers," e.g. *CIJ* 78, 92, 124, 141, 206, 217, 234, and 311. In general, however, Jewish women at Rome were not commemorated more often than males.

[71] The non-Jewish data are derived from Kajanto 1968.

Table 5: Survival rate in percentages until 40 years of age

Jewish	Male	Female
Rome	23.2	27.0
Teucheira	24.6	25.6
Tell el-Yehoudieh	25.5	16.0

Non-Jewish	Male	Female
Rome	17.2	8.3
Carthage, early Empire	39.9	29.3
Carthage, late Empire	39.7	33.2
Quattor Coloniae	56.0	52.7
Celtianis	79.6	75.1

Approach 3.c offers the advantage over the other approaches discussed so far that it excludes references to the elderly and thus eliminates the biases as they result from a disproportionate number of references to this particular group. Yet, like Approach 1, Approach 3.c does not take into account the high percentage of references to children under the age of ten as documented in some collections of inscriptions. As a result, the average age at death in inscriptions from North Africa are, once again, remarkably high in comparison to the percentages in inscriptions from other parts of the Empire. The high survival rate of Jewish women in Rome and Teucheira can be explained in much the same way. At both sites, fewer Jewish women than Jewish men under the age of ten were recorded. In short, then, Approach 3.c offers no reliable method for the study of ancient demography. Like the other approaches, it shows that commemorative practices determine to a large extent the rationale behind including or excluding references to age at death in inscriptions.

Approach 4: The Problem of Age-Rounding

Approach 4 focuses on still another factor that affects the calculations of average age at death, although its distorting effect is more limited than the distortions that result from the disproportional commemoration of certain age groups: a preference for rounded ages, that is, for ages divisible by five.[72]

Scholars have often wondered whether the ancients really knew how old they were. In the Jewish inscriptions from Tell el-Yehoudieh, as well as in Jewish inscriptions from Venosa and Naples, references

---

[72] Although ages ending in the number 6 were popular in Egypt and ages ending in 1 were widespread in Africa, multiples of five are by far the most common in inscriptions from the Roman Empire.

to age are often preceded by the word "approximately" (ὡς and *plus minus* respectively).⁷³ By the fourth century C.E. *plus minus* seems to have become a standard formula in inscriptions referring to age at death. That many Jews, like non-Jews, only had a rather vague idea of their age becomes particularly evident when we turn to an investigation of this preference for rounded ages.⁷⁴

In every population, 20% of all people should have an age that is divisible by five. To establish if the number of ages divisible by five contained in inscriptions is correct, it is therefore necessary to establish if any *excesses* in age rounding occur (that is, if the 20% that can theoretically be expected is surpassed). The following formula used by the US Census Bureau to determine excesses in age rounding for the US Census of 1910 may be employed to determine the eventual existence of such excesses: (sample-20) x 1.25.⁷⁵ In the next table, Jewish and non-Jewish examples of excess in age rounding have been arranged for people in the age group between 23 and 62 years of age.⁷⁶

Table 6: Excess in age rounding. In parentheses is the total number of inscriptions on which each calculation is based.

Jewish	Males	Females
Rome	43.5 (82)	9.0 (63)
Teucheira	43.9 (65)	77.9 (39)
Tell el-Yehoudieh	58.3 (44)	48.5 (25)
Non-Jewish	Males	Females
R. land-troops	37.9 (408)	-
R. civilians and *incerti*	48.4 (1271)	48.9 (1003)
R. freedmen/women	47.4 (295)	52.9 (279)
R. slaves	48.5 (132)	58.8 (89)
I. town-councillors	15.1 (75)	-
I. civilians and *incerti*	42.8 (904)	-
I. freed-troops	47.2 (299)	-
I. freedmen and slaves	49.5 (117)	-
A. councillors	17.5 (34)	-
C. citizens and *incerti*	42.8 (123)	33.1 (117)
C. freedmen/women	49.7 (141)	62.0 (87)
C. Christian tombstones	58.5 (38)	69.6 (38)

---

⁷³ Boyaval 1976, 224 and n.53 maintains that ὡς must sometimes be considered as synonym for ἐτῶν.

⁷⁴ In general, see W. Levison, "Die Beurkundung des Zivilstandes im Altertum. Ein Beitrag zur Geschichte der Bevölkerungsstatistik," *Bonner Jahrbücher* 102 (1898) 1–82; A. Mócsy, "Die Unkenntnis des Lebensalters im römischen Reich," *Acta Archaeologica Academiae Scientiarum Hungaricae* 14 (1966) 387–421; Clauss 1973, 396-97; Duncan Jones 1977, *id.*, "Age–Rounding in Greco–Roman Egypt," *Zeitschrift für Papyrologie und Epigraphik* 33 (1979)169–77; *id.*, 1990.

⁷⁵ See Duncan-Jones 1977, appendix I; see also Mócsy 1966, 395-96.

⁷⁶ The non-Jewish data derive from Duncan-Jones 1977, 338-39, 341, 343 and *id.* 1990, 84. My calculations differ from those suggested by van der Horst 1991, 82.

(Table 4 continued)

Zone 1	39.4 (560)	38.0 (440)
Zone 2	46.4 (689)	47.9 (627)
Zone 3	59.2 (1574)	60.4 (1200)
Italy without Rome	42.6 (1213)	41.8 (789)
Gaul	44.1 (311)	43.1 (232)
Rome	47.0 (2337)	50.2 (1371)
Africa and Numidia	51.4 (3110)	52.2 (2490)
Mauretania	51.6 (298)	54.1 (162)
Dalmatia	53.3 (358)	56.0 (229)
Spain	56.6 (721)	58.4 (615)
Moesia	57.2 (193)	73.3 (80)
Germany	57.3 (350)	20.7 (45)
Dacia	61.2 (134)	65.0 (87)
Pannonia	64.8 (489)	75.9 (211)
Noricum	82.1 (206)	77.3 (169)

Explanation of the abbreviations: R.= city of Rome, I.= Italy, city of Rome excluded; A.= Africa; C.= Carthage; zone 1 = towns 1-200 km. from Carthage; zone 2 = towns 201-300 from Carthage; zone 3 = towns 301-400 km. from Carthage.

These calculations are especially interesting from the point of view of Roman social history. In inscriptions commemorating (non-Jewish) town-councillors from Italy and North Africa, excess in age rounding is relatively low. This suggests not merely a greater age awareness among the higher echelons of Roman society than among the rest of the population of the Roman Empire; it also suggests that such people had access to a more reliable recording system and/or that literacy was more widespread among this group than among other groups of the Roman population.[77]

Without exception, excess in age rounding for people other than town-councillors is high. Although some differences in excess of age rounding can be seen in Carthage between citizens as opposed to slaves and freedmen/women, such differences are far from substantial. Therefore, it should not surprise that also in Italy no significant differences exist in age rounding in inscriptions relating to land-troops, citizens, freedmen/women, and slaves. One of the reasons why there was little difference between the free and unfree in terms of age awareness may have been that many of these citizens had originally been slaves who had been manumitted. Evidence from Africa indicates that age awareness decreased markedly in the countryside. In fact, there seems to be a causal relationship between the level of age awareness and distance to major urban centers like Carthage. In other parts of the Roman empire, as for example in the northern provinces including Moe-

---

[77] Discussion in Duncan-Jones 1977, 336-38.

sia, Germany, Dacia, Pannonia, and Noricum, a notable excess in age rounding should probably be ascribed to a lack of Romanization.[78]

The percentages recorded for the various Jewish sites indicate that in terms of excess in age rounding, the patterns documented in the Jewish inscriptions are close to the patterns documented for the general populace. Such an observation can be taken to imply that the great majority of people commemorated in the Jewish inscriptions did not belong to the class of town-councillors.

In Table 6 it can furthermore also be observed that in general, excess in age rounding is higher for women than for men. Yet, there are several exceptions to this pattern. In order to make these differences more easily identifiable visually, a separate table has been included here.[79]

Table 7: Differences between excess in age rounding in inscriptions referring to women as opposed to those referring to males.

Teucheira (J.)	+ 34.5
Moesia	+ 28.1
Terenuthis	+ 24.8
Pannonia	+ 17.1
Rome	+ 6.8
Dacia	+ 6.2
Dalmatia	+ 5.1
Mauretania	+ 4.8
Spain	+ 3.2
Africa and Numidia	+ 1.6
Italy outside Rome	- 1.9
Gaul	- 2.3
Noricum	- 5.8
Tell el-Yehoudieh (J.)	- 9.8
Rome (J.)	- 34.5
Germany	- 63.9

Explanation: J.= Jewish inscriptions

Table 7 illustrates that excess in age rounding is usually higher in inscriptions referring to women than in epitaphs relating to males. Yet, for some areas, including Italy, Gaul, Noricum, and Germany, and for the Jewish evidence from Rome and Tell el-Yehoudieh, excess in age rounding is higher for males than for females. Duncan-Jones believes that this discrepancy results from the relatively low number of samples on which the percentages preceded by a minus sign are based.[80]

---

[78] Mócsy 1966, 405f.

[79] The non-Jewish data derive from Duncan-Jones 1979, 170 and 1990, 86. Note that percentages in the later publication are higher than in the former. I have followed the latter publication.

[80] Duncan-Jones 1977, 340.

This explanation is not entirely convincing. In Africa (in zone 1, Table 6), for example, where the excess in age rounding is lower for women than for men, these percentages are based on a data-base consisting of hundreds of inscriptions. To a lesser extent, the same holds true in case of inscriptions from Germany and Gaul.

Another explanation, which has also been put forward by Duncan-Jones, is that a high excess in age rounding points to a low level of literacy. On the basis of evidence from underdeveloped countries, collected from the mid 1940s to the mid 1960s, it can be shown that there usually exists a close relationship between age-rounding and illiteracy. In the modern samples, an excess in age rounding of only 30% corresponds to a level of illiteracy that is as high as 70%.[81] Therefore, it is appealing to explain the generally higher excess of age rounding among women in the Roman world as reflecting a higher degree of illiteracy.[82] Yet, again, this explanation is not entirely satisfactory. It is based on the supposition that references to age at death in inscriptions relating to women derive exclusively from the information provided by such women. We do not know, however, if this was really the case. It is possible that males who had perhaps easier access to education (and thus to a more reliable recording system) may have been be responsible for the reference in inscriptions relating to women. After all, in a considerable number of cases the survivors rather than the person commemorated in an inscription are likely to have determined the references to age at death to be included in the epitaph.

In conclusion, Approach 4 shows that the phenomenon of age rounding influences to some extent the character of references to age at death in ancient inscriptions. References to age at death in Jewish inscriptions indicate that age rounding was common in Jewish inscriptions too, and that excesses in age rounding were not lower or higher than in most non-Jewish inscriptional materials. Although independent evidence suggests that some Jews may have had some general idea concerning their age,[83] Approach 4 shows indisputably that in terms of

---

[81] Convincing evidence to this effect can be found in Duncan-Jones 1977, Appendix III.

[82] E.g. Duncan-Jones 1977, 334-35 and *id.* 1990, 90, Table 33. That illiteracy was more widespread among women than among men is a view held by many scholars, e.g. O. Montevecchi, *La papirologia* (Torino: Società editrice internazionale, 1973) 400 referring to the archive of the family of Tebtynis and W. V. Harris, *Ancient Literacy* (Cambridge Mass: Harvard U. P., 1989) 312f., who estimates the level of literacy in the Roman Empire around 20-30% for males and less than 10% for females.

[83] E.g. the Jews who tried to enter the ephebeion in Alexandria, Cyrene, and in several cities in Asia Minor (*CPJ* 1, 39, n. 99 and no. 150, lines 92-93; S. Applebaum, *Jews and Greeks in Ancient Cyrene* [Leiden: Brill, 1979] 167-68; Lüderitz 1983, 11-21, 57-58, and 185) must have had some idea of how old they were, because in order to join this institution one had to prove that one was at least fourteen years of age (*P.Oxy.* 1202 = *Loeb Selected Papyri* [1977] vol. 2, no. 300). Note also that rabbinic law made legal status de-

age awareness, Jews did not stand out in any way among the general population. Conceivably, the level of literacy of the Jews of Rome was quite comparable to that of the non-Jewish population there. Approach 4 also seems to suggest that in Rome, as in other parts of the empire, a majority of Jews did not belong to leading strata of Roman society. Rather, they could be found in the heterogeneous class composed of citizens, freedmen/women and slaves.

Approach 5.a: The Pattern of Mortality

The question of the reliability of the data contained in inscriptions referring to age at death can be approached from still another angle. This approach analyzes the pattern of mortality, that is, the number of deaths occurring within a given period of time (for example, five or ten years).

Upon subdividing the references to age at death into five-year periods, a pattern can be observed that at first sight seems demographically probable. When divided into periods of five years, the Jewish inscriptions from Rome, for example, permit one to draw the following picture: A high mortality during the first five years for both males and females; then, for those who survived until their fifth birthday, a decrease in the chance of dying; from ages nineteen to thirty-five, a mortality that was again consistently high; and, finally, only few people that lived to see old age.

At first sight, such a pattern seems to correspond rather closely to what one would expect in this sort of society: high infant mortality, and a relatively low level life expectancy for the rest of the population that survived the hazardous years of weaning and childhood.[84] Despite its suggestiveness, this type of calculation does not lead to reliable results. In order to determine if the pattern of mortality as suggested by ancient inscriptions is reliable, it is necessary to study the percentage of deaths within every age group (in this case, within periods of five

---

pendent on age, see *Sifre Numbers* 153 (*P. Mattot*) ed. Horovitz p. 199; *Mishnah, Avoth* 5.21; *Mishnah, Niddah* 5.6f. and the discussion in the *Babylonian Talmud, Niddah* 46a. See, in general, L. Löw, *Die Lebensalter in der jüdischen Literatur* (Szegedin, 1875) 142-63. Note also that in *Mishnah, Avoth* 5:21 mostly rounded ages (5, 10, 13, 15, 18, 20, 30, 40, 50, 60, 70, 80 and 100) are preferred in describing what one should do at what age, see eventually J. Maier, "Die Wertung des Alters in der jüdischen Überlieferung der Spätantike und des frühen Mittelalters," *Saeculum* 30 (1979) 354–64. Finally, it should be noted that three Jewish inscriptions from Rome (*CIJ* 9, 368, and 408) that mention gerusiarchs also include references to age at death that are high.

[84] The data from Teucheira suggest a very similar pattern. At Tell el-Yehoudieh, however, child mortality cannot so easily be observed; here there is a steady increase in mortality among young adults.

years) rather than the total number of inscriptions.[85] The following table contains these percentages as they can be derived from the Jewish epitaphs from Rome.[86] Note that the mentioned percentages always stand for the percentage of deceased people *within* a specific age group, and not for absolute percentages.

Table 8: Percentages of recorded deaths within various age groups at Rome (Jewish)

	Males			*Females*		
x	l_x	$_nd_x$	$_nq_x$	l_x	$_nd_x$	$_nq_x$
0 - 1	82	1	1.2	62	0	0
1 - 4	81	17	30.0	62	12	15.4
5 - 9	64	8	12.5	50	6	12.0
10-14	56	6	16.7	44	3	6.6
15-19	50	6	12.0	41	10	24.4
20-24	44	11	25.0	31	5	16.1
25-29	33	2	6.1	26	6	23.1
30-34	31	4	12.9	20	2	10.0
35-39	27	8	25.6	18	1	5.6
40-44	19	2	16.5	17	4	23.5
45-49	17	1	5.7	13	2	15.4
50-54	16	5	31.3	11	0	0
55-59	11	1	5.1	11	2	16.2
60-64	10	2	20.0	9	2	22.2
65-69	8	1	12.5	7	0	0
70+	7	7	-	7	7	

Explanation: x: age
l_x: survivors at the beginning of an age interval
$_nd_x$: number of deaths recorded;
$_nq_x$: percentage.

To evaluate fully the percentages recorded in Table 8, it is necessary to compare them with the percentages that can be derived from other groups of inscriptions. These data are presented in coherent form in Table 9.

---

[85] Regarding non-Jewish materials, such an approach can be found in Durand 1959-60, 368 and in Hopkins 1966. The percentages recorded in Table 8 are arrived at by dividing the number of deaths by the number of survivors at the beginning of each interval.

[86] These percentages have also been calculated for the Jewish inscriptions from Teucheira and Tell el-Yehoudieh, but it is not necessary to present them separately at this point. They have been included in Table 9 below.

Table 9: Male and female chances of death in percentages.

Males

Age	1	2	3	4	5	6	7	8
0 - 1	1.2	0	0	1.6	-	0.4	0.5	33.2
1 - 4	30.0	10.8	9.5	17.5	11.6	2.8	3.7	26.8
5 - 9	12.5	17.2	16.5	20.4	12.2	2.9	3.7	8.9
10-14	16.7	8.3	2.5	14.2	10.0	4.5	4.3	5.6
15-19	12.0	13.6	6.1	20.9	15.7	11.2	6.6	6.8
20-24	25.0	26.6	22.6	22.4	17.7	15.0	8.6	8.6
25-29	6.1	10.7	26.8	21.4	16.3	14.6	8.7	10.0
30-34	12.9	20.0	26.3	21.6	16.0	13.8	9.2	12.0
35-39	25.6	20.0	14.3	23.4	18.2	12.5	8.9	14.7
40-44	16.5	6.3	16.6	22.4	20.8	13.8	9.5	18.5
45-49	5.7	6.7	50.0	17.7	16.9	13.2	9.2	22.4
50-54	31.3	21.4	40.0	17.3	24.8	16.0	11.4	25.7
55-59	5.1	0	0	17.8	19.9	18.7	11.5	30.3
60-64	20.0	45.5	66.6	26.6	28.7	24.6	14.8	35.0
65-69	12.5	0	0	20.9	15.1	22.5	15.7	-
70+								

Females

Age	1	2	3	4	5	6	7	8
0 - 1	0	0	0	1.3	3.5	0.2	0.4	30.7
1 - 4	15.4	2.6	4.4	13.6	8.2	2.4	3.8	27.3
5 - 9	12.0	13.2	4.5	15.3	10.0	4.6	3.7	9.3
10-14	6.6	5.1	4.8	13.1	10.9	5.1	4.2	6.3
15-19	24.4	10.0	10.0	23.8	20.2	13.7	7.4	7.7
20-24	16.1	22.2	22.2	30.3	24.8	17.5	9.3	9.5
25-29	23.1	4.8	35.7	34.3	25.1	22.0	11.0	11.3
30-34	10.0	20.0	44.4	31.7	22.8	18.5	11.6	13.2
35-39	5.6	31.3	20.0	28.9	19.8	18.1	11.6	15.3
40-44	23.5	16.2	25.0	26.0	22.7	18.0	11.7	17.0
45-49	15.4	0	66.6	25.8	18.0	16.9	9.2	19.1
50-54	0	22.2	-	21.9	21.6	26.0	11.0	21.9
55-59	16.2	26.6	-	19.1	16.2	24.7	11.8	25.3
60-64	22.2	20.0	-	26.0	30.6	30.0	16.8	26.0
65-69	-	-	-	26.6	22.0	23.4	16.5	38.0
70+								

Explanation: 1.Rome-Jewish; 2.Teucheira-Jewish; 3. Tell el-Yehoudieh, Jewish; 4. Rome, non-Jewish; 5. Italy, Rome excluded, non-Jewish; 6. Spain, non-Jewish; 7. Africa, non-Jewish; 8. UN Model Life Table.

Column 8 in Table 9 is of particular interest. It contains a Model Life Table published by the United Nations. This Model Life Table shows average age-specific death-rates that have been calculated on the basis of demographic data originating from 50 preindustrial societies and collected between 1900 and 1950. By comparing this modern UN Life Table with data as they can be inferred from ancient skeletal remains excavated in Hungary, Ery has shown that this life table can be taken as reliable indicator of the mortality patterns in preindustralized societies that predate the twentieth century.[87] Thus, the UN Life Table is an important tool for checking the reliability of the mortality patterns as they are suggested by ancient inscriptions.[88] Let us study the evidence provided by this life table in greater detail.

Column 8 of Table 9 (the UN Life Table) documents that during the first four years of life the chance of dying is very high for both males and females. Around 30% of all males and females in this age group do not survive their fourth birthday. In the following years, however, the chance of surviving increases dramatically. From age 20 onwards, again a higher percentage of people within the age group is likely to die. Such percentages increase steadily for both males and females as the years go by. The regularity of decrease and increase in mortality is particularly evident when the evidence is presented as in Chart 1.

Comparing the UN Life Table (Table 9, column 8) with the other seven columns in this table, the differences are striking. In *none* of the "inscriptional" columns can any regular patterns be observed. Irrespective of whether the percentages in the other seven columns relate to males or females, or to Jews or non-Jews, gradual decreases or increases are completely absent. In other words, ancient inscriptions that include references to age at death not merely contain unreliable data for estimating the average life span of past populations. They are also completely unreliable for reconstructing the ups and downs in chances of survival during specific periods in the course of human life. How unreliable the ancient inscriptions are in this respect becomes particularly evident when the dates are presented in a chart (Chart 2). It can easily be observed that the irregular pattern in this chart contrasts markedly with the regular pattern of Chart 1.

Approach 5.a thus illustrates from yet another angle that commemorative practices determine the character of references to age at death in ancient inscriptions. It may be clear that these inscriptions were never meant to provide statistically reliable information on the demographic structure of the ancient world. Jewish inscriptions from Rome, Teucheira, and Tell el-Yehoudieh do not form an exception to this phenomenon. Also, among the Jews of antiquity, epigraphic customs

---

[87] Ery 1967, 53-54.

[88] Of course, the UN Life Table only gives averages. Therefore, the chances of death may be different at different sites at different points in history. The 473 skeletons discovered in the Lankhills cemetery in Britain, for example, show that adult women were much likelier to die at a young age than would be expected on the basis of the UN data. For this cemetery, see Humphreys and King 1981, 19-20.

and norms, and not demographic concerns, determine the practices surrounding inscriptional references to age at death.

Chart 1: Male and female chances of death according to the UN Model Life Table (= Table 9, column 8)

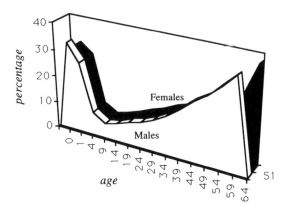

Chart 2: Male and female chances of death according to Jewish and non-Jewish inscriptions from Rome (= Table 9, columns 1 and 4)

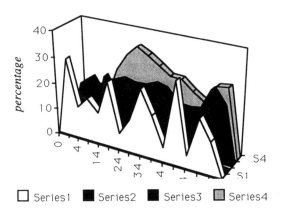

Explanation: Series 1: Jewish males - Rome; Series 2: Jewish females - Rome; Series 3: non-Jewish males - Rome; Series: non-Jewish females - Rome.

Approach 5.b: The Hypothetical Life Table and the Archaeological Remains

An approach that is somewhat similar to Approach 5.a contrasts life expectancy as it can be derived from a hypothetical life table and from archaeological remains with the life expectancy as it can be derived from ancient inscriptions. The hypothetical life table mentioned is based on literary evidence provided by Ulpian and has recently been reconstructed ingeniously by Frier.[89] In addition to these data, the following table also contains the life expectancy derived from Jewish inscriptions from Rome.[90]

Table 10: Life expectancy according to Frier's suggested Life Table, skeletal materials, and epigraphic sources.

Age	Life Table	Keszthely-Dobógo	Spain Males	Rome-Jewish Male and Female combined
0	21.11	-	(35.5)	27.4
1	31.70	-	39.1	27.3
5	37.13	37.38	37.0	26.7
10	34.50	36.17	33.1	26.1
15	31.12	32.34	29.5	23.2
20	28.41	28.85	27.5	23.2
25	25.75	25.51	26.7	20.9
30	23.13	22.08	25.7	16.6
35	20.56	18.40	24.6	16.0
40	18.05	15.98	22.4	15.7
45	15.63	13.13	20.9	13.7
50	13.13	11.47	18.2	11.9
55	11.23	9.93	16.4	10.1
60	9.14	8.61	13.7	8.7
65	7.38	6.11	13.0	7.5
70	5.82	3.85	10.1	7.0
75	4.53	2.02	9.9	6.5
80	3.77	-	8.1	4.3

Explanation: the evidence from Keszthely-Dobógo (near lake Balaton - Pannonia) is archaeological; the evidence from Spain (non-Jewish) and Rome (Jewish) is epigraphic.

Both the hypothetical life table and the archaeological evidence indicate that as people grow older, there is a regular decrease in life ex-

---

[89] B. Frier, "Roman Life Expectancy: Ulpian's Evidence," *Harvard Studies in Classical Philology* 86 (1982) 213–51; *id.*, "Roman Life Expectancy: The Pannonian Evidence," *Phoenix* 37 (1983) 328–344; Duncan-Jones 1990, 96f.

[90] For the non-Jewish data, see Frier 1983, 335. And see, in general, A. J. Coale and P. Demeny, *Regional Model Life Tables and Stable Populations* (New York: Academic Press, 1983).

130   CHAPTER THREE

pectancy. By contrast, the decrease in life expectancy from age thirty onwards as suggested by inscriptions from Spain is unrealistically small.[91] According to these inscriptions, life expectancy at age sixty five is twice as high as the life expectancy that is archaeologically documented. Thus, the Spanish evidence suggests figures that cannot be considered demographically reliable. Only in the twentieth century did life expectancy for the elderly rise to the levels suggested by the Spanish evidence.

Life expectancy as suggested by the Jewish inscriptions in Table 10 seems less deviant than the Spanish evidence, at least insofar as people over forty are concerned. In reality, however, the Jewish funerary inscriptions are not more reliable than the Spanish epitaphs. This can be deduced from the fact that in the first fifteen to twenty years, there is almost no decrease in life expectancy. The unreliability of the epigraphic evidence can best be seen when all these data are presented in the form of a chart (Chart 3).

Thus, Approach 5.b documents in a different manner than does Approach 5.a that the *pattern* of mortality/life expectancy as it can be inferred from inscriptions cannot serve as indicator of ancient chances of survival at any specific stage in life.

Chart 3: Life expectancy as documented in Table 10

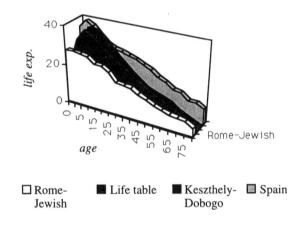

□ Rome-Jewish   ■ Life table   ■ Keszthely-Dobogo   ▨ Spain

---

[91] The same exceptionally low figures can be observed on inscriptional material from Egypt, Greece, and Illyricum, see Frier 1983, 335, Table IV.

Sociological Inferences: Jewish Women in Antiquity

In the previous pages it has been observed on various occasions that references to age at death in inscriptions relating to males differ from those in inscriptions relating to females. In the following table the total numbers of inscriptions relating to males and females have been illustrated in percentages insofar as such inscriptions include references to age at death. Added in parentheses are the percentages of all inscriptions which refer to males and females.[92]

Table 11: Percentage of inscriptions mentioning age at death of males versus females in percentages. In parentheses is the percentage of inscriptions mentioning males and females.

Jewish	Males	Females
Rome	56.6 (55.0)	43.4 (45.0)
Teucheira	62.5 (60.9)	37.5 (39.1)
Tell el-Yehoudieh	63.8 (62.5)	36.2 (37.5)
Joppa	- (82.4)	- (17.6)
Beth She'arim	- (65.5)	- (30.5)
Non-Jewish	Males	Females
Rome	56.7	43.3
Italy	61.2	38.8
Spain	55.7	44.3
Africa	56.3	41.7
Kom Abou Billou	48.6	42.3
Tehneh	55.9	37.0
Alexandria	55.5	30.1
Akhmim (early-Chr.)	55.6	32.9

Table 11 shows that at all three Diaspora sites, inscriptions containing references to age at death favor males slightly over females (35-45% women versus 55-65% males). At first sight, such percentages seem to

---

[92] Inscriptions that mention males or females in the capacity of dedicants have, of course, been excluded from Table 11. In Rome, the percentages per catacomb are the following: Villa Torlonia catacombs: 57.1% males versus 42.9% females; Vigna Randanini catacomb: 53.1% males versus 46.9% females; and Monteverde catacomb: 57.2% males versus 42.8% females. The non-Jewish data (Rome, Italy, Spain, and Africa) derives from Hopkins 1966, 252-55; the Egyptian data is based on Boyaval 1976, 223. Unfortunately, there is no information available concerning the total number of non-Jewish inscriptions that concern males versus females. Some of the issues raised in this section have been discussed by R. Kraemer, "Non–Literary Evidence for Jewish Women in Rome and Egypt," *Helios* 13 (1986) 85–101, but her treatment is not entirely satisfactory. Kraemer does not take into account recent epigraphic publications (e.g. Solin 1983; Lüderitz 1983) and, consequently, her percentages are different than the ones suggested here. She also excludes the data from the Jewish cemetery at Jaffa. Furthermore, it is misleading to distinguish between the Via Nomentana and the Villa Torlonia catacomb, inasmuch as this is one and the same catacomb. Finally, contrary to Kraemer, I do not believe that Jewish inscriptions from regions that are as far apart as Asia Minor and North Africa should be lumped together.

indicate that Jewish men were more likely to receive an inscription referring to their age at death than their female counterparts, but such an impression is misleading. A comparison of all funerary inscriptions at a given site, referring to either males or females, shows that in general Jewish men were more likely to receive an epitaph than were Jewish women (the percentages are roughly 40% for women versus 60% for males). Thus, the inclusion of references to age at death was not more common in inscriptions erected for males than in those commemorating women.

It is noteworthy, furthermore, that the Jewish data do not stand alone. In non-Jewish inscriptions that refer to age at death, a similar relationship can be observed between the inscriptions relating to males compared to those relating to females. In non-Jewish inscriptions, too, the general chance of being remembered with an inscription was some 20% higher for males than it was for females. That these differences should indeed be interpreted as resulting from epigraphic practice follows from an analysis of sex ratios, that is, from the number of documented males per 100 females. Table 12 illustrates these ratios as they can be inferred from Jewish and non-Jewish inscriptions and compares them with Jewish evidence collected in the nineteenth century.[93]

Table 12: Sex ratio (men per 100 women), based on inscriptions mentioning age at death. In parentheses: sex ratio based on the total number of inscriptions.

Jewish			
Rome		130	(122)
Teucheira		166	(155)
Tell el-Yehoudieh		176	(166)
Non-Jewish			
Rome		131	
Italy (Rome excl.)		158	
Spain		126	
Africa		135	
Gaul		147	
Germany		475	
Britain		220	
Danubian provinces		170	
Misenum		2100	
Jewish Nineteenth Century			
Algeria	1873	103	
Austria	1861-71	128	
Budapest	1868-75	114	
Budapest	1878-82	103	

---

[93] The non-Jewish data are based on Hopkins 1966, 261 and Hopkins in Hinard 1987, 118; the data for Misenum derive from Claus 1973, 405; the Jewish nineteenth-century data were taken from *The Jewish Encyclopedia* 3 (1903) 225.

(Table 12 continued)

Budapest	1898	104
France	1854-59	111
Hungary	1876-78	114
Hungary	1896	104
Posen	1819-73	108
Prague	1865-74	111
Prague	1879-80	105
St.Petersburg	1866-72	147
Vienna	1865-74	117
Vienna	1897	103

Modern parallels suggest that in a stationary population, approximately 102-106 males are born to 100 females, a rate that is then equalized during infancy.[94] Supposing that a similar ratio prevailed in antiquity for stationary populations,[95] the data as derived from inscriptions and presented in Table 12 suggest ratios that are far too high. Ratios start at 130, but often they are much higher.[96] This holds true for both Jewish and non-Jewish inscriptions.[97] Such evidence implies that inscriptions cannot serve to reconstruct past sex ratios.

A comparison of the sex ratios as reflected by the inscriptions with ratios documented for various Jewish populations during the nineteenth century likewise indicates that the ratios suggested by the ancient inscriptions are far too high. Admittedly, ancient and nineteenth century ratios sometimes correspond, yet this only happens when the ancient inscriptional data are exceptionally low and the nineteenth century demographic data are exceptionally high (compare, for example, the sex ratio of St. Petersburg 1866-72 with that of ancient Rome). In the nineteenth century, the sex ratio for Jews in Europe averaged 105-110—a figure that seems reliable in the light of recent collections of sex ratios for non-Jewish populations.[98] This is considerably lower than the approximately 150 suggested by Jewish inscriptions dating to the first few centuries of the common era. Thus Table 12 shows that the lower percentage of epitaphs dedicated to women as documented in Table 11 must indeed be ascribed to epi-

---

[94] Hopkins 1966, 260 and especially Keyfitz 1971, 307f.

[95] Blumenkranz 1961, 345 suggests that Jewish immigrants from other parts of the Empire and converts seriously affected the demographic structure of the Jewish community of ancient Rome, but such an assertion must remain entirely hypothetical.

[96] It should be noted that the evidence from *Misenum* is atypical (21 males for every female!). This must be explained by the fact that Misenum was a naval base (Suet. *Aug.* 49), which attracted males who, in their capacity of soldiers, could not marry officially-- at least not until the 160s C.E., see C. G. Starr, *The Roman Imperial Navy. 31 B.C.-A.D. 324* (New York: Barnes and Noble, 1960) 90-91. Thus, many epitaphs from the area concern unmarried sailors who had died before fulfilling their full term of military service.

[97] Note the almost identical ratios for Jewish and non-Jewish Rome.

[98] See the tables in Keyfitz 1971.

134   CHAPTER THREE

graphic practice and not to the particular demographic characteristics of ancient populations. Put differently, in Roman Palestine as well as in the Diaspora, Jewish men were more likely to receive an epitaph than Jewish women. Only in Jewish inscriptions that postdate Late Antiquity does this disparity appear to have disappeared.[99]

The relative lack of inscriptional references to women can be explained in at least two different ways (note that one explanation does not exclude the other). One explanation stresses the fact that upon marriage males were generally older than females. Therefore, males were likely to predecease their wives, who then commemorated their dead husbands with an inscription.[100] The implication of such an explanation is that these widows never received a proper epitaph because once they themselves passed away there was nobody left to commemorate them. A second explanation pays special attention to the content and function of funerary inscriptions. In the Jewish inscriptions from Beth She'arim, it can be observed that inscriptions do not so much commemorate individuals. Rather these inscriptions document the owner of a grave, a set of graves, or a burial room. Thus, even though many different people could be buried in a given area, the accompanying inscriptions often features the name of only one individual, namely the person who had paid for these graves.

The evidence compiled in Table 11 and 12 is especially interesting inasmuch as it compels one to reflect on several issues raised in recent work on Jewish women in antiquity. It has been argued that titles such as "mother of the synagogue," "elder" and the like should not *a priori* be considered as mere honorary epitaphs. Some scholars maintain that such titles could equally well point to a more active kind of leadership on the part of women in the Jewish communities of antiquity.[101] The concept of "women leaders" is certainly a useful one as long as one keeps in mind that female "leadership" in the ancient synagogue may have taken a variety of forms—just as in the early Church, where the donation of large sums of money by women did not automatically result in such patrons becoming an integral part of ecclesiastical leadership.[102] The actual number of inscriptions attesting to females

---

[99] E.g., in thirteenth and fourteenth century Paris 46.4% of the Jewish epitaphs relate to males and 53.6% to females; in Strasburg during the same period, figures are comparable (48.5% males and 51.5 % females). These percentages have been derived from the inscriptions published by Nahon 1986.

[100] Thus Ery 1967, 52-53 in reference to non-Jewish inscriptions.

[101] B. Brooten, *Women Leaders in the Ancient Synagogue* (Chico: Scholars Press, 1982). esp. 7-10; R. Kraemer, "An Inscription from Malta and the Question of Women Elders in the Diaspora Jewish Communities," *Harvard Theological Review* 78 (1985) 431-38.

[102] See, in general, E. A. Clark, "Women as Patrons in Late Ancient Christianity," *Gender and History* 2 (1990) 253–73.

playing roles in the synagogue is another factor that needs to be taken into account in discussions on women in the ancient synagogue. The fact that in Rome only five Jewish female dignitaries are attested as opposed to 117 Jewish male dignitaries (or one female per 23.4 male functionaries) certainly tells us that within this Jewish community, Jewish women in leadership positions were the exception rather than the rule.[103] In any attempt to define the role played by women in Jewish religious life, enough attention should finally also be paid to evidence other than titulature. That women may have played a somewhat limited role in the religious life of the community, at least insofar as the Jewish community of ancient Rome is concerned, is borne out by epithets in Jewish inscriptions from Rome—even though the evidence is rather scarce. In these inscriptions, men are sometimes called νομομαθής or μαθητής σοφῶν (most likely the Greek equivalent of the Hebrew תלמיד חכם or "student of the sages"), while for a Jewish woman the much more neutral "lover of the law" is used on one occasion.[104] The famous epitaph of Regina likewise talks about her *observantia legis* in rather general terms, but it does not specify what such observance of the Law actually entailed.[105] It is conceivable that the diversity in terminology is indicative of differences in communal and religious requirements for males as opposed to females (males study the law, females merely observe it). The fact that women are sometimes also referred to as ὅσια cannot necessarily be taken to indicate that women were active in the public religious life of the synagogues, for it can equally well point to scrupulous religious observance in a more private sphere.[106]

In this context it is interesting to turn briefly, again, to the relevant epigraphic practices. Even though women were less likely to be recorded in funerary inscriptions (above, Tables 11 and 12), this cannot be taken to mean that in the commissioning of funerary inscriptions, Jewish women were less active than their male counterparts.

---

[103] *Contra* Brooten 1982, who writes (p. 149) that "the point is not whether these women were the exception or not ... but whether their titles were merely titles or whether they implied actual functions, just as for the men."

[104] *CIJ* 193, 333, and 508 (males); *CIJ* 132 (female). The same expression is also to be found on *CIJ* 72, but the inscription is not Jewish, see Chapter 2, section "Jewish Sarcophagi from Rome." Kraemer 1986, 89 suggests that the *discipulina* mentioned in *CIJ* 215 should be taken as a female student of the law, but this suggestion must hypothetical: note also that in *CIJ* 190 a female μαθητής is mentioned, yet also in this case has the addition "of the law" been omitted. Juvenal's sixth *Satire* cannot be used as evidence to argue that Jewish women studied Jewish law at Rome, *contra* R. S. Kraemer, *Her Share of the Blessings. Women's Religions Among Pagans, Jews, and Christians in the Greco-Roman World* (New York and Oxford: Oxford U. P., 1992) 109. For male "lovers of the law," see *CIJ* 203 and 509.

[105] *CIJ* 476. Note that not all scholars consider this as a Jewish inscription.

[106] *Contra* Kraemer 1986, 89-90. On this term, see Chapter 5.

Thirty-two times Jewish women can be observed to have arranged independently for such an epitaph (as opposed to forty-five epitaphs erected by males).[107] Such evidence suggests, generally speaking, that in the private (as opposed to the public or religious) sphere Jewish women in Rome may have had the freedom to act as they saw fit. More specifically, this evidence indicates that the task of erecting a funerary inscription was not passed automatically to other male members of the family, but that Jewish women could take such matters into their own hands. In non-Jewish inscriptions, the inclusion of the commemorator's name was probably motivated by concerns of heirship.[108] Thus, by analogy, the evidence provided by the Jewish epitaphs from Rome might indicate that Jewish women in Rome could be heirs to the property of their spouses.

The epigraphical evidence from Rome is particular interesting when it is compared to the taxonomy of women as reflected in the Mishna. In both cases, participation of women in the study of sacred texts and other comparable public religious activities appears to have been limited, while in private affairs (oaths, vows, litigation, business), Jewish women enjoyed a considerable degree of freedom.[109] Of course, this is not to intimate that the Jewish epitaphs from Rome attest to a profound influence or even to the presence of a rabbinic type of Judaism in that city. The similarities only suggest that the differences between the role of Jewish women in this Diaspora community as reflected in epitaphs and the way rabbinic circles in Roman Palestine perceived that role may have been smaller than one would perhaps anticipate. Such similarities raise the possibility that Judaism of a rabbinic type was accepted in Diaspora communities because it built on traditions that had long been widespread among the Jewish communities of antiquity.[110]

Implications

In this chapter, we have seen that to refer to age at death in inscriptions is a common Roman practice that was taken over by Jews in the Diaspora and, to a lesser extent, also by the Jews of Roman Palestine. We have furthermore also seen that the same factors are operative in

---

[107] Sixteen times epitaphs were erected by males and females together.

[108] R. P. Saller and B. D. Shaw, "Tombstones and Roman Family Relations in the Principate: Civilians, Soldiers and Slaves," *Journal of Roman Studies* 74 (1984) 126; E. A. Meyer, "Explaining the Epigraphic Habit in the Roman Empire: The Evidence from Epitaphs," *Journal of Roman Studies* 80 (1990) 74-78.

[109] J. R. Wegner, *Chattel or Person? The Status of Women in the Mishnah* (New York: Oxford U. P., 1988) *passim*.

[110] See, for further details, Chapter 5.

both Jewish and non-Jewish inscriptions. Like non-Jewish inscriptions, Jewish inscriptions cannot be used to reconstruct ancient mortality and life expectancy patterns. Inasmuch as epigraphic practices determine the character of references to age at death in all ancient inscriptions (above, Approaches 1 through 5.b), inscriptions referring to age at death reflect the statistics of commemoration rather than the statistics of mortality.

A detailed analysis of references to age at death in Jewish epitaphs in comparison to such references in non-Jewish epitaphs has permitted us to see that similarities between Jewish and non-Jewish epigraphical materials are surprisingly strong. For example, the percentages of male and female children under the age of ten referred to in inscriptions (*supra*, Table 2) are practically identical for both Jewish and non-Jewish epitaphs. Likewise, the percentages of inscriptions relating to Jewish and non-Jewish male children are somewhat higher than those concerning female Jewish and non-Jewish children (Table 2). Again, the total number of inscriptions commemorating Jewish and non-Jewish children is higher than the number of inscriptions erected for Jewish and non-Jewish people over seventy years of age (Table 2). Comparably, no significant differences exist in the excesses in age rounding in Jewish and non-Jewish inscriptions from Rome, at least insofar as males are concerned (above, Table 6). Moreover, in Rome, the relationship between the total number of inscriptions referring to age at death relating to males as opposed to females is strikingly similar in Jewish and non-Jewish inscriptions (above, Tables 11 and 12). To all these similarities can finally be added that Jewish and non-Jewish inscriptions from Rome are identical in that in both cases about one third of the total number of inscriptions refers to age at death.[111]

Such a state of affairs raises important questions about the relationship between Jews and non-Jews in ancient Rome, or, for that matter, anywhere in the Diaspora. Given the fact that referring to age at death was a Roman epigraphic custom, one wonders how and why Jews copied it. Did such Jews copy it directly from Roman inscriptions they had seen, or did they avail themselves of Roman workshops in the same way as Jews who ordered wall painting and sarcophagi from common Roman workshops?[112] Did Roman stonecutters offer certain standard formulae from which their Jewish customers could then

---

[111] For the non-Jewish inscriptions, see Kajanto 1968, 10. Of the 430 Christian inscriptions studied by P. Saint-Roch, "Enquête 'sociologique' sur le cimetière dit "Coemeterium Sanctorum Marci et Marcelliani Damasique'," *Rivista di Archeologia Cristiana* 59 (1983) 421-22, 142 or 33% refer to age at death; 62 or 43.6% relate to children under 10 years of age.

[112] On workshop identity, see Chapter 2.

choose?¹¹³ To what extent were Jews who buried their dead in the Monteverde, Vigna Randanini, and Villa Torlonia catacombs tempted to use (Roman) standard phraseology? If they did, at what point did such standard phraseology lose its non-Jewish connotations to become an integral part of Jewish funerary inscriptions? Can the similarities between Jewish and non-Jewish inscriptions from Rome that refer to age at death be taken to imply that in terms of commemorating the dead, Jews and non-Jews had more in common than traditional scholarship has been willing to allow? If it is indeed true that Jews and non-Jews shared ideas as to who (male/female) and when (young/old) to commemorate someone by means of a funerary inscription, what are the implications of this common trait in terms of the social and intellectual relationship between Jews and non-Jews in antiquity?

It is most unfortunate that we cannot fully answer all of these questions. Despite such unsatisfactory results, one conclusion can nevertheless be drawn here: the relationship between Jews and non-Jews was more organic and dynamic in nature than some previous students of the Roman Jewish community have been prepared to admit. If the relationship between these two groups would have been otherwise, Jews would never have borrowed epigraphic formulae from non-Jews. In turn, such a conclusion raises another question: given the fact that Jews freely borrowed epigraphic formulae from the larger world surrounding them, what is the extent of such borrowing and how is one to evaluate it? To answer this question, other aspects of the Jewish epitaphs from Rome must be analyzed. Let us turn to the onomastic data provided by the Jewish funerary inscriptions from Rome, and then to the language of Rome's Jewish epitaphs.

---

[113] Note that in the two Jewish catacombs under the Villa Torlonia in Rome, references to age at death appear twice as often in marble inscriptions (which had to be carved by professional craftsmen) than in painted inscriptions (which could be painted by Jews on the spot). I have counted 19 references on a total of 29 marble inscriptions (56.6%) and 19 references to age at death on a total of 75 *dipinti* (26.6%). This suggests that the customs of referring to age at death originated in inscriptions. Note also that in the case of early Christian inscriptions referring to age at death, the influence of the *marmorarii* (as opposed to the family of the deceased) is believed to have been considerable, see H. Nordberg, *Biometrical Notes*. Acta Instituti Romani Finlandiae II:2 (Helsinki: Tilgman, 1963) 44.

CHAPTER FOUR

THE ONOMASTICON OF THE JEWISH COMMUNITY OF
ROME: JEWISH VIS-A-VIS NON-JEWISH ONOMASTIC
PRACTICES IN LATE ANTIQUITY

Introduction

Some scholars interpret the employment of non-Jewish names by the Jews of antiquity as indicative of their Hellenization or Romanization. Other scholars believe, however, that name-giving practices tell us little or nothing about the relationship between Jews and non-Jews in the ancient world. In this chapter we will study this old question afresh on the basis of the rich onomastic evidence contained in the Jewish funerary inscriptions from Rome.[1] Along with Jewish name-giving practices in Rome, Edfu (Upper Egypt), Beth She`arim (Galilee), and Venosa (southern Italy), non-Jewish onomastic evidence will be studied, too, to determine the characteristics of third- and fourth-century C.E. Jewish onomastic practices in Rome.

Inasmuch as it may serve as an introduction to this complicated subject, let us first turn to an investigation of Leon's analysis of Jewish onomastic evidence from Rome.

*A Critique of Leon's Interpretation of Jewish Onomastic Evidence
From Rome*

To date, Leon's investigation of the names contained in the Jewish epitaphs from the Jewish catacombs is the only substantial study to deal with this particular aspect of Jewish funerary epigraphy in Rome.[2] Although new inscriptions have come to light since Leon first published the results of his research, these new discoveries do not affect significantly the figures and percentages Leon originally calculated. Presenting most of the evidence in the form of tables, Leon discovered that among the Jews of ancient Rome Latin names were more popular than Greek and Greek more popular than Semitic names. He also noted that Latin names outrank Greek ones not merely in Jewish

---

[1] I have excluded from the following survey the names of Roman Jews that are mentioned in literary sources (see the list in Solin 1983, 658-61). The limited number of these references precludes using them for statistical purposes.
[2] Leon 1960, 93-121.

funerary inscriptions written in Latin, but also in Jewish epitaphs composed in Greek. Recent counts have confirmed Leon's conclusions. In Rome, the Jewish onomastic repertoire as recorded in inscriptions consists of 274 Latin names, 230 Greek names, and 79 names of Semitic origin.[3]

Leon then attempted to explain some of the patterns he observed in terms of a higher or a lesser degree of Romanization. After determining how often Semitic, Greek, and Latin names occur in the four major Jewish catacombs of Rome, Leon concluded that "the Appia [that is, Vigna Randanini] group included the most Romanized congregation, the Monteverde the most conservative, and the Nomentana [that is, the two Jewish catacombs under the Villa Torlonia] the most Hellenized and least Romanized."[4] This conclusion was based on the following table.

Table 1: Language of name by catacomb = Leon's Table IV.

Names	Vigna Randanini	Monteverde	Villa Torlonia
Some Latin	63.6%	55.0%	48.3%
Some Greek	38.1%	34.3%	45.0%
Some Semitic	11.3%	20.1%	11.7%

Reading the columns in Table 1 from left to right, it is clear why Leon believed that the names preserved in the three main Jewish catacombs in Rome reflect different degrees of Romanization. In the Vigna Randanini, or, to use Leon's terminology, in the most Romanized of all Jewish catacombs, Latin names occur more often (63.3%) than in the Monteverde (55.0%) or Villa Torlonia catacombs (48.3%). Comparably, Greek names are more common in the Villa Torlonia (or most Hellenized) than in the other two Jewish catacombs (45.0% versus 38.1% and 34.3%, respectively). Differences in the percentages of Semitic names further complement this picture. The fact that the percentage of Semitic names in the Monteverde catacomb is as high as 20.1% seems to justify an identification of this Jewish catacomb as the most conservative.[5]

Yet, there exist several other ways to interpret the data collected by Leon. In addition to pointing out which percentages are higher than others, it is no less essential to keep in mind *how much* higher some percentages are in comparison to others. For example, the percentage

---

[3] Solin 1983, 711.

[4] Leon 1960, 110.

[5] Note also that the percentage of double names made up of a Latin and a Semitic names is highest in case of the Monteverde catacomb. *Duo* and *tria nomina* in the Jewish epitaphs will be discussed *infra*, see "The Influence of Roman Name-Giving Practices on the Jewish Onomasticon in Late Ancient Rome: The Question of the Duo and Tria Nomina."

of Greek names recorded in the Vigna Randanini or most Romanized catacomb is practically identical to the percentage of Greek names in the conservative (as defined by Leon) Monteverde catacomb (38.1% versus 34.1%). Likewise, the differences between the percentages of Latin names in the two Villa Torlonia or most Hellenized of the three Jewish catacombs cannot be said to be substantially different from the percentages documented for the conservative Monteverde catacomb (48.3% versus 55.0%).

Similarly, Leon noted that Semitic names occur in one sixth of all Jewish inscriptions written in Greek, and that in Jewish inscriptions composed in Latin the total number of recorded Semitic names drops to one eighth.[6] Converted to percentages, 16.3% of all Jewish epitaphs in Greek contain Semitic names as opposed to 12.2% of the Jewish epitaphs in Latin. At best, such figures illustrate the general unpopularity of Semitic names. In my view the differences in recorded percentages are, however, really too insignificant to argue that traditional Semitic names were a remarkable feature of inscriptions written in Greek as opposed to inscriptions carved in Latin. These percentages suggest rather that no direct relationship existed between the language of the inscription and the name of the person commemorated in it.

Along similar lines, *duo* and *tria nomina* (double or triple names) do not occur in significantly higher numbers in the Vigna Randanini than in the Monteverde catacomb (45 versus 41 examples).[7] Put differently, there are no significant differences in the type of Latin names employed by users of the Monteverde as opposed to those of the Vigna Randanini catacomb. Finally, the percentage of Greek names documented in the Hellenized community buried in the Villa Torlonia catacombs is a mere 6.9% higher than those recorded for the Romanized community of the Vigna Randanini catacomb. It may thus be clear that Leon's exclusivist explanation of the data in terms of accommodation versus traditionalism (that is, the use of specifically Jewish names) highlights certain data at the expense of others.

Another way of studying the percentages contained in Table 1 is by reading the columns in Table 1 from top to bottom rather than from left to right. Thus, a different and, in my view, a more nuanced picture of onomastic practices within each individual Jewish catacomb emerges. We may observe, for example, that in the Villa Torlonia or the most Hellenized of catacombs the percentage of Latin names is higher (!) than the percentage of Greek names, even though the difference is only 3.3%. Comparably, in the Monteverde or most conserva-

---

[6] Leon 1960, 109.

[7] The percentages given by Leon 1960, 112, Table VII are flawed for reasons that will be explained in the section "The Influence of Roman Name-Giving Practices on the Jewish Onomasticon in Late Ancient Rome: The Question of the Duo and Tria Nomina."

tive catacomb, in which Semitic names occur more frequently than in any of the other Jewish catacombs of Rome, Latin names makes up more than half (!) of all epigraphically recorded names. If one would systematically apply Leon's terminology to the Monteverde catacomb, one would have to conclude that the people buried in this catacomb were conservative and Romanized at the same time. How minimal the differences between the different catacombs really are in terms of the kind of names employed (Semitic, Greek, and Latin) is particularly evident when the figures calculated by Leon are presented in the form of a chart (chart 1).

Chart 1: Leon's Table IV in chart-form.

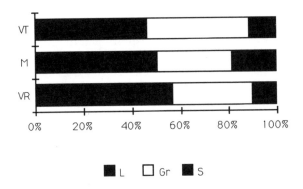

Still other problems arise from Leon's attempt to interpret differences in onomastic practices as reflecting variations in the degree of accommodation to contemporary Roman society. The language in which an inscription was carved is no less of a determining factor for evaluating the historical value of onomastic evidence. In the Vigna Randanini catacomb, Latin names are admittedly employed approximately twice as often than Greek names (63.6% versus 38.1%). Yet, such onomastic preferences contrast with the fact that the majority of epitaphs from this same catacomb was not composed in Latin but in Greek (63.1% inscriptions in Greek versus 36.9% in Latin).[8] In other words, although the Jews buried in this catacomb showed a predilection for Latin names, such a predilection was combined with a preference for the Greek language. In the Monteverde catacomb, these differences are even more striking. There, some 78% of all inscriptions were written in Greek, yet more than half of all inscriptions carry Latin names (55%).[9] Even in the two catacombs under the Villa Tor-

---

[8] For the Vigna Randanini catacomb, see *CIJ* 81-204 (in Greek) and 204-76 (in Latin).
[9] For the Monteverde catacomb, see *CIJ* 290-455 (in Greek) and 456-93 (in Latin).

Ionia in which only 6% of the inscriptions are in Latin, 48.3% of the names are in Latin. Put differently, in the "conservative" Monteverde and Villa Torlonia catacombs, Greek inscriptions are more likely to carry Latin names than in the "Romanized" Vigna Randanini catacomb. Clearly, to apply Leon's ill-defined and antithetically conceived concepts "Romanization," "Hellenization," and "conservatism" to such evidence is to create categorizations that are contradictory.

Also problematic about Leon's presentation of the evidence, finally, is that he represents the Jewish onomastic data from Rome as a chronologically-uniform corpus of evidence. He compares the total number of Greek names with the total number of Latin and Semitic names without paying any attention to the dating of individual inscriptions. Such an approach prevents one from detecting eventual shifts in onomastic adaptation on the part of the Roman Jewish community.[10]

It may be clear, in short, that the onomastic evidence does not support Leon's view that "the users of the Via Appia catacomb represent the most Romanized group, those of the Monteverde catacomb the most conservative, and those of the Via Nomentana catacomb the most foreign and least assimilated."[11]

In accommodating to non-Jewish practices, no identifiable differences between the various Jewish catacombs in Rome can be detected. Inasmuch as we know that people from the very same Jewish community are buried in different catacombs, such differences would really come as a surprise.[12] Let us see if there exist other ways to determine if (and if so, how) onomastic data can be used to determine the extent to which Jews adapted to non-Jewish name-giving practices.

Aspects of Jewish Onomastic Practices in Late Ancient Rome: Looking for a Pattern

Two methodological problems that are difficult to solve face the student of ancient Jewish onomastic practices: (1) our inability to determine if certain names are really Greek names or transliterations of Hebrew names and (2) our inability to date inscriptions except in the most general terms.

Leon has suggested that the Greek Σίμων is nothing but the Greek rendering of the Hebrew שמעון. Further examples of Greek names that

---

[10] Leon 1960, 87-92 dating the inscriptions to the second and third century C.E. The Jewish funerary inscriptions from Rome, however, date to the third and fourth, and possibly the early fifth century C.E., see Solin 1893, 656, 684, 694 n. 235, 701, 708 n. 287.

[11] Leon 1960, 258.

[12] This was the case with, for example, the members of the Jewish community of the Siburesians, see Leon 1960, 151.

are very similar to Hebrew names, and that were perhaps chosen by Jews because they were translations of Hebrew names, or because such names sounded similarly, can easily be added.[13] It is evident that in a study on the interaction of Jews and non-Jews it is crucial to know whether Jews either chose Greek names because of homonimity with Hebrew names or whether they chose such names because they were Greek names. Similarly, it would be crucial to know whether certain Latin names are real Latin names or whether they derive from Greek prototypes.[14] Only in exceptional cases does the transcription of a Hebrew name into Greek show that a Greek name is to be understood as a translation of a Hebrew original.[15] Unfortunately, in most other cases, we simply cannot know what considerations determined the choice of the names in question. As provisional such a solution may seem, in the following survey I have arranged names such as Σίμων under the heading "Greek names."

We have just seen that one of the criticisms that may be raised against Leon's treatment of the onomastic data is that his approach does not enable us to detect eventual shifts in onomastic practices. This brings us to the second methodological problem just referred to, namely our inability to date individual inscriptions. If one could date each inscription, it would be easy to arrange the onomastic data in periods of fifty and hundred years. Thus, fluctuations in onomastic practices could easily be registered, and much could be learned about chance or continuity in name-giving practices over the long term.

Yet, even though we cannot divide the onomastic history of the Jews of Late Ancient Rome into chronologically-defined periods, it is possible to study some cases in the development of Jewish onomastic

---

[13] Leon 1960, 105. For further examples, see J. Juster, *Les Juifs dans l'Empire Romain* (Paris: Geuthner, 1914) vol. 2, 230-32; H. Leclercq in *DACL* 12 (1936) 1552; Schwabe and Lifshitz 1974, 209; Lifshitz in Prolegomenon to *CIJ* (1975) 23; Solin 1983, 639-40; G. Mussies, "Jewish Personal Names in Some Non-Literary Sources," in J. W. van Henten and P. W. van der Horst (eds.), *Studies in Early Jewish Epigraphy* (Leiden: Brill, 1994) 244-49. N. Cohen, "Jewish Names as Cultural Indicators in Antiquity," *Journal for the Study of Judaism* 7 (1976) p. 112 attributes the the popularity of the name Σίμων among the Jews of the Roman period to the influence of Greek culture rather than to the fact that the name had biblical roots, but then contradicts this statement elsewhere in her article (n. 74).

[14] On foreign substrata in Latin names, see I. Kajanto, *Onomastic Studies in the Early Christian Inscriptions of Rome and Carthage = Acta Instituti Romani Finlandiae II.1* (1963) 55 and *id.*, *The Latin Cognomina = Commentationes Humanarum Litterarum* 36.2 (1965) 16f.

[15] E.g., a Jewish inscription from Gortyn on Crete transcribes the name Moses as Μωσης while Christians normally transcribe this name as Μωυσῆς, see S. V. Spyridakis, "Notes on the Jews of Gortyna and Crete," *Zeitschrift für Papyrologie und Epigraphik* 73 (1988) 173. Similarly, Jews at Corycos prefer Symoon and Iako where Christians appear to have favored Symeonis and Iakobos, see M. H. Williams, "The Jewish Community of Corycus. Two More Inscriptions," *Zeitschrift für Papyrologie und Epigraphik* 92 (1992) 251.

practices over shorter periods of time. Inscriptions often mention not only the name of the deceased, but also that of another family member, usually the father and/or mother of the deceased. By collecting these inscriptions it is possible to study Jewish name-giving practices over shorter periods of time, namely within individual families. The evidence normally spans two generations, or forty to sixty years. In Table 2.a and 2.b, the relevant data have been arranged graphically.

Table 2.a: Inscriptions from the Jewish catacombs of Rome that include the name of at least one parent and one child. Inscriptions in *Latin*.

	Parents		Children	
	Males	Females	Males	Females
Vigna Randanini				
208	-	Lat(dn)	-	Lat/Gr(dn)
209	Lat/Sem(dn)	Lat(dn)	Gr	-
212	Lat/Sem(tn)	Lat/Sem(dn)	-	Lat(dn)
213	-	Gr	-	Sem
230	-	Gr	Lat (2x)	-
232	-	Gr	Gr	-
240	Lat	Lat	-	Gr
241	-	Lat	-	Lat
248	Lat	Lat	-	Lat
252	Lat	Sem	-	Gr
259	Lat/Sem(dn)	-	Lat/Gr(dn)	-
263	Lat	-	Sem	-
264		Lat	Lat	-
267	Lat	Sem	-	Lat
269	Sem	-	Sem	-
Monteverde				
461	-	Lat/Gr(dn)	-	Lat(dn)
470	Lat(dn)	Lat/Gr(dn)	Lat(tn)(2x)	Lat/Sem(tn)(2x)
475	-	Lat	Lat	-
Possibly not Jewish				
30*	-	Gr	-	Lat
35*	Lat	Gr	Lat(dn)	-

Explanation: numbers refer to *CIJ*; Lat=Latin name; Gr=Greek name; Sem=Semitic name; dn=*duo nomina*; tn=*tria nomina*.

Table 2.b: Inscriptions from the Jewish catacombs of Rome that include the name of at least one parent and one child. Inscriptions in *Greek*.

	Parents		Children	
	Males	Females	Males	Females
Provenance Unknown				
1	Lat	Sem	Lat	-
3	Gr	-	Lat	-

146   CHAPTER FOUR

(Table 2.b continued)
Villa Torlonia

12	-	Lat	Sem(2x)+Gr	Sem(2x)
13	Lat	-	Lat	-
30	Sem	-	Gr	-
31	Gr	-	Gr	-
55	Gr	-	Gr	-
F25	Gr	-	Lat	-
F36	Gr	-	Gr	-

Vigna Randanini

84	Gr	-	Gr	-
85	Gr	-	Gr	-
88	Lat	-	Sem/Lat	-
92	-	Gr	Lat+Gr	-
95	Gr	Lat	Gr	-
102	Lat	-	-	Lat
106	Gr	-	Lat	-
108	Lat	-	-	Gr
112	Gr	-	Gr	-
119	Gr	-	Gr(2x)	-
124	-	Lat	Lat	-
125	Lat	-	Lat	-
129	Lat	-	-	Lat
137	Gr	-	Sem	-
140	Gr/Sem	-	Lat	-
141	Gr	-	-	Lat
144	-	Lat/Gr(dn)	Lat	-
145	Lat	-	Lat	-
146	Lat(son of 145?)	-	Lat	-
147	Lat	-	-	Lat
149	Lat	Lat	Lat	-
153	-	-	Lat	Sem
155	Lat	-	-	Sem
156	-	Lat	-	Sem
159	Gr	-	-	Lat
167	Lat	-	-	Lat
169	Gr	-	-	Gr
171	-	Gr	Lat	-
172	Lat(dn)	-	-	Lat(dn)
286	-	Gr	Lat	-

Monteverde

307	-	Lat/Gr(dn)	Lat	-
310	Gr	-	Lat	-
314	Lat	-	-	Lat
322	Lat	-	Gr	-
329	-	Gr	Sem	-
334	Gr	-	-	Gr
370	Gr	-	Gr	-
376	Sem	-	-	Lat
385	-	Lat	Gr(grandson)	-
389	-	Lat	-	Lat
412	Gr	-	Gr	-
413	-	Lat/Lat(dn)	-	Lat/Sem(dn)

(Table 2.b continued)

Provenance Unknown

497	Sem(dn)	-	Gr	-
501	Gr	-	Gr	-
502	Gr	-	Gr+Lat	-
504	Lat	-	Lat	-
510	Sem	-	-	Sem
511	Lat	Gr	Gr	Sem
732	-	Gr	-	Sem

Explanation: numbers refer to *CIJ*; numbers preceded by F. refer to Fasola 1976; Lat=Latin name; Gr=Greek name; Sem=Semitic name; dn=*duo nomina*; tn=*tria nomina*.

Tables 2.a and 2.b illustrate that within one single Jewish family in Rome, children could be given names of different linguistic origin. In one case, two sisters and two brothers all had Semitic names, while a third brother bore a name that was distinctively Greek.[16] Other inscriptions display a similar pattern. They indicate that in one family, some children bore Semitic names, while for other children in that same family, Greek or Latin names were preferred.[17] Such data suggest that some Jewish families in Rome chose freely from among the entire repertoire of available names (Semitic, Greek, and Latin), rather than confining themselves exclusively to one linguistic group of names.

The impression that the Jews of Late Ancient Rome switched back and forth almost indiscriminately between Semitic, Greek, and Latin names is furthermore confirmed when the data contained in Tables 2.a and 2.b are rearranged in a somewhat different manner. Tables 3.a and 3.b contain references to parents whose names differ linguistically from the names borne by their children. When parents bear Semitic or Greek names, but their children Greek or Latin names, one can consider such evidence as an indication of a continuing adaptation to non-Jewish onomastic practices. Conversely, when parents bear Latin or Greek names, while at the same time they have chosen Greek and Semitic names for their offspring, a reversion to more traditional practices of name-giving may be postulated. Here are the data.

---

[16] *CIJ* 12.

[17] E.g. *CIJ* 92: a Latin and a Greek name for children in the same family; and *CIJ* 153: a Latin and a Semitic name for children in the same family.

148   CHAPTER FOUR

Table 3.a: Inscriptions of parents and children bearing linguistically-differing names. Accommodation to non-Jewish onomastic practices in the case of children.

Name Parents	Name Children	No. of Inscriptions
1. Sem	Gr	2
2. Sem	Lat	3
3. Gr	Lat	12
4. Lat/Sem(dn)	Lat (dn)	1
5. Lat/Gr(dn)	Lat (dn)	1
6. Lat	Lat (dn)	1
Total		20

References: 1.: *CIJ 30*, 497; 2.:*CIJ* 263, 267, 376; 3.: F25, *CIJ* 92, 106, 140, 141, 159, 171, 230, 310, 503, 30*; 4.: *CIJ* 212; 5.: *CIJ* 461, 470; 6.: *CIJ* 35*.

Table 3.b: Inscriptions of parents and children bearing linguistically differing names. Reversion to Jewish onomastic practices in the case of children.

Name Parents	Name Children	No. of Inscriptions
1. Lat	Gr	8
2. Lat	Sem	2
3. Gr	Sem	5
4. Lat/Sem(dn)	Gr	1
5. Lat/Lat(dn)	Lat/Sem(dn)	1
6. Lat/Lat(dn)	Lat/Gr(dn)	1
Total		18

References: 1.: *CIJ 12*, 108, 125, 240, 252, 322, 385 (grandson), 511; 2.:*CIJ* 155, 156; *CIJ* 3, 137, 213, 329, 511, 732; 4.: *CIJ* 209; 5.: *CIJ* 413; 6.: *CIJ* 208.

Tables 3.a and 3.b show that in name-giving, the Jews of Late Ancient Rome displayed a considerable degree of flexibility. A total of twenty inscriptions indicate that segments of the Roman Jewish community named their children in ways that were customary in contemporary non-Jewish rather than in Jewish society (Table 3.a). Yet, eighteen inscriptions attest to the opposite tendency, namely, the employment of names for children that are more traditional than the names borne by the parents (Table 3.b). Thus, studying the surviving evidence in this way, there is no reason to suppose a consistent and continuous process of accommodation during which Latin names were increasingly replaced by Greek and Semitic names. Rather, Semitic, Greek, and Latin names all seem to have been used freely by the Jewish community in third- and fourth-century C.E. Rome.

It could be argued that this conclusion merely results from the specific research method employed. One could maintain that *if* the inscriptions were more precisely datable, a different, more consistent pattern of Jewish name-giving could perhaps be identified. Yet, interestingly, when the Jewish onomastic data are placed into their original archaeological context, a comparable mixture in the use of Semitic, Greek, and Latin names can be observed.

Only in the cases of the two Jewish catacombs under the Villa Torlonia can the exact provenance of Jewish funerary inscriptions be established.[18] Most inscriptions stem from two areas in the lower catacomb and were written in Greek.[19] A considerable number of them contain names that can be reconstructed and identified.[20] As has been observed in Chapter 2, this part of the Villa Torlonia catacomb consists of a regular network of underground galleries. Although we do not know how much time elapsed between the construction of one gallery and the next, the formal features of the galleries indicate that this part of the catacomb had been laid out systematically to accommodate a possibly high number of burials. One may presume that most, if not all burials in this region of the catacomb took place in one and the same period.

Looking at the distribution of Semitic, Greek, and Latin names in this part of the catacomb, it can be observed that no concentrations of any one group of names exist. Rather the opposite is the case. People bearing Greek names were buried alongside those bearing Latin and Semitic names. In fact, the amalgam of Semitic, Greek, and Latin names almost could not have been more complete. It thus becomes clear that Leon's contention that the people buried in the two catacombs under the Villa Torlonia were more Hellenized than other segments of the Roman Jewish community needs to be modified. The evidence from the lower catacomb under the Villa Torlonia suggests that among the people who were buried there over a relatively short period of time, Greek, Latin, and to a lesser extent Semitic names, all enjoyed a comparable degree of popularity.

Such evidence complements tables 2.a, 2.b, 3.a, and 3.b in that it shows that the Jews of Rome used Semitic, Greek, and Latin names interchangeably.

There is, however, still a third way to document that the pattern observed is characteristic of Roman Jewish onomastic practices in Late Antiquity, namely by studying the relationship between Semitic, Greek, and Latin names borne by various Jewish community officials in Rome. These data have been presented in Table 4. References to

---

[18] The excavators of the other Jewish catacombs of Rome did not bother to record exactly the original find spot of the inscriptions they had discovered.

[19] Region J and E on the map published by Beyer and Lietzmann 1930; on Fasola's 1976 map, the region is designated by the letter D.

[20] Add to the inscriptions published by Beyer and Lietzmann 1930, the inscriptions in Fasola 1976, 56 (three inscriptions). The following inscriptions are too mutilated to reconstruct the names they once contained: Beyer and Lietzmann 1930 (map) nos. 7, 9, 10, 12, 13, 25, 30, 36, 37, 39, 40, 53, 55, 57, 59, and 60. Note, however, that not all the inscriptions Beyer and Lietzmann published appear on their map. I have been unable to locate inscriptions nos. 38, 41-47, 49-52, 64, 66, and 67.

female community officials have not been included, because only four such officials appear in the Jewish funerary inscriptions from Rome.[21]

Table 4: Personal names used by Jewish community officials in Rome

	Sem	Gr	Lat/Sem(dn)	Lat	Lat(dn)	Lat(tn)
Archisyn.	1	4	-	1	-	-
Archon	6	18	1	16	5	3
Didaskalos	-	1	-	-	-	-
Hiereus	3	-	-	-	-	-
Hyperetes	-	-	-	-	1	-
Gerusiarch	-	7	-	6	1	-
Grammateus	2	3	-	12	1	1
Nomomathes	-	1	-	-	-	-
Pater Syn.	2	3	-	1	1	1
Phrontistes	-	1	-	1	-	-
Prostates	-	-	-	1	-	-
Totals	14	38	1	38	9	5
Totals		52			51	

Explanation: Sem=Semitic name; Gr=Greek name; Lat=Latin nam; dn=*duo nomina*; tn=*tria nomina*.

Table 4 illustrates that Jewish community officials in Rome bore Semitic, Greek, and Latin names indiscriminately. Particularly interesting is the numerical relationship between the more traditional names, such as the Semitic and Greek names, and the openly untraditional names such as the single, double, and triple Latin names. Fifty-two times, synagogue officials bear Semitic and Greek names. Latin names, on the other hand, can be associated with such officials on fifty-one occasions. Thus, among the Jewish community officials in Late Ancient Rome, Jewish ancestral names did not enjoy a higher degree of popularity than less conventional names.[22] In fact, the relationship of Semitic to Greek and Latin names in the 104 inscriptions relating to synagogue officials (14:38:52 or 13.5%:36.5%:50%) is practically identical to the relationship of Semitic to Greek and Latin names in *all* Jewish epitaphs from Rome (13.1%:31.8%:46.1%).[23] In other words, the naming practices for community officials reflect reliably the naming practices of the Roman Jewish community at large.

---

[21] The evidence has been derived from *CIJ* with the addenda in Fasola 1976 and Solin 1983, 633-37. Under the heading *archon* I have included variants such as *exarchon* and *archon alti ordinis*.

[22] Contrast this evidence with priests of Iuppiter Dolichenus, who often carry oriental names, see M. P. Speidel, *The Religion of Iuppiter Dolichenus in the Roman Army* (Leiden: Brill, 1978) 46. It is interesting to note that the three priests mentioned in the Jewish epitaphs from Rome all bear Semitic names. The only female priest, however, bears a Latin name (*CIJ* 315).

[23] The latter percentages are those given by Leon 1960, 107, Table I.

The evidence presented in Tables 2, 3, and 4 consistently points toward a community whose onomastic practices were not uniform. To describe this community on the basis of simplified categories such as "Romanized" or "Hellenized" is to impose distinctions on the onomastic data that are far too rigid. Tables 2, 3, and 4 show that Greek and Latin names were used by Jews interchangeably. As we will now see, the interchangeable use of Semitic, Greek, and Latin names is not characteristic of Jewish onomastic practices in third and fourth century C.E. Rome alone. Jewish onomastic evidence from other parts of the Roman Empire attest to very much the same phenomenon.

Onomastic Preferences Among Jews in Other Parts of the Roman Empire

In Apollinopolis Magna (Edfu, Upper Egypt) receipts of the yearly *fiscus Judaicus* have survived in the form of ostraca. They form a rich source of information concerning Jewish onomastic practices during the period of 70 to 116 C.E. In fact, the evidence is so rich that it is possible to reconstruct a number of family trees.[24]

The name-giving as reflected in several of these family trees is particularly interesting because it displays characteristics that are remarkably similar to those documented for Jewish Rome. It can be observed that alongside Greek names, Semitic, Latin, and Egyptian names were used among the Jews in this locality.[25] It can also be observed that among the non-Jewish names employed, Greek names were most popular. Put differently, Jews in late first and second century C.E. Edfu had in common with the Jews in third and fourth century C.E. Rome that they both underwent strong outside influences in onomastic practices and that both communities used names of different linguistic origin synchronically.

As was the case in Rome, the onomastic data from Edfu do not permit us to reconstruct onomastic trends over long periods of time. Despite this drawback, the editors of the *Corpus Papyrorum Judaicarum* nevertheless maintain that the Egyptian element was continuously on the rise in the Jewish onomasticon in Edfu. In fact, they believe Egyptian names won a "complete victory" in due time. To support their statement, they observe that in documents from Edfu dating to a slightly later period (151-165 C.E.), Egyptian names occur more often than in the earlier ostraca of 70-116 C.E.[26]

---

[24] *CPJ* 2, 108f. and esp. 117.
[25] See e.g. family-tree no. 11 in *CPJ* 2, 117. Other good examples include family-trees nos. 3 and 4.
[26] *CPJ* 1, 84 and vol. 2, 118; *CPJ* 2, nos. 375-403.

Upon closer investigation, such observations fail to convince. The increase of Egyptian names in the later documents is really immaterial for the question of onomastic change. Almost all these later documents refer to members of one single family only, more precisely to the three sons of one Achillas Rufus. To be sure, one cannot be certain that the naming practices in this one Jewish family were common among other Jewish families in the area as well during the middle of the second century C.E. In addition, the earlier and later evidence from Edfu is separated by a violent and ill-documented Jewish revolt in North Africa in the reign of Trajan. We do not know how these events affected the Jewish community at Edfu. Thus, there is no way to establish whether the later tax documents from Edfu reflect the onomastic practices of the original Jewish community in Edfu, or if these documents relate to new Jewish immigrants who had settled in the area following the suppression of the Jewish revolt of 115-117 C.E.

Although little information concerning changes in onomastic patterns among the Jews of Edfu can be deduced from the later tax documents, some interesting observations on onomastic patterns can be made on the basis of the earlier ostraca. Because reigning years of individual Roman emperors are often included in these inscribed artifacts, it is possible to date the ostraca very precisely. In the following table I have subdivided the available onomastic data according to the linguistic origin of the names. The names of women have been excluded, because, once again, there are too few of them. Since the Jewish tax had to be paid every year, some names occur several times. In the following table I have mentioned such names only once.[27] The names were then classified in chronological order. The entire period for which ostraca have survived (70-116 C.E.) has been divided in two, thus creating an earlier and later generation of twenty three years each.[28]

Table 5: Names borne by Jews in ostraca from Edfu.

	70-93 C.E.	93-116 C.E.
Semitic	5 (14.7%)	12 (29.3%)
Egyptian	1 (2.9%)	6 (14.6%)
Greek	25 (73.5%)	23 (56.1%)
Latin	3 (8.8%)	- -

---

[27] Sometimes people are referred to as X son of Y. In such cases I have only included X. Y belongs to an earlier generation that falls outside the chronological framework documented in Table 5.

[28] Note that on the basis of tax documents, papyrologists consider one generation to amount to approximately thirty five years, see R. Bagnall "Religious Conversion and Onomastic Change in Early Byzantine Egypt," *Bulletin of the American Society of Papyrologists* 19 (1982) 113.

Table 5 suggests that in the early second century C.E. Semitic and Egyptian names were more popular than in the late first century C.E. among tax-paying Jews in Edfu. Supposing that Edfu's Jewish population was stationary, Table 5 suggests that a slight shift in Jewish onomastic practices occurred early in the second century C.E. Yet, unfortunately, there is no coherent body of Jewish evidence from Egypt postdating the early second century C.E. to help put the data from Edfu into a broader historical perspective. As we have just seen, later tax documents are not very helpful in this respect.

The Jewish funerary inscriptions from Beth She'arim (Galilee) constitute another important source of information concerning Jewish onomastic practices in antiquity. The Beth She'arim catacombs were in use from approximately 200 C.E. to 350 C.E.[29] Thus they are roughly contemporary to the Jewish catacombs in Rome. Most inscriptions in Beth She'arim were written in Greek, but inscriptions in Hebrew, Aramaic, and Palmyrean have also been found. As in Rome, the Beth She'arim inscriptions cannot be dated with any degree of precision. Unlike Rome, however, in Beth She'arim developments in onomastic practices can sometimes be followed over three and, in exceptional cases, over four generations. Thus, entire family trees can be reconstructed. All the pertinent data are presented in Tables 6.a and 6.b.

Table 6.a: Greek inscriptions from Beth She'arim containing the names of parents and their offspring.

	Parents		Children		Grchildren		Grgrchildren	
	M	F	M	F	M	F	M	F
BS II								
16+17	Gr	-	Sem+Gr	-	-	-	-	-
48	-	Sem	Sem	-	-	-	-	-
18-42	Palm	Sem	Sem	Sem	Sem(2x)	-	Sem(3x)	-
48-51	-	Sem	Sem	-	-	-	-	-
60	Gr	-	Gr	-	-	-	-	-
61	Gr	-	Gr+Lat	-	-	-	-	-
62	Sem	-	Sem	-	-	-	-	-
66+68	-	Sem	-	Sem	-	Sem	-	-
75+76	Sem	-	Sem	-	-	-	-	-
80	Sem	-	Sem	-	-	-	-	-
85	Gr	-	-	Sem	-	-	-	-
91	Palm	-	-	Palm	-	-	-	-
94+96	Sem	-	Gr	-	Sem	-	-	-
99+100	Gr	-	Sem	-	Sem+Lat	-	Sem	-
105	Sem	-	Sem	-	-	-	-	-
107	Gr	-	Gr	-	-	-	-	-
116	Pal	-	Palm	Gr	-	-	-	-
125	-	Gr	Sem+Lat	-	-	-	-	-

---

[29] Mazar 1973; Schwabe and Lifshitz 1974; Avigad 1976.

(Table 6.a continued)

126	Sem	-	Sem	-	-	-	-	-
127	Gr	Gr	Lat	-	-	-	-	-
147	Gr	-	-	Gr	-	-	-	-
156	Lat	-	Gr	-	-	-	-	-
182	-	Gr	-	Gr	-	-	-	-
183	-	Gr	-	Palm	-	-	-	-
189	Lat	-	Lat	-	Sem	-	-	-
296	Lat	-	Sem	-	-	-	-	-
208	Gr	-	Sem	-	Lat	-	-	-
219	Sem	-	-	Sem	-	-	-	-
220	Gr	-	Sem	-	Gr	-	-	-

Explanation: Grchildren: grandchildren; Grgrchildren: great grandchildren; M: male; F: female; BS II: Schwabe and Lifshitz 1974; Palm: Palmyrean name.

Table 6.b: Inscriptions in Hebrew, Aramaic and Palmyrean from Beth She`arim containing the names of parents and their offspring.

	Parents		Children		Grchildren		Grgrchildren	
	M	F	M	F	M	F	M	F
BS I								
12	-	Gr	-	Gr	-	-	-	-
23	Palm	-	Sem	-	-	-	-	-
40	Sem	-	Sem	-	-	-	-	-
50	Sem	-	-	-	-	-	-	-
BS III								
2	Sem	-	Sem	-	-	-	-	-
5-6	Sem	-	Sem	-	-	-	-	-
13	Sem	-	Sem	-	Sem	Lat	-	Lat
16	Lat	-	Sem	-	Sem	-	-	-
21	Sem	-	Sem	-	-	-	-	-
22	Sem	-	Sem	-	-	-	-	-
24	Sem	-	Sem	-	-	-	-	-
26	Sem	-	Sem	-	-	-	-	-
28	Sem	-	Sem	-	-	-	-	-

Explanation: BS I: Mazar 1973; BS III: Avigad 1976.

Tables 6.a and b are interesting in two respects. First, both tables document a general preference for Semitic-Hebrew-biblical names.[30] Parents and children often carry Semitic names. In one case, Semitic names can be shown to have run in a Jewish family for as many as four generations or some hundred and twenty years, while in another case, Semitic names were used by a mother, her female child, and her female grandchild (Table 6.a nos. 18-42 and nos. 66+68). On an onomastic level, the influence of the Greco-Roman world was apparently rather limited in the northern part of Roman Palestine during the second and third century C.E. That Jewish onomastic preferences in

---

[30] This has also been observed by Schwabe and Lifshitz 1974, 207.

Palestine differed from those in the Diaspora has recently been reconfirmed by the inscriptions discovered in a first century C.E. tomb in Jericho. There too, all names are Semitic—except for one Greek theophoric name, which was probably a translation of a Hebrew theophoric name discovered in the same funerary complex.[31] Given the predominance of Semitic names in Roman Palestine, it does not come as a surprise to note that the Greek names adopted by the Hasmoneans (as opposed to their Hebrew names) never became popular there.[32]

Second, Tables 6.a and 6.b document that Jewish onomastic practices in Beth She`arim were strikingly similar to those in Jewish Rome in one respect: the indiscriminate use of names of different linguistic origin. For example, one father, himself bearing a Greek name, gave one of his sons a Semitic name, but then opted for a Greek name for his second son (Table 6.a, nos. 16-17). Similarly, another parent had a Semitic name, but he gave his son a Greek name. Yet, the name of the latter's grandchild was again Semitic (Table 6.a, nos. 94 and 96). Conversely, one father had a Greek name, his son a Semitic one and his grandson once again a Greek name (Table 6.a, no. 220). Finally, one inscription commemorates two brothers, one with a Semitic and the other with a Latin name. They were the sons of a mother whose name was distinctively Greek (Table 6.a, no. 125). Thus, even in late Roman Palestine itself, next to Semitic names, Jews freely employed names that belonged linguistically to non-Jewish cultures—although it must also be noted that this happened much less extensively than in the Jewish communities of the Diaspora and more often than not in Greek inscriptions. Again, the interchangeable use of names that were of different linguistic origin was not limited to the people buried in Beth She`arim only, but is also characteristic of Jewish onomastic practices in Roman Palestine during an earlier period. The list of names of the five *zugot* in *Mishnah Avoth* 1 may serve as a prime example to illustrate this point: it includes biblical, Hebrew, Aramaic, and Greek names.

Another small but onomastically important corpus of Jewish inscriptions comes from Venosa, a city in Basilicata (southern Italy). In its vicinity one and possibly two Jewish catacombs have been preserved. The Jewish inscriptions from Venosa can be dated only in very general terms. The majority most likely belongs to the fourth through sixth centuries C.E. Two thirds of the inscriptions were written in

---

[31] R. Hachlili, "The Goliath Family in Jericho. Funerary inscriptions From a First Century A.D. Jewish Monumental Tomb," *Bulletin of the American Schools of Oriental Research* 235 (1979) 31-66, Table 1 on p. 34-35.

[32] T. Ilan, "The Greek Names of the Hasmoneans," *Jewish Quarterly Review* 78 (1987) 14-15.

Greek, and one third in Latin. Isolatedly, also inscriptions in Hebrew occur. In contrast to earlier Jewish inscriptions from Italy in which Hebrew phraseology is usually confined to expressions such as "Peace" or "Peace on Israel," the inscriptions from Venosa attest to a renewed interest in the use of the Hebrew language. One inscription in particular documents that in terms of language, the Venosa materials constitute a transitional phase in (Italian) Jewish history: the epitaph in question was partly conceived in Greek and partly in Hebrew, yet the script used throughout was Hebrew.[33]

The "renaissance" of the Hebrew language in the inscriptions from Venosa is paralleled by an increase of names of Semitic origin. In Jewish Rome, Semitic names constitute no more than 13.5% of the entire onomasticon used by Jews, while 39.5% of the Jews there had Greek and 47% Latin names. In Venosa, the percentage of Latin names is still considerable. In fact, it is even higher than the percentage documented for Jewish Rome, namely 60.4%. Thus, while most inscriptions in Venosa are in Greek, most names are in Latin (a situation similar to that in Rome). On the other hand, and in contrast to the Jewish onomastic data from Rome, the numerical relationship between Greek and Semitic names has now been reversed. In Venosa, Greek names occur in only 14.3% of the inscriptions, while Semitic names are attested in 25.3% of all epitaphs carrying identifiable names.[34]

Such percentages indicate that in comparison to an earlier period, the use of Semitic names was on the rise towards the end of antiquity in southern Italy. Yet, despite an increased preference for Semitic names in general, even in Venosa names of different linguistic origin continued to be used freely. One Isaac, for example, gave his son the Latin *Faustinos* rather than a Hebrew name.[35] Similarly, another member of the Venosan Jewish community, a teacher by the name of Jacob, called his daughter *Severa*.[36] Clearly, the Hebraization of the onomasticon used by Jews in southern Italy was a gradual process that took a number of generations to complete—just as the change from

---

[33] H. J. Leon, "The Jews of Venusia," *Jewish Quarterly Review* 44 (1953-54) 267-84; Solin 1983, 734-35 with further references to the numerous publications of C. Colafemmina. For the second catacomb, see E. M. Meyers, "Report on the Excavations at the Venosa Catacombs 1981," *Vetera Christianorum* 20 (1983) 445-59 and C. Colafemmina, "Saggio di scavo in località 'Collina della Maddalena' a Venosa," *Vetera Christianorum* 18 (1981) 443-51. For the inscriptions, see now Noy 1993, nos. 42-116. On the dating of inscriptions, see Solin 1983, 734 and C. Colafemmina, "Nuove scoperte nella catacomba ebraica di Venosa," *Vetera Christianorum* 15 (1978) 369-81.

[34] Dividing these percentages over males and females, the figures are as follows. Males: Latin names: 58.7%; Greek names 19.1% and Semitic names: 22.2%; Females: Latin names: 64.3%; Greek names: 3.6% and Hebrew names: 32.2%.

[35] *CIJ* 600 = Noy 1993, 76.

[36] *CIJ* 594 = Noy 1993, 48.

Greek and Latin to Hebrew in Jewish inscriptions was a very gradual process.[37]

There exist still two other bodies of Jewish onomastic evidence, namely the Jewish funerary inscriptions from North Africa and the list of Jews recorded in an important inscription discovered in Aphrodisias. The North African inscriptions have been excluded from the present survey. They originate from a number of sites and are likely to date to very different periods. Therefore, they cannot be treated as a cohesive body of evidence.[38]

The relative predominance of biblical names for Jews in the inscription from Aphrodisias, on the other hand, is exceptional and so far without parallel among the Jewish Diaspora communities of antiquity. Inasmuch as we do not know whether there was a Jewish community in Aphrodisias during the period preceding the inscription, it is impossible to tell whether the names mentioned in the inscription belong to Jews who had always remained faithful to traditional Jewish onomastic practices, or to Jews who were in the process of onomastic Hebraization. Or should these names be considered as documenting the existence, among the Jews of Aphrodisias, of a considerable number of immigrants from Roman Palestine?[39] However this may be, in and by itself the evidence from Aphrodisias is certainly not enough to show that already in the third century C.E. there was a general upsurge in the use of biblical names among the Jewish communities of the Diaspora.[40]

This investigation of three internally-consistent Jewish bodies of onomastic evidence has clarified that two important characteristics of Jewish onomastic practices in Rome are not isolated occurrences: in the Diaspora (except for the problematic evidence from Aphrodisias), the non-Jewish onomasticon appealed to Jews more than in Roman Palestine; and, despite clear preferences for certain kinds of names, this preference did not exclude the simultaneous use of names of different linguistic origin.

---

[37] Jewish medieval inscriptions from southern Italy may be found in *CIJ* and in C. Colafemmina, "Archeologia ed epigrafia ebraica nell'italia meridionale," in *Italia Judaica. Atti del I convegno internazionale. Bari 18-22.V.1981* (Rome: Ministero per i beni culturali e ambientali, 1983) 199-210. Particularly interesting is the inscription published by C. Colafemmina, "Nota su di una iscrizione ebraico-latina di Oria," *Vetera Christianorum* 25 (1988) 648-50.

[38] *Contra* Y. Le Bohec, "Juifs et Judaïsants dans l'Afrique romaine. Remarques onomastiques," *Antiquités Africaines* 17 (1981) 209-29.

[39] Reynolds and Tannenbaum 1987, 91-115.

[40] *Contra* S. Honigman, "The Birth of a Diaspora: The Emergence of Jewish Self-Definition in Ptolemaic Egypt in the Light of Onomastics," in S. J. D. Cohen and E. S. Frerichs (eds.), *Diasporas in Antiquity* (Atlanta: Scholars Press, 1993) 118. H. Boterman, "Griechisch-jüdische Epigraphie. Zur Datierung der Aphrodisias-Inschriften," *Zeitschrift für Papyrologie und Epigraphik* 98 (1993) 184f. dates the Aphrodisias inscription to the fourth century, but her argument is based on circular reasoning.

The Influence of Roman Name-Giving Practices on the Jewish Onomasticon in Late Ancient Rome: The Question of the Duo and Tria Nomina

As we have just seen, Jewish onomastic practices in Late Ancient Rome did not develop in complete isolation. The influence of non-Jewish onomastic practices was, however, more profound than the mere borrowing of non-Jewish names. The following comparative study reveals that the Latin names borne by Jews in Rome follow surprisingly closely in both form and content trends that are characteristic of non-Jewish name-giving practices in Late Antiquity. In order to appreciate these similarities, it is necessary to sketch briefly the history of Roman onomastic practices.[41]

Towards the end of the Republic, free Romans normally bore names that consisted of three parts. First came the *praenomen* or first name of which there were only few. During this period, 20% of all Roman males bore the name *Gaius*; another 20% bore the name *Lucius*. The first name was followed by the *nomen* or family name, also known as the *nomen gentilicium*. To these two names a *cognomen* or surname was then added. To the standard triple name (*tria nomina*) system other information could still be appended. Sometimes one encounters filiation (for example *Lucius M[arci] f[ilius]*), while at other times a second surname (*agnomen, cognomen secundum*) was added to honor outstanding qualities or accomplishments (for example *Africanus, Cunctator*). References to one's *tribus* or tribe may also be found.[42]

The Roman triple name system as summarized here was not static, but evolved constantly over time. For example, in the second century B.C.E. the practice of bearing *cognomina* appears to have been confined to the Roman nobility. In the following centuries the use of *cognomina* then started to spread gradually. First *liberti* began using it, and then the custom was also copied by the *plebs ingenua*. By the time of Augustus, *cognomina* had become an integral part of the standard Roman name. Similarly, while references to tribe and filiation mostly occur in the early Principate, *agnomina* gained especial popularity from Severan times onwards.[43]

---

[41] In general, see B. Doer, *Untersuchungen zur römischen Namengebung* (Stuttgart, 1937; reprint New York: Arno Press, 1975); Kajanto 1963 and *id.*, 1965; G. Alföldy, *Die Personennamen in der römischen Provinz Dalmatia* (Heidelberg: Carl Winter, 1969); H. Solin, *Beiträge zur Kenntnis der griechischen Personennamen in Rom* (Helsinki: Societas Scientiarum Fennica, 1971); N. Duval (ed.), *L'onomastique latine = Colloques internationales du CNRS no. 564* (Paris: CNRS, 1977); O. Salomies, *Die römischen Vornamen. Studien zur römischen Namengebung* (Helsinki: Societas Scientiarum Fennica, 1987).

[42] Salomies 1987, 27 and 36.

[43] Solin 1971, 36 and 124; Salomies 1987, 277f.

As a result of intensive research by Finnish scholars in particular, the historical development of the Roman *tria nomina*-system can now be reconstructed as follows. Three stages can be distinguished. Towards the end of the Republic, Romans commonly had a personal *praenomen*. This was followed by a hereditary *gentilicium* and a *cognomen* that was generally (but not always) hereditary (stage 1). In early imperial times, Romans continued to carry personal *praenomina* and hereditary *gentilicia*. The *cognomina* however, ceased to be inherited from one generation to the next. Rather, *cognomina* were now being individualized and started to serve as personal names (stage 2). In the course of the second century C.E. further changes manifested themselves. *Praenomina* developed into hereditary names. *Gentilicia* remained hereditary and *cognomina* continued to function as personal names. Obviously, it was henceforth only the *cognomen* that could serve to identify people. *Praenomina* and *gentilicia* thus lost their original function of denoting someone's onomastic identity (stage 3).[44]

In Late Antiquity the triple name form described here as stage 3 disintegrated gradually. From the third century C.E. onwards, the *tria nomina* were no longer automatically included in funerary inscriptions. The Roman name-system continued to appeal in a more simplified form, namely as *duo nomina*. Yet even this simplified name-system soon lost its appeal. It was superseded almost entirely by an even simpler name-system, namely one consisting of single names only. Early Christian inscriptions dating to the early fourth century C.E. document the extent to which traditional onomastic practices changed in Late Antiquity. 80% of all names recorded in these inscriptions are single names. This percentage further increased as time went by.[45]

There is ample evidence to show that the developments that transformed Roman onomastic practices also affected the Latin names contained in the Jewish funerary inscriptions from Rome. For example, one of the few Jewish epitaphs from Rome that contains a collection of triple names (*CIJ* 470) reflects the third and final stage in the development of the Roman triple name system.[46] It records the names of two brothers and one sister. Besides having identical *nomina gentilicia*, they all bear the same (hereditary) *praenomen*. Thus, the personal *cognomen* provides the only means to distinguish between one person and the next: *Lucius Maecius <u>Constantius</u>, Lucius Maecius <u>Victorinus</u>* and *Lucia Maecia <u>Sabbatis</u>*.[47]

---

[44] Salomies 1987, 363-66 and 378f. For a discussion of the ancient literary sources relating to this phenomenon, see Doer 1937, 17f.

[45] Kajanto 1963, 5, 6 Table 12, and 9-10 Tables 3 and 4.

[46] Note, however, that this inscription may not be Jewish, see I. di Stefano Manzella, "L. Maecius archon, centurio alti oridinis. Nota critica su *CIL*, VI, 39084 = *CII*, I, 470," *Zeitschrift für Papyrologie und Epigraphik* 77 (1989) 103-112.

[47] A second sister bears the name *Maecia Lucianus* (daughter of) *Lucius*.

To be sure, this Jewish inscription is not the only inscription that testifies to this development. Other Jewish inscriptions from Rome, this time containing *duo nomina* rather than *tria nomina*, likewise document that by the time they were cut, *gentilicia* had ceased to be functional, and that *cognomina* were the only name that was used to identify a person.[48]

The list of Latin names published by Leon provides further evidence to suggest that the Latin names used by Jews belong to a late stage in Roman onomastic history.[49] This list is remarkable because (1) *gentilicia* and especially *cognomina* predominate; (2) some, although not all, of these *cognomina* had become hereditary;[50] and (3) *praenomina* are almost entirely absent.[51] Furthermore, such changes went hand in hand with a marked preference for single names. In fact, between 70% and 80% of all names in the Jewish funerary inscriptions from Rome are single names.[52]

That such developments are typically Late Ancient rather than uniquely Jewish becomes particularly evident when we look at non-Jewish inscriptions dating to the third and fourth century C.E. In these inscriptions, the use of *praenomina* was discontinued too. Instead, *cognomina* increasingly replaced the old family names.[53] As in the Jewish inscriptions, some *cognomina* in non-Jewish inscriptions now developed into hereditary names.[54] The preference of Latin over Greek names in Jewish inscriptions is paralleled by the onomastic data contained in early Christian inscriptions dating to the fourth and early fifth century C.E. In such inscriptions, Latin *cognomina* outnumber Greek *cognomina* two to one.[55] In the light of all these similarities, it should come as no surprise that the percentage of triple versus double

---

[48] *CIJ* 209, 213, 234, 267, and 461).

[49] Leon 1960, 95-101.

[50] E.g. *CIJ* 147, 172, and 248. Note that in other cases the old practice of inheriting a *gentilicium* from one generation to the next persisted e.g. *CIJ* 172, 212, 259, 413, 461, and 504.

[51] There are only few examples: *Lucius* in *CIJ* 155, 212, 470 (twice), an inscription which may not be Jewish, see Stefano Manzella 1989; *Gaius* in *CIJ* 57(?), 100, 101, 236 and 465; *Publius* in Moretti 1974, 215; *Marcus* in *CIJ* 284 and Solin 1983, 656 no. 9a. Leon 1960, 95-99 collects all single names into one list, but this is misleading inasmuch as many of these names are either *cognomina* or *gentilicia*. In other cases, it is not always possible to identify a name as a *gentilicium* rather than a *cognomen*. Some *gentilicia* are clearly used as *cognomina*, e.g. *CIJ* 123 (*Iulia Aemilia*), 216 (*Aurelia Flavia*), 266 (*Aelia Patricia*). For the latter phenomenon in non-Jewish inscriptions, see Kajanto 1963, 18f. who notes that this practice was more common for women than for men.

[52] See Leon 1960, Table 4.

[53] Kajanto 1963, 16-18 with further explanations. For the onomastic evidence from the catacomb of Ss. Marcellino e Pietro, see Guyon 1987, 60 and fig. 46.

[54] Kajanto 1963, 22 and 52f.; see also Kajanto 1965, 20.

[55] Kajanto 1963, 57 and esp. Table 16.

and single names in Jewish and non-Jewish inscriptions is also practically identical.[56]

In addition to formal similarities, the Latin names used by Jews have much in common with the names borne by non-Jews in terms of the kinds of names chosen. A list of Latin names published recently by Solin enables us to determine the history of individual Roman names from Republican times through the fourth century C.E. Comparing Solin's list of Roman names with the list of names used by Jews and published by Leon, it can be observed that the names preserved in the Late Ancient Jewish epitaphs from Rome have parallels in the (non-Jewish) Late Ancient rather than in the Republican onomastic materials.[57] While twenty of the names preserved in the Jewish funerary inscriptions seem to have already been popular during the Republic, as many as seventy-seven Latin *gentilicia* and *cognomina* carved in the Jewish funerary inscriptions fall into the category of typically Late Ancient names.[58] This is important evidence to document that Jews in third and fourth century C.E. Rome having Latin names did not choose just any Latin name. Rather, they specifically chose names that were popular in contemporary non-Jewish society at large.

A comparison of *gentilicia* used in Jewish and non-Jewish inscriptions further complements this picture. In Late Antiquity, many people

---

[56] Compare the data in Leon 1960, 111 (Jewish) with Kajanto 1963, 9 Table 3 (non-Jewish). Note that the use of *duo nomina* is higher among Jewish and early Christian women than it is among Jewish and early Christian men. Note also that in case of inscriptions, single names are often used by the dedicant, and that the *tria nomina* are cited in full only in case of the deceased. Of course, the use of a single name by a dedicant cannot be taken to mean that such dedicants never carried triple names. It only means that in their capacity as dedicants they did not care to record such triple names, see Salomies 1987, 393. For a good Jewish example of this phenomenon, see *CIJ* 149 and 150: *Petronia* (single name) when commemorating her son; and the same *M[..]a Petronia* (double name, although fragmentary) when commemorated herself. And see L. Moretti, "Iscrizioni greco-giudaiche di Roma," *Rivista di Archeologia Cristiana* 50 (1974) 213-19, at 215-18. In the Jewish inscriptions from Rome, *duo nomina* relating to the deceased occur in sixty cases, while *duo nomina* relating to the dedicant occur in only twenty nine cases. Leon failed to take this epigraphic practice into account when he compiled his table V on p.111 in which he catalogued the occurrence of double and triple names. When the above considerations are taken into account the relationship between double names in Greek and Latin inscriptions is not 39 versus 50, as Leon suggests, but rather 31 versus 30. Note finally that single names were also preferred in the commemoration of children under twenty years of age, see Solin 1971, 45. And note that the name of the commemorator was primarily included in Roman inscriptions to indicate that he or she was the heir of the deceased, see Meyer 1990, 74-78.

[57] For non-Jewish names, see Solin in Duval 1977, 105-38. The Republican names have been culled from *CIL* 1; the Late Ancient names are based on *ICVR* 1 and 5. For Jewish names, see Leon 1960, 95f.

[58] The most popular *cognomina* are listed by Kajanto 1965, 29-30: they can be encountered in the Jewish epitaphs from Rome as well. Note also that in Late Antiquity there was an increased preference for suffixed name forms, see Kajanto 1963, 61f. Jewish inscriptions also reflect this phenomenon, e.g. *CIJ* 129: *Ursacius* (a derivative of *Ursus*) and *CIJ* 110: *Eutychianus* (a Greek name to which the Latin suffix *-ianus* was added).

had the same *gentilicium*. In fact, no more than eight *gentilicia* appear in 51% of all inscriptions! Most of these *gentilicia,* especially in the provinces, were of imperial origin.[59] Significantly, the four most popular of these eight *gentilicia—Aurelius, Flavius, Iulius,* and *Aelius*—are exactly the *gentilicia* that we encounter most frequently in Jewish funerary inscriptions from Rome—albeit in a slightly different order of popularity.[60]

The fact that contemporary non-Jewish onomastic practices provided a model for Jewish onomastic habits in Late Antiquity explains some peculiarities in the Jewish onomasticon in Rome. Since the Flavian emperors were responsible for the destruction of the Second Temple in Jerusalem in 70 C.E. and for the subsequent institution of the *fiscus Judaicus,* the relatively frequent occurrence of the *gentilicium Flavius-Flavia* among Jewish males and females in Rome is surprising.[61] Yet, as has just been pointed out, this *gentilicium* was one of the eight most popular *gentilicia* in Late Antiquity. Thus in the third or fourth century C.E. some Jews simply assumed a name that was *en vogue* without bothering about the original negative connotation this name had carried.[62]

The use of pagan theophoric names must be explained along similar lines. The practice is attested among the Jews in Rome, Egypt, and in Palestine itself. As in non-Jewish inscriptions, pagan theophoric names usually appear in inscriptions in Greek.[63] Malaise has observed with regard to theophoric names referring to Isis, that such names do not necessarily reflect on the religious affiliation of its bearer, but that they may also serve as indicator of someone's ethnic origin.[64] The use

---

[59] Kajanto 1963, 16; Kajanto in Duval, 1977, 426 referring to the work of A. Mócsy. And see the *index nominum* of *CIL* 6 under the respective family names of the individual Roman emperors. The *gentilicium Aurelius* outnumbers all others.

[60] In Jewish inscriptions, the *gentilicia Aurelius* and *Iulius* both occur sixteen times; *Flavius* occurs eleven, and *Aelius* six times. In comparison to the *gentilicia* in non-Jewish inscriptions the *gentilicia Iulia* and *Flavius* have traded places.

[61] E.g. *CIJ* 172 (2x), 216, 234 (2x), 235, 299, 361, 416, 417 and 463.

[62] For the use of names of one's enemies, see Hachlili 1979, 52 (see the sons of Sisera in Nehemiah 7:55). On the Roman use of names that had originally had a negative connotation, see W. Kubitschek, "Spurius, Spurii filius, sine patre filius usw.," *Wiener Studien* 47 (1929) 130-43.

[63] *CIJ* 533: referring to *Livius Dionysius.* The name Dionysius enjoyed particular popularity among pagans, see Solin 1971, 108-9, 111. The name derives from the name of the Greek god Dionysus to which a suffix was added, giving the name a dedicatory meaning ("belonging to Dionysus"); *CIJ* 91 and 92: *Asclepiodote; CIJ* 232: *Afrodisia* and *CIJ* 229 and 291: *Isidora/us* (the most popular of all Isiac theophoric names), see the percentages given in M. Malaise, *Les conditions de pénétration et de diffusion des cultes égyptiens en Italie* (Brill: Leiden, 1972) 46; Egypt: *CPJ* 1, 29 and 228; Beth She`arim: Schwabe and Lifshitz 1974, no. 52. For the non-Jewish evidence, see Kajanto 1965, 53f.

[64] Malaise (previous note) 26-34.

of pagan theophoric names by Jews can perhaps be explained among similar lines: such names had lost their original pagan connotation.[65]

To summarize. By comparing the Latin names in third and fourth century C.E. Jewish funerary inscriptions with contemporary onomastic data in non-Jewish inscriptions, it can be shown that in this respect the Jewish epitaphs reflect remarkably closely onomastic trends that are general in Late Antiquity. The gradual disintegration of the traditional *duo* and *tria nomina* system, formal changes in the traditional Roman naming-system, as well as the preference of specific *gentilicia* and *cognomina* over others are all typically Late Ancient developments. Jewish inscriptions moreover reflect these developments with remarkable accuracy. Thus, in Rome, Jewish accommodation to non-Jewish onomastic practices went further than the arbitrary choice of Latin names alone. Rather, Jewish onomastic practices tended to follow general trends in contemporary non-Jewish onomastic practice exceedingly closely. In the onomastic sphere, therefore, the interaction of Jews and non-Jews in Late Ancient Rome can be said to have been particularly intense.

The Semitic Names in the Jewish Epitaphs from Rome

Even though Latin names determine to a significant extent the onomasticon used by Jews in Late Ancient Rome, their importance should not be exaggerated. It is surely significant that in Late Antiquity, Semitic names never disappeared completely. In fact, if certain Greek and Latin names were indeed conceived as translations of Hebrew names, as has often been suggested, the share of Semitic names in the total number of names attested for Roman Jews would rise considerably. It is most regrettable therefore, that the Semitic origins of certain Greek or Latin names cannot be proved in individual cases.

The use of Semitic *agnomina*, not only in some Jewish inscriptions from Rome, but also in Jewish inscriptions from Edfu and Beth She`arim, suggests that some Jews bore Greek and Latin names when dealing with the outside world, and Hebrew ones when dealing with the Jews themselves.[66] In the Roman world it was rather customary to

---

[65] See also the detailed discussion in Mussies in Henten and van der Horst 1994, 245-49.

[66] An *agnomen* or *supernomen* is a second name, appended to the first name by the formula ὁ καὶ - ἡ καὶ or *qui/quae et*. When this name is found detached, however, or when it is preceded by the word *signo* such a second name is called *signum*, see Kajanto 1963, 31, and 47f. For Jewish examples, see *CIJ* 108, 140, 206, 282 (problematical), 362, 379 (Rome); *CPJ* 2, nos. 223, 248, 298 = 304 = 311 = 321(Edfu); Schwabe and Lifshitz 1974, nos. 88, 101, 191, and 199 and see no. 121, on which the pattern is reversed (Beth She`arim); Reynolds and Tannenbaum 1987, a.25, b.20 and b.28 (Aphrodisias). Note,

use local and "international" names at one and the same time. The graffiti discovered in La Graufesenque (France), for example, show that potters there used Rutenian or local names for local transactions, but that they adapted their names to Latin onomastic practices when signing products destined for export.[67]

Still another influence of Hebrew on Jewish name-giving in Latin is perhaps reflected by the formal characteristics of filiation. Le Bohec has recently suggested in reference to Jewish inscriptions from North Africa that filiations such as *Asterius, filius Rustici* (as opposed to the "classical" *Asterius, Rustici filius*) may be the result of direct translations of the typically Hebrew filiation-form *X ben/bath Y*.[68] Unfortunately, filiation appears in Roman Jewish inscriptions only in exceptional cases. Yet, in one such epitaph, a bilingual inscription in Greek and Latin carrying Semitic names, filiation follows the Hebrew rather than the Latin pattern.[69] The same seems to have been the case in other parts of Italy too.[70]

Both the names and filiation-forms thus suggest that the Semitic practices may have loomed larger on the minds of individual Jews than we are able to ascertain today. Differing fundamentally in both form and structure, in the eyes of their Roman contemporaries, Semitic names were unmistakable indicators of the ethnic background and/or religious affiliation of the people that bore these names. That in the eyes of the Jews themselves, too, biblical names were considered typically Jewish follows, for example, from the confusion Christians caused among the Jews of Caesarea (in Roman Palestine) around 309 C.E. when they started using biblical names.[71]

The importance of the Semitic names undoubtedly lies in their presence rather than in their actual number and in the fact that Semitic names occur in all the major Jewish underground cemeteries in Rome. The presence of these names indicates that not all Roman Jews adapted readily to non-Jewish onomastic practices. Thus, despite the strong influence of non-Jewish onomastic practices, Latin or Greek names never completely replaced Semitic names. Onomastic variety rather than complete adaptation to non-Jewish name-giving trends

---

however, that there exist also Jewish inscriptions in which the *agnomina* are Greek or Latin rather than Hebrew, e.g. *CIJ* 47, 86, 140, and 32*.

[67] B. and H. Galsterer, "Romanisation und einheimische Traditionen," in H.-J. Schalles et al. (eds.), *Die römische Stadt im 2. Jahrhundert n.Chr. Der Funktionswandel des öffentlichen Raums* (Cologne and Bonn: Rheinland and Habelt, 1992) 378.

[68] Le Bohec 1981, 226.

[69] *CIJ* 497.

[70] E.g. *CIJ* 558, 613, 614, and 629.

[71] G. Stemberger, *Juden und Christen im Heiligen Land. Palästina unter Konstantin und Theodosius* (Munich: Beck, 1987) 259 n. 168 with further references.

characterizes the Jewish onomasticon in third and fourth century C.E. Rome.

The Onomastic Practices of Jewish Women in Late Ancient Rome

While inscriptions relating to Jewish women in Late Ancient Rome differ in some aspects from inscriptions relating to men, no substantial differences exist insofar as onomastic practices are concerned. Leon has pointed out that among Jewish women in Rome, Latin names were more common than Greek names. Semitic names were least widespread. Thus, in third and fourth century C.E. Rome onomastic preferences for Jewish women generally corresponded to name-giving practices among Jewish males.[72]

The practices surrounding name-giving for Jewish women in Rome are similar to that observed in connection with Jewish males in that names of different linguistic origin are used interchangeably. In five inscriptions mothers bear either Semitic or Greek names, and their daughters Greek or Latin names.[73] But in five other inscriptions the daughters bear Greek or Semitic names, and the mothers Latin or Greek names.[74] Sometimes daughters were called after their mothers, while at other times their father's name was adopted.[75] Occasionally, however, daughters bore names different from those of their parents. In one inscription husband and wife bear the same *gentilicium*.[76] This could be due to chance or it could indicate that both were *liberti* from the same household.[77] In another inscription, a mother and her son have the *gentilicium* in common.[78]

---

[72] Leon 1960, 109, Table III.

[73] *CIJ* 212, 252, 267, 461, and 30*.

[74] *CIJ* 12, 156, 240, 511, 732.

[75] *CIJ* 147, 172 and 248: a father and daughter bearing the same *cognomen* (*CIJ* 172: this name may also be a *gentilicium* rather than a *cognomen*). The opposite can also be observed. For a mother and son having the *cognomen Severus-Severa* in common, see *CIJ* 264.

[76] *CIJ* 219.

[77] On this phenomenon in non-Jewish inscriptions, see Doer 1937, 214f.

[78] *CIJ* 149. For the same phenomenon in non-Jewish inscriptions, see Doer 1937, 218f referring to *CIL* 6.16436 and 23612. The Jewish inscription, however, is difficult to interpret because it only contains single names (i.e. *Petronius-Petronia*). It is possible that these names, which had originally been *gentilicia* functioned as *cognomina* in this specific Jewish inscription (in the same way as can also be observed in early Christian inscriptions, see Kajanto 1963, 22-23). *Petronius* could also have been the son of a former marriage (see e.g. *CIL* 5.4000 and 6.4627) or of a marriage of two parents with different legal status (Gaius, *Inst.* 1.84: the son of a slave father and a freeborn mother follows that of the *ingenua*). See also, in general, Salomies 1987, 409-10.

Leon has stressed that the use of Latin names was more widespread among Jewish women than among Jewish men.[79] Similarly, in Rome, the use of *duo nomina* was more common among women, but the use of *tria nomina* was not.[80] Such assertions do not come as a complete surprise. Scholars have often observed that the onomasticon of Jewish women tends to follow contemporary non-Jewish naming practices to a greater extent than was the case with Jewish males.[81] It is not correct to assume, however, that in the Diaspora the names of Jewish women always and by definition reflect the onomastic customs of non-Jewish society. In Venosa, for example, the percentage of Jewish females with Hebrew names is higher than that of Jewish males (32.2% versus 22.2%).

Despite small differences, I believe that the importance of Latin names among the Jewish women of Late Ancient Rome should not be exaggerated. The preponderance is merely a reflection of the general propensity, also documented for Jewish males, to prefer Latin over Greek and Semitic names. In antiquity, the names of women are largely linguistically identical to those borne by males in a given locality. In Roman Palestine, the majority of such names was Hebrew; in Rome it was Latin.[82]

Names Borne by Roman Jews as Indicators of Social Status?

The formal characteristics of the names in the Jewish epitaphs from Rome inform us not only about onomastic trends but also about the social status of the Jews bearing them. Although it has recently been suggested that the Jewish epitaphs are difficult to evaluate in terms of social history, the information provided by the inscriptional evidence is in reality quite unambiguous.[83]

In Late Antiquity a marked tendency towards polynomy characterizes the onomastic practices of Rome's upper classes. It resulted in

---

[79] Leon 1960, 109.

[80] Leon 1960, 109-111. Note that also in early Christian inscriptions, females bear double names more often than males, see Kajanto 1963, 9 Table 3. Note also that in Jewish inscriptions from Carthage, triple names occur more often in the case of Jewish males than Jewish females, see Le Bohec 1981, 216-17.

[81] S. D. Goitein, *A Mediterranean Society. The Jewish Communities of the Arab World as Portrayed in the Documents of the Cairo Geniza*. Vol. 3. *The Family* (Berkeley: University of California Press, 1978) 314-15; Le Bohec, 1981, 220; G. Nahon, *Inscriptions hebraïques et juives de France médiévale* (Paris: Corlet, 1986) 36.

[82] T. Ilan, "Notes on the Distribution of Jewish Women's Names in Palestine in the Second Temple and Mishnaic Periods," *Journal of Jewish Studies* 40 (1989) 186-200. The percentages in G. Mayer, *Die jüdische Frau in der hellenistisch-römisch Antike* (Stuttgart: Kohlhammer, 1987) 33-34 are not reliable.

[83] *Contra* G. Fuks, "Where Have All the Freedman Gone? On an Anomaly in the Jewish Grave-Inscriptions from Rome," *Journal of Jewish Studies* 36 (1985) 25-32.

long sequences of personal names. Sometimes, these sequences are so long that it is impossible today to identify a family name as opposed to a *cognomen*.[84] By contrast, Late Ancient onomastic practices of the *plebs urbana* and of the population living in the provinces of the Roman empire show an ever increasing preference for the use of single names in general and for the use of *cognomina* in particular.[85] Thus, in Late Antiquity, the name-giving practices of the upper and lower classes in society developed in opposite directions.

We have just seen that the use of *cognomina* and single names is a characteristic feature of Jewish onomastic practices in Late Ancient Rome. Polynomy, on the other hand, is not attested in Jewish funerary inscriptions. Evidently, the Jews buried in the Jewish catacombs of Rome did not belong to the upper crust of late Roman society. Additional support for the contention that Jews belonged to the same social class as most of Rome's city may be found in the occurrence of *agnomina* or *supernomina* in Jewish inscriptions from Rome. Kajanto has argued that such *agnomina* originated in the eastern Mediterranean and that they were mostly used by classes other than senatorial or equestrian.[86] Cumulatively, such evidence demonstrates that the assertion that the names borne by Jews indicate that "they were also at home in the upper-class gentile Roman world" is too general to be of much value.[87]

The previous analysis of Jewish onomastic practices in Rome also explains a supposed anomaly in the Jewish funerary inscriptions. In a recent article, Fuks described this anomaly as follows. Freedmen (*liberti*) usually included the sigla *lib.* or *l.* in their epitaph. In addition such freedmen often bear "servile names." Inasmuch as a significant part of the Roman Jewish community is known to have consisted of slaves that had been manumitted, it is strange that not one single reference to servile status can be detected in the Jewish funerary inscriptions from Rome. The lack of such references cannot have been accidental. Rather, it must be regarded as a deliberate attempt to avoid references to servile status. According to Fuks the phenomenon is to be interpreted on the basis of Josephus' description of the Fourth Philosophy (the Zealots): "They have a passion for liberty that is almost unconquerable, since they are convinced that God alone is their leader

---

[84] A. Cameron, "Polynomy in Late Roman Aristocracy: The Case of Petronius Probus," *Journal of Roman Studies* 75 (1985) 164-82, esp. 171-75.

[85] See Salomies 1987, 390f., 400 and 404.

[86] For Jewish examples, see (*CIJ* 47, 86, 108, 140, 206, 282 (problematic) 362, 379, 32*). In general, see I. Kajanto, *Supernomina. A Study in Latin Epigraphy* = *Commentationes Humanarum Litterarum* 40:1 (1966), 14-15.

[87] *Contra* N. Cohen, "Jewish Names as Cultural Indicators in Antiquity," *Journal for the Study of Judaism* 7 (1976) 121.

and master."[88] Does the evidence really justify maintaining that part of the Roman Jewish community consisted of Zealots or people holding comparable convictions?

A closer look at Fuks's arguments reveals that his line of reasoning is flawed. It is true that part of the Jewish community in Rome originated among Jewish slaves who had been shipped there on various occasions during the first centuries B.C.E. and C.E.[89] It is also correct to assert that freedmen/women can be recognized on the basis of their names and certain epigraphic formulae. Upon manumission slaves often took on the *praenomen* and *gentilicium* of their former master. To these two names a *cognomen*—usually the slave's original personal name—was added. Further additions indicating servile origin such as *libertus, conlibertus, verna*, or a reference to one's *patronus* were commonly appended too. Finally, the fact that filiation was not included in the names of former slaves also made it possible to distinguish them onomastically from freeborn—as did the presence of a Greek *cognomen*.[90] Thus, despite the fact that in the first century C.E. the *tria nomina* connoted free status, former slaves with triple names could easily be recognized as such.[91] As many as 70% of the inscriptions from Rome that carry triple names relate to freedmen/women rather than to freeborn. Apparently, the recording of the *tria nomina* mattered especially to those who had newly won this privilege. The wish to integrate as quickly as possible also explains why the lower classes so avidly copied the names popular among the upper classes of Roman society.[92]

Fuks has failed to realize that all these points hold true for the first two to three centuries of the Common Era only. Yet, as we have seen earlier, Roman onomastic practices were far from static. "The age of Constantine the Great marked the turning point in the history of the

---

[88] Josephus, *AJ* 18.23; Fuks 1985, *passim*.

[89] Josephus, *AJ* 14.78; Plutarchus, *Vita Pomp.* 45.1-2; Appian, *Mith.* 117.571; Eutropius, *Breviarium a.U.c.* 6.16 (under Pompey) and Josephus, *BJ* 7.118 and *CIJ* 556 (70 C.E.).

[90] On filiation and the names of slaves, see H. Solin, "Riflessioni sull'esegesi onomastica delle iscrizioni romane," *Quaderni Urbinati di Cultura Classica* 18 (1974) 122 n. 32. See also Kajanto 1963, 5, Salomies 1987, 229f. and G. Alföldy, "Notes sur la relation entre droit de cité et la nomenclature dans l'Empire romain," *Latomus* 25 (1966) 37-57. On the servile connotation of Greek *cognomina*, see T. Frank, "Race Mixture in the Roman Empire," *American Historical Review* 21 (1916) 693, Kajanto 1963, 58-60, and Solin 1971, 49, 123f., 137. Note that Greek names do not always point to servile origin, but may may indicate that its bearer originated in the eastern part of the Mediterranean, see Solin 1974, 117f. and M. L. Gordon, "The Nationality of Slaves under the Early Roman Empire," *Journal of Roman Studies* 14 (1924) 93-111, esp. 101f.

[91] See Juvenal, *Sat.* 5.125-27; Suetonius, *Claud.* 25.3. And see Kajanto 1963, 6-7 and Alföldy 1966, 47-55.

[92] L. R. Taylor, "Freedmen and Freeborn in the Epitaphs of Imperial Rome," *American Journal of Philology* 82 (1961) 113-32, esp.117f.; Solin 1971, 51, 135-36.

Latin name system."[93] Among other things, it affected the epigraphic formulae that had originally denoted servile or freedmen/woman status. From the third century C.E. onwards, inscriptions lack references to freedmen. In addition, Greek names appear to have lost their one-time servile connotation.[94]

The onomastic data they contain indicate that the Jewish funerary inscriptions from Rome belong to a relatively late stage in Roman epigraphic history. Without exception they date to the third and fourth century C.E. Since Jewish tombstones follow contemporary non-Jewish epigraphic practice in a number of respects, it should come as no surprise that references to freedmen/women are lacking on these inscriptions—just as such references are lacking on non-Jewish inscriptions.

The *only* possible clue to indicate that some people recorded in the Jewish tombstones may have been descendants of slaves is provided by the imperial *gentilicia*. The interpretation of this evidence is, however, extremely ambiguous. As has already been pointed out, in Late Antiquity, this particular class of *gentilicia* was used by practically everyone, including people who were certainly not descendants of imperial slaves. Thus, even if a Jewish male or female bears an imperial *gentilicium* one cannot be sure whether he or she was a freedman/woman or a descendant of such *liberti*.

Whatever the case may be with regard to freedmen/women, there is, in short, nothing anomalous about the Jewish funerary inscriptions from Rome once they are seen within the context of Late Ancient epigraphic practices and trends. They were carved in a time when the Zealots of first century Palestine had long vanished from history. In terms of social history, the onomastic evidence preserved in the Jewish epitaphs indicates what Jews in Late Ancient Rome were not: leading personalities in the rapidly changing political landscape of Late Antiquity. Onomastically, Jews had much in common with the non-Jewish urban masses.

---

[93] Kajanto 1963, 13.

[94] See Taylor 1961, 120f.; Kajanto 1963, 6-9; Solin 1971, 97 and 137; Solin 1974, 122-23; W. Eck, "Römische Grabschriften. Aussageabsicht und Aussagefähigkeit im funerären Kontext," in von Hesberg and Zanker 1987, 61-83, esp. 73f. and *id.*, "Aussagefähigkeit epigraphischer Statistik und die Bestattung von Sklaven im kaiserzeitlichen Rom," in P. Kneissl and V. Losemann (eds.), *Alte Geschichte und Wissenschaftsgeschichte (FS K. Christ)* (Darmstadt: Wissenschaftliche Buchgesellschaft, 1988) 130-39.

Onomastic Practices as an Indication of Interaction

Can onomastic data be used as evidence of interaction between Jews and non-Jews in antiquity? Or are onomastic data merely the accidental result of unreflective behavior or fashion?

Scholars have given different answers to this question. The editors of the *Corpus Inscriptionum Judaicarum* and M. Hengel assume that changes in Jewish onomastic practices indicate the influence of Hellenistic civilization on the Jews. L. H. Feldman has criticized this assumption. In a recent book on interaction he writes that "Again the adoption by Jews of Greek names turns out not to be a very meaningful criterion of their degree of assimilation." Elsewhere in the same book, however, Feldman, inconsistently, uses names such as *Sabbatis* and *Sabbatia* to argue that non-Jews observed the Sabbath. Similarly inconsistent is O. Rössler when he uses Libyan names to identify Libyans, but then argues that no importance should be attached to the non-Libyan names in these inscriptions, because there are nothing but a "modische Mimikry." Studying the names preserved in a building in Aquileia that was once believed to have been a synagogue, F. Vattioni concluded that "the Jews...have never been fanatic in giving names to their children; on the contrary, they have displayed such an indifference or caution in using clearly pagan theophoric names, according to some in order to camouflage themselves better." Similarly, Y. Le Bohec has explained the Latinization of the onomasticon of the Jews of North Africa as resulting from a "fear of isolation"—an idea that had already been posited by H. Leclercq in 1936. R. MacMullen has observed that name change in Roman Egypt among the more progressive elements of the population points towards a "slackening loyalty to their own traditions." R. J. Rowland, surveying the name-giving practices on Roman Sardinia, has observed that Sardinian names make up no more than 2.5% of the names found in Sardinian inscriptions, adding that such names invariably appear in inscriptions from inland—that is the least Romanized—sites. And E. J. Bickerman has maintained that in antiquity people did not easily give up traditional onomastic practices and argues that certain names can inform about the hopes and beliefs of the namegiver.[95]

---

[95] See *CPJ* 1, 27; Hengel 1974, 61f.; L. H. Feldman, "Hengel's *Judaism and Hellenism* in Retrospect," *Journal of Biblical Literature* 96 (1977) 377-78; Feldman 1993, 418 and 359-60; Rössler in *Beihefte der Bonner Jahrbücher* 40 (1980) 282; F. Vattioni, "I nomi giudaici delle epigrafi di monastero di Aquileia," *Aquileia Nostra* 43 (1972) 128; Le Bohec 1981, 227; Leclercq in *DACL* 12 (1936) 1551; R. MacMullen, "Nationalism in Roman Egypt," *Aegyptus* 44 (1964)189; R. J. Rowland, "Onomastic Remarks on Roman Sardinia," *Names* 21 (1973) 97; E. J. Bickerman, *The Jews in the Greek Age* (Cambridge Mass. and London: Harvard U.P., 1988) 46-47.

Inasmuch as these statements are all based on assertions and unproven assumptions, they are not very useful for answering the question of whether onomastic data tell us much about interaction. It is much more useful to investigate what the ancients themselves had to say about name-giving practices. In addition, we also need to explore whether further information on this issue can be gained from the epigraphic data that have been studied in this chapter.

Ancient literary sources indicate that in the Roman world, people were able to identify the linguistic origin and the etymology of certain names. Philo, for example, discusses extensively several passages from the Hebrew Bible that relate how the names of several well known figures of early Jewish history including Abram (Abraham), Sarai (Sarah), Jacob (Israel) were changed. Without exception, Philo interprets the change of names as "signs of moral values, the signs small, sensible, obvious, the values great, intelligible, hidden."[96] Somewhat more than a century later, Justin, in his *Dialogue with Trypho,* explains the significance of the name Israel as "a man who overcomes" (*Isra*) "power" (*El*).[97] In his turn, Tertullian observes that the Creator "figuratively employs names of places as a metaphor derived from the analogy of their sins" and then gives several examples of this practice.[98] The author of the pseudo-Cyprianic treatise *De Montibus Sina et Sion* dwells extensively on the etymology of a series of Hebrew biblical names and also enlarges upon the numerical value of the letters that make up individual names.[99] In the same general period, Origen notes that in his day, Jewish names were taken from the Hebrew Bible or from words "the meaning of which is made clear by the Hebrew language."[100] Early in the fourth century C.E., Lactantius cites Cicero's *De Natura Deorum* in explaining the etymology and meaning of the names Kronos and Saturnus, while elsewhere he explains the meaning of the geographical name Latium through wordplay.[101] In these same years, Eusebius refers to a certain Theodotus, whom he describes as a person "who by his very acts proved his own name." A second person by the name of Theotecnus is typified by Eusebius as "a clever unprincipled trickster who belied his name."[102] Towards the end of the fourth century, Jerome observed that

---

[96] Philo, *Mut.* 59-65. And see, in general, L. Grabbe, *Etymology in Early Jewish Interpretation: The Hebrew Names in Philo* (Atlanta: Scholars Press, 1988).

[97] Justin, *Dialogue with Trypho* 126 (ed. Goodspeed 1914, 245-46).

[98] Tertullian, *Adversus Marcionem* 3.13.9-10 (*CCL* 1, 525-26).

[99] A short discussion and examples in J. Daniélou, *The Origins of Latin Christianity* (London and Philadelphia: Darton, 1977) 41-42 and 44.

[100] Origen, *Contra Celsum* 4.34 (*PG* 11, 1081).

[101] Prudentius, *Divinae institutiones* 1.12.9 (*SChr* 326, 140) and 1.13.8 (*SChr* 326, 144).

[102] Eusebius, *Historia ecclesiastica* 7.32.23 (*GCS* 9.2, 726) and 9.2.2 (*GCS* 9.2, 808).

Iovinian's name was a *nomen quod de idolo derivatum est*.[103] Comparably, Jerome's contemporary Zeno of Verona is well aware of the fact that the meaning of the Hebrew name Isaac is *laetitia*.[104] And another contemporary, Sulpicius Severus, calls the name change of Sarah and Abraham a "not unimportant mysterium," but then refrains from explaining why this is so.[105] Ammianus reports that in the early 370s Romans chose certain names because they were considered distinguished. Ammianus also displays the same sensitivity to the meaning of individual Latin names as does Eusebius for Greek names.[106] In early Christian literature with its strong tendency to explain the Hebrew Bible (Old Testament) allegorically, etymological studies gained popularity primarily because, in Augustine's words, they helped "solving the enigmata of the Scriptures."[107] Later rabbinic sources disapprove of the use of pagan theophoric names, exactly because of the pagan, idolatrous associations inherent in such names.[108] In mystical and magic circles the belief that a deeper meaning was attached to certain names was quite widespread.[109] Speculations on the etymology of the names of angels were particularly widespread, as both literary sources and amulets indicate.[110] Awareness of the linguistic origin or religious affiliation inherent in certain names appears to have been especially high among people who converted to either Judaism or Christianity[111] or who were initiated into Oriental cults.[112] Last but not

---

[103] Jerome, *Adversus Jovinianum* 2.384 (*PL* 23, 352).

[104] Zeno, *Tractatus* 1.59.5 (*CCL* 22, 135).

[105] Sulpicius Severus, *Chronicorum liber* 1.6.2 (ed. Lavertujon 1896, 13).

[106] Ammianus Marcellinus 28.4.

[107] See the short discussion in F. Wutz's otherwise purely descriptive *Onomastica Sacra. Untersuchungen zum Liber Interpretationis Nominum Hebraicorum des Hl. Hieronymus* (Leipzig: Hinrichs'sche Buchhandlung, 1914) 347-53.

[108] *Babylonian Talmud, Gittin* 11b.

[109] E.g. Kajanto, 1963, 93; I. Gruenwald, *Apocalyptic and Merkavah Mysticism* (Leiden: Brill, 1980) 104f.

[110] S. M. Olyan, *A Thousand Thousands Served Him. Exegesis and the Naming of Angels in Ancient Judaism* (Tübingen: Mohr, 1993); R. Kotansky, "Two Inscribed Jewish Aramaic Amulets from Syria," *Israel Exploration Journal* 41 (1991) 275-80.

[111] E.g. *CIJ* 523: *Beturia Paulla* from Rome converted to Judaism and changed her name to *Sara*. Another proselyte who had changed her name to Sara in Lüderitz 1983, no. 12. Note however, that not all proselytes appear to have changed their names, see *CIJ* 21, 68, 202, 222, and 256. See, in general, A. Harnack, *Die Mission und Ausbreitung des Christentums in den ersten drei Jahrhunderten* (Leipzig: Hinrichs'sche Buchhandlung, 1902) 307.

[112] L.Vidman, *Isis und Serapis bei den Griechen und Römern. Epigraphische Studien zur Verbreitung und zu den Trägern des ägyptischen Kultes* (Berlin: de Gruyter, 1970) 133; R. Bagnall, "Religious Conversion and Onomastic Change in Early Byzantine Egypt," *Bulletin of the American Society of Papyrologists* 19 (1982) 105-123; G. H. R. Horsley, "Name Change as an Indication of Religious Conversion in Antiquity," *Numen* 34 (1987) 1-17.

least, *cognomina* had at first been adopted by the Romans specifically for the meaning inherent in such surnames.[113]

It would be an overstatement, of course, to maintain that everybody always realized the implications or associations of the names they bore. Yet, the above-mentioned literary sources clearly document sensitivity in respect to onomastic practices. One literary source in particular shows that in Late Antiquity the use of certain names was considered by some as profoundly indicative of one's attitude towards society and culture. In *Leviticus Rabbah*, a homiletic midrash, dating (apparently) to the fifth century C.E., Rabbi Huna states in the name of Bar Kappara that "Israel was redeemed from Egypt on account of four things, namely because they did not change their names, they did not change their language, they did not go tale-bearing and none of them was found to have been immoral. They did not change their name, having gone down as Reuben and Simeon and having come up as Reuben and Simeon. They did not call Judah 'Leon,' or Reuben 'Rufus,' nor Joseph 'Lestes,' nor Benjamin 'Alexander'."[114] A similar sensitivity probably underlies the fact that those responsible for the Septuagint almost never translated Hebrew names into Greek, but rather prefered to transliterate the indeclinable originals.[115]

That Jewish onomastic data must indeed be interpreted in terms of interaction with non-Jews (or the absence thereof) finds further confirmation when we turn, once again, to the inscriptional evidence. In Beth She'arim, a city located in a region with a strong Jewish presence, the majority of names are of Semitic origin. Names of supra-regional origin such as Palmyrean names can be attributed to the Jews who had been brought from the Diaspora to Beth She'arim for final burial. Similarly, in first century C.E. Jericho, a city located in the Jewish heartland, Hebrew names predominate. Such data contrast markedly with the Jewish onomastic evidence from the Jewish Diaspora. In Rome, a city which housed a Jewish community that was small in comparison to the city's non-Jewish populace, the percentage of Semitic names constitutes no more than a small fraction of the total number of names used by Jews. Instead, local non-Jewish onomastic practices set the example for the onomasticon used by Jews.[116] A similar phenomenon can be observed in other Jewish Diaspora communi-

---

[113] E.g. *Coriolanus*, captor of the Volscan town of *Corioli*, Livy, 2.33. Late ancient examples in Ammianus Marcellinus 15.13: Constantine changing the name of *Strategius* to *Musonian*; and *Julian*, jestingly called *Victorinus* by the courtiers of Constantius after the former's victory over the Alamanni in 357 C.E., see Ammianus Marcellinus 16.12.68.

[114] *Leviticus Rabbah* 32:5. Translation by J. Israelstam and J. Slotki, *Midrash Rabbah, Leviticus* (London: Soncino, 1939).

[115] See Bickermann 1988, 113.

[116] Note that Roman Jews who can be shown to have originated in the eastern Mediterranean, all bore Greek names, see *CIJ* 25, 296, 362, 370, 501, 502, Fasola 1976, 22 (a person who supposedly came from Lindos), and 47.

ties such as Edfu or Venosa. There too, the influence of non-Jewish on Jewish naming-practices was considerable.

The use of Latin names cannot always or automatically be taken to indicate fluency in the Latin language. In some cases it does not even prove familiarity with Roman culture—as G. Alföldy has maintained in his study of name-giving practices in Dalmatia.[117] Studying the Jewish epitaphs from Rome, we have seen that there is no clear relationship between the language in which an inscription was written and the linguistic origin of the names chosen: Latin names not only predominate in inscriptions written in Latin, but also in those composed in Greek; Semitic names are used in Greek and Latin inscriptions alike. The Jewish onomastic evidence from Rome, however, shows such strong parallels in structure and content with the contemporary non-Jewish onomasticon in this city that one cannot but reach the conclusion that this evidence is indicative of more than a superficial acquaintance with Roman name-giving practices.

One would very much like to know exactly how Late Ancient names were transmitted to and appropriated by the Jews of Rome. Perhaps, Jews were exposed to Latin names in the streets of Rome and in the city's marketplaces. One would also like to know why the Jews of ancient Rome prefered Latin and Greek names over more traditional, that is, Jewish ones. An element of free choice certainly played a role here. The onomastic data in question derive from catacombs in which only Jews were buried. In contrast to Roman monumental tombs constructed above ground and designed publicly to commemorate the accomplishments of their owners, such catacombs constitute a more private environment. In such an environment, Jews were free to use epigraphic formulae without worrying about the reaction of the non-Jewish outside world to such formulae. Rather than being the result of direct outside influence (such as the wish of public representation or a Roman master giving a Roman name to his or her Jewish slaves), names encountered in inscriptions from the Jewish catacombs must therefore be assumed *a priori* to reflect the onomastic practices and preferences of the Jews themselves.[118]

Whatever the answer to these questions, the corollary of the Greek and in particular of the Latin names preserved in the epitaphs from the Jewish catacombs is regular contact with non-Jews in daily life. When seen in this light, the question of whether the use of non-Jewish names by Jews results from fashion loses much of its relevance. Even if Jews chose non-Jewish names because they simply liked these names and did not attach deeper meaning to them, in and of itself such a choice reflects a situation of regular contacts between Jews and non-Jews.

---

[117] Alföldy 1969, 17-18.
[118] *Contra* Berliner 1893, 56 and Vogelstein and Rieger 1896, 59.

How deep such contacts went and what evidence may be used to measure the nature of such contacts is, of course, another question. The Jewish epitaphs from Rome show in any event that in Late Antiquity, Jewish onomastic practices do not attest to the Jews' isolation, but rather to a lively interaction between Jews and non-Jews.

CHAPTER FIVE

THE JEWISH FUNERARY INSCRIPTIONS FROM ROME:
LINGUISTIC FEATURES AND CONTENT

Introduction

Much can be learned about how Roman Jews viewed themselves and about how these Jews interacted with their non-Jewish contemporaries by studying the language used in the Jewish funerary inscriptions from Rome. To evaluate the languages of these inscriptions properly, it is necessary to investigate separately the formal aspects and the content of these inscriptions. A brief analysis of Leon's work on the languages of the Jewish epitaphs from Rome will serve as an introduction to the larger question of how Roman Jews did and did not adapt to the larger non-Jewish world of Late Antiquity.

The Languages of the Jewish Funerary Inscriptions from Rome

To date, some 595 Jewish mostly funerary inscriptions have been discovered in and around Rome. Of these inscriptions 467 (or 79%) have been written in Greek and 127 (or 21%) in Latin. One further inscription was composed in Aramaic, one is a bilingual in Greek and Aramaic, and six Greek inscriptions carry a short concluding formula in Hebrew.[1] Studying the languages of the inscriptions per catacomb, Leon observed that clear differences exist between the various Jewish catacombs in terms of languages used. He calculated the following figures (Table 1).[2]

Table 1: Language per catacomb according to Leon

	Greek	*Latin*	*Semitic*	*Bilingual*
Vigna Randanini	124 (63.6%)	71 (36.4%)	0	0
Monteverde	161 (78.2%)	41 (19.9%)	3 (1.4%)	1 (0.5%)
Villa Torlonia	63 (92.6 %)	4 (6.0%)	1 (1.4%)	0

---

[1] Solin 1983, 701 and n. 248. And see Fasola 1976, 22. The figures suggested by Leon 1960, 76 are slightly different because Leon could not take into account the (mostly Greek) inscriptions that were discovered by Fasola in the upper and lower Villa Torlonia catacombs in 1973-74.

[2] Leon 1960, 77.

Upon closer inspection, the percentages presented by Leon are not entirely correct. Most problematically, in Table 1 Leon inconsistently included inscriptions he himself did not regard as Jewish. At the end of his book, Leon published an appendix of Jewish inscriptions that was based on an earlier collection of Jewish inscriptions, J.-B. Frey's *Corpus Inscriptionum Judaicarum*. The main differences between Leon's and Frey's corpus of inscriptions is that Leon did not accept some inscriptions that Frey had included in his collection, either because these inscriptions were too fragmentary or because Jewish provenance could not be ascertained. Yet, when it came to calculating the relationship between inscriptions written in Greek as opposed to those composed in Latin, Leon used Frey's collection. Thus Leon maintained, for example, that a total of 124 Greek inscriptions were found in the Vigna Randanini (Frey's figure), but he then failed to notice that the total number of Greek inscriptions he himself ascribed to the Vigna Randanini catacomb was no higher than 106.[3]

Leon's presentation of these data is also problematical for another reason. Seventeen Jewish funerary inscriptions from Rome are in Latin, yet the character of the letters used is Greek. Conversely, three inscriptions were carved in Latin characters, yet the language employed was Greek. To gain a better understanding of the relationship between the Greek and Latin in the Jewish epitaphs from Rome, it is necessary to list inscriptions separately. Taking into account the shortcomings in Leon's presentation of the data, and adding the inscriptions that have been discovered in the upper and lower Villa Torlonia catacombs during the excavations of 1973-74, a more precise picture can now be drawn of the relationship between Jewish epitaphs in Greek and Latin (Table 2).[4]

Table 2: Language of inscription per catacomb

	Greek	Latin	Bilingual Gr.Lat
Vigna Randanini	121 (64.7%)	44 (23.5%)	22 (11.8%)
Monteverde	150 (79.0%)	25 (13.2%)	7 (3.7%)
VT-Upper	24 (100.0%)	0	0
VT-Lower	85 (98.8%)	0	1 (1.2%)

---

[3] Leon 1960, 274-92 follows Frey's enumeration, but has no reference to inscriptions such as *CIJ* 164, 174-75, 177-78, 181-84, 186-88, 194, 196-200.

[4] I have excluded a few inscriptions that refer to proselytes, because such inscriptions do not necessarily inform about the language situation of the Jews. Four Latin inscriptions that supposedly belong the Villa Torlonia catacombs (*CIJ* 68-71) have not been included in Table 2 because they were not actually found in the catacomb. I have included fragmentary inscriptions (such as *CIJ* 174 and 175), but I have excluded inscriptions whose Jewish provenance cannot be proved (e.g. *CIJ* 251 and 272).

(Table 2 continued)

	Bil Gr./Aram./Hebr.	Hebr.
Vigna Randanini	0	0
Monteverde	5 (2.6%)	3 (1.6%)
VT-Upper	0	0
VT-Lower	0	0

Explanation: Gr.=Greek; Lat.=Latin; Aram.=Aramaic; Hebr.=Hebrew.

Despite some minor differences, Table 2 provides us with a picture that is very similar to the one originally drawn by Leon: in all Jewish catacombs, Greek inscriptions are more popular than Latin ones; and Latin inscriptions are more frequently to be found in the Vigna Randanini than in the Monteverde or Villa Torlonia catacombs.

According to Leon, such figures provide indisputable evidence to show that the users of the Vigna Randanini catacomb belonged to the most, and those using the Villa Torlonia catacombs to the least, Romanized segment of the Roman Jewish community.[5] Inasmuch as 35.5 % (that is 23.5% + 11.8%) of the inscriptions from the Vigna Randanini catacomb suggest use of or familiarity with Latin, and inasmuch as, on the other hand, Latin played a negligible role in inscriptions from the upper and lower Villa Torlonia catacombs, Leon's conclusion seems, at first sight, self-evident. It is undeniable that the total number of Latin inscriptions from the Vigna Randanini, as opposed to the Monteverde and Villa Torlonia catacombs, is higher. Yet, to conclude from this that the users of the Vigna Randanini catacomb were therefore more Romanized than their Jewish contemporaries who were laid to rest in other Jewish catacombs is problematic, because such a conclusion is based on the unproven assumption that the Greek and Latin inscriptions constitute comparable bodies of evidence. To prove that the users of some catacombs were more Romanized than the users of other catacombs, one would have to prove that Greek and Latin were used during the same general period. If, however, one could show that the Latin inscriptions generally postdate the Greek ones, then the relative predominance of Latin in inscriptions from the Vigna Randanini catacomb would not document that some Jews were more Romanized than others, but would simply point to a period when Greek had ceased to be the most popular language for funerary inscriptions.

In Chapters 2 and 4 we have seen that archaeological and onomastic evidence does not support the idea that the Jews buried in one catacomb were generally more Romanized than those buried in another. Nor is this very surprising. We know that members of one and the

---

[5] Leon 1960, 77.

same Jewish community were sometimes buried in one catacomb and at other times in another.[6] Rather than supposing that the membership of individual Jewish communities consisted of an amalgam of more Romanized, more Hellenized, and more conservative elements (to use Leon's terminology) who were buried in different catacombs because they wanted to be separated in death when they had been united while still alive, it makes perhaps more sense to suppose that the members of a Jewish community were buried in different catacombs simply because sometimes there was room available in one catacomb, while at other times a burial slot could more easily be procured in another catacomb. Inasmuch as we know that catacombs were used during relatively long periods of time, the prominence of Latin inscriptions among the epigraphical materials from the Vigna Randanini catacomb may simply mean that this catacomb was in use when burial in the Monteverde and Villa Torlonia catacombs had stopped. Is there any evidence to support this view?

It is very unfortunate that the exact find spot of most Jewish funerary inscriptions from Rome remains unknown.[7] It is not possible to determine, therefore, whether in Rome the same general Latinization took place as in Venosa, where Greek inscriptions predominate in the earliest parts of one of the Jewish catacombs there, and where Latin inscriptions predominate in the more recent galleries of this underground complex.[8] Although we cannot really prove, then, that in Rome the use of Latin in Jewish epitaphs reflects a relatively late stage in the history of this community, such a hypothesis makes sense. Inasmuch as Greek was the most common non-Semitic language of Roman Palestine—at least insofar as inscriptions are concerned[9]—it is reasonable to suppose that Greek (as opposed to Latin) was the language most likely to be used by the first Jewish immigrants to Rome.

---

[6] This happened with members of the Subura community, who were buried in the Monteverde, Vigna Randanini, and Villa Torlonia catacombs, see *CIJ* 18, 22, 36, 37, 140, and 380. See, eventually, the evidence collected by Williams 1994.

[7] See the remarks by J. B. Frey, "Inscriptions inédites des catacombes juives de Rome," *Rivista di Archeologia Cristiana* 5 (1928) 280.

[8] G. Bognetti, "Les inscriptions juives de Venosa et le problème des rapports entre les Lombards et l'Orient," *Comptes Rendus de l'Académie des Inscriptions et Belles Lettres* 1954, 195; C. Colafemmina, "Gli Ebrei in Basilicata," *Bolletino Storico della Basilicata* 7 (1991) 13.

[9] On the use of Greek by the Jews of Roman Palestine, see G. Mussies, "Greek in Palestine and the Diaspora," in S. Safrai *et al.* (eds.), *The Jewish People in the First Century. Historical Geography, Political History, Social, Cultural, and Religious Life and Institutions* (Assen and Amsterdam: Van Gorcum, 1976) vol. 2, 1040-64; Schürer 1976-86, vol. 2, 20-28; H. B. Rosén, "Die Sprachsituation im römischen Palästina," in G. Neumann and J. Untermann (eds.), *Die Sprachen im römischen Reich. Beihefte der Bonner Jahrbücher* 40 (Cologne and Bonn: Rheinland und Habelt, 1980) 215-39; van der Horst 1991, 22-39 (inscriptions); F. Millar, *The Roman Near East 31 BC - AD 337* (Cambridge Mass. and London: Harvard U.P., 1993) 337-86.

We know that descendants of immigrants other than Jews continued to use Greek long after immigration had taken place.[10] We also know that the few Jewish inscriptions from Rome that commemorate people who can be shown to have originated in the eastern part of the Mediterranean are invariably written in Greek.[11] Having long been accustomed to Greek and immigrating into a bilingual city, upon immigrating, there was no good reason for Jews to give up the Greek they had used for such a long time already.[12]

When, at some later stage, Roman Jews switched to Latin, this shift must have occurred rather gradually. Studying references to age at death and onomastic practices in Chapters 3 and 4, we have seen that Latin epigraphic practices started to manifest themselves among Roman Jews at a time when the inscriptions were still written in Greek. Latin inscriptions written in Greek characters (classed as bilingual in Table 2) probably constitute the next step in this process of gradual Latinization. Besides documenting the continued importance of Greek forms, these inscriptions suggest that a time of experimentation preceded the period during which Latin gained full acceptance as language for inscriptions. The final step was the use of inscriptions in which both the language and the letters were Latin. The predominance of Latin inscriptions in the Vigna Randanini catacomb can be explained by supposing that this catacomb remained in use until after burial in the Monteverde catacomb had ceased. Such a hypothesis is shown to be likely by the fact that a particular type of abbreviations (for details, see next section below) appears somewhat more frequently in Latin inscriptions from the Vigna Randanini catacomb than in those from the Monteverde catacomb (51% versus 40%).

Before we can draw any further conclusions, let us first turn to the two catacombs under the Villa Torlonia. These catacombs present a case all to themselves. Of the five Latin inscriptions that have been found in the area of the Villa Torlonia catacombs, only one was actually found in the (lower) catacomb (hence the exclusion of the other four inscriptions from Table 2). The general lack of Latin inscriptions from the Villa Torlonia catacombs may be due to the fact that many of the Jews buried in the Villa Torlonia catacombs belonged to the lower strata of society. Therefore they may have had less access to Latin, or, perhaps, they were less interested in it. In an earlier chapter we have already seen that the Villa Torlonia catacomb differed from all other Jewish catacombs in that it was planned systematically to accommo-

---

[10] For the non-Jewish evidence, see I. Kajanto, *A Study of the Greek Epitaphs from Rome*. Acta Instituti Romani Finlandiae II:3 (Helsinki: Tilgmann, 1963) 5.

[11] *CIJ* 25, 362, 370, 502. And see *CIJ* 296 and 501.

[12] For good examples of the bilingual nature of Rome see, e.g. the dedications in *IGUR* 1, 60-61.

date large groups of people who were then buried in simple burial slots.[13] In contrast to the Monteverde and Vigna Randanini catacombs, where the use of inscriptions engraved on marble plates was customary, people buried in the upper and lower Villa Torlonia catacomb were likely to receive an inscription that was painted or incised into the stucco sealing off the graves. Such inscriptions, and the graffiti in particular, were incised on the spot by those present at the deceased's burial. Yet, even though these inscriptions are in Greek, they all display the same tendency towards Latinization as do Greek inscriptions from the Monteverde and Vigna Randanini catacombs. Not only do the inscriptions from the Villa Torlonia catacombs contain references to age at death—a typically Latin epigraphical custom—54% percent of all names used in Greek inscriptions from the lower Villa Torlonia catacomb are Latin.[14] In Greek inscriptions from the upper Villa Torlonia catacomb, all names are Latin.

Such evidence indicates that the relationship between Greek and Latin in the Jewish epitaphs from Rome cannot be understood properly merely by contrasting the inscriptions written in one language with those written in another. The Greek inscriptions from the Villa Torlonia catacombs (or, for that matter, from the Vigna Randanini and Monteverde catacombs) show a community open to linguistic influences and in the process of linguistic change. This process must probably to be viewed as follows. The earliest inscriptions that were set up by the Jewish community of ancient Rome were all in Greek. The high percentage of Latin names used in these inscriptions suggests that at a relatively early stage a tendency towards Latinization started to manifest itself (late second and third century C.E.). No differences existed between the various Jewish catacombs of Rome insofar as the influence of Latin on Greek was concerned: references to age at death, which were included in Greek inscriptions in direct imitation of Latin epigraphic practices, appear in Greek inscriptions from all Jewish catacombs equally frequently, just as Latin names invariably predominate in the Greek inscriptions from all Jewish catacombs.[15] In the course of the fourth century, the influence of Latin became ever more palpable. In this period the first inscriptions in Latin started to be carved. That Latin probably never entirely replaced Greek as the language for inscriptions and that it became more and more popular only gradually is borne out by the relatively large number of bilingual inscriptions, as well as by onomastic evidence. In Chapter 4 we have seen that of all Semitic names preserved in Jewish inscriptions from

---

[13] For details, see Chapter 2.

[14] Note that this percentage is surprisingly close to the percentage of Latin names in the Greek inscriptions from the Vigna Randanini catacomb (58%).

[15] For details and percentages, see Chapters 3 and 4.

Rome, 79% of them occur in inscriptions in Greek and 21% in inscriptions in Latin. Such figures suggest that the adoption of the Latin language cannot be taken as an indicator of complete Latinization, but that it could go hand in hand with more traditional onomastic practices.

That the shift from Greek to Latin was gradual is most evident, however, when we look at the content of these inscriptions. Inasmuch as they often provide exactly the same kind of information, it is usually impossible to ascertain any differences between the inscriptions carved in Greek as opposed to those written in Latin except for that language difference.[16] That the Jewish inscriptions in Latin indeed follow the ones in Greek quite closely is particularly clear from the commemorative patterns documented on both types of inscriptions. In Greek inscriptions, as, for example, in Asia Minor, the name of the dedicator traditionally appears first and is followed by the name of the deceased. In Latin epitaphs, by contrast, this pattern is normally reversed so that the name of the dedicator follows that of the deceased.[17] As might be expected, the Jewish epitaphs from Rome composed in Greek follow the Greek pattern.[18] Yet, interestingly, the Jewish epitaphs from Rome written in Latin do not follow the Latin pattern, but rather the Greek one.[19] Nor did people who used Latin as opposed to Greek consider themselves less Jewish. While 34% of all Greek inscriptions from the Monteverde catacomb carry an incised menorah, 35% of the Latin inscriptions from this catacomb have been adorned with this typically Jewish symbol. In case of the Vigna Randanini catacomb the percentages are somewhat different, but still quite comparable. There a menorah appears in 22.5% of all Greek inscriptions, while it has been included in 15.5% of all inscriptions in Latin. Cumulatively, such evidence shows that the change from Greek to Latin in the Jewish epitaphs from Rome is not as revolutionary as a first look at the language of these inscriptions might suggest.

An explanation of the language of the Jewish epitaphs that stresses the chronological differences between inscriptions in Greek and Latin also helps to solve an apparent discrepancy between these inscriptions and the early Christian inscriptions from Rome. To date, some 25,500

---

[16] It is certainly telling that the Jewishness has been questioned of exactly those two inscriptions that differ markedly in both phraseology and nomenclature from the other Jewish inscriptions in Latin. On *CIJ* 470, see Stefano Manzelia 1989; on *CIJ* 476, see A. Ferrua, "Epigrafia ebraica," *La Civiltà Cattolica* 1936, 308.

[17] Kajanto 1963, 21 and 24.

[18] In 66.6% of these inscriptions the name of the dedicator is mentioned first.

[19] In 65% of these inscriptions the name of the dedicator is mentioned first. Note that in non-Jewish inscriptions in Latin, the name of the dedicator is mentioned first in 32% of the inscriptions, while the name of the deceased is mentioned first in 68% (these percentages have been calculated on the basis of the evidence provided by Kajanto 1968, 24).

early Christian inscriptions from Rome have been published.[20] According to my estimates, 2,298 or 9% of these inscriptions were written in Greek, while 23,100 or 91% were composed in Latin. Of course, the relationship between Greek and Latin is not the same in all Christian catacombs. Even within single catacombs, considerable differences exist. In the lower part of the Domitilla catacomb, for example, 15.5% of all inscriptions were in Greek, but this percentage is reduced to 7.5% in inscriptions found in the upper galleries of this same catacomb.[21] In the Priscilla catacomb, another catacomb that was used early in the history of Rome's early Christian community, 19% of all inscriptions were in Greek.[22] Usually, however, in the epigraphical materials from the Christian catacombs, Greek inscriptions form only a minor fraction of the total "harvest" of inscriptions: percentages normally hover between 0 and 10 or 15%. In general, then, these percentages contrast markedly with the percentages documented for the Jewish catacombs of Rome. We may recall that there, 79% of all inscriptions are in Greek. Can such differences be taken to imply that the Jewish community of ancient Rome was isolated linguistically after all?

Taking into account the location of the Greek inscriptions found in the early Christian catacombs, it seems likely that in case of these catacombs, too, a shift occurred that led from Greek to Latin as the main language for inscriptions. In early Christian catacombs that were used in the third century, Greek was still relatively common. It quickly disappeared, however, from the inscriptions found in catacombs that were dug during the fourth century—that is, during the period that the construction of catacombs seems to have reached its peak.

To explain the numerical difference between the Jewish and early Christian inscriptions in terms of language used, we have to consider, furthermore, the character of the communities that produced them. We have already seen that there existed, within the Jewish community, a tradition of producing inscriptions in Greek. Greek owed its prominence not only to the importance attached to it by Jews who had originated in the eastern part of the Mediterranean, but also to the fact that it served as a vehicle to express someone's Jewishness (this thesis will be dealt with in greater detail below). As time went on, Roman Jews continued to erect inscriptions in Greek. The Latin influence mani-

---

[20] *ICVR* 1-10.

[21] *ICVR* 3. Similarly "high" percentages can be found in other early catacombs such as the Callixtus catacomb, see *ICVR* 4 (24.5% [*inscriptiones cryptarum ad S. Cornelium*]; 17% [*inscriptiones coemeterii inferiori*]; 13.5% [*inscriptiones coemeterii superioris*]) in the catacomb of Praetextatus, see *ICVR* 5 (15.3%) and in the *coemeterium Pamphili*, see *ICVR* 10 (18.5%).

[22] *ICVR* 9.

fested itself only gradually, for it had to compete with an epigraphic tradition that was deeply rooted.

In case of Rome's early Christian community, by contrast, the situation was rather different. While people of very different social and cultural backgrounds felt drawn to this community, sizable increases in membership happened only at a relatively late point in history, namely in the course of the fourth century, when the Greek element in the early Church in Rome was quickly disappearing. When the mass production of early Christian epitaphs finally got underway, Latin immediately assumed an importance which it could assume in Jewish inscriptions only after it succeeded in supplanting existing epigraphical traditions.[23]

That the use of Greek in Jewish inscriptions can indeed not be seen as testifying to an isolated position of the Jews in late Roman society is particularly evident when one studies the linguistic features of these inscriptions—a subject to which we will now turn.

The Linguistic Features of the Jewish Epitaphs from Rome

Even a brief look at the Jewish epitaphs from Rome suffices to show that the Greek in these inscriptions closely follows the Greek in non-Jewish inscriptions in terms of phonology, morphology, and syntax. The extent to which phonological changes that affected post-Classical Greek in general also affected the Greek of the Jewish funerary inscriptions is evident, for example, from the way the word κεῖται ("he" or "she lies," commonly used in constructions such as ἐνθάδε κεῖται, "here lies") has been rendered in these inscriptions (Table 3).[24]

Table 3. Κεῖται and its variants in Jewish and early Christian funerary inscriptions from Rome.

	Vigna Randanini	*Monteverde*	*V.Torlonia*	*ICVR 1-7*
Κεῖται	6	14	5	13
Totals	6 (17.1%)	14 (15.9%)	5 (13.2%)	13 (25%)
Κεῖτε	21 (60%)	42 (47.7%)	11 (28.9%)	6 (11.5%)
Κεῖτη	—	1	—	1
Κεῖτι	—	1	—	—
Κεῖθαιν	—	1	—	—

---

[23] M. Guarducci, *Epigrafia greca*. 4. *Epigrafi sacre pagane e cristiane* (Roma: Istituto Poligrafico dello Stato, 1978) 302-3.

[24] Fragmentary inscriptions have been left out.

(Table 3 continued)

Κητε	—	—	1	—
Κίται	1	3	2	1
Κίτε	7 (20%)	17 (19.3%)	18 (47.4%)	22 (42.3%)
Κιετε	—	1	—	—
Κίκε	—	—	1	—
Κίτεν	—	1	—	—
Κίθε	—	1	—	—
Κάτακιτε	—	1	—	9 (17.3%)
Χεῖθε	—	2	—	—
Χίτε	—	1	—	—
Εκειθεν	—	1	—	—
Εκιθεν	—	1	—	—
Totals	29 (82.9%)	74 (84.1%)	33 (86.8%)	39 (75%)

Table 3 shows that the properly Classical κεῖται appears in only a fraction of either the Jewish or the early Christian funerary inscriptions from Rome. Instead, one encounters phonological changes that are generally characteristic of post-Classical Greek, including the substitution of ε for αι (as in κεῖτε), ι for ει (as in κίτε), and less common substitutions such as χ for κ (as in χίτε). Further morphological changes including the adding of a ν (as in κίτεν) or the transfer of an augmental ε to a verb that is already in the past tense (as in ἔκειθεν).[25]

Likewise typically late antique is the ἐνθάδε κεῖται formula as a whole. In pagan inscriptions in Greek that contain this formula, the subject usually precedes the verb. In inscriptions dating to Late Antiquity, this order is normally reversed so that the subject follows the ἐνθάδε κεῖται which, in such cases, is put at the beginning of the inscription.[26] This is the pattern one frequently encounters in the Jewish epitaphs from Rome: of all Jewish epitaphs, 89% follow the Late Ancient and 11% follow the Classical pattern.[27]

Perhaps most interesting aspect of Table 3 is that no major differences exist between the individual Jewish catacombs insofar as the Greek used in inscriptions is concerned. In all catacombs, κεῖται is used only occasionally, namely in less than one fifth of all cases. Users of these catacombs all seem to have preferred κεῖτε and κίτε instead. Such written evidence suggests that the users of different Jewish catacombs all spoke the same kind of Greek. A comparison of the syntactical errors contained in the inscriptions from the different Jew-

---

[25] For parallels, see F. T. Gignac, *A Grammar of the Greek Papyri of the Roman and Byzantine Periods* (Milan: Cisalpino, 1976-1981) vol. 1, 76-77, 113, 185, 191; vol. 2, passim.

[26] Kajanto, 1963, 21-22.

[27] These are the precentages per catacomb: in the Monteverde catacomb, 97% of the inscriptions follow the Late Ancient pattern, in the Vigna Randanini catacomb 79%, and in the Villa Torlonia catacombs 87%.

ish catacombs shows that the Jews buried in one catacomb were not generally better educated than those buried in another.[28] Inasmuch as other linguistic features of the Jewish epitaphs in Greek were studied in detail by Leon, it is not necessary to repeat his observations here, except to note that these researches led Leon to conclude, correctly, "that the Jews of ancient Rome spoke essentially the same Greek that was prevalent among the lower classes during these centuries."[29]

Such a conclusion also holds true for the Jewish inscriptions in Latin. At first sight, the presence of vulgar Latin in Jewish inscriptions in Latin appears to have been minimal.[30] While the preposition *cum* is followed by an accusative in one case only, the other inscriptions have the regular *cum* with ablative.[31] Finally, in Jewish epitaphs in Latin, the final "m" is never dropped as is often the case in inscriptions in vulgar Latin.[32] Similarly, the influence of vulgar Latin in the spelling of words such as *benemerenti* is not very large.[33]

Upon closer investigation, however, the influence of vulgar Latin turns out to have been considerable. It is evident, for example, from the spelling of the word *vixit*, especially where it concerns inscriptions from the Monteverde catacomb.[34] Vulgar Latin influences are particularly evident in the rendering of the *pronomen relativum*. In 70% of the Latin inscriptions from the Vigna Randanini catacomb *que* has been substituted for the Classical *quae*—percentages that are surprisingly close to those calculated for early Christian inscriptions in Latin.[35] That the Jewish inscriptions in Latin belong to a relatively late

---

[28] Compare, for example, the substitution of ὅς or ἥ by οιτος (in the Villa Torlonia catacomb, *CIJ* 16), by ὅσιος or ἵτις (in the Vigna Randanini catacomb, *CIJ* 100 and 102) or ὅστις (in the Monteverde catacomb, *CIJ* 299).

[29] H. J. Leon, "The Language of the Greek Inscriptions from the Jewish Catacombs of Rome," *Transactions of the American Philological Society* 58 (1927) 233. And see the evidence collected by van der Horst 1991, 25-31.

[30] On vulgar Latin in general, see V. Väänänen, *Introduction au Latin vulgaire* (Paris: Klincksieck, 1963); A. Acquati, "Il vocalismo latino-volgare nelle iscrizioni africane," *Acme* 24 (1971) 155-84; *ead.*, "Il consonantismo latino-volgare nelle iscrizioni africane," *Acme* 27 (1974) 21-56; *ead.*, "Note di morfologia e sintassi latino-volgare," *Acme* 29 (1976) 41-72; E. Löfstedt, *Il latino tardo. Aspetti e problemi* (Brescia: Paideia, 1980).

[31] *Cum* with accusative: *CIJ* 262; *cum* with ablative: *CIJ* 220, 236, 242, 262, 276, and 457.

[32] See *CIJ* 210, 220, 236, 242, 262, 276, and 457.

[33] For examples of *benemerenti* according to the Classical spelling, see *CIJ* 206, 207, 220, 221, 225, 235, 236, 237, 242, 256, 260, 265, 266, 270, 456, 457, and 474. For examples of a spelling in vulgar Latin, see *CIJ* 213, 233, 257, and 462.

[34] For examples of a Classical spelling, see *CIJ* 210, 217, 230, 237, 241, 242, 247, 259, 276, 457, 462, 463, 465, and 482. For renderings in vulgar Latin, see *CIJ* 206, 254, 260, 456, 466, 468, and 480.

[35] For *que*, see *CIJ* 217, 230, 237, 240, 241, 242, and 267; for *quae*, see *CIJ* 247, 254, and 257. For *que-quae* in early Christian inscriptions, see H. Zilliacus, *Sylloge inscriptionum Christianorum veterum musei Vaticani*. Acte Instituti Finlandiae. Vol. I:2 (Helsinki: Tilgmann, 1963) 9, 21, and 24.

phase in the history of this language is also evident, finally, from the gradual disappearance of the *ablativus durativus* which was traditionally used to indicate the number of years, months, and days the deceased was supposed to have lived. Instead of *vixit* x *annis, y mensibus, z diebus* (all ablatives), Late Ancient inscriptions prefer *vixit* x *annis, y menses, z dies* (one ablative and two accusatives).[36] In referring to age at death, Jewish inscriptions in Latin usually only include the number of years. Conforming to Late Ancient practice, references to years have been rendered in the ablative. In the few inscriptions which include a reference to the number of months and days, one follows the Classical pattern, and three the Late Ancient one.[37]

When one contrasts the total number of Jewish inscriptions in Latin that display elements of vulgar Latin with the total number of inscriptions that lack such elements, roughly three quarters of the Latin inscriptions from both the Vigna Randanini and the Monteverde catacombs show the influence of vulgar Latin.[38] Frequently, two or more elements characteristic of vulgar Latin appear in one and the same inscription.[39] Such evidence suggests that in their use of Latin Roman Jews did not distinguish themselves from non-Jews. A comparison of the Latin in inscriptions from the Vigna Randanini with those from the Monteverde catacomb suggests, furthermore, that it is impossible to draw a distinction between the users of both catacombs on linguistic grounds.

That Jewish inscriptions in Latin follow rather closely contemporary (non-Jewish) patterns may also be evident when we turn to study a characteristic that has not received proper attention in studies on these Jewish inscriptions: contractions. In Latin inscriptions, three types of contraction occur.[40] For the present purpose, we need to investigate the history of only two of these contractions, namely the contraction proper and the abbreviation.[41] Studying a collection of 1,632 mostly funerary inscriptions from Rome, 27% of which were pagan and 73% of which early Christian, Hälvä-Nyberg observed that

---

[36] G. Konjentzny, "De idiotismis syntactis in titulis latinis urbanis (*CIL* vol.VI) conspicuis," *Archiv für lateinische Lexicographie und Grammatik* 15 (1908) 297-351, esp. 331; Zilliacus (previous note) 28; H. Nordberg, *Biometrical Notes*. Acta Instituti Romani Finlandiae II:2 (Helsinki: Tilgmann, 1963) 21 and 25.

[37] For the Classical pattern, see *CIJ* 237; for the Late Ancient pattern, see *CIJ* 242, 262, and 276.

[38] In calculating this relationship, I have excluded the Latin inscriptions written in Greek characters. I have also excluded Latin inscriptions that consist of names only.

[39] See, for example, *CIJ* 213, 217, 219, 230, 234, 237, 242, 260, 264, 267, 456, 466, and 468.

[40] See U. Hälvä-Nyberg, *Die Kontraktionen auf den lateinischen Inschriften Roms und Afrikas bis zum 8.Jh.n.Chr.* (Helsinki: Suomalainen-Tiedeakatema, 1988) 17-19.

[41] A word like *v(i)x(i)t* constitutes a proper contraction; a word like *Kal(endas)* constitutes an abbreviation.

in Late Antiquity there was a general increase in the use of all three types of contractions in Latin inscriptions. Hälvä-Nyberg has also observed that these different kinds of contractions enjoyed popularity at different points in time. While abbreviations were most popular during the fourth and fifth centuries, the proper contraction did not start to become more widely used until late in the fifth century, and even then it never became as popular as abbreviations had been before.[42]

When we turn to abbreviations and contractions in Jewish epitaphs in Latin, the following picture emerges.[43] Of the Latin inscriptions found in the Vigna Randanini catacomb, one inscription displays a proper contraction (2.5%), nineteen inscriptions contain abbreviations (48.7%), and, finally, nineteen further inscriptions contain neither proper contractions nor abbreviations. Six Latin inscriptions from the Monteverde catacomb (40%) carry abbreviations. Proper contraction does not appear in any Latin inscription from this catacomb. The remaining nine Latin inscriptions from Monteverde (60%) do not carry abbreviations.

The absence of proper contraction in Latin inscriptions from the Vigna Randanini and Monteverde catacomb suggests that the Jewish epitaphs generally predate the late fifth century when proper contractions first started to become more widely used. A dating of these inscriptions to the fourth or perhaps to the early fifth century seems more likely, especially because abbreviations are rather common in the Jewish funerary inscriptions in Latin. The fact that abbreviations appear somewhat more frequently in the inscriptions from the Vigna Randanini catacomb than in those from the Monteverde catacomb (51% as opposed to 40%) could indicate that the inscriptions found in the former catacomb slightly postdate those discovered in the latter. Independent of what else such differences might mean, however, it is clear that formally the Jewish epitaphs in Latin testify to the same phenomenon as those composed in Greek: in Late Antiquity Jews did not employ any special kind of Latin, they used exactly the same kind of Latin that non-Jews used in their inscriptions.

Still another point merits our attention in this context. Leon observed that fewer "aberrations" from correct pronunciation and grammar occur in inscriptions written in *Latin* as opposed to those carved in Greek. Such observations led Leon to argue that the Jews who ordered epitaphs in Latin were more prosperous and better educated than those using Greek.[44] Inasmuch as other scholars have argued in reference to non-Jewish inscriptions that the persons recorded in *Greek*

---

[42] Hälvä-Nyberg 1988, 39 and 108.

[43] From this survey I have excluded incomplete inscriptions and Latin inscriptions that have been written in Greek characters.

[44] Leon 1960, 87, 90-92; repeated by Solin 1983, 706-7.

epitaphs were generally of higher social status than those represented in the Latin epitaphs, it is necessary to review the evidence bearing on Jewish epitaphs in Greek and Latin.[45]

That Leon's conclusion is rather problematic may become evident when it is seen against the previous discussion of vulgar Latin in the Jewish inscriptions from Rome. Taking into account that the influence of spoken (vulgar) Latin on literary Latin was tangible, especially in Late Antiquity, one may wonder whether it is correct to speak of "aberrations" at all. What is the norm from which these Late Ancient inscriptions are supposed to show aberrance? From the Latin as documented in literary sources dating to the first century B.C.E.? The idea of "aberration" presupposes that there existed a standardized Latin, that people ackowledge the standard as such, and that, in Late Antiquity, people wanted to have their epitaphs carved in this standardized type of Latin. The presence of vulgar Latin in Jewish and non-Jewish epitaphs alike suggests, however, that in Late Antiquity vulgar Latin became the standard language of the populace. Thus, to speak of "aberrations" makes no sense; it amounts to saying that these inscriptions deviated (or did not deviate) from a standard that had ceased to be the standard. The same argument holds true for the Jewish inscriptions in Greek. In this case too, non-Classical Greek is used most frequently in Late Ancient inscriptions. Thus, to say that Jewish inscriptions in Latin contain fewer mistakes than those composed in Greek is incorrect on at least two counts. First, the influence of the spoken language ("incorrect" Latin or Greek) is not less tangible in Latin as opposed to Greek inscriptions. Second, it is useless to study Late Ancient inscriptions on the basis of standards that belong to a different period.

There are also other problems with Leon's thesis that contrasts the Jewish inscriptions in Latin with those composed in Greek. Inscriptions inform, in the very first place, about the level of education of the person who carved these inscriptions. They inform only indirectly, however, about the level of education of the person ordering an inscription. Of course, someone who knew his or her Latin well is not likely to have accepted an inscription full of syntactical flaws. Yet the opposite scenario is very well imaginable: someone who did not know Latin well is likely to have accepted an inscription written in proper Latin. Put differently, even if an inscription is written in correct Latin,

---

[45] For the non-Jewish evidence, see Kajanto 1963, 6. This view is not shared by everyone. For the view that the users of Greek belonged to the lower strata of society, see E. C. Polomé, "The Linguistic Situation in the Western Roman Provinces of the Roman Empire," *Aufstieg und Niedergang der römischen Welt* II.29.2 (1983) 516; J. Kaimio, *The Romans and the Greek Language*. Commentationes Humanarum Litterarum 64 (1979) 24.

this cannot automatically be taken to mean that the person who ordered it knew Latin perfectly. To determine whether the users of Latin inscriptions were really so much better educated than those using Greek ones, it would be necessary to study the Latin used in graffiti, that is, in inscriptions that the deceased's family or friends (as opposed to professional stonecutters) incised into the wet stucco directly after burial had taken place. Unfortunately, this is not possible because practically all graffiti in Latin concern names only, and do not include other information.[46]

Therefore, linguistic arguments alone are certainly not sufficient for concluding that the Jews who used Latin were generally more prosperous than those using Greek. As has already been observed, inscriptions in Greek and Latin are remarkably similar insofar as their content is concerned. Also, a comparison of the stones into which these inscriptions have been cut reveals that no major differences exist between the Greek and Latin inscriptions in terms of artistic quality: Latin inscriptions are not generally more carefully carved than Greek ones, nor are the stones that carry such inscriptions generally larger or of better quality. In fact, studying the collection of Jewish epitaphs preserved in the Musei Vaticani, it may be observed that most Jewish funerary inscriptions in Latin can hardly be called key monuments of art.[47]

That the users of Greek were not generally poorer than those using Latin is borne out by the fact that the relationship between the more expensive carved inscriptions and the certainly much cheaper painted and engraved inscriptions (the so-called dipinti and graffiti) is more or less identical in both cases. While of all Greek inscriptions from the Vigna Randanini catacomb, 14% are either graffiti or dipinti, the percentage of painted or engraved inscriptions rises to 20% in case of the Latin inscriptions from this catacomb. In the Monteverde catacomb, 18% of the Greek inscriptions are graffiti or dipinti, while 16% of the Latin inscriptions belong to this class. It may finally be noted that the relationship of Greek to Latin on (expensive) sarcophagi used by Jews is 10:1—a relationship that differs fundamentally from that documented for non-Jewish sarcophagi on which inscriptions in Latin

---

[46] For names in the Latin graffiti from the Vigna Randanini catacomb, see *CIJ* 223, 226, 227, 238, 244, 253, 258, 261, the only exception being *CIJ* 211. For the Monteverde catacomb, see *CIJ* 473. Contrary to S. V. Tracy. "Identifying Epigraphical Hands," *Greek, Roman, and Byzantine Studies* 11 (1970) 321-33 I do not believe that in the case of inscriptions it is not possible to establish common workshop identity. Tracy's argument is in any event based on circular reasoning (e.g. p. 325 n. 30).

[47] E.g. *CIJ* 466, 467, and 477. Note also that the illustrations published in *CIJ* are particularly misleading in this respect. *CIJ* 353, for example, is one of the most impressive Jewish epitaphs in Greek from Rome. It measures 53 x 87 cm. Yet, the picture in Frey gives one the impression that this inscription is but a small and insignificant burial plaque.

by far outnumber inscriptions in Greek.[48] Again, we should be careful not to draw conclusions too quickly: some inscriptions mentioning synagogue officials are far from impressive from an artistic point of view.[49] Cumulatively, such evidence suggests that there is no reason to conclude that a strong dichotomy existed between the users of Greek and Latin in terms of social position or cultural background. The Greek and Latin inscriptions from the Jewish catacombs do not point to two different groups within the Roman Jewish community, but rather document two chronologically different stages in the history of this community.

The Content of the Jewish Funerary Inscriptions

Having much in common linguistically with non-Jewish inscriptions, the inscriptions from the Jewish catacombs of Rome differ in one important aspect from their non-Jewish counterparts: in content.

Sometimes, Jewish funerary inscriptions carry formulae that occur much more frequently in these than in pagan or early Christian inscriptions. The expression ἐν εἰρήνη ἡ κοίμησίς αὐτοῦ - αὐτῆς ("in peace his/her sleep") is one such formula. According to my estimates, in Rome it occurs in 27.8% of all Jewish and in only 1% of all early Christian inscriptions in Greek.[50]

At other times, Jewish inscriptions provide information that is lacking in non-Jewish inscriptions. The differences between Jewish and non-Jewish inscriptions from Rome are particularly evident from the use of epithets and from references to offices related to the Jewish community.

To show the differences in epithets used by Jews, Christians and pagans in Rome, I have included the following two tables (Tables 4 and 5). In them, I have compared the epithets that were collected by I. Kajanto while studying a collection of 800 Greek and 2,000 Latin (pagan) inscriptions from Rome with evidence I have collected from the *Corpus Inscriptionum Judaicarum* and the ten volumes of the *Inscriptiones Christianae Urbis Romae*.[51]

---

[48] These percentages have been calculated on the basis of the evidence provided by Konikoff 1986. For the non-Jewish evidence, see Koch and Sichtermann 1982, 26.

[49] E.g. *CIJ* 433.

[50] This figure is based on the evidence culled from *CIJ*, Fasola 1976, Solin 1983 and *ICVR* 1-10). Early Christian inscriptions sometimes prefer variants such as ἐν θεῷ (*ICVR* 9.26052) or ἐν ἁγίῳ (*ICVR* 10.27233).

[51] Kajanto 1963, 30-39. The word θεοσεβής has been excluded from this survey. Only complete inscriptions have been included.

Table 4. Epithets in pagan, Jewish and early Christian epitaphs from Rome. Inscriptions in Greek. Epithets that occur in Jewish and not in pagan inscriptions have been underlined. Epithets that occur only in early Christian inscriptions have been italicized. Epithets that occur in both Jewish and early Christian, but not in pagan inscriptions have been underlined and italicized.

Epithet	Pagan	Jewish	Early Christian
1 ἀβλαβής	—	1	—
2 ἀγαθός	12	—	4
3 ἀγαθώτατος	2	—	—
4 *ἀγαπητός*	—	3	8
5 ἁγνός	2	—	1
6 ἁγνοτάτη	1	—	—
7 ἀείμνηστος	9	1	22
8 *ἄκακος*	—	—	1
9 ἄλυπος	3	—	—
10 ἄμεμπτος	2	2	—
11 ἀμίαντος	—	1	—
12 ἀμίμητος ˜ ἀμείμητος	1	—	—
13 ἀναμάρτητος	1	—	1
14 ἄξιος	11	3	5
15 ἁπλουστάτη	1	—	1
16 ἄριστος	1	—	—
17 ἀστομάχητος	1	—	—
18 ἀσύνκριτος	11	3	13
19 ἄφθορος	1	—	—
20 ἄωρος	3	—	—
21 *γλυκύς*	—	—	2
22 γλυκύτατος	112	10	163
23 δικαία	—	1	—
24 ἔντιμος ˜ ἔντειμος	1	—	—
25 ἐπιζήτητος	1	—	—
26 εὐδιδακτή	—	1	—
27 εὐλογημένη	—	1	—
28 εὐνοήσας	1	—	—
29 εὐπρόσδεκτος	1	—	—
30 εὐσεβής	3	—	—
31 εὐσεβέστατος	8	—	—
32 εὐτυχεστάτη	1	—	—
33 *εὐφροσύνη*	—	—	3
34 *εὐχάριστος*	—	—	1
35 ἡδύς	—	1	—
36 ἥρως	8	—	—
37 θαυμαστός	1	—	—
38 θεοφιλεστατός	1	—	2
39 καλή	1	—	1
40 καλόμαλλος	1	—	—
41 κομψός	1	—	—
42 κυρία	1	—	—
43 *λαμπρότατος*	—	—	1

(Table 4 continued)

44	μακαρία	1	—	—
45	μόνανδρος	1	—	—
46	ὅσιος	—	14	—
47	πανάρετος	1	—	—
48	πανφίλος	—	—	1
49	πασιφίλος	—	—	1
50	πιστός	1	—	11
51	πιστοτατή	1	—	—
52	ποθεινότατος	1	—	—
53	σεμνός	4	—	3
54	σεμνοτάτη	2	—	7
55	σοφός	—	—	1
56	σπάταλος	1	—	—
57	σώφρων	1	—	1
58	σωφρονεστάτη	1	—	—
59	τίμιος	1	—	—
60	τιμιώτατος	4	—	—
61	φειλητός (πασι)	—	1	—
62	φιλάδελφος	—	3	1
63	φίλανδρος	4	1	8
64	φιλάνθρωπος	1	—	—
65	φιλέντολος	—	3	—
66	φιλόγονευς	—	1	—
67	φιλόθεος	—	—	1
68	φιλίλιης	—	1	—
69	φιλόλογος	1	—	—
70	φιλόνομος	—	1	—
71	φιλοπάρθενος	—	—	1
72	φιλοπάτωρ	—	1	1
73	φιλοπένης	—	1	—
74	φιλόστοργος	2	—	—
75	φιλοσυνάγωγος	—	1	—
76	φιλότεκνος	—	1	2
77	φιλοχήρα	—	—	1
78	φίλτατος	3	—	—
79	χρηστός	8	—	—
80	χρηστοτάτη	1	—	—
81	ψυχή ἀγαθή	2	—	1
	ψυχή ἀείμνηστος	2	—	—
	ψυχή ἄκακος	1	—	—
	ψυχή ἀσύνκριτος	1	—	—
Totals		250	57	270

A look at Table 4 reveals some highly interesting differences between pagan, Jewish, and early Christian inscriptions insofar as the use of epithets is concerned. Through Late Antiquity, there seems to have been a general decrease in the use of epithets. While of all pagan inscriptions that predate the Late Ancient period, 31% percent carry an epithet, this percentage is reduced by more than half (12.2%) in Jewish inscriptions. In the case of early Christian inscriptions the percentage is even slightly lower, namely 11.8%.[52] The decrease in the number of epithets goes hand in hand with what appears to be a general decrease in the variety of epithets. In pagan inscriptions we find as many as 52 different epithets. In Jewish and early Christian inscriptions the number of epithets is reduced to 24 and 31, respectively. A closer look at the epithets reveals that in Late Antiquity there was also a shift in the adjectives that served as epithets. This shift can be observed most easily in Jewish inscriptions. Of the 24 different epithets contained in these inscriptions, 18 (75%) do *not* occur in pagan inscriptions. In early Christian inscriptions, the shift is somewhat less pronounced, but still quite tangible: 29% of the epithets in early Christian inscriptions do not occur in pagan inscriptions.[53] Even more interesting than studying the shift itself is to review which epithets replaced the ones used earlier.

Pagan and early Christian inscriptions have in common that one epithet outranked by far all others in popularity: γλυκύτατος ("sweetest"). In early Christian inscriptions the epithet is even more popular than in pagan ones (60.4% versus 44.8%). In Jewish inscriptions γλυκύτατος is fairly popular (17.5%), yet it is less popular than ὅσιος ("holy," "devout"—a term used in the Septuagint to translate the Hebrew חסיד). This word occurs in 24.5% of all Jewish epitaphs carrying an epithet. It is, therefore, the single most popular epithet in Jewish funerary inscriptions.[54] No less significant is the fact that ὅσιος *never* appears in either pagan or early Christian inscriptions.[55]

Further differences between the epithets in Jewish as opposed to pagan and early Christian inscriptions are equally telling. Jewish inscriptions show a clear preference for epithets that consist of the word φιλο- followed by a noun (61% of the epithets used in Jewish inscrip-

---

[52] The pagan percentages have been calculated on the basis of the evidence provided by Kajanto 1963, 31. Of the 467 Jewish inscriptions 56 carry epithets. Of the 2298 early Christian epitaphs in *ICVR* 1-10, 270 carry epithets.

[53] Early Christian inscriptions carry eleven epithets that do not occur in pagan inscriptions, yet two of these (namely ἄκακος and γλυκύς) appear in pagan inscriptions in a somewhat different form. They have therefore been excluded from the calculation used to arrive at the precentage mentioned in the main body of the text.

[54] On ὅσιος in the LXX, see *Theologisches Wörterbuch zum Neuen Testament* 5 s.v. And see N. Glueck, *Hesed in the Bible* (Cincinatti: The Hebrew Union College Press, 1967) 20-21. On ὅσιος in the Jewish inscriptions of Beth She`arim, see Lifshitz and Schwabe 1974, 34, 35,,126, 157-58, 163, 173, and 193. Note that in a bilingual inscription in Greek and Hebrew from Otranto the μετὰ τῶν ὁσίων formula has been rendered as מכשבם עים צדי קים, see C. Colafemmina, "Di una iscrizione greco-ebraica di Otranto," *Vetera Christianorum* 12 (1975) 131-37.

[55] Only one ὁσία-inscription has been published in *ICVR*, see *ICVR* 1.1399. As the name of the person referred to indicates, this inscription is, however, to be regarded as Jewish, as Leon recognized, see Leon 1960, 344, no. 733.

tions fall into this category). Although compounds of this type also occur in pagan inscriptions, Jewish inscriptions show a much greater variety in their use of φιλο-compounds. In addition to their variety and their numerical importance,[56] Jewish φιλο-compounds differ from pagan ones in still another respect: while pagan φιλο-compounds normally denote someone's affectionate qualities (such as φιλάνθρωπος or φιλόστοργος), Jewish φιλο-compounds not only include terms of endearment (φιλάδελφος, φιλόγονευς, φιλοπάτωρ, and φιλότεκνος), but also words that point to someone's love for the Jewish community (φιλοσυνάγωγος, φιλόλαος, φιλοπένης) and for "the law"—a general term that should perhaps be understood as reference to Jewish religious practices (φιλέντολος, φιλόνομος). In early Christian inscriptions φιλο-compounds are not unusual either,[57] yet in these inscriptions the usage of such compounds is closer to pagan than to Jewish practices. While terms of endearment predominate in such Christian inscriptions, terms that are religious in nature are encountered only twice (φιλόθεος, φιλοχήρα, that is, "someone who is kind to widows"). Similarly significant is that, concerning epithets other than φιλο-compounds, Jewish inscriptions display a preference for terms that describe someone's moral qualities or intellectual capabilities. Next to ὅσιος, we encounter terms such as ἀβλαβής ("innocent"), ἀμίαντος ("pure," "undefiled"), δικαία ("righteous"), εὐδιδακτή ("well instructed") and εὐλογημένη ("blessed"). Although it is true that these terms occur only occasionally, it is significant that they occur, and that they occur in Jewish inscriptions only. This suggests that although they were using the same language everyone used in Late Antiquity, Roman Jews employed this language to give expression to feelings and ideas that non-Jews did not share. Everyone who knew Greek could understand words such as φιλέντολος, φιλόνομος, or φιλοσυνάγωγος. Yet the point is that these words were neologisms, and that we find them in a Jewish context only. If one contrasts the epithets in early Christian inscriptions with those used in pagan inscriptions, one quickly realizes that there is much greater continuity in their use of epithets than is the case, with Jewish inscriptions. 75% of the epithets in Jewish inscriptions being unattested in pagan inscriptions, Jews clearly distinguished and wanted to distinguish themselves from non-Jews.

Finally interesting is the division of epithets per catacomb. Epithets occur in 25% of the Greek inscriptions from the Vigna Randanini catacomb. In Greek inscriptions from the Monteverde and the Villa Torlonia catacombs, on the other hand, the percentage of epithets drops to

---

[56] Of all pagan inscriptions carrying an epithet, only 3.2% displays a φιλο-compound. Of all Jewish inscriptions that fall into this category, 26.8% carries such a compound.

[57] They occur in 5.5% of all early Christian inscriptions carrying epithets.

5%. Inasmuch as the use of epithets was more common in Latin than in Greek inscriptions (for details, see below), the relatively widespread use of epithets in the Greek funerary inscriptions from the Vigna Randanini catacomb should probably be ascribed to the influence of Latin. That the Greek inscriptions from this catacomb underwent a strong Latin influence is also made likely by the formal appearance of references to age at death. While 50% of the references to age at death in Greek inscriptions from the Vigna Randanini catacomb follow the Latin pattern, in the Monteverde 35% and in the Villa Torlonia catacombs 15% conform to the Latin pattern.[58]

The differences between the epithets used in pagan, Jewish, and early Christian inscriptions are particularly interesting when they are seen against the background of the epithets used in pagan and Jewish inscriptions in Latin.[59] Here are the data (Table 5).

Table 5: Epithets in pagan and Jewish inscriptions. Inscriptions in Latin. Epithets that occur in Jewish inscriptions only have been underlined.

Epithet	Pagan	Jewish
1 Amantissima	2	—
2 Anima -		
benemerens	1	—
dulcis	1	—
dulcissima	1	—
bona	—	1
bona et benedicta	1	—
innox	—	1
3 Benedicta	2	1
4 Benemerens	456	35
5 Benemerita	8	—
6 Bonus	1	1
7 Bona Iudaea	—	1
8 Cara	1	—
9 Carissimus	140	4
10 Castissima	4	—
11 Desiderantissima	1	1
12 Dignissimus	5	—
13 Dulcis	2	1
14 Dulcissimus	110	6
15 Fidelis	1	—
16 Fidelissimus	3	—
17 Frugalissimus	1	—
18 Inconparabilis	14	2
19 Indulgentissimus	3	—

---

[58] On the Latin and Greek patterns in general, see Kajanto 1963, 13: Greek inscriptions normally refer to age at death; Latin inscriptions, on the other hand, include reference to the number of months, days, and hours the deceased is supposed to have lived.

[59] The epithets in early Christian inscriptions in Latin have been excluded because an analysis of the more than 23.000 Latin inscriptions in *ICVR* 1-10 is beyond the scope of this chapter. For the pagan data, see Kajanto 1963, 34 (based on *CIL* 6.24321-26321).

(Table 5 continued)

20 Ingenuosissimus	1	—
21 Innocentissima	2	—
22 Iucundissimus	1	—
23 Merens	2	—
24 Merentissima	1	—
25 Merita	1	—
26 Obsequentissimus	1	—
27 Optimus	52	1
28 Pius	11	—
29 Pient/pi/issimus	107	—
30 Rara	1	—
31 Rarissimus	10	—
32 Sancta	4	—
33 Sanctissimus	25	1
34 Univira	2	—
Totals	979	56

Most striking about Table 5 is that pagan and Jewish inscriptions in Latin are surprisingly similar insofar as the use of epithets is concerned. In pagan inscriptions in Latin, epithets are more frequently used than in pagan inscriptions composed in Greek (41.3% as opposed to 31%). At the same time, however, there is less variety among epithets used in pagan inscriptions in Latin than in pagan inscriptions in Greek (33 as opposed to 52). Jewish inscriptions in Latin largely follow this pattern. While epithets occur more frequently in Jewish inscriptions in Latin than in those composed in Greek (44.1% versus 12.2%), the use of such epithets goes hand in hand with a reduction in the number of epithets (14 different epithets are attested in Jewish inscriptions in Latin, as opposed to 24 in Jewish inscriptions in Greek). Yet, where there was a decrease in the total number of epithets in Jewish inscriptions in Greek as opposed to pagan inscriptions in Greek, in Jewish inscriptions in Latin the percentage of epithets used is practically identical to that in pagan inscriptions in Latin (44.1% as opposed to 41.3%). Similarly, while there were clear differences between Jewish and pagan inscriptions in Greek in terms of epithets used, such differences have disappeared in the case of the Jewish and pagan inscriptions in Latin. In both types of Latin inscriptions, the epithet *benemerens* outranks all other epithets. In pagan inscriptions, *benemerens* is followed, in order of popularity, by *carissimus* and *dulcissimus*. In Jewish inscriptions in Latin the same pattern can be observed, except for the fact that *dulcissimus* and *carissimus* have now traded places.[60] Most significantly, in their use of epithets, Jewish in-

---

[60] In pagan inscriptions in Latin, *benemerens* appears as epithet in 46.6% of all inscriptions, *carissimus* in 14.3% and *dulcissimus* in 11.2%. In Jewish inscriptions in Latin, *benemerens* appears as epithet in 62.5% of all inscriptions, *dulcissimus* in 10.7% and *carissimus* in 7.1%.

scriptions in Latin lack exactly that feature that characterizes Jewish inscriptions in Greek: the use of epithets that do not occur in non-Jewish inscriptions. Except for *bona Iudaea*, the two other epithets that occur only in Jewish inscriptions in Latin, namely *anima bona* and *anima innox* are expressions that are not typically Jewish and that could have appeared equally well in pagan inscriptions in Latin. Having noted the preference for moral or religious epithets in the Jewish inscriptions in Greek, and for the epithet ὅσιος in particular, one would perhaps expect to encounter frequently epithets such as *sanctus*. This, however, is not the case. Differences between individual Jewish catacombs exist only insofar as the use of epithets being somewhat more common in Latin inscriptions from the Vigna Randanini catacomb than in those from the Monteverde catacomb, namely 57.5% and 33% respectively.

It is in the use of epithets, then, that we can make, for the first time, a distinction between Jewish epitaphs composed in Greek and those written in Latin. While Jewish epitaphs in Latin tend to follow the standard Latin vocabulary, Jewish epitaphs in Greek tend to diverge from the epithets commonly used in non-Jewish Greek inscriptions. Inasmuch as the epithets in question are compounds that are made of existing Greek words, there is nothing specifically Jewish about epithets such as φιλόνομος, ἀμίαντος or εὐδιδακτή. Yet it is conceivable that for the people who ordered the epitaphs that carry these epithets, these inscriptions denoted qualities that had typically Jewish connotations. The word ὅσιος in particular must have had special meaning for Roman Jews. Perhaps even non-Jews recognized that there was something distinctively Jewish about the word. It is certainly intriguing to note that while Jewish inscriptions from Rome freely use the formula μετὰ τῶν ὁσίων, early Christian inscriptions from Rome never use this expression and prefer μετὰ τῶν ἁγίων instead.[61]

In addition to using epithets which are absent from non-Jewish inscriptions, Jewish epitaphs stand out for still another reason: they indicate that Jewish communal activities held a place of central importance in the lives of many Roman Jews.[62] According to my estimates,

---

[61] For Jewish inscriptions carrying the μετὰ τῶν ὁσίων formula, see *CIJ* 55, 340; Fasola 1976, 57; for early Christian inscriptions carrying the μετὰ τῶν ἁγίων formula, see e.g. *iCVR* 4.10552; 6.16851; 8.22792 and 23372 and 9.26162A variant of this formula, μετὰ τῶν δικαίων, appears in Jewish and early Christian inscriptions alike, see *CIJ* 78, 110, 118, 150, 194, 281 and see 210; *ICVR* 2.4433. On the special significance of the term ἅγιος for and its use by Christians, see *Reallexikon für Antike und Christentum* 2 (1954) 1116 s.v. "Christennamen" (H. Karpp). For the formula μετὰ εὐλογίας, see Fasola 1976, 24.

[62] The point was first observed by J. Z. Smith, "Fences and Neighbors: Some Contours of Early Judaism," in W. S. Green (ed.), *Approaches to Ancient Judaism* (Chico: Scholars Press, 1980) 1-25. Note, however, that the figures suggested by Smith are incorrect.

20.5% of the roughly 595 Jewish epitaphs from Rome refer to various offices people held within the Jewish community.[63] That this figure is significant becomes especially evident once it is seen against the wider background of Late Ancient Jewish and non-Jewish epigraphic practices.

One of the most striking features of the Jewish funerary inscriptions from Rome is the contrast between the relatively-frequent occurrence of references to community-related functions and the absence of other types of reference, such as references to occupational status. Apparently, Roman Jews preferred to be remembered as, for example, ἀρχισυνάγωγος rather than as, say, "painter." Such evidence suggests that within the Roman Jewish community of Late Antiquity, status depended to a significant extent on one's involvement in the Jewish community. Involvement in the Jewish community may have entailed a number of activities, but it certainly included participation in religious practices. In antiquity, "loyalty to the community was inseparable from loyalty to the deity who called it into being."[64]

It is exactly in this attachment to the community that Roman Jews differed from their pagan and Christian contemporaries in that city—at least if we accept that epigraphic evidence can serve as a trustworthy indicator in such matters. Since the inception of the series *Inscriptiones Christianae Urbis Romae* in the last century, at least some 25,500 early Christian inscriptions from in and around Rome have been published. Studying what I believe to be a representative sample of around 5,500 early Christian inscriptions from the Christian catacombs of Callixtus (including the *cripta dei Papi*), Sebastiano, Praetextatus, Agnese, the coemeterium Majus, and various catacombs along Via Portuensis, it became evident that references to ecclesiastical functions such as *presbyter*, *acolythus* and so forth are almost entirely absent in this corpus. They occur in less than 1% of the inscriptions.[65]

An analysis of the pagan evidence, finally, provides us with comparable results. *Collegia* or voluntary associations may have given a sense of belonging and pride to people of low social status. As members of *collegia*, people were able to fulfill all sorts of functions. Not

---

[63] Note that of all inscriptions on sarcophagi used by Jews, 80% refer to synagogue-related functions.

[64] E. P. Sanders, *Judaism. Practice and Belief. 63 B.C.E. - 66 C.E.* (London and Philadelphia: SMC and Trinity Press International, 1992) 144.

[65] This figure is based on the evidence culled from *ICVR* 2 (1935), *ICVR* 4 (1964), *ICVR* 5 (1971), *ICVR* 8 (1983). For lack of space, I refrain here from listing these inscriptions separately. To judge on the basis of the consular data, these materials date from the period of the 270s C.E. to the 540s C.E. I have specifically selected the Christian catacombs mentioned in the text because they are located along the same *viae consulares* as the Jewish catacombs of Rome.

infrequently, people with little social standing in society succeeded in rising within an association's strictly hierarchical structure to positions of power that they would never attain in Roman society.[66] Because of the specific social role these *collegia* fulfilled (as alternative *cursus honorum*), it is all the more surprising to observe that such functions are almost never mentioned in the epitaphs. Out of a collection of around 4,500 funerary inscriptions from various *columbaria*, not more than 2.5% refer explicitly to functions *collegia*-members held within individual associations.[67]

Studying references to occupational status in a collection of 10,523 inscriptions published in the *CIL* 6 (mostly pagan and probably largely predating the third century C.E.), P. Huttunen observed that references to someone's occupation are not altogether absent from this corpus of evidence: they occur in 9.5% (13.9%) of the inscriptions included in this survey.[68] That such references occur at all is due to the fact that the upper classes of Roman society are included in this sample. Roughly two thirds of the inscriptions which mention occupation refer to people belonging to the senatorial and equestrian orders, or to employees of the imperial service. The rest (that is, the majority) of the population in Rome was, however, less accustomed to include references to occupational and/or religious status in inscriptions. They occur in 3.5% (5%) of all inscriptions in *CIL* 6.[69]

In short, then, the profusion of references to community-related offices, together with the absence of references to Jews holding positions within the Roman administrative system, shows the importance Roman Jews attached to their Jewish community. Again, it is not possible to argue that the users of one catacomb were less involved in Jewish communal affairs than those using another catacomb. While 19% of all Jewish inscriptions from the Monteverde catacomb refer to community related offices, 17% of all Jewish inscriptions from the Vigna Randanini catacomb also carry this type of reference. People whose inscription carried a reference to the function they held within the Jewish community seem to have had a slight preference for Greek

---

[66] E.g., slaves could become members of *collegia* as could women; see J. P. Waltzing, *Étude historique sur les corporations professionelles chez les Romains depuis les origines jusqu'à la chute de l'Empire d'Occident* (Louvain: Peeters, 1895-1900), vol. 1, 346-48; 368 (slaves as presidents of *collegia*). On the different functions that could be held within a *collegium*, see *ibid.* 358f.; 385f. (note the great variety of offices).

[67] This evidence is based on the inscriptions collected by Waltzing (previous note). In this particular case, the absence to such references may in part be due to the fact that inscriptions from *columbaria* are usually not very verbose.

[68] P. Huttunen, *The Social Strata in the Imperial City of Rome. A Quantative Study of the Social Representation in the Epitaphs Published in the Corpus Inscriptionum Latinarum, Vol. V* (Oulu: University of Oulu, 1974) 48.

[69] Huttunen (previous note). And see his remarks on the equestrian order pp. 77f. and on *collegia* and the oriental cults on pp. 110-111.

as opposed to Latin. Of all Jewish inscriptions in Greek, 19.1% refer to community-related offices. Of all Jewish inscriptions in Latin 12.6% carry this type of reference. Such figures are hardly surprising in the light of what has been observed earlier in connection with the epithets used in Jewish inscriptions, namely that Greek inscriptions are more likely than Latin ones to express someone's attachment to Judaism. Interestingly, the use of certain epithets and references to community-related functions often go hand in hand. Of the seven references to males who are called ὅσιος, one is to a child and another to a certain, otherwise-unknown Ῥωμανος Ἀσκλήπιος. The five other references concern people who are known to have been active in Jewish communal affairs.[70]

Implications

Like no other ancient source, the Jewish funerary inscriptions from Rome inform us about how Roman Jews viewed themselves. There is one reason in particular why the Jewish epitaphs can be considered as a reliable guide in these matters. Sealing off graves in the dark underground galleries of the catacombs, these epitaphs were hardly visible. Thus, rather than being the result of the wish to impress passersby, the content of these inscriptions was determined, in the very first place, by the way the person ordering a funerary inscription wanted the person commemorated in each inscription to be remembered.[71]

In this chapter we have seen that, linguistically, the Jews of ancient Rome can hardly be called isolated. From a phonological, morphological, or syntactical point of view, the Greek and Latin in Jewish inscriptions did not differ in any way from the Greek and Latin commonly used in non-Jewish inscriptions. Nor is it possible to view the users of one catacomb as more Romanized than the users of another catacomb. In fact, the formal characteristics of the Jewish epitaphs in Greek and Latin rather suggest that the choice of catacomb was not determined by someone's cultural background and preferences.

A brief analysis of the content of the Jewish funerary inscriptions in general, and of the use of epithets and references to community-related functions in particular, has shown, on the other hand, that while using the language of the city in which they were living, Jews expressed ideas and ideals for which no parallels exist in non-Jewish inscriptions. Although such ideas and ideals find expression, above all,

---

[70] *CIJ* 111 and Solin 1983, 657 no. 67; *CIJ* 93, 100, 103, 145, and 321.

[71] The person commissioning the inscription need not necessarily be identical to the deceased, but in a number of cases he or she certainly was.

in Greek inscriptions, it has been noted that they are not altogether absent from Latin inscriptions, either. We can only guess why Greek enjoyed such popularity in Jewish epitaphs: perhaps Greek lent itself more than did Latin to the formation of the "typically-Jewish" epithets we encounter in these inscriptions; Greek was valued perhaps because it was used in the liturgy, or because the Septuagint or other translations of the Hebrew Bible into Greek were popular among Roman Jews;[72] Greek perhaps also reminded Jews of their home-country, Greek conceivably being the language in which they conversed with the Palestinian rabbis and representatives of the Patriarchate who are known to have visited Rome on various occasions. The idea that Greek continued to be popular among Jews because the use of this language enabled them to attract proselytes is, however, less attractive, primarily because it is based on a complex of unproven assumptions.[73]

It is difficult to determine why Hebrew is absent from the Jewish epitaphs except for short fomulaic phrases such as שלום על ישראל. Such short phrases in Hebrew always occur as additions to inscriptions that have been written in Greek.[74] Most of these inscriptions were found in the Monteverde catacomb. The rudimentary character of these Hebrew phrases and the way in which they have been rendered (usually as graffiti or painted) suggests that they were added to underline someone's attachment to the Jewish people. But these inscriptions can probably not be taken as evidence for the gradual Hebraization of Jewish epigraphy as documented in Late Ancient and early medieval Jewish inscriptional remains from Southern Italy. Nor would it be correct to conclude from the general absence of Hebrew in the Jewish inscriptions from Rome that Roman Jews hardly knew any Hebrew or that Hebrew meant little to them. Among Roman Jews who buried in the catacombs, Hebrew simply never became a standard inscriptional language. We can only guess why this was so.

The problems surrounding Hebrew in Jewish epitaphs from Rome raises a question that has concerned us already earlier, although in a somewhat different form: How did Roman Jews relate to the Jews of Roman Palestine in general, and to rabbinic Judaism in particular?[75]

---

[72] Note that *CIJ* 201 follows the LXX and *CIJ* 370 Aquila's translation. *CIJ* 86 is a mixture of LXX and Aquila.

[73] *Contra* M. Simon, *Verus Israel. A Study of the Relations between Christians and Jews in the Roman Empire (135-425)* (Oxford: Oxford U.P., 1986) 294-95 and *contra* Solin 1983, 710. The unproven assumptions are: 1) that Judaism was a missionary religion—which is disputable—and 2) that the linguistic preferences of the Jews were essentially determined by the wish to gain converts—which is a *reductio ad absurdum* of a complicated and multifacetted problem.

[74] *CIJ* 283, 296, 319, 349, 397, and 497.

[75] See Chapter 3.

In the centuries following the destruction of the Second Temple in 70 C.E., Jewish life in Palestine changed profoundly in a number of respects. The main area of Jewish settlement shifted from the Jewish heartland in Judaea to the Galilee and the Golan in the North. In the course of time, a Patriarch came to represent the Jews vis-à-vis the Roman authorities. Together with his staff, the Patriarch also regulated matters that were to have an increasingly profound bearing on the daily lives of the Jews themselves. In modern scholarship, the various stages along which a rabbinic class rose to prominence are as disputed as the rise of this class itself is beyond question.[76]

For lack of evidence, it is usually difficult to tell, how these developments in Roman Palestine affected the Jewish communities in the Diaspora. In rabbinic literature, accounts of Palestinian scholars visiting Jewish communities in different parts of the Roman world are not uncommon. We have every reason to suppose that such visits indeed took place. It is much more difficult, however, to measure the effects such visits had on a Jewish community in a given locality. It is equally difficult to evaluate the reliability of rabbinic literature when it relates events that happened at places that were far removed geographically and intellectually from the world in which the redactors of this literature were living. The discussions surrounding the Roman *yeshivah* of Mattiah ben Heresh may serve as an example to illustrate our problem.

According to a *baraitha* preserved in the *Babylonian Talmud*, Mattiah ben Heresh's *yeshivah* was regarded as one of the most outstanding of such institutions of his day.[77] Previous scholars have interpreted such literary evidence as pointing to a strong rabbinic presence in Rome as early as the second century C.E. Others have gone even further, maintaining that already in the first century C.E., Rome became the seat of an influential rabbinic academy.[78] It may be clear that such interpretations pass in silence over the many interpretational problems raised by the evidence contained in the Babylonian Talmud. Where did redactors of the Babylonian Talmud get their information?

---

[76] See, in general, P. Schäfer, *Studien zur Geschichte und Theologie des rabbinischen Judentums* (Leiden: Brill, 1978); G. Stemberger, *Juden und Christen im Heiligen Land. Palästina unter Konstantin und Theodosius* (Munich: Beck, 1987); L. I. Levine, *The Rabbinic Class of Roman Palestine in Late Antiquity* (Jerusalem: Ben-Zvi, 1989).

[77] *Babylonian Talmud, Sanhedrin* 32b. For a disucssion of other references to M. ben Heresh in rabbinic literature, see A. Toaff, "Matia ben Cheresh e la sua accademia rabbinica di Roma," *Annuario di Studi Ebraici* 2 (1964) 69-80 and L. A. Segal, "R. Matiah Ben Heresh of Rome on Religious Duties and Redemption: Reacting to Sectarian Teaching," *Proceedings of the American Academy for Jewish Research* 58 (1992) 221-41. On the meaning of the word *yeshivah*, see D. M. Goodblatt, *Rabbinic Instruction in Sasanian Babylonia* (Leiden: Brill, 1975) 69.

[78] Most explicit in this respect is U. Cassuto, "La *Vetus Latina* e le traduzioni giudaiche medioevali della Bibbia," *Studi e Materiali di Storia delle Religioni* 2 (1926) 153 citing Philo, *Leg.* 23 (!) as proof for the existence of such academies.

204   CHAPTER FIVE

Are not these redactors known to have taken little interest in historiographical matters, to have lived under very different political conditions and far away from Rome, and to have been active many centuries after these events are supposed to have taken place? Even more important: How can one be sure that in reporting about Mattiah ben Heresh, these redactors did not read their own concerns into the traditions they ascribed to Mattiah ben Heresh? Did Mattiah ben Heresh's academy ever exist at all? If so, what was his influence on the Roman Jewish community at large?

Comparable uncertainties exist concerning the interpretation of the rabbinic traditions relating to another famous Roman Jew, Todos (or Theudas). According to rabbinic sources, Todos was much appreciated in Palestine for his financial support of the rabbis there. The fact that he roasted and ate a lamb during the festival of Passover in a time when such a practice had been abolished, however, drew sharp criticisms from the same Palestinian circles.[79]

At first sight, such traditions seem to document both the involvement of the rabbis of Palestine with the Jewish community of Rome (and vice versa).[80] Yet, a closer reading of the redactional history of the Todos story indicates that it went through various phases. B. M. Bokser has argued that in its earliest "layers," the Todos story documents the *lack* of influence of rabbinic Judaism on Jews in Rome. He points out that over time the discussions surrounding Todos increasingly became a vehicle for the rabbis to express concerns of a decidedly religious rather than of a strictly historiographical nature.[81] Thus, ultimately, the Todos story does not inform us in a very reliable way about the relationship between Roman Jewry and the rabbis of Palestine.

Although it is often difficult to reconstruct the historical reality that lies behind rabbinic texts, it seems rather likely that Roman Jews will have learned sooner rather than later about the developments that shaped and changed the lives of their co-religionists in Roman Palestine. Even though the relevant evidence comes from other parts of Italy and not from Rome itself, there is reason to suppose that the Roman Jewish community was keenly aware that in Roman Palestine the Patriarchate had turned into an institution that was highly respected by both Jews and Roman officials. A bilingual Jewish inscription from Catania (Sicily) that was erected in 383 C.E. contains,

---

[79] *Tosefta Yom Tov (Besah)* 2:15; *Palestinian Talmud, Moed Qatan* 3:1 and parallels.

[80] Examples in B. M. Bokser, "Todos and Rabbinic Authority in Rome," in J. Neusner, et al. (eds.), *New Perspectives on Ancient Judaism. 1. Religion, Literature, and Society in Ancient Israel, Formative Christianity and Judaism* (New York and London: Lanham, 1987) 117-30, at 117 n.2.

[81] Bokser (previous note) *passim*.

among other adjurations, the phrase "I adjure you likewise by the honors of the patriarchs...let no one open the tomb [etc]."[82] A law promulgated in Rome in the early fifth century and addressed to the *praefectus praetorio* of Italy and Africa, indicates that during the fourth century, it was not uncustomary for Jews from the Western provinces of the Roman Empire to send money to the Patriarch.[83] Given the size and importance of the Roman Jewish community, it is conceivable that Late Ancient legislators had Roman Jews in mind when they first formulated this law—even though, upon promulgation, they then preferred to phrase this regulation in more general terms. Jewish immigrants such as a certain Ionius who originated from Sepphoris and who was buried in the Monteverde catacomb could inform Roman Jews about the latest developments in Roman Palestine.[84] No less significant is the appearance of the term of rabbi in inscriptions from Italy dating to the fourth and fifth centuries.[85] In the present context, it is irrelevant to determine whether the rabbis attested in inscriptions represent the same class of people as those responsible for rabbinic literature or whether, in these inscriptions, the term rabbi was used as *terminus technicus* rather than as epithet for people whose status depended on qualities other than having been ordained.[86] What matters here is that this typically Semitic title occurs at all; it shows the gradual emergence of an element that had previously been absent from the Greek and Latin speaking Jewish communities of Italy.

Meager though this evidence might appear, it helps us to draw a more organic picture of the relationship between rabbinic and Diaspora Judaism. Traditionally, scholarship has conceived of this relationship in terms of power and control—or the absence thereof. Most recently, some scholars have argued, for example, that an important inscription from Aphrodisias attests to "the beginning of the imposition of rabbinical orthodoxy on occidental Jewry."[87] Lacking evidence to support this claim, other scholars have used evidence from Aphrodisias to argue that in this city Jews "were still a long way from

---

[82] *CIJ* 650 = Noy 1993, no. 145.

[83] *CTh* 16.8.7.

[84] *CIJ* 362. Sepphoris was a city that housed the Patriarchate for some time.

[85] In inscriptions from Naples, see E. Miranda, "Due iscrizioni greco-giudaiche della Campania," *Rivista di Archeologia Cristiana* 55 (1979) 338 and *CIJ* 568 (= Noy 1993, no. 36); in an inscription from Venosa, see *CIJ* 611 (= Noy 1993, 86). An inscription from Oria referring to a Rabbi *Giulius* and published by C. Colafemmina, "Note su di una iscrizione ebraico-latina di Oria," *Vetera Christianorum* 25 (1988) 648 is probably medieval.

[86] See on this question S. J. D. Cohen, "Epigraphical Rabbis," *Jewish Quarterly Review* 72 (1981) 1-17, who appositely concludes (p. 12) that "not all rabbis were Rabbis."

[87] J. Reynolds and R. Tannenbaum, *Jews and Godfearers at Aphrodisias* (Cambridge: Cambridge U.P., 1987) 82.

206   CHAPTER FIVE

behaving as the rabbis of Palestine would have liked."[88] Comparably, the author of a recent study on the Jews of Rome concluded that Judaism in Late Ancient Rome was virtually identical with "rabbinic Judaism of the pharisaic type," while elsewhere he observes that "hellenistic Judaism" continued to exist in Rome—very different, if not mutually exclusive, claims that have in common, however, that the author fails to back up either the one or the other.[89] All these assertions show that recent scholarship on Diaspora Judaism tends to view rabbinic and pre-rabbinic Judaism not merely as very different but as antithetically-opposed entities.

Upon closer reflection, one may doubt whether such a view is convincing. It fails to take into account that rabbinic Judaism itself evolved over time. It also fails to do justice to the fact that in rabbinic literature itself contradictory views often coexist. It furthermore fails to consider that rabbinic Judaism built on traditions that were recognizable by many Jews, or, to put it differently, that in many areas, the shift from pre-rabbinic to rabbinic Judaism must have been a gradual one. It finally sees the emergence of rabbinic Judaism in terms of influence from Roman Palestine, and thus fails to consider the possibility that in the Diaspora rabbinic Judaism possibly owed its popularity to the efforts of the Diaspora Jews themselves. I therefore believe that to contrast a supposedly full-fledged, orthodox rabbinic Judaism of Palestine with a consciously un-rabbinic Judaism of the Diaspora is to impose categorizations on the evidence that are far too rigid.

Similarly misleading is the tendency in recent studies on Diaspora Judaism to stress a "stunning diversity" as the most outstanding characteristic of Jewish life outside Roman Palestine. A. T. Kraabel has argued, for example, that synagogues discovered in various parts of the Diaspora reflect diverse forms of organization and religiosity.[90] Taking into account these observations, J. A. Overman has even pro-

---

[88] M. H. Williams, "The Jews and Godfearers Inscription from Aphrodisias-A Case of Patriarchal Interference in Early 3rd Century Caria?" *Historia* 41 (1992) 304.

[89] Solin 1983, 716 and 709. And see *id.*, "Gli Ebrei d'Africa: una nota," in A. Mastino (ed.), *L'Africa romana. Atti del'VIII convegno di studio Cagliari, 14-16.XII 1990* (Sassari: Gallizzi, 1991) 616 and 622.

[90] A. T. Kraabel, "Social Systems of Six Diaspora Synagogues," in J. Gutmann (ed.), *Ancient Synagogues. The State of Research* (Chico: Scholars Press, 1981) 79-91, reprinted in J. A. Overman and R. S. MacLennan (eds.), *Diaspora Jews and Judaism. Essays in Honor of, and in Dialogue with A. Thomas Kraabel* (Atlanta: Scholars Press, 1992) 257-67. See also Kraabel 1982, 457.

posed that we should henceforth speak of Judaisms (plural) rather than of Judaism (singular).[91] Other scholars would agree.[92]

Upon closer investigation, views such as these are not persuasive because they stress some evidence at the expense of other evidence. It is true that synagogue buildings in different parts of the Diaspora look different, but neither is this surprising, nor can differences in building technique or formal appearance be taken to reflect profound differences between one Jewish community and another in terms of religious practices and beliefs. The primary importance of these buildings is not the way they look, but the fact that they exist at all. Along similar lines, the fact that titles such as ἀρχισυνάγωγος occur during a period of several centuries among Jewish communities in many parts of the Roman world suggests that such communities may have had much more in common that the notion of "diversity" is capable of allowing for.[93] Finally, the fact that Todos ate a roasted lamb on Passover (if we take the Todos-story at face value) can hardly be interpreted as indicative of a Judaism that was ideologically at odds with the Judaism represented by the rabbis who criticized Todos for his act. The issues that separated Todos and his rabbinic friends were of far less importance than the practices that united them: did not all agree on the central importance of the festival of Passover?

Such assertions are not to deny, of course, that local differences in Jewish practices and beliefs existed. In fact, the existence of complete uniformity between communities that were as far apart as the Rhineland, Roman Spain, Asia Minor, and Mesopotamia would really be surprising. Yet, on the other hand, differences of this kind do not allow us to speak of Judaisms (in the plural). Such a categorization is based on the presupposition that, just as in the first few centuries of the Church, Christianity manifested itself in a variety of forms, there likewise existed many different kinds of Judaism. Judaism, however, was not Christianity: it had different historical roots, it developed differently, and it held very different ideas as to what constituted schism and heresy.

Inasmuch as, in my view, the similarities between the various Jewish communities in the Roman world far outweigh the differences, at least generally speaking, I believe that there exists some justification in applying the term "normal" or "common Judaism" to the Late

---

[91] J. A. Overman, "The Diaspora in the Modern Study of Ancient Judaism," in id. and R.S. MacLennan (previous note) 63-78. Comparable is Smith 1980, 15 and 19 and M. White, *Building God's House in the Roman World. Architectural Adaptation among Pagans, Jews, and Christians* (Baltimore and London: Johns Hopkins, 1990) 61.

[92] See, for example, the remarks of J. Neusner in *Journal of Jewish Studies* 24 (1993) 317-23.

[93] See the survey of T. Rajak and D. Noy, "*Archisynagogi*: Office, Title and Social Status in the Greco-Jewish Synagogue,' *Journal of Roman Studies* 83 (1993) 75-93.

Ancient period. In a recent study on first century C.E. Judaism, E. P. Sanders has introduced the term.[94] Arguing against the view that holds that Judaism in first century Palestine was divided into a variety of parties, including the Pharisees, the Sadducees, and the Essenes, Sanders maintains that such parties played a very subordinated role in the daily life and religious practices of most Jews. In his view, the different parties were too peripheral and not powerful enough to be able to impose their particular views on sizable portions of the Jewish population. Sanders believes, therefore, that rather than following partisan views, a majority of Jews, including ordinary priests, agreed on basic theological points and on observing a series of basic religious practices: the celebration of the Sabbath and of major Jewish annual festivals (which could entail, once in a while, a trip to Jerusalem), as well as the observance of a number of other rites, including circumcision, purity in the home, food laws, etc. Thus, according to Sanders' definition, the term "common Judaism" is a convenient concept to indicate that in first century Palestine (and probably also in the Greek-speaking Diaspora during this period) most Jews agreed what were the most fundamental characteristics of their religion.[95]

The surviving evidence suggests that there are good reasons to view Diaspora Judaism in Late Antiquity in terms of "common Judaism" too. In third and fourth century Rome, as in first century Palestine, Roman Jews distinguished themselves by agreeing on a clearly defined set of ritual practices. Paramount among these was the observance of Jewish holidays as they were celebrated in the synagogue. Although it is true that there there are only few sources available, the surviving evidence is amazingly consistent in stressing the centrality of the synagogue as a religious-communal institution among Roman Jews. In the first century Philo and Persius attest to it, as do Juvenal and Hippolytus at the beginning and towards the end of the second century respectively.[96] Then, from the third century onwards, the Jewish epitaphs document the importance of communal life among the Jews of Rome. That "Judaism" and "the synagogue" were considered by contemporaries as interchangeable also follows from events leading to the burning down of a Roman synagogue in the late 380s C.E.[97] Given the importance of the synagogue as an institution among Ro-

---

[94] Sanders 1992, 11f. and *passim*.

[95] Neusner's criticisms of the concept of common Judaism in *Journal of Jewish Studies* 24 (1993) 317-23 are not valid. Neusner misunderstands the main thrust of Sanders' argument.

[96] Philo, *Leg.* 155 and 158; Persius 5.176-84; Juvenal 3.296 and 14.96; on Callixtus breaking into a Roman synagogue, as described by Hippolytus, see H. Gülzow, "Kallist von Rom. Ein Beitrag zur Soziologie der römischen Gemeinde," *Zeitschrift für die Neutestamentliche Wissenschaft* 58 (1967) 102-21.

[97] Ambrose, *Epistula* 40 (*PL* 16, 1109).

man Jews, then, it is, finally, not surprising to note that synagogue buildings could serve Jews and non-Jews alike as a point of reference in the city's complicated topography.[98]

In the light of the evidence presented above, it is fair to say that Roman Jews practiced a "common Judaism" that Jews from Palestine or from other parts of the Mediterranean could easily recognize. Involvement in the Jewish community was one of the central features of this type of Judaism. "Proper Jewish behavior" was another. Far from being isolated linguistically, the Jews of ancient Rome adapted to their own purposes the two languages of the society in which they were living. Thus they succeeded in expressing an identity that was unmistakably Jewish.

---

[98] *CIJ* 531.

CHAPTER SIX

THE LITERARY PRODUCTION OF THE JEWISH COMMUNITY OF ROME IN LATE ANTIQUITY

Introduction

Without exception, all existing monographs on the Jewish community of ancient Rome pass in silence over the literary production of Roman Jews in Late Antiquity. This is most unfortunate. Two highly interesting fourth-century works in Latin whose authorship may be assigned to (Roman) Jews, still survive. The first, a treatise, written in Latin and known as the *Collatio Legum Mosaicarum et Romanarum* (also known as *Lex Dei quam praecepit Dominus ad Moysen*), has baffled scholars since its first publication by P. Pithou in 1573. The other, known after its *incipit* as the *Letter of Annas to Seneca* (*Epistola Anne ad Senecam*), has attracted scholarly attention only fairly recently, namely since 1984 through a publication of B. Bischoff.

The *Collatio* consists of a systematic comparison of Mosaic law with Roman law, more specifically with the works of the Roman jurists of the second and early third century and selected constitutions from the Gregorian and Hermogenian Codes. Since E. Volterra's seminal study of 1930, a number of scholars including E. Levy, I. Osterzeter, F. Schulz, B. Blumenkranz, C. Pietri, J. Gaudemet, L. Cracco Ruggini, D. Daube and A. M. Rabello believe that the author of the *Collatio* was an (anonymous) Jew who lived in Rome while composing this work.[1] In the wake of ideas developed by N. Smits, others,

---

[1] E. Volterra, "Collatio Legum Mosaicarum et Romanarum," *Atti della Reale Accademia Nazionale dei Lincei. Memorie* 6.3.1 (1930); E. Levy, Review of Volterra 1930, *Zeitschrift der Savigny-Stiftung für Rechtsgeschichte. Romanistische Abteilung* 50 (1930) 698-705, at 700-701; I. Osterzeter, "La 'Collatio Legum Mosaicarum et Romanarum.' Ses origines-son but," *Revue des Études Juives* 97 (1934) 65-96; F. Schulz, *History of Roman Legal Science* (Oxford: Clarendon, 1946) 311-314 and in various articles that will be cited *infra*; B. Blumenkranz, *Die Judenpredigt Augustins. Ein Beitrag zur Geschichte der jüdisch-christlichen Beziehungen in den ersten Jahrhunderten* (Basel: Helbing and Lichtenhahn, 1946) 56-57; C. Pietri, *Roma Christiana. Recherches sur l'église de Rome, son organisation, sa politique, son idéologie de Miltiade à Sixte III (311-440)* (Rome: École Française, 1976) vol. 2, 1469; J. Gaudemet, *La formation du droit séculier et du droit de l'église au IVe et Ve siècles* (Paris: Sirey, 1979) 96-98; L. Cracco Ruggini, "Intolerance: Equal and Less Equal in the Roman World," *Classical Philology* 82 (1987) 187-205, at 202 as well as in all her other articles on Jews in Late Antiquity, see esp. "Ebrei e Romani a confronto nell'Italia tardoantica," in *Italia Judaica. Atti del I convegno internazionale, Bari 18-22.V.1981* (Rome: Ministero per i beni culturali etc., 1983) 38-65; D. Daube, "*Collatio* 2.6.5.," in *id., Collected Studies in Roman Law* I, edited by D. Cohen and D. Simon (Frankfurt on the Main: Klostermann 1991) 107-122, at 107; A. M.

including F. Triebs, C. Hohenlohe, L. Wenger, H. Chadwick, H. Schreckenberg, M. Lauria, and E. Schrage maintain, however, that the author of the *Collatio* was not Jewish at all. They argue that the *Collatio* is to be considered as a fourth-century Christian product. Recently, this latter suggestion has once again been forcefully advocated by D. Liebs in an important study on Roman law in Italy in Late Antiquity. [2]

In the following pages, I will analyze the *Collatio* from a historical perspective and pay special attention to the question of its authorship. I will argue that once the *Collatio* is placed within the larger framework of Jewish and early Christian attitudes towards the Torah, there can be little doubt that the author of the *Collatio* was Jewish. This identification has several important implications that will also be explored.

The *Letter of Annas to Seneca* is a treatise that has been preserved only partially. It is difficult, therefore, to identify its author and to establish its precise purpose. Despite such difficulties, this letter provides interesting evidence for determining social relations between Jews and non-Jews in Late Antiquity.

Several other literary sources including the *Codex Theodosianus*, the works of Ambrosiaster, a treatise known as *Fides Isaacis ex Iudaeis*, and the *Actus Sylvestri* also bear on the history of the Roman Jewish community in Late Antiquity, at least seemingly so. That I have nevertheless refrained from analyzing these works separately in the following pages, needs to be explained briefly.

Contrary to Vogelstein and Rieger, I do not believe that much can be learned about the Roman Jewish community from the *Codex Theodosianus*, except in a very general way; the laws in question relate to

---

Rabelo, "Alcune note sulla 'Collatio Legum Mosaicarum et Romanarum' e sul suo luogo d'origine," in *Scritti sull"ebraismo in memoria di Guido Bedarida* (Florence, 1966) 177-86; *id.* "Sul l'ebraicità dell'autore della 'Collatio Legum Mosaicarum et Romanarum'," *La Rassegna Mensile di Israel* 33 (1967) 339-49 and *id.*, "Sul Decalogo 'Cristianizzato' e l'autore della 'Collatio Legum Mosaicarum et Romanarum'," *La Rassegna Mensile di Israel* 55 (1989) 133-35.

[2] N. Smits, *Mosaicarum et Romanarum Legum Collatio* (Haarlem: Tjeenk Willink, 1934); F. Triebs, *Lex Dei sive Collatio Legum Mosaicarum et Romanarum* (Wratislava: Nischkowsky, 1902) 7; for Hohenlohe, see the review by B. Kübler in *Zeitschrift der Savigny-Stiftung für Rechtsgeschichte. Romanistische Abteilung* 56 (1936) 361-62; L. Wenger, *Die Quellen des römischen Rechts* (Vienna: Holzhausens, 1953) 547-48; H. Chadwick, "The Relativity of Moral Codes: Rome and Persia in Late Antiquity," in W. Schoedel and R. L. Wilken (eds.), *Early Christian Literature and the Classical Intellectual Tradition. In honorem Robert M. Grant* (Paris: Beauchesne, 1979) 135-37; H. Schreckenberg, *Die christlichen Aversus-Judaeos-Texte und ihr literarisches Umfeld (1.-11.Jh.)* (Frankfurt on the Main-Bern: Lang, 1982) 303; M. Lauria, "Lex Dei," *Studia et Documenta Historiae et Iuris* 51 (1985) 257-75; D. Liebs, *Die Jurisprudenz im spätantiken Italien (260-640 n. Chr.)* (Berlin: Dunker and Humbolt, 1987)162-74; E. J. H. Schrage, "La date de la 'Collatio Legum Mosaicarum et Romanarum' étudié d'après les citations bibliques," in J. A. Ankum *et al.* (eds.), *Mélanges Felix Wubbe* (Fribourg: U. P. Fribourg, 1993) 401-17.

the Jews in the later Roman empire as a whole rather than to the Jewish community of Rome specifically.³

Comparably, the works of Ambrosiaster tell us little or nothing about the Jews of Rome in Late Antiquity. Ever since Erasmus discovered that the *Quaestiones Veteris et Novi Testamenti 127 (150/115)*, as well as a collection of commentaries to most of the letters of Paul had not been authored by Ambrose, but by an anonymous writer ("Ambrosiaster") who was active in the second half of the fourth century, scholars have tried to determine the religious background of this author.⁴ Because he displays some familiarity with Jewish customs, G. Morin suggested at the turn of the century that Ambrosiaster was perhaps identical with a converted Jew by the name of Isaac, who is known to have been active during the pontificate of Pope Damasus (366-384 C.E.). While Morin himself abandoned this identification in later years, C. Martini in particular has shown that there are considerable differences in style and language between the surviving writings of Isaac and those of Ambrosiaster. Most likely, Ambrosiaster was a pagan convert to Christianity.⁵

The *Fides Isaacis ex Iudaeis* falls into the same category as the works of Ambrosiaster. It is a short treatise that was composed by the aforementioned Isaac. Being a typically Christian theological treatise, the work is, however, too stereotypical to be of much evidential value.⁶

Particularly interesting are the traditions surrounding a legendary disputation with, and supposed conversion of many Roman Jews by, Sylvester, events said to have taken place in the reign of Constantine (described in the so-called *Actus Sylvestri*).⁷ In the Middle Ages, the

---

³ Vogelstein and Rieger 1896, 108f.

⁴ His works have been published in *CSEL* 50 (*Quaestiones*) and in *CSEL* 81, 1-3 (Commentaries). A commentary on Hebrews is lacking.

⁵ For the thesis that Ambrosiaster was Jewish, see G. Morin, "L'Ambrosiaster et le Juif converti Isaac, contemporain de Damase," *Revue d'Histoire et de Littérature Religieuses* 4 (1899) 97-120; H. J. Vogels in *CSEL* 81.1, XII-XIV; M. Zelzer, "Zur Sprache des Ambrosiaster," *Wiener Studien* 83 (1970) 196-213, esp. 212-13. For the thesis that Ambrosiaster was a pagan convert to Christianity, see C. Martini, *Ambrosiaster. De auctore, operibus, theologia* (Rome: Cuggiani, 1944) 147f., and esp. 154-160. Martini's conclusions were intimated by A. Souter, *The Earliest Latin Commentaries on the Epistles of St. Paul* (Oxford: Clarendon, 1927) 45; A. Stuiber, "Ambrosiaster," *Jahrbuch für Antike und Christentum* 13 (1970) 119-23, at 119-20. And see A. Souter, *A Study of Ambrosiaster* (Cambridge: Cambridge U. P., 1905) 166-74 (chronological evidence). L. Speller, "Ambrosiaster and the Jews," *Studia Patristica* 17 (1982) 72-77 (knowledge of Jewish customs).

⁶ *CCL* 9, 336-48.

⁷ For a comprehensive study, see W. Levison, "Konstantinische Schenkung und Sylvesterlegende," *Studi e Testi* 38 (1924) 159-247 (= *Miscellanea F. Ehrle* II). More recent studies of the *Actus Sylvestri* include A. Ehrhardt, "Constantine, Rome and the Rabbis," *Bulletin of the John Rylands Library* 42 (1959-60) 288-312 and F. Parente, "Qualche appunto sugli *Actus Beati Sylvestri*," *Rivista Storica Italiana* 90 (1978) 878-97.

story of Sylvester's public confrontation with twelve learned rabbis gained widespread popularity as a result, among other things, of its inclusion in Jacob a Voragine's *Legenda Aurea*. Wall paintings in a chapel of the basilica of the Santi Quattro Coronati in Rome and the apse of the church of San Silvestro in Tivoli testify to the supposed irresistibility of Sylvester's arguments. Yet, despite its popularity in later ages, at present the study of the hundreds of manuscripts containing an account of these events has simply not advanced far enough to determine whether a disputation involving twelve rabbis and Sylvester (as well as Constantine and Helena) ever took place in Rome. The two versions of this event most easily available reflect decidedly Christian concerns, but contain frustratingly little concrete information concerning the Roman Jewish community during the period under discussion in this book.[8] It appears that overall the *Actus Silvestri* are more indicative of the theological (Christian) concerns of the fifth century than of the historical reality of the fourth century. It is obvious, then, that an analysis of these traditions is beyond the scope of this book.

The Collatio: General Characteristics

The *Collatio* comprises 16 *tituli* (titles) in which individual laws taken from the Pentateuch are compared to Roman laws.[9] Except for the last *titulus* which deals with private law, all other titles concern criminal law. The comparison consists of a juxtaposition of Mosaic and Roman law, rather than of an analytical discussion in which the similarities between the two legal systems are treated in detail.

With the exception of *Collatio* 6.7.1, all titles start with the rendering of a Pentateuchal law. The phrase *Moyses dicit* (Moses says) normally introduces such a law. Following the introductory citation of Pentateuchal law is a rich selection of Roman laws. Although these Roman laws deal with approximately the same subject matter as the Mosaic laws preceding them, they usually treat individual legal issues in greater detail. In fact, more often than not, the injunctions selected from the Torah contrast markedly with the laws derived from the Roman legal tradition in terms of their concise and even terse legal phraseology.

---

[8] *PG* 110, 596-604 and *PG* 121, 520-540. And see also *PL* 8, 814.

[9] For an edition of the text, see J. Baviera, *Fontes Iuris Romani Antejustiani* (Florence: Barbèra, 1940) 544-89. An edition of the Berlin or oldest manuscript, with English translation, can be found in M. Hyamson, *Mosaicarum et Romanarum Legum Collatio* (London: Oxford U. P., 1913). On the manuscript traditions, see the detailed discussion by F. Schulz, "The Manuscripts of the *Collatio Legum Mosaicarum et Romanarum*," in M. David *et al.* (eds.), *Symbolae ad jus et historiam antiquitatis pertinenetes, Julio Christiano van Oven dedicatae (Symbolae van Oven)* (Leiden: Brill, 1946) 313-32; and see Volterra 1930, 8f.

While Pentateuchal laws have been chosen freely from the books Exodus through Deuteronomy, passages taken from Exodus and Deuteronomy numerically outweigh those derived from Leviticus and Numbers.[10] The Roman laws integrated in the *Collatio* have been extracted from several juristic works as well as from collections of constitutions. In the sequence of their occurrence, they include: Paulus' *Sententiae*, Ulpian's *De officio proconsulis*, Ulpian's *Regulae*, the *Codex Gregorianus*, Papinianus' *De adulteriis*, Paulus' *Liber singularis de adulteriis*, Ulpian's *Ad Edictum*, the *Codex Hermogenianus*, Paulus' *Liber singularis de iniuriis*, Gaius' *Institutiones*, Papianus' *Definitiones* and *Responsa*, Paulus' *De poenis paganorum liber singularis*, his *Responsa*, and the *De poenis omnium legum liber singularis* by the same author, and, finally, Modestinus' *Differentiae*. This list is especially remarkable in that it contains the names of exactly those five jurists whose works were later declared authoritative without further need of checking in the so-called "Law of Citations" of 426 C.E.[11]

The *Collatio* was composed in Latin. As Volterra's comparison of the various text traditions has shown, the biblical passages contained in the *Collatio* display little similarity in style and content with Jerome's Latin Bible translation of the late fourth and early fifth centuries. In general, the biblical passages of the *Collatio*, rather, tend to follow the text of the Septuagint, or, more precisely, of a Latin translation (or translations) of the latter work. Some deviations from this particular text tradition occur, but in general they seem to be due to the particular needs ensuing from the structural characteristics of the *Collatio* (namely the juxtaposition of Jewish and Roman law).[12]

Although the author of the *Collatio* always indicates the provenance of the Roman legal texts he cites, it seems rather unlikely that he himself bothered to consult all these works separately.[13] Analyzing the order in which the juristic works and imperial constitutions included in the *Collatio* appear, Schulz has pointed out that the author of the *Collatio* used excerpts in which earlier Roman legal materials had been arranged in groups according to their authors.[14] Furthermore, both Schulz and Niedermeyer have suggested that textual differences between the *Collatio* and the original juristic writings should not be

---

[10] For a list of the passages in question, see below, Table 1.

[11] *Codex Theodosianus* 1.4.3.

[12] Although other scholars such as T. Mommsen and P. Jörs had been aware of this question (see the latter in *RE* 4 [1901]] 368 s.v. *Collatio legum;* Hyamson 1913, XXXIII), Volterra 1930, 54-80 was the first to study it in depth. See also the extensive discussion in Smits 1934, 31f.

[13] Note that, by contrast, the author of the *Collatio* never indicates the provenance of the Pentateuchal passages he includes.

[14] F. Schulz, "Die Anordnung nach Massen als Kompositionsprinzip [etc.]," *Atti del congresso internazionale di diritto romano (Bologna e Roma 17-27.IV.1933)* (Pavia: Successori, 1935) vol. 2, 11-25, esp. Table 1 on p. 22-23.

attributed to the author of the *Collatio* himself, but rather resulted from the activities of previous (post-Classical) revisors.[15] Overall, from a Roman legal point of view, the author of the *Collatio* does not appear to have been a very original thinker. In fact, Schulz goes so far as to maintain that the *Collatio* initially constituted a collection of Roman legal materials similar to the *Fragmenta Vaticana* and that the Mosaic laws were added only at a later stage by someone with strong theological interests.[16] Although it cannot now be determined whether the organizing principle of the author of the *Collatio* indeed consisted of skillfully appending biblical materials to an already existing collection of Roman juristic works,[17] it is certainly true that his capacities as a jurist are far from impressive. Given the inconsistencies between the legal materials he includes under single headings,[18] it seem rather unlikely that he was a jurist at all.[19]

The exact date of the *Collatio* is a complicated question. The inclusion of references to the *Codex Gregorianus* and *Codex Hermogenianus* as well, as a possible allusion to a law known to have been promulgated by Constantine, provide an indisputable *terminus post quem*.[20] The two aforementioned codices were composed in the reign of Diocletian, or, more precisely, in the last decade of the third century.[21] Constantine's constitution dates to 315 C.E. The inclusion of these materials throughout the *Collatio* shows that the *Collatio* could not have been compiled before the early fourth century.

A *terminus ante quem,* on the other hand, cannot so easily be established. It seems beyond doubt that the *Collatio* predates the year 438 C.E., because the *Codex Theodosianus* is never referred to. To be more specific as to when exactly the *Collatio* came into being is, however, exceedingly difficult. In the fifth title of the *Collatio* an imperial constitution has been included that had been promulgated by Valen-

---

[15] Schulz 1946, 312. And see H. J. Wolf, "Ulpian XVIII ad Edictum in Collatio and Digest and the Problem of Postclassical Editions of Classical Works," in *Scritti in onore di Contardo Ferrini pubblicati in occasione della sua beatificazione* (Milan: Vita e pensiero, 1948) 64-90.

[16] Schulz 1946, 313-14.

[17] See the sensible criticisms voiced by Liebs 1987, 171-72.

[18] A systematic and detailed analysis of these inconsistencies may be found in Smits 1934, 75f.

[19] *Pace* Smits 1934, 172 and Liebs 1987, 170, 172. See also the important remarks by Levy 1930, 704.

[20] The allusion to Constantine's constitution may be found in *Collatio* 14.3.6. Note that some scholars do not believe that this passage alludes to a constitution of Constantine, e.g. A. Masi, "Ancora sulla datazione della 'Collatio Legum Mosaicarum et Romanarum'," *Studi Senesi* 77 (1965) 418-19.

[21] H. Honsell, T. Mayer-Maly and W. Selb, *Römisches Recht. Aufgrund des Werkes von P. Jörs, W. Kunkel, L. Wenger* (Berlin: Springer, 1987) 38; Liebs 1987, 134-43: The *Codex Gregorianus* probably dates to 291 C.E., and the *Codex Hermogenianus* to 295 C.E.

tinian II, Theodosius and Arcadius in the short interval between 390 and 392.[22] Later, this same constitution appeared in the *Codex Theodosianus*, albeit in a slightly different form.[23] Important though such evidence is in helping us to determine the *Collatio*'s chronology, it does not, however, directly solve the question of its precise dating. Because this particular constitution occurs in virtual isolation and because most other legal materials included in the *Collatio* do not seem to postdate the early fourth century, some scholars maintain that the second half of title five of the Collatio is to be regarded as an interpolation.[24] Arguing in favor of a dating between 314 and 324, Volterra has furthermore pointed out that crucifixion, a punishment referred to in the *Collatio* on several occasions, was abolished under Constantine. Similarly, the title on adultery (*Collatio* 6) describes practices that were outlawed in 342 C.E.[25] Consequently, passages such as these reflect the penal practices of a period predating the early fourth century. In Volterra's view, their inclusion in the *Collatio* makes a late fourth century dating of this work in its entirety highly unlikely, if not impossible.[26]

Such evidence notwithstanding, other scholars disagree with this "modische Schichtentheorie" and continue to prefer a late fourth dating for the *Collatio* as a whole.[27] Liebs, repeating an observation first made by Smits in 1934, argues that the fifth title of the *Collatio* would have been remarkably short and conceivably incomplete had the reference to the imperial constitution of Valentinian II, Theodosius, and Arcadius of 390-392 C.E. not been included from the outset. Believing that this imperial constitution was an integral part of the *Collatio* from the beginning, Liebs maintained that the *Collatio* in its entirety came into being around the year 400 C.E.[28]

A glance at the relevant secondary literature shows that even today the question of an early or late dating of the *Collatio* has not been settled. Despite frequent controversies among Roman legal historians, neither those supporting an early fourth century dating nor those advocating a dating later in the fourth-century have succeeded in putting forward completely conclusive arguments.[29] For reasons I shall ex-

---

[22] *Collatio* 5.3.1-2; discussion in Volterra 1930, 52 and Liebs 1987, 163, 165-70.

[23] *Codex Theodosianus* 9.7.6.

[24] This was first argued by Volterra 1930, 96; Schulz 1946, 314; G. Cervenca, "Ancora sul problema della datazione della 'Collatio Legum Mosaicarum et Romanarum'," *Studia et Documenta Historiae et Iuris* 29 (1963) 273; Rabello 1966, 181.

[25] *Codex Theodosianus* 3.12.1.

[26] *Collatio* 1.2.2; 7.4.4, and 14.2.2.; Volterra 1930, 102.

[27] This expression is used by Liebs 1987, 164.

[28] Smits 1934, 162; Liebs 1987, 163.

[29] Supporters of an early date (admitting the existence of later interpolations) include Triebs 1902, 7; Volterra 1930, 100f.; Levy 1930, 703; Schulz 1946, 314; H. Niedermeyer, "Voriustinianische Glossen und Interpolationen etc." *Atti del congresso internazionale di*

plain in detail below, I believe that a dating in the later fourth century is to be preferred.[30] Whatever the exact dating of the *Collatio* may be, there can no longer be any doubt, however, that the work essentially dates to the fourth and not to the fifth or sixth century, as an earlier generation of scholars had assumed.[31] No less important, if it is indeed true that the *Collatio* was first composed in the early fourth century and that various additions to it where made in the course of that century, as many scholars maintain,[32] this would imply that the purpose for which the *Collatio* had originally been assembled (see *infra*) continued to be relevant for a period of at least some eighty to hundred years.[33]

We do not know where the *Collatio* was composed. Since the work is written in Latin and because the three surviving manuscripts were all found in western Europe, the *Collatio* most likely originated in the western part of the later Roman Empire. It has been argued that the *Collatio* was assembled in Rome. *Collatio* 5.3.2 contains an imperial constitution which concludes with the statement "issued the fourteenth of May *in atrio Minervae.*" We know that this atrium of Minerva was a public space in Rome that served the promulgation of laws. It was located on the Imperial fora near the temple of Augustus. It most likely also housed an archive, and we may suppose that it was there that the author or a later interpolator of the *Collatio* found the materials which are now part of the fifth title of the *Collatio*.[34] Liebs has furthermore pointed out that the *novellae constitutiones* concerning kidnappers and mentioned in passing in *Collatio* 14.3.6 are probably identical to two constitutions preserved in Roman legal sources other than the *Collatio*. These sources indicate that one of these so-called "new constitutions" was in force in the city of Rome only, while the other was first promulgated there.[35] Taken together, such evidence does not necessarily prove, of course, that the Collatio was indeed written in Rome; it merely shows that the author of the *Collatio* had access to legal materials that were first promulgated in that city. The

---

diritto romano (Bologna e Roma 17-27.IV.1933) (Pavia: Successori, 1935) vol. 1, 353-384, at 369; Cervenca 1963, 272-73 (an early fourth-century "Grundschrift," but substantial later additions, contra Masi); Masi 1965, *passim*. Supporters of a late date usually follow Mommsen; they include: O. Karlowa, *Römische Rechstgeschichte* (Leipzig: Von Veit, 1885) vol. 1, 967; Jörs in *RE* 4 (1901) 369; Hyamson 1913, XLVI-XLVIII; Smits 1934, 162-63; Lauria 1985, 261; Liebs 1987, 136, 163.

[30] See the section "The Collatio as a Late Ancient Jewish Treatise" *infra*.

[31] See Volterra 1930, 50.

[32] E.g. Schulz 1934, 12-13; *id.* 1946, 313-14; Cervenca 1963, 273-75. For a convenient summary of these interpolations, see Cracco Ruggini 1983, 47 n. 20.

[33] Contra Niedermeyer 1934, 369.

[34] See Liebs 1987, 165-66.

[35] Liebs 1987, 166.

evidence does suggest, however, that a western origin is more likely for the *Collatio* than an eastern one.³⁶

Finally, it is also impossible to determine whether the *Collatio* available today is complete or whether it is only part of what was originally a much-larger work. As was the case with the dating of the *Collatio*, scholarly opinion is divided over the issue and no definitive solution has yet been found.³⁷

The Collatio: A Christian or a Jewish Work?

Like many other issues concerning the *Collatio,* the question of its authorship has been the subject of much speculation. In the nineteenth century, several renowned Roman legal historians believed that the anonymous author was a Christian.³⁸ In more recent years a considerable number of scholars have abandoned this contention, primarily as a result of Volterra's groundbreaking study published in 1930. Many students of Roman law now maintain that the *Collatio* was originally composed by a Jew. To be sure, Volterra's arguments have not convinced all Roman legal historians. D. Smits, L. Wenger, and more recently M. Lauria and D. Liebs, all adamantly oppose the idea of a Jewish authorship of the *Collatio*. Liebs in particular has marshalled all the available evidence to argue that the author of the *Collatio* was a Christian after all.³⁹

Can we determine whether the author of the *Collatio* Jewish or Christian? In order to answer this question, I believe that the *Collatio* must be placed into the larger framework of Late Ancient Jewish and Christian attitudes towards the Hebrew Bible in general, and towards the Pentateuch in particular. Volterra was the first to realize the potential and methodological usefulness of this approach, but he did not exploit it fully.⁴⁰ In the following pages we will, therefore, pursue this

---

³⁶ *Contra* Hyamson 1913, XLVIII. Schrage 1993, 416 suggests that the *Collatio* originated in North Africa because it supposedly uses a Latin Bible translation that was made in North Africa. Such a suggestion remains hypothetical because it is based on the assumption that this Latin translation was only used in North Africa. Besides, the hypothesis does not address the fact that similarities between the biblical materials of the *Collatio* and Latin Bible translations are practically inevitable: after all, all these biblical citations are based on the same Hebrew or Greek text.

³⁷ See Schulz 1946 (*Symbolae van Oven*), *passim*. The following scholars believe the *Collatio* is incomplete: Jörs in *RE* 4 (1901) 369; Hyamson 1913, XXXI, who gives references to earlier literature (Huschke, Rudorff, Mommsen); Volterra 1930, 7; Smits 1934, 149 and 159-60 (hypothetical); Baviera 1940, 589; Rabello 1966, 180. Those who maintain that the *Collatio* is complete include: Smits 1934 159-60 (unconvincing); Liebs 1987, 172 (useful observations).

³⁸ E.g., Karlowa 1885, 967-68; Jörs in *RE* 4 (1901) 367.

³⁹ For references, see the list in the section "Introduction."

⁴⁰ Volterra 1930, 86f.

approach in a more systematic fashion than has been attempted hitherto.

The Pentateuch in Early Christian Thought

The single most consequential statement to determine early Christian thought about the Pentateuch was Jesus' assertion that he had come not to abolish the law and the prophets, but to fulfill them (Matthew 5:17). In the present context, it is not necessary to establish exactly what Jesus meant by this statement or even whether he actually ever said such a thing.[41] What matters here is that early Christian authors assumed the statement to be genuine. For all its concision, it was a phrase that was to attract the attention of many early Christian writers.

The question of which biblical laws had ceased to be binding and which parts of the Torah continued to be relevant posed itself in full force in many early Christian communities in the course of the first century—as can be inferred from the writings of Paul.[42] Again, it is not necessary for the present purpose to explore in any detail the complex issue of Paul's thought on these matters.[43] In this discussion it will be more pertinent to study the ways in which Paul's writings were received and interpreted by later generations of Christian thinkers. In the western part of the Mediterranean the letters of Paul initially failed to make a significant impression in Christian circles. Only in the second half of the fourth century C.E. did writers start to take his writings more seriously. As we will see shortly, the increase in the number of commentaries on the letters of Paul during this period is in fact such that it may rightly be called explosive. But first let us turn to a brief investigation of early Christian literature predating the Late Ancient period.

In early Christian literature a rich variety of attitudes towards the Torah and the Hebrew Bible can be found. Yet, for all their variety, early Christian authors often propose very similar solutions to the inherently-ambiguous idea that Pentateuchal law was, at one and the same time, valid and invalid (not abandoned but fulfilled). In an attempt to salvage the Hebrew Bible for the Christian cause, early Christian writers shared the common trait that they no longer accepted the laws of the Pentateuch *in toto*. Rather, in much of early Christian literature, a clear distinction was drawn between various parts of the

---

[41] See the discussion in E. P. Sanders, *Jesus and Judaism* (Philadelphia: Fortress, 1985) 262-63; and see his conclusions on pp. 267-68.

[42] Another good example is Trypho's *Dialogue*. In it, the discussion of Mosaic Law comprises at least one quarter of the entire *Dialogue*.

[43] Here it suffices to refer to one of the many studies on the subject, E. P. Sanders, *Paul, the Law and the Jewish People* (Philadelphia: Fortress, 1983) esp. 143f.

Law as they corresponded to different stages in human history: (1) before the Law (that is, before the giving of the Law on Mount Sinai); (2) under the Law (after the receipt of the Law on Sinai); and, finally, (3) under Christ.

The work of Tertullian may serve as a prime example to show that some early Christian writers regarded the Law as an entity that was in a continuous state of transformation. Tertullian held that parts of the Law had first emerged at the time of the creation of the world. Pointing out that the Law had been modified subsequently by Moses, he then argued that even Mosaic law represented no more than an intermediate state of a transitory nature. In Tertullian's eyes the process of the ongoing revelation of the Law continued unabated also after Moses' "intervention." It would not cease until the advent of Jesus.[44]

In early Christian thought, the earliest phase in human history ("before the Law") comprised the period from creation to the dissemination of the Law on Mount Sinai. Although Mosaic law did not yet exist during this period of time, Christian thinkers describe the world of the primordial fathers as a time in which natural law prevailed. As Tertullian put it in reference to Noah, Abraham, and Melkizedek, "there was a law unwritten, which was habitually understood naturally."[45] Similar ideas also surface in Eusebius' *Praeparatio evangelica,* and in the works of Ephrem Syrus and Ambrose, to give but a few other representative examples.[46] Describing the sacrifice of Isaac, Zeno of Verona observed that Abraham fulfilled the Law even though he was not under it, and elsewhere added that Abraham simply did not need the Law because he himself incorporated it.[47]

Because the many injunctions of Mosaic Law were seen as punishment the age of the patriarchs is often painted with the traits of a golden age in early Christian discussions of the Law. These men, including Enoch, Noah, Abraham, and Melkizedek were not only seen as having served God faithfully because of their natural instinct for justice; they were also seen as exemplary figures because they lacked one of the more distinctive marks of Judaism: circumcision.[48] It is for

---

[44] Tertullian, *Adversus Iudaeos* 2 (*CCL* 2, 1341-44); see also Aphraat, *Demonstration* 11.11 (*PS* 1, 499-500) who speaks not so much of continuously-changing laws but rather of continuously-changing covenants (and see *Demonstration* 12.11 [*PS* 1, 533-34] which refers to Jeremiah 31:31-32).

[45] Tertullian, *Adversus Iudaeos* 2.7 (*CCL* 2, 1342). In *Adversus Iudaeos* 2.2-3 (*CCL* 2, 1341) Tertullian calls the Law current during this period the *lex primordialis*; Novatian, *De cibis Iudaicis* 3.2-3 (*CCL* 4, 93-94).

[46] Eusebius, *Praeparatio evangelica* 7.6 and 7.8.20 (*SChr* 215, 168-72 and 186); Ephrem, *Hymnen de Ecclesia* 43.3 (*CSCO* 198, 108 and see *CSCO* 199, 103; Ambrose, *Epistula* 63.2-3 (*Maur.* 73 - *CSEL* 82.2.143).

[47] Zeno, *Tractatus* 1.43.8 and 1.62.1 (*CCL* 22, 116, and 141).

[48] E.g. Eusebius, *Praeparatio evangelica* 7.8.20 (*SChr* 215, 186); Zeno, *Tractatus* 1.3.5 (*CCL* 22, 25).

exactly that reason that in early Christian thought, Abraham quickly assumed the role of Christianity's forefather *par excellence*. Like no other, it was he, *Abraham patriarcha noster*, who came to symbolize the conversion of the gentiles, as can be seen for example in the *Epistle of Barnabas*, and in the works of Justin, Irenaeus, Eusebius, Aphraat, Zeno, Ambrose, Eucherius, and Prudentius.[49] Such writers realized, of course, that Abraham had proceeded to circumcise himself, but they assured their audience on more than one occasion that this happened only late in life and long after he had first embraced monotheism and given extensive testimony of his great faith in God.[50] In the view of many early Christian authors, the earliest period in human history thus offered incontestable proof to show that a true relationship between God and the human race did not depend on circumcision or on the Law but rather on a more spiritualized sense of what was right and wrong.

The next phase in human history ("under the Law") started with the receipt of the Law by Moses on Mount Sinai. Conforming to the account in Exodus 19, early Christian writers distinguished two stages in this process: the first and the second ascension to Sinai.

Upon his first descent from Mount Sinai Moses had brought with him the Ten Commandments only. In early Christian thought, these commandments were not considered innovative by any means: they were believed to merely document in written form the natural law of the preceding period.[51] The Ten Commandments therefore comprised all injunctions necessary to guarantee the proper ethical behavior of humankind.[52] Had not Jesus himself pointed out that by keeping just these few commandments one would be able to attain the eternal

---

[49] *Epistle of Barnabas* 13.7a (*SChr* 172, 176); Justin, *Dialogue* 11.4; 23.4; 25.1, asserting that the Jews are not the sons of Abraham (ed. Goodspeed 1914, 103, 117, and 118); Irenaeus, *Adversus haereses* 4.8.1, fulminating against Marcion (*SChr* 100, 464-65) and ibid. 4.16.2 (*SChr* 100, 562-63); Eusebius, *Praeparatio evangelica* 7.8.22 (*SChr* 215, 188); id., *Historia ecclesiastica* 1.4.11f. (*GCS* 9.1.42f.); Aphraat, *Demonstrations* 2.3 (*SChr* 349, 238-39) 11.1 (*PS* 1, 467-68) and 16.8 (*PS* 1, 783-84) arguing that the Christians are the children of Abraham and the Jews the children of Cain; Zeno, *Tractatus* 1.3.6 (*CCL* 22, 25) who regards Abraham as forefather of both Jews and Christians; Ambrose, *Epistula* 65.2 (*Maur.* 72 - *CSEL* 82.2, 179 and 188); Eucherius, *Instructionum liber* 1.3.2. (*CSEL* 31, 77); Prudentius, *Liber Apotheosis* 361-68 (*CCL* 126, 89).

[50] Tertullian, *Adversus Iudaeos* 3.1 (*CCL* 2, 1344); Aphraat, *Demonstrations* 11.3-4 (*PS* 1, 475-78).

[51] E.g., Trypho, *Dialogue* 45.3-4 (ed. Goodspeed 1914, 142-43); Irenaeus, *Adversus haereses* 4.14.3 and 4.16.2-4 (*SChr* 100, 548-49 and 564-71); Tertullian, *Adversus Iudeos* 2.2-3 (*CCL* 1, 1341) and Origen, *Contra Celsum* 5.37 (*GCS* 2, 41). The same idea also surfaces in medieval Canon Law. See also the remarks of Eusebius in his *Demonstratio evangelica* 1.5. (*GCS* 23, 20, 23-27). And see the discussion in A. Lenox-Conyngham, "Law in St. Ambrose," *Studia Patristica* 23 (1989) 149-52.

[52] E.g., Eucherius, *Instructionum liber*1, 38 (*CSEL* 31, 78). And see Julian, *Contra Galilaeos* 152B-D.

life?⁵³ And had not Jesus observed that the entire law and the prophets depended in essence on the observance of but two commandments: to love God and to love one's neighbor (that is, commandments taken from the Decalogue)?[54] Early Christian authors did not grow tired of stressing that Jesus had sanctioned the Decalogue.[55]

In turn, Jesus' sanctioning of the Decalogue determined early Christian views of Moses. Since Moses was responsible for making known publicly the commandments that were later to be endorsed by Jesus, he is often represented in early Christian literature and art as *typus Christi*, or prefiguration of the Christian Messiah.[56] Again, as early Christian authors realized full well, it was Jesus himself who had first stressed this connection when he stated that "if you believe Moses, you would believe me, for he wrote of me."[57]

Perhaps most striking about descriptions of Moses in early Christian literature, however, is the fact that Moses is frequently represented as everything but a Lawgiver. That such descriptions resulted from familiarity with Hellenistic Jewish descriptions of Moses as statesman, general, and philospher is possible.[58] Yet, I believe that descriptions of Moses as universal leader rather than as legislator resulted above all from Christianity's inherently ambivalent attitude towards biblical law. Aphraat, for example, described in glowing terms Moses' eight major achievements, including the latter's ascent to Mount Sinai, but never even once mentioned Moses' role as disseminator of God's law.[59] Similarly, around 390, Gregory of Nyssa wrote a "Life of

---

[53] Matthew 19: 18-19, which focusses on the commandments in the second table of the Decalogue. See also Luke 18:18-20 and Mark 10:17-19.

[54] Matthew 23:37-40.

[55] On Matthew 19:18-19, e.g. Tertullian, *Adversus Iudaeos* 2.2-3 (*CCL* 2, 1341) and Irenaues, *Adversus haereses* 4.12.5 and 4.14.3 (*SChr* 100, 520-21; 548-51); on Matthew 23:37-40 see Irenaeus, *Adversus haereses* 4.12.2-3 (*SChr* 100, 512-15), Aphraat, *Demonstration* 2.7 (*PS* 1, 61-62), and Ephrem, *Sermo* 3.273-75 (*CSCO* 212, 28 and 213, 41). These ideas may also have reached non-Christians, see, e.g., the way Pliny described the baptismal oath taken by Christians in his *Epistula* 10.96.7.

[56] *Epistle of Barnabas*12.2f. (*SCh* 172, 166f.); Tertullien, *Adversus Iudaeos* 10.10 and 13.12 (*CCL* 2, 1377 and 1387); *id.*, *Adversus Marcionem* 2.26.4 (*CCL* 1, 505); Clement, *Stromateis*1.168 (*GCS* 15, 104-5); Eusebius, *Demonstratio evangelica* 3.2.6f. (*GCS* 23, 97f.), on which, see the discussion by M. J. Hollerich, "Religion and Politics in the Writings of Eusebius: Reassessing the First 'Court Theologian'," *Church History* 59 (1990) 309-25, esp. 320; Eusebius, *Historia ecclesiastica* 1.3 (*GCS* 9.1, 28). For art, see J. J. M. Timmers, *Christelijke symboliek en iconografie* (Haarlem: De Haan, 1981) 38-39; 63. Note also that of all biblical figures in the New Testament, Moses is the person most frequently mentioned.

[57] John 5:46. Origen, *In Jesu Nave* 2 (*SCh* 71, 96); Irenaeus, *Adversus haereses* 4.2.3 and 4.12.1 (*SCh* 400-1 and 510-11).

[58] That early Christian authors were familiar with such traditions follows from, e.g., Clement, *Stromateis* 1.151f. (*GCS* 15, 93f) and Eusebius, *Praeparatio evangelica* 8.6 (*GCS* 43.1, 427-29), 9.7 (*GCS* 43.1, 493) and 9.26f. (*GCS* 43.1, 519f.). Gregory of Nyssa's *Life of Moses* depends on Philo's *Vita Mosis*.

[59] Aphraat, *Demonstration* 1.14 (*SCh* 349, 226).

Moses," which, somewhat surprisingly, carries the subtitle "Concerning Perfection in Virtue." It largely ignored the legal materials of the Pentateuch. Instead it stressed the ascetic aspects of Moses' life in an attempt to find biblical support for the monastic movement.[60] In the West, it was likewise customary to downplay Moses' role as Lawgiver. In the first book of his "History of the Franks," Gregory of Tours summarized the history of the world from its creation down to the death of St. Martin. Dwelling extensively on the passage through the Red Sea, Gregory passed over the events at Sinai in complete silence. He only remarked obliquely that the forty years in the desert served for the Israelites "to familiarize themselves with *their* laws [italics mine]."[61] Thus, Aphraat's, Gregory of Nyssa's, and Gregory of Tours' writings all document a general tendency in early Christian thought, namely either to gloss over biblical law or to interpret it spiritually rather than literally whenever possible. In Christian eyes, it was Moses himself who had first opened the way to such interpretations.[62]

Much the same attitude can be seen when we now turn to early Christian attitudes toward Moses' second ascent of Sinai. Christian writers such as Irenaeus were quick to point out that the incident involving the golden calf demonstrated that in the particular case of the Jews, the Decalogue alone would not be sufficient to keep them in check.[63] To control the Jews, another, more stringent law was called for. For that reason (and not merely to replace the broken tables of the Law), Moses had to ascend Sinai a second time. He brought back new laws, or, as Aphraat called them in reference to Ezekiel 20:25, "the commandments which are not excellent."[64] Thus the Jews were pun-

---

[60] A. J. Malherbe and E. Ferguson (transl.), *Gregory of Nyssa. The Life of Moses* (New York: Paulist Press, 1978). Note, for example, that Gregory devotes only two paragraphs to the giving of the Law on Sinai (1.47-48), while his description of the Tabernacle comprises no less than six paragraphs (1.49-55) (comparable is 2.165f.)! Similarly, 77 paragraphs concern the *historia* or account of Moses' life, while the *theoria* or second part, interpreting Moses' life allegorically, is made up of 321 paragraphs.

[61] Gregory, *Historia francorum* 1.10-11.

[62] Deuteronomy 30:6: "And the Lord your God will circumcise your heart;" see, e.g., Zeno, *Tractatus* 1.3.6.13 (*CCL* 22, 27).

[63] Irenaeus, *Adversus haereses* 4.14.3 (*SChr* 100, 548-51). One often encounters the idea that the Jews are inherent transgressors of the Law (even of the second law) in early Christian literature, e.g., Tertullian, *Adversus Marcionem* 2.18.3 (*CCL* 1, 496); Irenaeus, *Adversus haereses* 4.12.1 and 4.14.3 (*SCh* 100, 510-11 and 546-47); Commodian, *Instructionum liber*1.38 (*CCL* 128, 32). And see Ephrem, *Contra Julianum* 1.16-19 (*CSCO* 174, 74-75 = 175, 68) who maintains that Jews recognized in the bull represented on the coins of Julian the golden calf of Exodus! B. L. Visotzky, "Anti-Christian Polemic in Leviticus Rabbah," *Proceedings of the American Academy for Jewish Research* 56 (1990) 89-94.

[64] Aphraat, *Demonstration*15.8 (*PS* 1, 755-56). See, in general, P. W. van der Horst, "I gave them laws that were not good. Ezekiel 20:25 in Ancient Judaism and Christianity," in J. N. Bremmer and F. García Martinez (eds.), *Sacred History and Sacred Texts in Early Judaism (FS van der Woude)* (Kampen: Kok Pharos, 1992) 94-118.

ished.⁶⁵ Yet, at the same time they were also offered a tool to help them improve their ways.⁶⁶

As might be expected in the light of the above, early Christian authors did not consider this second law to be incumbent on Christians.⁶⁷ In fact, in their view of human history, observance of the second law represented an entirely unnecessary step between a period in time governed by natural law (as made explicit in the Decalogue) and an age determined by the presence of Christ. Yet, paradoxically,⁶⁸ even though the second law was regarded as resulting from specific historic events in which Christians prided themselves not to have taken any part;⁶⁹ some of the religious customs described in the second law were not per se considered superfluous. Rather than rejecting such customs outright, they were now spiritualized. Thus, the fathers of the Church succeeded in arguing that Christians, like their Messiah, were not abolishing the law of old, but that they were actually fulfilling it daily.

Passages illustrating the spiritualization of biblical law abound in early Christian literature. Because it is representative in this respect, it will suffice here to cite only part of the second chapter of Augustine's "Against the Jews." He writes that:

> It would take too long, however, to dispute these charges one by one; how we are circumcised by putting off the old man and not in despoiling our natural body; how their [i.e. of the Jews] abstinence from certain foods of animals corresponds to our modification in habits and morals; how we present our bodies, a living sacrifice, holy and pleasing to God before whom we intelligently pour forth our souls in holy desires, instead of blood...when we find rest in Him [Jesus] we truly observe the Sabbath, and the observance of the

---

⁶⁵ In general: Galatians 3:19; Lactantius, *Divinae institutiones* 4.10, 11-13 (*SCh* 377, 88); Hilary, *Commentarius in Matthaeum* 13.22 (*PL* 9, 922); *Consultationes Zacchaei et Apollonii* 2.7 (*PL* 20, 1119); Ephrem, *Hymnen de Ecclesia* 43.1 (*CSCO* 198, 108 and *CSCO* 199, 103); Ambrose, *Epistula* 65.2 (*Maur.* 75 - *CSEL* 82.2, 157). And see Trypho, *Dialogue* 16.1-3 and 19.1 (Goodspeed 1914, 108-9 and 111): circumcision is a punishment; Trypho, *Dialogue* 19.5-6 and 22.1 (Goodspeed 1914, 112 and 114), Aphraat, *Demonstration* 15.6 (*PS* 1, 747-48), and John Chrysostom, *Adversus Iudaeos* 4.6 (*PG* 48, 880): sacrifice was instituted to save the Jews from worse forms of idolatry; Tertullian, *Adversus Marcionem* 2.18.2 (*CCL* 1, 495-96) and Aphraat, *Demonstration* 15.3 (*PS* 1, 733-36): the food laws are meant to check lust and to encourage continence; Irenaeus, *Adversus haereses* 4.15.2 (*Sch* 100, 554-55) and Tertullian, *Adversus Marcionem* 4.34 (*CCL* 1, 634f.): discussing Matthew 19:7-8 and interpreting the Mosaic laws of divorce as punishment.

⁶⁶ E.g., Novatian, *De cibis Iudaicis* 3.2-3 (*CCL* 4, 93-94) and Ambrose, *Epistula* 64.3 (*Maur.* 74 - *CSEL* 82.2, 150): *Lex paedagogus est*.

⁶⁷ E.g., Eusebius, *Praeparatio evangelica* 8.1 (*GCS* 43.1). The expression *lex prima* and *lex secunda* may be found in, e.g., Commodian, *Instructionum liber*1.25 (*CCL* 128, 20).

⁶⁸ See, e.g., the discussion in Julian, *Contra Galilaeos* 305D-314E; 351A; 354A.

⁶⁹ Julian, *Contra Galilaeos* 319D.

new moon is the sanctification of our new life. Christ is our Pasch; our unleavened bread is sincerety of truth without the leaven of decay.[70]

This passage from Augustine shows how the Church came to regard itself as *Verus Israel*, a spiritual, yet, in Christian eyes, a real heir to the whole of the Jewish tradition. The passage is interesting in another respect too. It illustrates appropriately a further characteristic of early Christian discussions of the second law, namely that the legal materials from the Hebrew Bible are usually elaborated upon in a rather reductionistic way. No longer was the second law with its 613 positive and negative commandments represented as a coherent body of legally and socially structuring principles. Quite the contrary. In the works of many early Christian thinkers, the second law now came to equal precisely those few Jewish religious customs, the practice of which they disliked most: the celebration of the Sabbath, the circumcision of the males of the community on the eighth day after birth, the observance of laws regulating purity in respect to food, and, finally and somewhat less frequently, Temple sacrifice and the celebration of other Jewish religious festivals (Passover, new moon etc.). A mere glance at early Christian literature suffices to show that in its discussions of the second law, it in fact hardly ever pays attention to anything but the Sabbath, circumcision and food laws.[71] The difference between early Christian thinkers and their Jewish counterparts was thus fundamental. While for Jews—or at least for some Jews—study of the implications of the legal materials in the Torah had always been a major concern and continued to be so in Late Antiquity, a majority of early Christian writers never felt compelled to discuss individual biblical laws with the same painstaking philological precision and eye for detail as their Jewish contemporaries.[72]

It has already been observed that according to early Christian writers, the third and final phase in human history started with the min-

---

[70] *Tractatus adversus Iudaeos* 2 (*PL* 42, 52). For a "spiritualized" interpretation of Jewish religious customs, see also *Epistle of Barnabas* 9.1f. and 10 (*SChr* 172, 140f., 148f.); Tertullian, *Adversus Marcionem* 1.20 (*CCL* 1, 461-62); *id., Adversus Iudaeos* 3.4 and 5 (*CCL* 2, 1345 and 1349-52); Irenaeus, *Adversus haereses* 4.16.1 (*SCh* 100, 558-63); Irenaeus, *Adversus Haereses* 4.17.1-5 and 4.18 and 14.9.1 (*SChr* 574-617); Novatian, *De cibis Iudaicis* 2.1 and 3.1 (*CCL* 4, 90 and 93); Ambrose, *Epistula* 69 (*Maur.* 72 - *CSEL* 82.2, 188).

[71] See e.g. the *Epistle of Barnabas*, Trypho, *Dialogue*; Tertullian, *Adversus Marcionem*; Irenaeus, *Adversus haereses* 4; Novatian, *De cibis Iudaicis*; Commodian, *Instructionum Liber*; Eusebius, *Praeparatio evangelica*; Ephrem, *Sermo* III; Aphraat, *Demonstrations* 11, 12, 14, and 16; Prudentius, *Liber apotheosis*; *Consultationes Zacchaei et Apollonii*; Augustine, *Adversus Iudaeos*. Compare also Julian's *Contra Galilaeos*.

[72] The difference in approach can be illustrated by contrasting the treatment of legal materials in *halakhic midrashim* with the treatment of such materials in early Christian commentaries on the Hebrew Bible.

istry of Jesus. In discussing the changes brought about by Jesus, early Christian authors including Justin, Irenaeus, Tertullian, Aphraat, and the anonymous author of the *Consultationes Zacchaei and Apollonnii* pointed out that the prophecies of Isaiah were now becoming reality (Isaiah 2:3-4: "From Zion shall go forth the Law and the word of the Lord from Jerusalem").[73] In Christian exegesis, this specifically meant that the injunctions of the Decalogue remained in full force.[74] The hundreds of other commandments recorded in the second law, on the other hand, now lost their relevance for good. According to the early Christian point of view, rather than strict and scrupulous adherence to the Law, faith and grace henceforth shaped human existence.[75] As Chromatius, bishop of Aquileia, was to put it succinctly in the late fourth century: "Jews seek [truth] through their law...but they seek in vain, because they do not follow the way of truth."[76] From such a perspective, Jesus came to symbolize *finis Legis*.[77] Benevolently freeing Christians from "the cloud of the Law and the yoke of the statutes," he quickly came to be represented as mankind's true saviour.[78]

In the Latin Christianity of the second half of the fourth century (that is, of the same general period that the *Collatio* was in use), several important early Christian writers turned their attention to the old question of the validity/invalidity of Pentateuchal law. Early Christian interest in this question was above all sparked by a renewed concern with the letters of Paul, as this concern came of age in the sixties of the fourth century in the western part of the Mediterranean, and in the city of Rome in particular. Where previous generations of Christians theologians in that city had paid only limited attention to the Pauline corpus, the second half of the fourth century witnessed a profound change in this respect.[79] In a little more than half a century, a number of major commentaries on Paul's writings were composed in Latin.

---

[73] Trypho, *Dialogue* 24.1 (Goodspeed 1914, 117); Irenaeus, *Adversus haereses* 4.34.4 (*SCh* 100, 856-57); Tertullian, *Adversus Iudaeos* 3.9 (*CCL* 2, 1346); Aphraat, *Demonstration* 16.1 (*PS* 1, 761-62); *Consultationes Zacchaei et Apollonii* 2.7 (*PL* 20, 1119); Augustine, *Adversus Iudaeos* 7-8 (*PG* 42, 58-59) and, of course, the pseudo-Cyprianic treatise *De montibus Sina et Sion*.

[74] And see Irenaeus, *Adversus haereses* 4.16.4 (*SCh* 100, 570-71).

[75] The idea that faith replaces the Law is found, for example, in Ambrose, *Epistula* 64.1 (*Maur.* 78 - *CSEL* 82.2, 160) Augustine, *Adversus Iudaeos* 7.9 (*PL* 42, 58) who equates the law of Jesus to the *lex fidei*, and Paulinus, *Carmen* 31, 351-54 (*CSEL* 30, 319). Grace plays a significant role in Augustine, see below.

[76] *Sermo* 128, see the discussion in L. Cracco Ruggini, "Il vescovo Cromazio e gli Ebrei di Aquileia," in *Aquileia e l'Oriente Mediterraneo* 1 = *Antichità Altoadriatiche* 12 (1977) 353-81, at 376.

[77] Novatian, *De cibis iudaicis* 5.2 (*CCL* 4, 97); Clement, *Stromateis* 6.11.94 (*GCS* 52, 479); John Chrysostom, *Adversus Iudaeos* 2.2 (*PG* 48, 858-59).

[78] Paulinus, *Carmen* 24.653-65 (*CSEL* 30, 228).

[79] E.g., E. Dassmann, *Der Stachel im Fleisch. Paulus in der frühchristlichen Literatur bis Irenäeus* (Münster: Aschendorff, 1979).

They include commentaries on Galatians, Ephesians, and Philippians written by Marius Victorinus,[80] commentaries on all of Paul's letters by Ambrosiaster,[81] commentaries on Philemon, Galatians, Ephesians, and Titus by Jerome,[82] several commentaries on Romans and Galatians by Augustine,[83] an anonymous commentary dating to the years 396-405,[84] and commentaries on all of Paul's letters except Hebrews by Pelagius.[85] Rufinus' early fifth-century translation into Latin of Origen's commentary on Romans likewise documents the centrality which Paul's writings assumed in Christian thinking during this period.[86] Also illustrative of the renewed interest in Paul is, finally, the apocryphal correspondence between Seneca and Paul. This collection of letters was composed sometime between 324 C.E. and 392 C.E. By forging a correspondence between two figures that were held in high esteem by pagans and Christians respectively, its anonymous author wanted to interest the pagan intelligentsia in the letters of Paul.[87]

It is not entirely clear what caused this sudden interest in Paul in the later fourth century. Anti-Manichaean polemics may have led to a greater sensibility to Pauline writings in more orthodox Christian circles, but, according to W. Geerlings, this is only part of a more complex picture. He points out that the upsurge in commentaries on Paul in the second half of the fourth century went hand in hand with a renewed interest in the life and sufferings of Job and maintains that both these developments reflect an intensified preoccupation with the question of theodicy. According to this view, Paul's writings could give people guidance in a period characterized by a general anxiety vis-à-vis the omnipotence of evil and the unpredictability of *fatum*.[88] Although additional research remains to be done to explore the further implications of Geerlings' views, his explanation seems somewhat

---

[80] *CSEL* 83.1-2 and the Teubner edition of 1972 by A. Locher. Victorinus' commentaries date to the 360s C.E.

[81] *CSEL* 81, 1-3 (they were written around 366-84 C.E.).

[82] *PL* 26, 331-656 (dating to the year 386 C.E.).

[83] Augustine, *Expositio quarundam propositionum ex Epistola ad Romanos* (394-95 C.E.): *CSEL* 84; *Epistolae ad Galatas expositionis liber unus*, *CSEL* 84; *Epistolae at Romanos inchoata expositio*, *CSEL* 84; *De diversis quaestionibus ad Simplicianum* (396-97- C.E.): *PL* 40, 101-48, and *De spiritu et littera* (412-13 C.E.): *PL* 44, 201-46.

[84] H. J. Frede, *Altlateinische Paulus-Handschriften* (Freiburg: Herder, 1964); and see id., *Ein neuer Paulustext und Kommentar* (Freiburg: Herder, 1973).

[85] *PL Supplementum* 1, 1110-1374 (410 C.E.).

[86] See C. P. Hammond Bammel, *Der Römerbrief des Rufin und seine Origenes-Übersetzung* (Freiburg: Herder, 1985) 505-37.

[87] The texts are edited by L. Bocciolini Palagi, *Il carteggio apocrifo di Seneca e San Paolo* (Florence: Olschki, 1978) esp. 7-9 and 50f.

[88] W. Geerlings, "Hiob und Paulus. Theodizee und Paulinismus in der lateinischen Theologie am Ausgang des vierten Jahrhunderts," *Jahrbuch für Antike und Christentum* 24 (1981) 56-66.

more attractive than one that attributes the rediscovery of Paul primarily to influence of the person of Simplician alone.[89]

Whatever the background of this renewed fascination with Paul may have been, it is not necessary here to analyze further the factors that might have contributed to the rapid formation of a series of Latin commentaries on his works in the course of the fourth century. In the present context it suffices to observe that the preoccupation with and exegesis of Paul's letters led to substantial discussions of some of the more important issues raised in his writings. Foremost among such issues was the question that had threatened to rip apart early Christian communities as they sprang up in various parts of the Mediterranean during the first century: To what extent is Mosaic law incumbent on non-Jews who believed that Jesus was the Messiah?[90]

In their commentaries on Paul, the aforementioned exegetes addressed this particular question on various occasions. A brief review of their views is in order here, for it will help to uncover some fundamental differences between the treatment of Pentateuchal law by early Christian writers as opposed to the way it was treated contemporaneously by the author of the *Collatio*.

Marius Victorinus' exegetical remarks on the ideas expressed in Paul's writings are usually short and to the point. At the beginning of his commentary on Galatians he summarized the contents of this letter, and in so doing, he explained succinctly the essence of the Law as he, Victorinus, saw it: Paul wrote his letter to the Galatians to preach to them the *evangelium fidei*, to instruct them that one cannot be saved by works of the Law, and thus to correct them and to "call them back" from Judaism.[91] Clearly, Victorinus' most important distinction is between the Law on the one hand and faith on the other.

Further remarks on individual Pauline passages (in which Victorinus hardly ever departed far from the text he commented upon) merely repeated the views presented by his introductory observations. Elaborating on Galatians 3:9-10, for example, Victorinus once again stressed the all-importance of *fides* as opposed to the works of the Law and, like Paul, refered to Abraham to prove his point.[92] Continuing his commentary on these passages, he then divided the "works of the Law" into two distinct categories. The first category, called *opera*

---

[89] See B. Lohse, "Beobachtungen zum Paulus-Kommentar des Marius Victorinus und zur Wiederentdeckung des Paulus in der lateinischen Theologie des vierten Jahrhunderts," in A. M. Ritter (ed.), *Kerygma und Logos. Beiträge zu den geistesgeschichtlichen Beziehungen zwischen Antike und Christentum (FS Andresen)* (Göttingen: Vandenhoeck und Ruprecht, 1979) 351-66.

[90] See especially Galatians 2-3 and Romans 3-4.

[91] *CSEL* 83.2, 95.

[92] One may recall here the special place Abraham assumed in early Christian thinking: he was a man of God, his faith was proverbial, but he was not under the Law.

christianitatis, is not very well defined, but it includes care for the poor and is considered to be of continuing importance for Christians. The second category, by contrast, is not incumbent on Christians, the reason being that it had been fulfilled by Christ. Victorinus' specification of what this "other law" encompassed should not surprise us: the sacrifice of a lamb on Passover, circumcision, and the laws regarding food.[93] In short, then, Victorinus' ideas about the Law are roughly in line with much of earlier Christian literature as it has been discussed in previous pages: while the Law divides into ethical precepts that still need to be kept and ceremonial precepts that can now be discarded, faith replaces the Torah as the most important ingredient in the proper worship of God.

Ambrosiaster, like Victorinus, held well defined ideas about the Law. One of the most important passages to illustrate Ambrosiaster's view on the Law is his commentary on Romans 3:20.[94] There he proposes a division of the Law into three parts. Despite some textual problems resulting from different manuscript traditions, Ambrosiaster's ideas can nevertheless be reconstructed with the help of his second major work, the *Quaestiones veteris et novi testamenti*. In Ambrosiaster's view of things, the so-called *lex divinitatis* constituted the first law. This law comprised the first four commandments of the Decalogue.[95] Then came a second law, or *lex naturalis*. It contained the remaining six commandments of the Decalogue and served to "prohibit sin."[96] The second law was in turn followed by third law, which Ambrosiaster names the *lex factorum*. It encompasses all further laws, and, more particularly, the ritual laws of the Jews.[97] Throughout his writings, Ambrosiaster was at pains to stress that this last law did not belong to the original set of laws, that it had been promulgated for a specific purpose (that is, to punish the Jews), and that it was valid for a limited time only (namely until the advent of mankind's *salvator*).[98]

At first sight, Ambrosiaster's tripartite treatment of Pentateuchal law seems to differ somewhat from Victorinus' discussions of this matter. The difference, however, is in degree and not in nature. Like Victorinus, Ambrosiaster considered as essential the contrast between faith, "which belongs to God," and the works of the Law, "which be-

---

[93] *CSEL* 83.2, 130-31.
[94] *CSEL* 81.1, 114-16.
[95] See *Quaestio* 7.1 (*CSEL* 50, 31).
[96] See *Quaestio* 7.2 (*CSEL* 50, 31); *Quaestio* 19 (*CSEL* 50, 435-36) and *Quaestio* 75.2 (*CSEL* 50, 468-69).
[97] In *Quaestio* 69.4 (*CSEL* 50, 121) Ambrosiaster refers to a fourth law, comprising the injctions in Leviticus 24:20 (an eye for an eye, etc.), but this fourth law does not seriously affect his treatment of the law as essentially tripartite.
[98] *Quaestio* 19 (*CSEL* 50, 436).

long to man."⁹⁹ Similarly, while Ambrosiaster subdivided the Decalogue into two parts, the basic distinction, as in Victorinus, remained between the ethical as opposed to the ceremonial aspects of Pentateuchal law. Both Victorinus and Ambrosiaster agreed that the ethical precepts (that is, the Decalogue) had not lost their original significance since Jesus had manifested himself.[100] They also agreed that it was because of the advent of Jesus that the observance of Jewish religious practices (the ceremonial law) had now become superfluous.[101]

The enormous influence of Paul on Ambrosiaster's thinking is no less evident in Ambrosiaster's other major work, in which he addresses the question of the relationship between Judaism and Christianity, his *Quaestiones veteris et novi Testamenti 127*. One *quaestio* in this work, no. 44, which was entitled *adversus Iudaeos*, deals specifically with the Jews.[102] In it, Ambrosiaster presents what at first seems a random selection of passages selected from the Hebrew Prophets to argue that the Jews have ceased to be the people of God. Closer inspection of the individual passages reveals that most, albeit not all, of the passages in question had been previously used by Paul for very similar purposes.[103] When seen against the larger background of early Christian writings on biblical law, other ideas expressed in *Quaestio* 44 likewise fail to strike us as revolutionary. As had been the case in his commentary on Romans, Ambrosiaster maintained in his *Quaestiones* that the display of *fides* was far more important than the observance of Pentateuchal law.[104] Not surprisingly, it was once again the figure of Abraham—*patriarcha autem noster, fidelissimus Abraham*—who served as exemplar to underscore that point.[105] Yet, where previous early Christian authors had followed Paul and had always associated Abraham's faith with the period before the latter's circumcision,[106] Ambrosiaster went so far as to interpret Abraham's circumcision itself as symbol of the latter's faith.[107]

---

[99] E.g., in the passage under consideration *CSEL* 81.1, 115, or in *Quaestio* 44.5 (*CSEL* 50, 74).

[100] *Quaestio* 69.4 (*CSEL* 50, 121).

[101] *Quaestio* 69.2 (*CSEL* 50, 119).

[102] *Quaestio* 44 (*CSEL* 50, 71-81).

[103] E.g. *Quaestio* 44.3 (*CSEL* 50, 73): Isaiah 59:20-21 see Romans 11:26-27; Habakkuk 2:4 see Romans 1:17; *Quaestio* 44.5 (*CSEL* 50, 75): Isaiah 10:22-23 see Romans 9:27-28; *Quaestio* 44.6 (*CSEL* 50, 75): Genesis 15:6 see Romans 4:3 and *Quaestio* 44.11(*CSEL* 50, 78): Hosea 2:23 see Romans 9:25.

[104] *Quaestio* 44.5 (*CSEL* 50, 74).

[105] *Quaestio* 44.6-8 (*CSEL* 50, 75-76). Similarly in *Quaestio* 117 (*CSEL* 50, 351-55) (citation).

[106] The passage in Paul is Romans 4:10. For other early Christian authors, see *supra*. Jerome agrees with the earlier interpretation (see *PL* 26, 372), but Augustine holds the same view as Ambrosiaster (see *CSEL* 84, 79).

[107] *Quaestio* 12 (*CSEL* 50, 36).

In line with early Christian literature on the matter, Ambrosiaster characterized Jewish law as consisting of a few religious customs only, to wit, the observance of the Sabbath and of food laws, the observation of the new moon, and, above all, the practice of circumcision.[108] These had been given to the Jews not for enjoyment, but as a punishment.[109] After all, had the Jews not always been rebellious troublemakers?[110] To Ambrosiaster it was obvious that such laws could never apply to Christians. In his eyes, the *novum testamentum* itself had made it clear that after Jesus' manifestion in the world, only natural law (Ambrosiaster's *lex naturalis*) continued to matter, and that the observance of Jewish ceremonial law (Ambrosiaster's *lex factorum*) was henceforth superfluous. The *lex divinitatis* was fulfilled by Christ.[111]

These then are the central ideas that permeate Ambrosiaster's writings insofar as they concern the Law. Even the fact that the *Quaestio adversus Iudaeos* is part of the *Quaestiones veteris testamenti* rather than of the *Quaestiones novi testamenti* exemplifies its author's views; Ambrosiaster did not doubt that except for a few isolated commandments, Judaism and Pentateuchal law were, overall, things of the past.

Further discussions of the Law in other late fourth century commentaries on Paul come to conclusions very similar to those first proposed by Victorinus and Ambrosiaster. Explaining in detail the exact meaning of Paul's letter to the Galatians verse by verse, Jerome stressed the importance of faith over against the Law, and asserted that Jewish ceremonial law—or, as he polemically called it, the *lex Pharisaeorum*—need no longer be observed on a daily basis.[112] The only law not made obsolete by the advent of Jesus is natural law; the rest of the Law, by contrast, was henceforth to be understood in purely spiritual terms only.[113]

Like Jerome, Augustine exploited, in his commentaries on Paul, the contrast between *fides* and *opera legis*.[114] In Augustine's view, the

---

[108] *Quaestio* 44.5 -8 (*CSEL* 50, 74-76). Comparable is *Quaestio* 69 (*CSEL* 50, 118f.).

[109] *Quaestio* 44.9 (*CSEL* 50, 77). And see *Quaestio* 8.1 (*CSEL* 50, 32); *Quaestio* 69.2 (*CSEL* 50, 119); *Quaestio* 75.2 (*CSEL* 50, 469) and *Quaestio* 76 (*CSEL* 50, 470).

[110] *Quaestio* 44.4 and 44.9 (*CSEL* 50, 73, and 76). Comparable is also *Quaestio* 4.1 (*CSEL* 50, 24-25).

[111] *Quaestio* 44.7 (*CSEL* 50, 77) and, comparably, *Quaestio* 69.4 (*CSEL* 50, 120) and *Quaestio* 19 (*CSEL* 50, 436). In *Quaestio* 44.7 (*CSEL* 50, 75), Ambrosiaster suggests in reference to Jeremiah 4:4 that the only type of circumcision that is desirable is spiritual circumcision ("of the heart"), but he does not develop this idea any further, but see *Quaestio* 69.3 (*CSEL* 50, 119-20). Comparable finally is also *Quaestio* 80-81 (*CSEL* 50, 138): *circumcisio enim signum Iudaismi, non Iudaismus*.

[112] *PL* 26, 368; 359-60; 367; 374.

[113] *PL* 26, 370, and 374.

[114] E.g., *Expositio ad Galatas* (*CSEL* 84, 84). On Augustine and the Jews in general, see Blumenkranz 1946.

works of the law consisted of two parts. The first part comprised Jewish ceremonial observances, such as circumcision, the Sabbath and so forth, *quorum utilitas in intellectu est*. The second part encompassed *mores*, that is, the laws that prohibit murder, adultery, and the like. Not surprisingly, only the laws belonging to this second category were said to have more than mere spiritual significance. Even those who have embraced Christianity are obliged to observe them.[115]

In his commentary on Romans 3:20 and in *Ad Simplicianum*, Augustine once again discussed the Law, but this time his focus was on the historical development of the Law rather than on its contents. He distinguished four phases: before the Law, under the Law, under grace, and in peace. In this scheme, the Law is "good,"[116] essentially because it fulfills a particular and clearly-defined function; the Law caused people to see sin and was given to help them refrain from it.[117] Yet, the Law can only contain sin; it cannot abolish it. Only grace can do that. Thus, grace is essential to reach the final stage *in pace*.[118] What had started as a commentary on Romans 3:20 thus turned into a discussion of Augustine's views on grace.

As a result of the Pelagian controversy, Augustine, in his later works, abandoned the idea that man could reach a state of complete peace in this life.[119] He also refined his doctrine of grace.[120] In this context, it is not necessary to elaborate further on these issues, except to observe that Augustine's basic views on the Law as sketched above did not change over time. When, in 412, he wrote his *De spiritu et littera*, he still believed that the Decalogue contained all the laws necessary to guarantee the well-being of society. Nor did he hesitate to discredit all "the other ordinances" (that is, Jewish ceremonial law) by equating them to "the letter that killeth."[121] Such ideas also surface in his voluminous *Enarationes in Psalmos*, completed in 416 C.E. There Mosaic law is called the *lex Hebraeorum* and is said to postdate natural law; only the New Testament deserved to be regarded as the *lex veritatis*.[122]

In conclusion, then, the "standard" fourth-century Christian attitude towards the legal materials in the Pentateuch as it was formulated against the background of exegetical activity involving Paul's writings can easily be grasped. Most commentators made explicit and recon-

---

[115] *Expositio ad Galatas* (*CSEL* 84, 76-77).
[116] *Expositio ad Romanos* (*CSEL* 84, 7): *Bona ergo lex*.
[117] Comparable is *De spiritu et littera* 8 (*PL* 44, 207).
[118] *Expositio ad Romanos* (*CSEL* 84, 6-9); *Ad Simplicianum* 2f. (*PL* 40, 110f.).
[119] See, e.g., *De peccatorum meritis et remissione etc.* of 412 C.E.
[120] E.g., in his *De spiritu et littera* of 412 C.E.
[121] *De spiritu et littera* 23-24 (*PL* 44, 223-24).
[122] *Enarationes in Psalmos* 57.1 (*CCL* 39, 708).

firmed ideas that had already started to emerge in early Christian literature of the preceding centuries. In much of this literature, fulfillment of the Law in the sense used by Jesus was taken to mean observance of the Decalogue only. Although said to encompass natural law, the Decalogue was ultimately believed to contain a collection of ethical precepts rather than of actual laws. Such an interpretation served to highlight the supposed conceptual differences between a Christian, more spiritual, and a Jewish, practical, tediouss, law. Except for occasional stereotypical references to Jewish religious holidays, circumcision, or the laws regarding food, legal materials other than the Decalogue did not really interest early Christian writers. In their view, the remaining portion of Pentateuchal law had been given specifically to the Jews as punishment. Therefore it did not need to bother non-Jews, especially not since a more sophisticated way of salvation had become available with the advent of Jesus. Thus, in early Christian thinking about Pentateuchal law, the most basic distinction was between the Ten Commandments and the second law. This distinction may strike modern readers as somewhat forced, and, in fact, even in antiquity it did not convince everyone.[123] Although it is true that on some occasions Christians would refer to practices documented in the Hebrew Bible to justify certain decisions, even in these cases, the generally-felt ambivalence towards the "Old Testament" was never far below the surface. In an attempt to escape ordination as a priest, Ammonius, a fifth century Egyptian monk, mutilated his ear and then referred to the Levitical regulations that specify that priests may not have blemishes. That Ammonius should have referred to the Hebrew Bible at all is significant. Yet, it is the words of Ammonius' bishop, Timotheus, that illustrate that few Christians concurred with Ammonius' defense: "That law prevails among the Hebrews. As for me, even if you would bring me a man with his nose cut off, I will ordain him, when he deserves it on the basis of his morals."[124]

Why is the Collatio not a Christian Work?

In the previous section, it has been argued that, except for the Decalogue, most early Christian writers did not elaborate systematically on the legal materials contained in the Hebrew Bible. It has also been pointed out that this selective approach was paramount especially in

---

[123] E.g., Julian (*Contra Galilaeos* 351a) who specifically discusses Matthew 5:17, and see Cyril, *Contra Julianum* 351a-b (*PG* 76, 1041a-b). On Celsus, Origen, and Josephus, see now L. H. Feldman, "Origen's *Contra Celsum* and Josephus' *Contra Apionem*," *Vigiliae Christianae* 44 (1990) 105-35.

[124] Palladius, *Lausiac History* 11, 2-3 (ed. Bartelink, 52). See Leviticus 21:17-23.

the second half of the fourth century, that is, in the same general period that the *Collatio* was in use. As a result of the publication of several volumes in a new series entitled *Biblia Patristica*, it is now possible to further document statistically the fundamental differences between early Christian authors and the author of the *Collatio* in their approach to the legal materials in the Torah.

The *Biblia Patristica* consists of long lists that indicate when passages from the Hebrew Bible are referred to in patristic literature.[125] In the following table, I have indicated the frequency with which the biblical passages cited in the *Collatio* occur in early Christian literature.

Table 1: Biblical passages in the *Collatio* and in early Christian literature

Collatio	1	2	3	4	5	6	7
Ex. 20:13	27	29	22	15	9	3	*19*
Ex. 20:16	(27)	3	1	9(8)	1	2	*7*
Ex. 21:16	—	—	—	1	1	—	*3*
Ex. 21:18-19	—	—	—	1	3	—	*1*
Ex. 21:20-21	—	—	—	(1)	—	—	*3*
Ex. 22:1-3	3	—	6	1	—	—	*13*
Ex. 22:2-3	—	—	(2)	1	—	—	*(9)*
Ex. 22:6	—	—	(1)	—	—	—	*1*
Ex. 22:7-8	—	—	(1)	—	—	—	*1*
Ex. 22:16-17	—	—	(1)	—	—	—	*4*
Lev. 20:10	5	7	9	5	—	—	*7*
Lev. 20:11-12	3	(2)	1	2	—	—	*2*
Num. 27:1-11	—	3	17	—	—	—	*12*
Num. 35:16,17, 20-21	—	—	2	—	2	—	*4*
Num. 35: 22-25	1	—	—	—	—	—	*10*
Deut. 18:10-13	—	2	5	1	2	—	*7*
Deut. 19:14	—	1	2	1	1	—	*2*
Deut. 19:16-20	—	1	2	—	—	—	*1*
Deut. 22:18	1	—	—	—	—	—	*1*
Deut. 24:7	—	—	1	1	—	—	*3*
Deut. 27:20,21, 22-23	1	1	2	1	—	—	*3*

---

[125] J. Allenbach *et al.*, *Biblia patristica. Index des citations et allusions bibliques dans la littérature patristique. 1. Des origines à Clement d'Alexandrie et Tertullien* (Paris: CNRS, 1986); iid., *Biblia patristica. Index des citations et allusions bibliques dans la littérature patristique. 2. Le troisième siècle (Origène excepté)* (Paris: CNRS, 1986); iid., *Biblia patristica. Index des citations et allusions bibliques dans la littérature patristique. 3. Origène* (Paris: CNRS, 1980); iid., *Biblia patristica. Index des citations et allusions bibliques dans la littérature patristique. 4. Eusèbe de Césarée, Cyrille de Jérusalem, Épiphanie de Salamine* (Paris: CNRS, 1987); iid., *Biblia patristica. Index des citations et allusions bibliques dans la littérature patristique. 5. Basile de Césarée, Grégoire de Nazianze, Grégoire de Nysse, Amophiloque d'Iconium* (Paris: CNRS, 1991); iid., *Biblia patristica. Supplément. Philon d' Alexandrie* (Paris: CNRS, 1982).

Explanation: 1. Early Christian authors from Clement to Tertullian (*Biblia Patristica* 1); 2. Early Christian authors of the third century (*Biblia patristica* 2); 3. Origen (*Biblia Patristica* 3); 4. Eusebius, Cyril and Epiphanius (*Biblia Patristica* 4); 5. Basil of Caesarea, Gregory of Nazianzus, Gregory of Nyssa and Amphilochius (*Biblia Patristica* 5); 6. Ambrosiaster, *Quaestiones* (*CSEL* 50); 7. Philo (*Biblia Patristica, supplément*). Since early Christian authors sometimes discuss a long passage rather than the specific passage referred to in the *Collatio* (e.g., Ex. 20:1-17 instead of Ex. 20:13) the same early Christian citation has sometimes been included twice in Table 1. In such cases, figures have been bracketed.

Table 1 illustrates that of all the biblical passages included in the *Collatio* only two passages sparked real interest on the part of early Christian authors: Exodus 20:13 and Exodus 20:16. Both these passages concern commandments from the Decalogue, namely "You shall not kill" and "You shall not bear false witness against your neighbor." One further passage included in the *Collatio* is likewise referred to relatively often in patristic literature, and in particular in the works of Origen: Leviticus 20:10. Leviticus 20:10 determines than an adulterer and an adulteress shall be put to death. Thus this passage from Leviticus must be seen as elaborating on one of the commandments that is also included in the Decalogue.[126] What is perhaps most striking about Table 1 is that all the other Pentateuchal passages that together determine the basic structure of the *Collatio*, on the other hand, are hardly ever referred to in early Christian literature. The paucity of references is particularly evident when one keeps in mind that the figures in Table 1 are based on a cross-section of a representative number of early Christian writings. No less important, even when occasional references to Pentateuchal passages occur in this literature, a short discussion of these materials commonly serves as an illustration of typically Christian ideas or views. The laws referred to are never studied in their own right, and there is no early Christian author among those included here who would ever have taken a selection of regulations belonging to the second law to determine the structure of the specific argument he had in mind. Thus, to maintain, as do Smits, Wenger, Lauria, and Liebs, that Christians regarded the Hebrew Bible as a holy book and that, therefore, the author of the *Collatio* may very well have been Christian simplifies matters too much.[127] The wealth of evidence presented here shows that in the fourth century, as in earlier centuries, Christians were far from accepting wholeheartedly every single legal guideline specified in the Hebrew Bible. The only legal

---

[126] Origen's interest in this particular passage resulted from his preoccupation with the theological reasons underlying early Christian ascetic practice.

[127] *Contra* Smits 1934, 171, Wenger 1953, 547, and Liebs 1987, 164-65. Liebs refers, among other writings, to Sulpicius Severus' treatment of Moses in his *Chronicles*. As A. Lavertujon, *La chronique de Sulpice Sévère. Texte, critique, traduction et commentaire* (Paris: Hachette, 1896) 195 already observed, however, the distance between Sulpicius and the *Collatio* is "incommensurable."

materials from the Hebrew Bible that really attracted their attention were the Ten Commandments.

The highly-selective character of Christian attitudes towards the legal materials in the Torah is the main argument that makes a Christian authorship of the *Collatio* unlikely. But there are other arguments too that militate against Christian involvement in the composition of the *Collatio*. Most significant among these is the fact that the *Collatio* shows no trace of an allegiance to or even knowledge of Christianity. Especially when placed against the larger background of Late Ancient legal writings, it is hard to believe, that if the author of the *Collatio* had been a Christian, he would have dispensed with references to Christian ideas about the legal implications of Jesus' appearance among humankind. Two examples, both from the eastern part of the Mediterranean, may serve to illustrate this point.

From the second half of the fifth century onwards, a collection of laws was used in the eastern part of the Mediterranean which is known to the modern scholarly world as the Syrian-Roman Law Book. It has survived in at least three recensions. It is not entirely clear what standing this law book enjoyed. To judge from the language in which it was written (Syriac) and from the laws it contains, it seems that it served primarily on a local level to help solve the common everyday legal problems as they would arise in the daily lives of inhabitants of the eastern Roman provinces. Roman law was clearly a major source of inspiration for the author(s) of the Syrian-Roman Law Book, yet, curiously enough, the most significant Roman law codices of Late Antiquity (*Codex Theodosianus*, the Gregorian and Hermogenian codes) are never referred to. We do not know who was responsible for the composition of the Syrian-Roman Law book. In view of the steadily-increasing involvement of the Church and particularly of Christian bishops in legal matters in Late Antiquity, however, it is conceivable that the Syrian-Roman Law book originated in ecclesiastical circles.[128]

Like the *Collatio*, the Syrian-Roman Law Book addressed questions of criminal law, among other things. Despite this rather superficial similarity, there are, however, significant differences between these collections of law. As can best be seen in the introductory statements preserved in the second recension, the Syrian-Roman Law Book takes a decidedly Christian view of history and of law.[129] According to this introduction, Moses was the first real Lawgiver on earth. His leg-

---

[128] K. G. Bruns and E. Sachau, *Syrisch-Römisches Rechtsbuch aus dem fünften Jahrhundert* (Leipzig: Brockhaus, 1880) and E. Sachau, *Syrische Rechtsbücher* 1 (Berlin: Reimer, 1907); for the dating of this document, see Bruns and Sachau 1880, 318-19 and Sachau 1907, VIII-IX. And see now W. Selb, *Sententiae Syriacae* (Vienna: Verlag der österreichischen Akademie der Wissenschaften, 1990).

[129] Sachau (previous note), 46-49, who also provides a German translation.

islative efforts were believed to have profoundly influenced all subsequent attempts by others to develop legal systems. Then came Jesus, the Messiah. He abolished all existing laws, and he gave the world only one other law instead: the law of the Messiah. The author of the Syrian-Roman Law Book never tells us what exactly this messianic law is supposed to entail. Assuming that his or her readers understood, the author merely asserted that the Christian Church was instrumental in keeping alive "the law of the Messiah." He or she also asserted that the Church was responsible for transmitting this new law to Christian emperors and that, henceforth, it was to be the task of these "victorious and Christian kings" to guarantee it. The anonymous author of these short and somewhat puzzling remarks was at pains to stress that several emperors, beginning with Constantine, had indeed lived up to that obligation already.[130]

It is hard to imagine that the author of the *Collatio* would have liked the introductory remarks of the Syrian-Roman Law Book very much. The *Collatio* never even once refers to Christianity, let alone acknowledges openly that Jesus is the world's Messiah. The Syrian-Roman Law Book, on the other hand, takes a very different approach. It explicitly includes a discussion of the most basic tenets of Christianity, even though such matters have little or no bearing on the actual legal discussions included in this collection.[131]

In other legal writings from this part of the Mediterranean world composed in the sixth century, ideas such as those expressed in the Syrian-Roman Law Book found an ever-stronger expression. Mar Abha, a patriarch who lived in Mesopotamia around the middle of the sixth century, for example, wrote a treatise dealing with marriage.[132] It was essentially based on the marriage laws of Leviticus 18 and 20. That Mar Abha had a great respect for biblical law indeed follows not only from the passages selected, but also from the phraseology permeating his little work. Yet, significantly, Mar Abha did not follow the injunctions in Leviticus slavishly. In fact, the main reason for dealing with this issue in this way at all was that he intended to "purify" Christian marital laws by isolating Jewish, pagan, and magical elements that had crept into traditional Christian practice long ago.[133] Thus, ultimately, Mar Abha's treatise reflects the same ambivalence towards biblical law we have encountered in much of early Christian literature: the "Old Testament" contains many interesting regulations

---

[130] Sachau (previous note), 46-49.

[131] In fact, no references to biblical law can be found in the actual laws of any of the three recensions. See also Volterra 1930, 41-44, 88.

[132] E. Sachau, *Römische Rechtsbücher* 3 (Berlin: Reimer, 1914), esp. 258f.

[133] Sachau 1914, XXII.

worth looking at, but none of these regulations can automatically or by definition be regarded as the final word on certain matters.

Mar Abha's approach, then, differs fundamentally from that exemplified by the *Collatio*. For the author of the *Collatio* there was no other law than the Pentateuch (except, of course, for Roman law). In addition, identifiably Christian concerns cannot be detected anywhere in the *Collatio*. Mar Abha's treatise, on the other hand, reflects an entirely different world. His decision to write on marriage was triggered precisely by his interest in the well-being of his fellow Christians. The Pentateuch served as a point of departure, but, to be sure, it was used only selectively and under great scrutiny. For Mar Abha, it would not have been conceivable to compose a work along the lines enunciated by the *Collatio*.

Another argument that has often been adduced to support a Christian authorship of the *Collatio* is based on questionable assumptions concerning the purpose of the *Collatio*. Both Smits and Liebs regard the *Collatio* as a piece of Christian propaganda, aimed at the conversion of the pagan population of the empire, and in particular of those with strong interests in Roman law. According to this view, the *Collatio* was meant to remove prejudices of a juristic nature by showing that Mosaic law had anticipated Roman law, and by documenting that the same legal principles were operative in both legal systems.[134] It is obvious that such an explanation is based on the presupposition that the jurists' professional interests and legal training would somehow frustrate their conversion to Christianity. Does such a supposition make sense?

In Late Antiquity, jurists—like anyone else—converted to Christianity for any number of reasons. For some it may have been a matter of political expediency, while for others, religious or other motifs may have been paramount (one reason does not, of course, automatically exclude one or more other reasons). As far as their legislative efforts were concerned, however, Christian jurists and lawgivers remained heavily indebted to traditional Roman legal practice. To speak of a *diritto romano cristiano* is misleading. Even under Justinian the Roman legal traditions of old remained the single most determining factor.[135] I have not found any evidence to suggest that this continued centrality of Roman non-Christian law in Christian Late Antiquity brought people in a state of complete religious confusion or prevented

---

[134] Smits 1934, 177-78; Liebs 1987, 170. Less explicit, but arguing along similar lines, is Wenger 1953, 547.

[135] For a recent discussion concerning the *status quaestionis* of the ideas proposed by B. Biondi, see G. Grifò, "Romanizzazione e Cristianizzazione. Certezze e dubbi in tema di rapporto tra Cristiani e istituzioni," in G. Bonamente and A. Nestori (eds.), *I Cristiani e l'impero nel IV secolo* (Macerata: Università degli studi, 1988) 75-106. For the importance of traditional Roman law in the late fourth century, see also Liebs 1987, 104-19.

them from conversion to Christianity. More likely, it was a *fait accompli* that Christian lawgivers took for granted. In addition, even if we try to regard the *Collatio* as a product of Late Ancient Christian propaganda aimed at pagans with a strong interest in law, the work fails to make sense. After all, with all its little inconsistencies, from a legal point of view, the *Collatio* was far from an impressive work.[136] Unless the reader of the *Collatio* already took a positive view of Pentateuchal law, it is hard to believe that he or she would have been very impressed by the concisely formulated Pentateuchal materials, as they contrast markedly with the more detailed regulations contained in Roman juristic works.

A final argument that has sometimes been adduced to argue that the author of the *Collatio* was a Christian is based on a few medieval references to the *Collatio*. Hincmar, archbishop of Reims, for example, refers to the *Collatio* twice in his *De divortio Lotharii regis et Thetbergae reginae*, dating to the year 860. A few other references from the same general period can also be added.[137] Interesting though such references may be, the fact that medieval writers occasionally used the *Collatio,* and, more specifically, the Roman legal materials transmitted by the *Collatio*, does not at all prove, in my view, that the *Collatio* was originally a Christian work. Not only do I disagree in principle with this part of Liebs' method of solving the authorship question of the *Collatio* as it consists of casual references to isolated medieval sources that reflect a very different historic reality;[138] I simply fail to see how medieval usage of the *Collatio* can tell us anything about either the author or the original purpose of this Late Ancient legal treatise.

In conclusion, it is fair to say that all the evidence presented thus far does not make Christian authorship of the *Collatio* very probable. Not only does the *Collatio* lack explicit references to Christianity, the entire approach to Pentateuchal law as reflected by the structural characteristics of this work differs substantially from contemporary Christian attitudes towards Pentateuchal law. In order to understand better the particular treatment of legal materials in the *Collatio* and to explain convincingly the rationale behind the work as a whole, it is necessary to consider another possibility, namely that the *Collatio* is a Jewish work.[139]

---

[136] For an analysis of the differences between the Pentateuchal law and Roman law by individual title, see Smits 1934, 75f.

[137] E.g. Wenger 1953, and Liebs 1987, 174, who has the most important references.

[138] *Contra* Liebs 1987, 165, 170, and 174.

[139] In his dicussion, of why the *Collatio* could never have been composed by a Christian, Volterra 1930, 96f. points out that only laws promulgated under pagan, and no laws promulgated by Christian, emperors have been included. It should be stressed that in Volterra's case, this argument is circular: If one maintains, as does Volterra, that the

The Pentateuch in Jewish Thought

There are two main reasons, in my view, why the *Collatio* must be regarded as a Jewish work. First, the treatment of legal materials in the *Collatio* reflects an attitude towards biblical law that is reminiscent of contemporary Jewish attitudes towards such materials. Second, only by regarding the *Collatio* as a Jewish work does it become possible to understand why this unique collection of Mosaic and Roman laws was written.

Information concerning Jewish attitudes towards the Torah can be found in the great variety of Jewish writings dating to the first few centuries of the Common Era. For the present purpose it is not necessary to embark on a systematic description of such attitudes. To show the plausibility that the *Collatio* was written by a Jewish author, we have merely to show that the strong dichotomy which early Christians believed existed between the Decalogue and the rest of the Torah is not a feature in the different kinds of Jewish writings composed during the first few centuries of the Common Era. Here, a brief discussion of this issue in Philo, Josephus, and rabbinic literature will serve to illustrate Jewish attitudes of Pentateuchal law.[140]

Philo of Alexandria's commentaries on Scripture are limited to the first five books of the Hebrew Bible. Of all the legal materials contained in these five books, Philo attributed the most special significance to the Decalogue. Philo's interest in the Decalogue was in fact such that it resulted in a separate, monographic discussion of the Ten Commandments, known as *De Decalogo*.[141] A combination of considerations may have triggered Philo's interest in the Decalogue. One reason why the Decalogue held such a special place in his conceptualization of the Torah can be attributed to his belief that the entire Decalogue had been given by God in person—in contrast to all the other

---

Collatio dates to the years 314-324 C.E., one cannot attach much importance to the absence of laws that, by and large, were promulgated after this period.

[140] It is true that both Philo and Josephus predate the *Collatio* by several centuries. Short discussions of their treatment of the Torah have been included to give greater depth to an account that would otherwise be based exclusively on attitudes derived from rabbinic literature. Hellenistic Jewish authors wrote extensively on Jewish law and on Moses, but they tend to focus on the non-halakhic parts of the Torah, see A. Terian, "Some Stock Arguments for the Magnanimity of the Law in Hellenistic Jewish Apologetics," *Jewish Law Association Studies* I (1985) 141-49 and P. W. van der Horst, "The Interpretation of the Bible by the Minor Hellenistic Jewish Authors," in M. J. Mulder and H. Sysling (eds.), *Mikra. Text, Translation, Reading and Interpetation of the Hebrew Bible in Ancient Judaism and Early Christianity* (Assen and Philadelphia: Van Gorcum-Fortress, 1988) 519-44.

[141] Y. Amir, "Die Zehn Gebote bei Philon von Alexandrien," and *id.*, "Mose als Verfasser der Tora bei Philon" in *id.*, *Die hellenistische Gestalt des Judentums bei Philon von Alexandrien* (Neukirchen-Vluyn: Neukirchener Verlag, 1983) 77- 106 and 131-63.

divine commandments that had reached humankind only indirectly, through Moses.[142] Another likely reason for Philo's fascination with the Commandments may be sought in his preoccupation with allegory and numerical symbolism. Further reasons could still be added.[143]

What matters most in the present context, however, is to observe that, engrossed as was Philo in his study of the Decalogue, his attraction to the commandments of the Decalogue never led him to question the validity of the other legal materials included in the Torah. Rather, the opposite was the case. In Philo's view, each Commandment of the Decalogue represented *in nuce* a whole series of legal injunctions that were specified in greater detail in other parts of the Torah. In an attempt to support this contention, Philo set out to link systematically laws from all parts of the Torah to those Commandments in the Decalogue most appropriate for such a linkage. The result of his efforts has survived in his lengthy treatise *De Specialibus Legibus*. Even more than the *De Decalogo*, this corpus provides us with an eloquent testimony to Philo's conviction that the Decalogue was a summary of all the other laws of the Pentateuch. Thus, Philo's approach to the legal materials in the Torah differed fundamentally from early Christian attitudes towards such materials. For Philo, the Decalogue was a *pars pro toto*, a summary of a much longer and much more detailed divine law, which permeated the entire Torah. Early Christian writers, by contrast, regarded the Decalogue as the only legally binding segment of the entire Torah. In their view, the Decalogue was nothing but a straightforward formulation of a few basic behavioral rules lacking more ramified consequences.

That Philo and early Christian authors were motivated by very different concerns in their discussions of the Torah can finally also be seen in Table 1 (above). In this table, I tried to establish how often early Christian writers referred to or discussed the Pentateuchal passages included in the *Collatio*. It was shown that there were only two or three legal passages of the Torah in which early Christian writers were really interested, and that all of them pertained to the Decalogue. When we now look at the Philonic references to the biblical passages included in the *Collatio* (Table 1, column 7), a different picture emerges. Where there are whole stretches in early Christian literature which lack any reference whatsoever to the passages in question,

---

[142] Philo, *Decal.* 18. See also Amir (previous note) 88-89.

[143] In his analysis of *De specialibus legibus*, E. R. Goodenough, *The Jurisprudence of the Jewish Courts in Egypt. Legal Administration by the Jews under the early Roman Empire as Described by Philo Judaeus* (New Haven: Yale U. P., 1929) 14f. maintains that Philo's interest in the laws of the Torah was triggered by a practical concern, namely to make the Torah conform to foreign jurisprudence, but Goodenough never presents evidence to support his hypothesis that Philo's concern was to show that (p. 214) "the Mosaic code was the supreme code for practical administration."

Philo, a single Jewish author, discusses every one of these passages in one or another of his writings—and he often does so on more than one occasion. On the basis of these data our previous observation can be reconfirmed: it was the Torah in its entirety that mattered to Philo, and not just part of it.[144]

Although Josephus was a man with a different background, different interests and, generally speaking, someone who was writing for a different kind of audience, he did not differ very much from Philo insofar as his basic views on the Law are concerned. On three occasions, Josephus inserted extensive discussions of the Law in his works. They include books three and four of his *Jewish Antiquities* and the second book of *Against Apion*.[145]

That these discussions should appear at all in Josephus' writings is highly significant. As L. H. Feldman has recently reminded us, while Josephus' *Jewish Antiquities* has much in common with the works of the great Greek and Roman historiographers, non-Jewish historiographers never dwell in much detail on questions of legal history.[146] Thus, the inclusion of substantial surveys of legal materials in Josephus' writings *in and by itself* indicates the importance Josephus attached to the Law when he set out to describe the features he believed to be most characteristic of the Jewish people.

Individual passages as well as the treatment of biblical laws in general further confirm that Josephus, like Philo, believed that the laws of the Torah were relevant in their entirety.[147] According to Josephus' rather apologetic scheme of things, the laws of the Torah ruled every aspect of daily life, and great care was being taken to ensure that the Law should not be forgotten.[148] Even the Torah scrolls were kept in special reverence, as may be inferred from his mention of the public execution of a Roman soldier who had been unwise enough to destroy deliberately a copy of the Torah.[149] Considering scrolls as sacred be-

---

[144] See, e.g., Philo, *Migr.* 89-93.

[145] Josephus, *AJ* 3.224-86; 4.67-75 and 199-301; *Ap.* 2.151-295.

[146] L. H. Feldman, "Use, Authority and Exegesis in the Writings of Josephus," in M. J. Mulder and H. Sysling (eds.), *Mikra. Text, Translation, Reading and Interpretation of the Hebrew Bible in Ancient Judaism and Early Christianity* (Assen and Philadelphia: Van Gorcum-Fortress, 1988) 455-518, at 510-13. And see D. A. Altschuler, *Descriptions in Josephus' Antiquities of the Mosaic Institutions* (Ph.D. Dissertation, Hebrew Union College-Jewish Institute of Religion: Cincinatti, 1977) and D. Goldenberg, *The Halakhah in Josephus and in Tannaitic Literature: A Comparative Study* (Ph.D. Dissertation, Dropsie College: Philadelphia, 1978).

[147] Note, for example, Josephus' insistence on the fact "that nothing has been added to the law for the sake of embellishment," in *AJ* 4.196.

[148] Josephus, *Ap.* 2.173-74; Josephus, *Ap.* 2.175.

[149] Josephus, *AJ* 20.115.

cause the writings they contained were considered sacred was a typically Jewish concept. The Romans did not know it.[150]

Given Josephus' approach to Pentateuchal law, it should not surprise us that some of the Pentateuchal passages included in the *Collatio* also appear in Josephus' writings.[151] As has been pointed out in connection with Philo, such evidence can and must be used to argue that the Pentateuchal materials contained in the *Collatio* represent passages that mattered to Jewish authors in ways that left early Christian authors indifferent.

In short, then, it is characteristic of Josephus' treatment of Pentateuchal law that he, like Philo before him, never subdivided such laws into separate groupings of relevant and irrelevant laws. For all his rearranging of individual laws,[152] Josephus never doubted that the laws of the Torah represented a coherent whole. The Torah was to be observed *in toto*. In Josephus's eyes, it was exactly this kind of observance that made Jews Jews.[153]

Rabbinic attitudes towards the Torah can finally be said to have much in common with the ideas expressed by Philo and Josephus, and in at least one respect all three express unusual commonality: they considered the Torah as a coherent complex of laws in which nothing was superfluous; significance inhered in even the minutest of details.[154] Passages that illustrate both the sanctity and the paramount importance attributed to the Torah (and the Hebrew Scriptures in general) abound in rabbinic writings. One tannaitic tradition held, for example, that the Torah was from heaven, and added that everyone who denied it would have no share in the world to come.[155] Another, roughly contemporary, rabbinic tradition stressed that the Torah had been an instrumental tool in the creation of the world.[156] But rather than by citing such individual instances, the deep respect for the Torah on the part of generations of rabbinic sages can perhaps best be illustrated by the general fact that, as Stemberger has put it, "rabbinic literature arose mostly out of the attempt to adapt the Torah as the Jew-

---

[150] This has been pointed out recently by M. Goodman, "Sacred Scripture and 'Defiling the Hands'," *Journal of Theological Studies* 41 (1990) 99-107, at 103.

[151] Josephus, *AJ* 3.274 = *Collatio* 4; *AJ* 4.219 = *Collatio* 8; *AJ* 4.225 = *Collatio* 13; *AJ* 4.277 = *Collatio* 2; and *AJ* 4.279 = *Collatio* 4.

[152] See Josephus, *AJ* 4.196.

[153] E.g., Josephus, *AJ* 3.222-23.

[154] Volterra has argued on the basis of *Collatio* 4.1 that its author was familiar with rabbinic traditions, but this is not correct; see the discussion in Smits 1934, 166 and see F. Schulz, "Die biblischen Texte in der Collatio legum Mosaicarum et Romanarum," *Studia et Documenta Historiae et Iuris* 2 (1936) 20-43, at 29-30.

[155] *Mishnah, Sanhedrin* 10:1.

[156] *Mishnah, Avot* 3:14.

ish rule of life to changing conditions."[157] Concern with and deep respect for the Hebrew Bible is evident in the redactional methods (the prooftexting) employed in both Talmuds (*Gemara*), as well as in the exegetical concerns that stand at the basis of *Midrash*.

In only one document in the large corpus of rabbinic writings, that is the *Mishnah*, the Hebrew Bible is used in a limited and very specialized way (the same holds true for the *Tosefta*). The limited use of the Hebrew Scriptures either as structuring principle or as authoritative source to justify the *Mishnah's* often apodictically-formulated *halakhot* has given rise to a variety of interpretations concerning the exact relationship of the *Mishnah* to the Hebrew Bible.[158] Whatever the reason(s) for this relative independence from the Hebrew Bible may be, it is important to note here that even the *Mishnah* never explicitly questions the authority of Scripture in a way that would even vaguely remind us of the ambivalence towards the Hebrew Bible reflected in contemporary early Christian writings. In the intellectual world of the framers of the *Mishnah*, the events at Sinai remained a milestone in human history, during which written and oral traditions had been revealed and to which they, the anonymous compilers of the *Mishnah*, consciously presented themselves as direct heirs.[159]

The differences in attitudes towards the Hebrew Bible in rabbinic as opposed to early Christian writings can finally also be illustrated by a short discussion of one passage in rabbinic literature that focuses on the Decalogue. In this passage, two different views are being advanced. R. Brooks, who discusses only the evidence from the Jerusalem Talmud, characterizes them as "two apparently contradictory themes," while E. Urbach, whose discussion includes passages from the Babylonian Talmud as well, tends to regard them more as different stages in an ongoing discussion.[160]

The first argument in the passage under discussion elaborates on the Decalogue in conjunction with the *Shema*, the prayer that proclaims the oneness of God.[161] In a dispute with R. Simon, R. Levi asserted

---

[157] H. L. Strack and G. Stemberger, *Introduction to the Talmud and Midrash* (Edinburgh: T&T Clark, 1991) 18.

[158] Discussion and further references in Strack and Stemberger 1991, 141-45. And see D. Weiss Halivni, *Midrash, Mishnah, and Gemara. The Jewish Predeliction for Justified Law* (Cambridge Mass. and London: Harvard U. P., 1986). And see what the *Mishnah* itself has to say in *Hagigah* 1:8.

[159] *Mishnah, Avot* 1:1-12.

[160] R. Brooks, *The Spirit of the Ten Commandments. Shattering the Myth of Rabbinic Legalism* (San Francisco: Harper & Row, 1990) 30f. and E. E. Urbach, "The Role of the Ten Commandments in Jewish Worship," in B. Segal and G. Levi (eds.), *The Ten Commandments in History and Tradition* (Jerusalem: Magness Press 1990) 161-189 and see E. E. Urbach, *The Sages. Their Concepts and Beliefs* (Cambridge Mass. and London: Harvard U. P., 1975) 360f.

[161] *Jerusalem Talmud, Berakhot* 1:5 (see *Babylonian Talmud* 9a-b).

that Jews should recite the *Shema* twice daily "because the Ten Utterances are contained in them [that is in the *Shema*]." According to this view, the *Shema* and the Ten Commandments were seen as organically related. Reciting the one implied that one was simultaneously reciting the other. The result of presenting the *Shema* and the Decalogue in this way was that the Decalogue received emphasis to such an extent that it appeared as foundational to Judaism.[162]

The second rabbinic argument, contained in the same Talmudic passage, presents a very different line of reasoning. It also focusses on the Decalogue, but this time the discussion centers on intentionally removing the Ten Commandments from the center of attention and devotion. An indication for the rationale behind this move is also included: if the Decalogue were recited, *minim* (heretics) could argue that only the Ten Commandments were given to Moses on Mount Sinai.[163] While the Decalogue had thus appeared as set of regulations that was quintessential to Judaism in the first half of this discussion, the second half proceeds seriously to downplay its centrality.

Brooks and Urbach explain differently the reasons underlying the rabbinic explanation of why the importance of the Decalogue was deemphasized. According to Brooks, the Decalogue was excluded from daily worship to guarantee the integrity of the belief in the dual Torah (that is the belief that there was a written as well as an oral Torah, that had both been revealed synchronically at Sinai): the explicit eulogizing of the written Torah could easily result in a general undervaluation of the oral Torah.[164] Urbach also explains the passage as reflecting internal Jewish concerns, yet he maintains that the Decalogue was removed from the center of attention because too great an emphasis on it alone might lead to the neglect of many *mitzvot* not explicitly mentioned in the Decalogue.[165]

The difficulty in explaining this passage is that we do not really know who these *minim* were. Contrary to what Brooks and Urbach maintain, it is quite possible, however, that the argument put forward by the rabbinic sages did not merely arise out of disputes that were carried on in exclusively Jewish circles. In the light of the above discussion of early Christian attitudes towards the Pentateuch, I believe that we have every reason to identify the *minim* in question with Christians rather than with some vaguely defined group of Jewish "dissidents."[166] It was characteristically Christian to downplay the im-

---

[162] Here I follow Brooks (above), 34.
[163] *Jerusalem Talmud, Berakhot* 1:5 (see *Babylonian Talmud* 9b).
[164] Brooks 1990, 35.
[165] Urbach 1990, 176-181.
[166] *Contra* Urbach 1990, 176 who denies that this is a possibility.

portance of the legal materials of the Torah and to focus on the Decalogue instead.

Whatever the identification of these *minim* may be, the above-mentioned rabbinic dispute regarding the Decalogue and the fact that discussions of the Decalogue are sporadic in rabbinic literature at large, document, once again, the essential difference between Jewish and early Christian attitudes towards the Hebrew Bible in general, and towards the legal materials in the Torah in particular. While Christians glossed over most of the materials except for the Decalogue, Jews refused, consciously and unconsciously, to subdivide their scriptures into relevant and redundant sections. For example, the author(s) of the *Midrash Yelamdenu* purposely equated those who maintained that only the Decalogue had been revealed at Sinai with the sons of Korah, thus suggesting that everyone who dared to hold such a view would invariably perish.[167] A typical early Christian reaction to contradictions between individual biblical laws was that such contradictory statements showed that biblical law in its entirety was defunct.[168] In rabbinic exegesis, by contrast, contradictions between scriptural passages were not seen as indicating the futility of the Hebrew Bible as a whole. Rather, such contradictions were made to yield a deeper meaning on the basis of a highly sophisticated set of hermeneutical rules.[169] According to the rabbinic point of view, even at the end of times, the Torah would not be abolished entirely. Rather, while several changes would be made in it, this last stage in human history held above all the promise that human beings would finally gain greater insight into the principles underlying the Torah's commandments.[170]

In this short survey of Jewish attitudes towards the Torah, I have shown that Philo, Josephus, and rabbinic literature agree in their discussions of legal materials in the Hebrew Bible in at least one essential aspect: their respect for the integrity of Scripture and their conscious or unconscious refusal to reduce Judaism to a simple set of a few isolated dogmas only. In my view, the choice of passages from the Torah and the importance attached to them indicate that the author(s) of the *Collatio* must have held views very similar to those ar-

---

[167] For references and discussion, see Urbach 1990, 176-78. The dating of this *midrash* is problematic, see Strack and Stemberger 1991, 332-33.

[168] Early Christian authors were eager to point out that circumcision on the eighth day and the observance of the sabbath could not always both be observed, and concluded that both practices were therefore mutually exclusive, e.g., Ephrem, *Hymnen in Ecclesia* 43.12-20 (*CSCO* 198, 109-10 and *CSCO* 199, 104-5); also typical is Zeno, *Tractatus* 1.3.2.3-4 (*CCL* 22, 24): "because one of them [i.e. the sabbath or circumcision] cannot be accomplished, they are both worthless."

[169] On the various kinds of *middot*, see the concise presentation in Strack and Stemberger 1991, 17-34.

[170] P. Schäfer, *Studien zur Geschichte und Theologie des rabbinischen Judentums* (Leiden: Brill, 1978) 198-213.

ticulated more fully by Philo, Josephus, and by the rabbinic sages. I believe, therefore, that it is not merely accidental that the author of the *Collatio* saw the Pentateuch as a *lex divina* or a *scriptura divina*.[171] Also in this respect, he (or she, or they?) had much in common with the rabbinic sages, who customarily referred to scripture as *kitvei ha-qodesh*.[172]

In contemporary early Christian literature, by contrast, the kind of respect for the Hebrew Bible as expressed in Jewish writings is by and large absent. Given the strong interest in Paul in the later fourth century, it seems in fact practically inconceivable that Christian writers would have ever thought of composing a treatise structured systematically and exclusively around "Old Testament" law. Compared to the interests expressed by the author(s) of the *Collatio*, writers such as Victorinus, Ambrosiaster, Jerome, Augustine, and Pelagius were living in a very different intellectual world indeed.[173]

The Collatio as a Late Ancient Jewish Treatise

In addition to using a number of Pentateuchal passages in a way that has previously been identified as typically Jewish, there is yet another reason why, in my view, the *Collatio* must be seen as the product of a Jewish author: the purpose of the *Collatio* can be explained most convincingly when it is seen as the work of a Jewish writer.

Only once the author(s) of the *Collatio* hints explicitly at the purpose of the work. In *Collatio* 7.1 the author gives a short description of the different ways in which thieves were to be treated according to the *lex duodecim tabularum*. Then, before citing the appropriate parallel passage in Pentateuchal law the authour adds: "know ye jurists, that Moses had previously so ordained." Although the author(s) of the *Collatio* does not elaborate any further on this observation, his systematic juxtaposing of Pentateuchal with Roman law throughout the *Collatio* expresses a very similar concern: the *Collatio* was composed to stress the primacy of Mosaic law and to show that the injunctions of Mosaic law were not at variance with the ordinances of Roman law. That there were good reasons, in the later fourth century, to stress that

---

[171] *Lex divina*: *Collatio* 6.7.1; *scriptura divina*: *Collatio* 16.1.1.

[172] E.g., already in *Mishnah, Shabbath* 16:1 and *Yadayim* 3:2. On one occasion (*Collatio* 1.1) Moses is called *dei sacerdos*. Smits 1934, 75-76 and 176 argues that a Jew would never have called Moses a priest. It is true that both Josephus and rabbinic literature refrain from connecting Moses with the priesthood, but Philo does not, see W. A. Meeks, *The Prophet-King. Moses Traditions and the Johannine Christology* (Leiden: Brill, 1967) 113-15, 117-20, 136-37. I do not think that this reference to Moses as priest makes a Jewish authorship of the *Collatio* difficult.

[173] Schrage 1993, 41 / misses this point completely

traditional Jewish and Roman law were not mutually exclusive becomes evident when we place the *Collatio* against the larger background of theological and legal discussions of the period.

We have seen that the *Collatio* was used (or composed, if one accepts a dating in the late fourth century) in a period, that was characterized, among other things, by an intense debate about Mosaic law. This debate gained momentum in the second half of the fourth century when several Christian writers turned their attention to the exegesis of the works of Paul. It also attracted the attention of people other than those directly involved in studying Paul. In his *Contra Galilaeos*, the emperor Julian, for example, addressed the issue, attacking his Christian contemporaries for their ambivalent treatment of Mosaic law.

Given such an intellectual climate, it is not surprising that a Jewish author would have joined in this debate by composing a work in which the author demonstrated the extent to which Mosaic regulations Christians believed to be redundant reflected exactly the same penal practices that would later be developed in Roman law. Like Josephus before him, who had represented the Jews as a law-abiding people in the face of a recently defeated revolt of considerable dimensions and in the face of long-standing Roman prejudices against the Jews, the author(s) of the *Collatio* was at pains to stress the legitimacy of traditional Jewish law in a period when this law had come increasingly under attack by Christian theologians. The author(s) of the *Collatio* furthermore had in common with Josephus that he too made minor changes in his rendering of biblical texts in order to represent the commonalities between Jewish and Roman law as convincingly as possible.[174]

Besides participating in a debate that preoccupied the author's non-Jewish contemporaries, there was still a second reason why a Jewish author would have written a work like the *Collatio* precisely in the fourth century, namely Roman interference with internal Jewish jurisdiction.

When the Jews had first come under Roman rule they were permitted to live according to the laws of their forefathers.[175] Such a regulation did not imply that Jews could not also avail themselves of the Roman legal system. The recently-discovered marriage contract which is part of the so-called archive of Babatha shows that there was nothing unusual about Jews having recourse to Roman law. It also

---

[174] These changes were studied in detail by Volterra 1930, 56-77; see also Smits 1934, 75f. and Schulz 1936, 20-43. As Levy 1930, 704 has pointed out, the author of the *Collatio* handles the biblical text freely, but treats Roman law "um so sklavischer." The author of the *Collatio* does not seem to have noticed that there are considerable discrepancies between the various Roman laws he cited in a single title.

[175] T. Rajak, "Was there a Roman Charter for the Jews?" *Journal of Roman Studies* 74 (1984) 107-23.

shows the extent to which, in the provinces of the Roman Empire, regulations characteristic of different legal systems (Jewish, generally "oriental-local," Hellenistic Greek, and Roman) could be intertwined.[176] The fact that there existed an organic relationship between Jewish and Roman legal practices in the case of Babatha's archive cannot be taken to mean, of course, that Jewish and Roman law were always the same. We know from incidents that took place in 204 C.E. and 213 C.E., respectively, that Jewish and Roman law differed in regulations concerning the redeeming of stolen property and in designating the beneficiary to a bequest or an inheritance.[177] In such cases, Roman law took precedence over Jewish law. Nor did the general rule that Jews could live according to their own laws prevent Roman legislators from trying to regulate aspects of internal Jewish jurisdiction. In 393 C.E. they interfered with Jewish marriage customs, prohibiting Jews "to contract nuptials according to their law, or enter into several matrimonies at the same time."[178] A few years later, in 398 C.E., Roman lawgivers determined furthermore that Jewish judicial powers would henceforth be limited to deciding exclusively those cases that were strictly religious—except that they continued to allow litigation in civil matters to take place in Jewish courts as long as both parties agreed on such an arrangement.[179]

Although we do not know why Roman authorities decided to regulate these matters only in the late fourth-century instead of earlier, the promulgation of the two aforementioned laws within a short period of time seems to suggest that in Late Antiquity Roman lawgivers were stepping up their efforts to limit Jewish jurisdiction. Considering such developments, I believe that we have every reason to see in the *Collatio* the work of a Jewish author. Despite the omnipresence of Roman legal materials, the *Collatio* was not conceived by a Roman jurist to serve as a manual of Roman law. Nor was the *Collatio* yet another Christian contribution to the debate on Mosaic law. The *Collatio* was written by a Jew (or Jews) and the decision to include or exclude Ro-

---

[176] See N. Lewis, R. Katzoff and J. C. Greenfield, "*Papyrus Yadin 18*. I. Text, Translation and Notes. II. Legal Commentary. III. The Aramaic Subscription," *Israel Exploration Journal* 37 (1987) 229-50; A. Wasserstein, "A Marriage Contract from the Province of Arabia Nova: Notes on Papyrus Yadin 18," *Jewish Quarterly Review* 80 (1989) 93-130; H. Cotton, "The Guardianship of Jesus son of Babatha: Roman and Local Law in the Province of Arabia," *Journal of Roman Studies* 83 (1993) 94-108.

[177] See the discussion with further references in D. Daube, "Jewish Law in the Hellenistic World," in B. S. Jackson (ed.), *Jewish Law in Legal History and the Modern World* = *The Jewish Law Annual, Supplement* 2 (Leiden: Brill, 1980) 56-58.

[178] *Codex Justinianus* 1.9.7. That Jewish and Roman marriage customs differed in certain respects is also evident in the *Collatio* 4.1; see the discussion in Volterra 1930, 62.

[179] *Codex Theodosianus* 2.1.10, see also A. Linder, *The Jews in Roman Imperial Legislation* (Detroit and Jerusalem: Wayne State U. P. and the Israel Academy of Sciences and Humanities, 1987) 204-11.

man legal materials from it rested on apologetic considerations alone.[180]

Regarding the *Collatio* as a work of fourth-century Jewish apologetics may also bring us a step closer to solving the question of the exact dating of the *Collatio*.[181] Volterra, who first proposed that the *Collatio* be seen as a Jewish apologetic work, suggested that the *Collatio* was composed early in the fourth century. He maintained that certain penalties referred to (such as crucifixion) make impossible a dating of the *Collatio* later in the fourth century.[182]

I disagree with such a line of reasoning for the following reason. As has already been pointed out, the inclusion of a law dating to the years 390-392 C.E. (*Collatio* 5.3.1-2.), shows indisputably that by the early fourth century the *Collatio* had not yet reached its final form. Rather than conceive of this particular law as an interpolation, it would perhaps make more sense to regard the *Collatio* in its entirety as a work of the late fourth century. As we have seen, it was exactly in the late fourth century that the (Christian) debate on Mosaic law was most intense. Moreover, it was only in this same period that it became more and more apparent that the legal position of the Jews was about to change for good—as, in fact, it did in the early fifth century. Finally, it was precisely at this particular period of time—and not earlier—that Church fathers, including Ambrose, Augustine, and Chrysostom, and imperial legislators had come to agree that Jewish and Roman law were categories that were mutually exclusive.[183] Such developments suggest, I believe, that by the 390s good reasons had emerged to compose a work like the *Collatio*. In the 310s, by contrast, such reasons were still largely absent. That the *Collatio* included "outdated" laws including crucifixion is not really surprising, as long as we keep in mind that the author of the *Collatio* was not a trained jurist himself, but rather someone who freely borrowed from whatever Roman juristic writings he could lay his hands on. The author of the *Collatio* was not an expert in Roman law, nor did he want to be. His concerns were purely apologetic.[184]

---

[180] *Contra* Hyamson 1913, XLII who maintains that the *Collatio* "served as an introduction to study Roman law and for the instruction of clerics."

[181] For discussion and references, see *supra*, the section "The Collatio: A Christian or a Jewish Work?"

[182] A convenient summary of the reasons why various scholars have accepted an early date may be found in Cracco Ruggini 1983, 46-47.

[183] Cracco Ruggini 1983, 57-58 for a short discussion and references.

[184] *Contra* Levy 1930, 702-703; Smits 1934, 22-23, Rabello 1967, 348-49 and Cracco Ruggini 1983, 46. These authors maintain that the *Collatio* cannot possibly postdate the early fourth century because it would no longer be affective as an Apologetic treatise; readers would immediately notice that some laws included in the *Collatio* had in the meantime been abolished or replaced by other laws. It should be stressed, however, that this line of reasoning is based on the *assumption* that all such readers had a good knowl-

One hypothesis concerning the authorship of the *Collatio* should finally also be discussed briefly. Although, to my knowledge, this suggestion has never been made, there exists a theoretical possibility that the *Collatio* was written by a Samaritan rather than by a Jewish author. The biblical passages cited in the *Collatio* were all taken from the Pentateuch, that is, from exactly those books considered, in the Samaritan tradition, as the only authoritative part of Scripture.[185] Samaritans regarded themselves as the quintessential keepers of Mosaic law; their writings as well as the name they chose for themselves all bear witness to that conviction. Furthermore, as is evident from the so-called Decalogue inscriptions found in various parts of Roman Palestine, the Ten Commandments held a place of special importance in Samaritan theology. Besides, as has been noted earlier, individual Pentateuchal passages in the *Collatio* often start with the short phrase "Moses says...." Again, this seems to be in line entirely with Samaritan traditions. Moses was a figure of central importance in Samaritan thought; the fact is indeed too well known to require further explanation here.[186]

It is very unfortunate that the Samaritans constitute one of the most elusive ethnic groups of Late Antiquity insofar as their communities outside Samaria are concerned. The Theodosian Code refers to Samaritans on three occasions.[187] That such references exist at all is significant. It suggests that in Late Antiquity Samaritans, like Jews, constituted an identifiable group that had to be mentioned separately in order for imperial constitutions regarding them to take effect. These three laws are also important in that they were all promulgated in different parts of the Roman world. This can be taken to imply that by the Late Ancient period, Samaritan communities had sprung up in dif-

---

edge of the latest developments in Roman law. I am not convinced that we can assume that many of the people who were able to read had such knowledge.

[185] Relevant evidence contemporary to the *Collatio* may be found in J. M. Cohen, *A Samaritan Chronicle. A Source Critical Analysis of the Life and Times of the Great Samaritan Reformer, Baba Rabbah* (Leiden: Brill, 1981). Baba Rabbah was active around 308-328 C.E. (Cohen, p. 225) See especially Baba Rabbah's speeches in which he stresses the centrality of the reading of the Torah (§ 8.1-3 = Cohen p. 17) and his criticisms of Roman rule (§ 1.15 = Cohen p. 6): the Romans are uncircumcised, they practice idolatry, and they repress the law of Moses. On Baba Rabbah's religious activities, see § 10.12 (Cohen p. 22).

[186] J. MacDonald, *The Theology of the Samaritans* (London: Bloomsbury, 1964) 147-222, e.g. p. 150 citing *Memar Marqa* 4.7: "no man can please God unless he believes in Moses;" W. A. Meeks, *The Prophet-King. Moses Traditions and the Johannine Christology* (Leiden: Brill, 1967) 216-57. The centrality of Moses is also evident in inscriptions, e.g., in a Samaritan inscriptions from Thessaloniki, *CIJ* 693a , see also G. H. R. Horsley, *New Documents Illustrating Early Christianity* 1 (North Ryde: Macquarie University, 1981) 108-10.

[187] *Codex Theodosianus* 13.5.18 (of 390 C.E., promulgated in Constantinople); 16.8.16 (of 404 C.E., promulgated in Rome) and 16.8.28 (of 426 C.E., promulgated in Ravenna).

ferent parts of the Roman empire.[188] Scattered literary references to Samaritans and some Samaritan inscriptions can still be added to this very patchy picture, with the dissatisfying result that we cannot tell, with any degree of reliability, where in the Samaritan Diaspora there were major communities, how Samaritans made a living, or in what ways these people related to the Pentateuch.[189]

Yet, despite the not-so-remote possibility that the *Collatio* was composed by a Samaritan author, I nevertheless believe that there is at least one compelling reason for regarding the *Collatio* as a Jewish work after all. The question can best be resolved by taking into account, once again, the historical developments of the later fourth century.

Fourth-century Christian literature suggests that on a theological level the flourishing Jewish communities of the Mediterranean posed a threat to Christians in a way that the Samaritan communities did not. Unlike such Jewish communities, Samaritan communities were either too insignificant or their lifestyle too unappealing to attract either converts (that is, Christians who forsook Christianity) or "sympathizers." In addition, Samaritans professed a religion which excluded all the prophetic books of the Hebrew Bible. Consequently, Samaritan theology interested Christian authors far less than did the theology of the Jews. Having accepted the prophetic books into their canon, the Jews disputed the Christian contention that this part of the Hebrew Bible contained evidence to prove the legitimacy of Christian Messianic claims.

It is in this atmosphere of Jewish-Christian fourth-century polemic that we must also place the *Collatio*. The role Samaritans played in these polemics was negligible. The *Collatio* shows the way in which a Roman Jew participated in the late antique debate on the Law. The anonymous author of this work wanted to defend his own religion and that of his co-religionists. In so doing, the author refrained from openly and polemically attacking Christian dogmas. He rather chose to follow a method that had long been succesfully used by Jews in defending Jewish religious practices against pagan and early Christian critics: to stress the great age of Mosaic law and to emphasize its essential conformity to the legal systems of other, non-Jewish peoples.

---

[188] This inference remains hypothetical because we do not know if the promulgation of these laws in question were at all triggered by the Samaritan community in the city were the law was promulgated.

[189] H. G. Kippenberg, *Garizim und Synagoge* (Berlin: de Gruyter, 1971) 145f.; F. Hüttenmeister and G. Reeg, *Die antiken Synagogen in Israel.* II. *Die samaritanischen Synagogen* (Wiesbaden: Reichert, 1977); P. W. van der Horst, "De Samaritaanse Diaspora in de oudheid," *Nederlands Theologisch Tijdschrift* 42 (1988) 134-44, containing, among other things, a sensible evaluation of Justin's *Apology* 1.26 (pp.137-38); A. D. Crown, *The Samaritans* (Tübingen: Mohr, 1989) 194-217.

Composed in Latin and therefore accessible to educated people other than Jews, the *Collatio* was the last major Jewish apologetic work to be written in antiquity. Whether the *Collatio* ever reached any of the jurists to whom it addressed itself or whether non-Jews who read it were at all convinced by the idea that Mosaic law stood as the basis of all subsequent law, it is impossible to say. Given the century-old Christian ambivalence towards Pentateuchal law, the author of the *Collatio* was most likely fighting for a cause that, in Christian eyes, had been lost long before. Yet, in a society in which old age and respectability went hand in hand, the surprising number of convincing parallels between Mosaic and Roman law cannot have failed to impress anyone who studied the *Collatio*. In that respect, most would have agreed that the *Collatio* constituted a worthy (although unusual) contribution to a debate that engaged some of the major intellectual figures of the later fourth century.

The Letter of Annas to Seneca

In 1984 B. Bischoff published a fragmentary letter in Latin which has become known after its *incipit* as the *Letter of Annas to Seneca*. Although the letter itself does not give any clues as to where, when, or by whom it was originally composed, it nevertheless merits a brief discussion within the framework of this chapter on Jewish literary production in Late Ancient Rome.[190]

The *Letter of Annas to Seneca* has been preserved in an early ninth-century manuscript (MS 17) that is now in the archepiscopal library in Cologne. In its present state, the text of the letter is not complete. It comprises approximately three full folio-pages (99 r.- 102 r.) before it breaks off. The letter carries the heading *Incipit Epistola Anne ad Senecam de superbia et idolis*. Although this heading gives a fairly good idea of the letter's contents, it has been suggested that it is not original. The actual text lacks further subject headings, yet it may be subdivided into three parts on the basis of the subjects it addresses. The letter starts with a panegyric to God, "*pater omnium mortalium, suorum amator multumque misericors.*" Then it continues with an attack on those who believe that they can understand the mysteries of

---

[190] B. Bischoff, *Anecdota Novissima. Texte des vierten bis sechzehnten Jahrhunderts* (Stuttgart: Anton Hiersemann Verlag, 1984) 1-9; A. Momigliano, "The New Letter by 'Anna' to 'Seneca'," *Athenaeum* 63 (1985) 217-19; L. Cracco Ruggini, "La lettera di Anna a Seneca nella Roma pagana e cristiana del IV secolo," *Augustinianum* 28 (1988) 301-25 and W. Wischmeyer, "Die Epistula Anne ad Senecam. Eine jüdische Missionsschrift des lateinischen Bereichs," in J. van Amersfoort and J. van Oort (eds.), *Juden und Christen in der Antike* (Kampen: Kok, 1991) 72-93, who provides a translation of the letter into German.

the world without knowledge of God; in that context it also criticizes what it sees as deviant views concerning the immortality of the soul. The third and last surviving part of the letter focuses on idol worship.[191]

Bischoff, who discovered the letter and published the *editio princeps*, regards the letter as a "Jewish-apologetic missionary treatise" of the fourth century. Other scholars have agreed with this interpretation.[192]

The identification of the letter as Jewish rests on several arguments. First, the author of the *Letter of Annas to Seneca* never refers to Jesus. Similarly, typically Christian theological ideas are entirely absent. Like the author of the *Collatio*, the author of this letter never even once engages in anti-Christian polemics. Another factor suggesting Jewish authorship is the fact that the *Letter of Annas to Seneca* develops ideas that are very similar to those expressed in *Wisdom of Solomon* (Chapters 13 and 14) and in the *Sibylline Oracles* (books three to five). In addition, views that seem to recall passages from the Hebrew Bible (Genesis through Job) have on occasion also been included.[193] Direct citations of biblical materials, however, are absent.

Furthermore, Bischoff has argued that the Annas referred to in the letter's *incipit* is to be identified with a (first century) Jewish contemporary of Seneca the Younger, namely Annas, who was the high priest of 6-15 C.E., or, more likely, another Annas who officiated as high priest during the years 62-68 C.E.[194] That the *incipit* is apocryphal follows from the fact that neither Annas nor Seneca are ever mentioned in the letter itself; rather, its anonymous author addresses the letter's projected recipients twice as *fratres*.[195] Another possibility, proposed by Momigliano, is that Annas was an otherwise unknown Jewish propagandist not identical to either of the high priests mentioned earlier. Such a suggestion opens the way to regard Annas as author of this letter after all.[196] While this name was not very common among the Jews of antiquity, it may be pointed out in support of Momigliano's hypothesis that a certain Annas is referred to twice in Late Ancient imperial legislation on the Jews.[197] According to this reconstruction, then, only the name Seneca would have to be regarded as apocryphal.

---

[191] For further subdivisions, see Bischoff 1984, 4.

[192] E.g. Momigliano 1985, 218: "*fratres* ...must be potential proselytes;" similarly, Cracco Ruggini 1988, 302.

[193] For references, see the footnotes in Bischoff 1984, 6-9.

[194] Bischoff 1984, 2-3.

[195] Lines 21 and 45.

[196] Momigliano 1985, 218-19.

[197] *Codex Theodosianus* 16.9.3 of 415 C.E. and 16.8.23 C.E., on which see Linder 1987, 273 n. 1.

That the *Letter of Annas to Seneca* was composed in Rome is certainly possible, but we cannot be sure.[198]

The commonly-accepted dating of the *Letter of Annas to Seneca* to the fourth century rests on assumptions concerning the letter's purpose. As has already been noted, Bischoff regards the letter as a missionary treatise. Pointing out that from Constantinian times onwards imperial law tried to ban Jewish missionary activities, Bischoff has therefore tentatively suggested that this letter is likely to have been composed before 325 C.E.[199] Wischmeyer has suggested that the *Letter of Annas to Seneca* must predate Jerome's translation of the Bible because the citation of Genesis 2:7 in line 37 of the *Letter of Annas to Seneca* differs from Jerome's rendering.[200]

Neither of these arguments is, however, entirely convincing. We have no evidence to suggest that Jews used Jerome's translation at all. Jews may very well have made their own translation directly from the Hebrew text or from one of the Greek translations that were in circulation. Besides, there is no reason to suppose that Jews would have discontinued to use a Latin translation other than Jerome's even after the latter's translation started to replace previous Latin translations in Christian circles. As for Bischoff's argumentation, it should be noted that although the author of the *Letter of Annas to Seneca* refers explicitly to the *veritas nostra* and presents his views as the only correct ones, he never speaks of actual conversion. Rather than seeing in the *Letter of Annas to Seneca* a work designed at winning proselytes, it should be regarded in the very first place as a treatise designed to curry the sympathy and favor of a pagan audience. The biblical passages included in the letter address issues that are, after all, not typically Jewish, but rather raise questions recognizable by non-Jews interested in popular philosophic discourse.[201] In fact, some of the very same issues that turn up in the *Letter of Annas to Seneca* had been previously addressed by Lactantius in his *Divinae institutiones* also.[202] When seen in this light, it is possible to maintain that the *Letter of Annas to Seneca* was composed after 325 C.E.; in Late Antiquity, a generalized interest in Judaism was not uncommon among pagans and Christians. Even though some Church fathers fulminate against it, sympathizing with Judaism was forbidden by no law. The dating of the *Letter of Annas to Seneca* can probably best be solved by a detailed study of its linguistic characteristics. Such an analysis is not in-

---

[198] Cracco Ruggini 1988, 305; Wischmeyer 1991, 91-92.
[199] Bischoff 1984, 3 and 5 and comparably Wischmeyer 1991, 81-82.
[200] Wischmeyer 1991, 81.
[201] *Pace* Wischmeyer 1991, 80-81; 90-91.
[202] E.g., *Divinae institutiones* 2.4 (against images); 7.8-7.12 (about the immortality of the soul) 7.14 (attack on Chaldei).

cluded here, for it would take us too far beyond the scope of this chapter.[203] If, however, the *Letter of Annas to Seneca* were indeed written to win the sympathy of non-Jews, one could argue that there were good reasons for doing so in the later fourth century (as opposed to the third or early fourth century). Perhaps the author of the letter of *Annas to Seneca* tried to accomplish what the author of the *Collatio* tried to accomplish by substituting philosophical for legal discourse.

Implications

The *Collatio* and the *Letter of Annas to Seneca* are very different kinds of documents. Yet, both provide evidence of a Judaism that was far from passive or isolated. Both were written not in Hebrew, Aramaic, or Greek, but in the dominant literary language of the western part of the Mediterranean, Latin. Both addressed issues that concerned not merely Jews, but that also interested the non-Jewish educated public of Late Antiquity. Finally, both reflect a level of education that is unusual when seen against the background of our previous observations concerning the social status of Jews in Late Ancient Rome.[204]

The evidence provided by the *Collatio* and the *Letter of Annas to Seneca* is particularly important because it helps us to draw a more nuanced picture of the Jewish Diaspora in the western Mediterranean during the late antique period. Recent scholarship has stressed that well into the fifth century, the Jewish communities of the eastern Roman provinces were full of self-confidence. The ubiquitous remains of synagogues in Late Ancient Palestine as well as Jewish and non-Jewish literary evidence such as rabbinic literature, Libanius' correspondence, or John Chrysostom's eight homilies against the Jews of 386-387 C.E. testify to the existence of a Judaism that was rather dynamic and that continued to manifest itself strongly. Despite increasingly restrictive imperial legislation and in the face of sudden outbreaks of popular violence against individual communities, in the fourth century Near East, the Jewish communities represented a force to be reckoned with both religiously and politically. Thus these Jewish communities related to contemporary non-Jewish society in a manner that contrasts markedly with the way in which the Jewish communities in the western part of the Mediterranean are believed to have defined their relationship with the world surrounding them. In Chapter 1 we saw that

---

[203] Bischoff 1984, 5; Wischmeyer 1991, 92, suggesting a third century date for the *Letter of Annas to Seneca* on the basis of the language used to which the *incipit* was added in the early fourth century.

[204] See Chapters 3 and 4.

Roman Jews in particular have been portrayed as an unintegrated minority which lived in virtual intellectual and cultural isolation.

Did the Jewish communities in the western part of the Mediterranean—as opposed to those in the eastern Mediterranean—really live in splendid isolation? Or is the supposed contrast between intellectual achievements and social position of Near Eastern as opposed to Roman Jews based on preconceived ideas that do not find support in the surviving evidence?

The analysis of the *Collatio* and the *Letter of Annas to Seneca* suggests that it is not correct to draw a sharp distinction between the Jewish communities in the eastern and the western part of the Mediterranean in terms of their interaction with contemporary non-Jewish society.[205] The author of the *Collatio*, for example, was a Jew with enough education and background in non-Jewish contemporary culture to be able to read Roman law in Latin.[206] In addition, the author was someone who knew how to get access to the place(s) where copies of these Roman laws were kept. He was also aware of the great respect in which his pagan and Christian contemporaries held Roman law, and thus was able to perceive the great potential of his particular treatment of Roman and Mosaic law.[207] The author of the *Collatio* acquired a fair knowledge of Roman law. He certainly did not belong to the class of people disparagingly described by Ammianus as "so totally uneducated that they cannot remember ever having possessed a law book, and if the name of an early writer is mentioned in cultivated company they think it is a foreign name for a fish or some other comestible."[208] Furthermore, by composing the *Collatio*, its author showed himself to be familiar with some of the issues that were debated in contemporary non-Jewish circles. Finally, the existence of *Collatio* in its present form also suggests that the person who wrote it realized that the rather theoretical debate on the status of the Law would ultimately affect deeply the actual sociopolitical position of his Jewish contemporaries.

---

[205] Although we do not know whether the *Collatio* or the *Letter of Annas to Seneca* originated in Rome, a western origin is probable. It seems likely that both works were produced in a city that could provide a setting for the intellectual debates sketched earlier and that housed a Jewish community. Cities that fall into this category include Rome, and perhaps also Carthage.

[206] On the study of Roman law in Late Antiquity in general, see Schulz 1946, 275-76; on the social status of the students, see Schulz 1946, 271-72.

[207] E.g., Eusebius *Historia Ecclesiastica* 10.8. On the knowledge of, attitudes toward and appropriation of Roman law in fourth-century patristic literature, see J. Gaudemet, "L'apport du droit romain à la patristique latine du IVe siècle," *Miscellanea Historiae Ecclesiasticae* 6 (1983) 165-81. For the great importance attached to Roman law in pagan circles of the fourth century, see Liebs 1987, 119 (analysis of the evidence in the *SHA*).

[208] Ammianus Marcellinus 30.4.

The *Letter of Annas to Seneca* reflects a very similar intellectual background. Its author wrote with a sophistication that far surpasses both the language and the arguments put forth in the roughly contemporary apocryphal correspondence between Paul and Seneca. Even more important, this author addressed issues of a religious-philosophical nature that interested not only his Jewish contemporaries exclusively, but also his non-Jewish contemporaries. Thus the author of the *Letter of Annas to Seneca* was able (at least theoretically) to reach an audience consisting of people who would have little appreciated a more literal exposition of biblical concepts.[209]

To be sure, the authors of the *Collatio* and the *Letter of Annas to Seneca* were not alone. They were part of a much larger phenomenon, the extent and character of which we can no longer determine very precisely. That in the early fifth century separate laws had to be promulgated to exclude Jews from the various branches of the empire's administration shows clearly that Jews other than the two authors mentioned here were educated enough to have been accepted in such functions. The existence of such laws also suggests that Jews themselves aspired to and were interested in accepting such functions—a fact which is not stressed enough in secondary literature on the subject.[210] Literary and epigraphical sources provide further evidence to show that until well into the fifth century C.E. there was nothing unusual about Jews in administrative positions.[211] We would like to know more about such people, about their social status within Roman society at large, about their education, and about their attachment to Judaism. How did their non-Jewish contemporaries recognize that these people were Jewish? How did their Jewishness affect their functioning as part of the Roman administrative system? How did their position within the late Roman administration affect the ways in which they interacted with non-Jews? It is extremely unfortunate that the surviving sources are extremely reticent in this respect. Whatever their intellectual and social background may have been, people of this kind cannot, however, have been completely exceptional: the *Codex*

---

[209] See Cracco Ruggini 1988, 318-23 who likewise evaluates the *Letter of Annas to Seneca* in these terms.

[210] *Codex Theodosianus* 16.8.16 of 404 C.E.; *Codex Thedosianus* 16.8.24 of 418 C.E.; *Constitutio Sirmondiana* 3 of 425 C.E.; Theodosius II's third *Novella* of 438 C.E.; *Codex Justinianus* 1.5.12) of 527 C.E. See also G. Kisch, "Zur Frage der Aufhebung jüdisch-religiöser Jurisdiktion durch Justinian," *Zeitschrift der Savigny-Stiftung für Rechtsgeschichte. Romanistische Abteilung* 77 (1960) 395-401.

[211] On Caecilian and Theodor, both *defensores civitates* on the island of Menorca in 418 C.E., see *PL* 20, 731-48 and *PL* 41, 821-32; on Jewish senators in the western provinces of the Empire, see Jerome, *In Esaiam* 18.66.20 (*CCL* 73A, 792); on *maiures cibitatis* (sic) on inscriptions from Venosa, see *CIJ* 611 (supposedly sixth century); and see *CIJ* 619 b-d (= Noy 1993, nos. 86, 114-16).

Theodosianus simply accepts it as a given that there were Jews who were "educated in liberal studies."[212]

In conclusion, the *Collatio* and the *Letter of Annas to Seneca* are important because they help us to refine reconstructions of a variety of aspects of Jewish life in the western Diaspora in Late Antiquity. They inform about the intellectual curiosities and concerns of Jews during this period. They also indicate the level of "general education" some Jews were able to attain. They show that there were Jews eager to get a training in law. They also show that non-Jews did not prevent Jews from receiving such a training. Most important of all, however, these two unique documents show that it is not correct to assume a strong dichotomy between the Jewish community of Rome and Jewish communities in other parts of the later Roman world in terms of social position or intellectual creativity.

---

[212] *Codex Thedosianus* 16.8.24 of 418 C.E. and promulgated in Ravenna.

CHAPTER SEVEN

CONCLUSIONS

Surveying the history of scholarship on the Jews of ancient in Rome in Chapter 1, we have seen that over the last three hundred and fifty years researchers with very different interests and backgrounds have come to agree on one fundamental point, namely that the Jews of ancient Rome lived in virtual isolation. In Chapter 1 I argued that earlier scholars reached this conclusion because of the particular ways in which they studied the evidence relating to the Roman Jewish community. We saw that until fairly recently, early Christian archaeologists did not take Jewish archaeological materials seriously, either because they did not want to or could not take such materials seriously. We have also seen that students of Jewish history did not place Jewish archaeological and epigraphical finds from Rome into the larger context of Late Ancient (non-Jewish) material culture—at least not systematically—because they believed that this would not help them in their attempt to gain a deeper understanding of the daily lives of the Roman Jews themselves. Perhaps the most outstanding feature of previous scholarship on the Jews of ancient Rome, however, is that earlier scholars never elaborated on how they came to their conclusions, or on what they actually meant by saying that Roman Jews were isolated. Remarkably enough, even in modern scholarly literature scholars often refrain from reflecting on this question. Thus E. M. Smallwood wrote that Jews characteristically resist assimilation "except in the superficial matter of language assimilation for everyday contact."[1] And Feldman observed that "the adoption by Jews of Greek names turns out not to be a very meaningful criterion of their degree of assimilation."[2]

A closer look at these statements reveals why the traditional approach to the question of how Jews related to non-Jews is unsatisfactory. First, scholars often fail to back up their claims. Thus one may wonder whether the use of Greek was just a superficial matter for those Jews who used the Septuagint, who recited the *Shema* in Greek,

---

[1] Smallwood 1976, 123.
[2] Feldman 1993, 418.

or who preferred to have their liturgy to be celebrated in Greek rather than in Hebrew. More fundamentally, if we believe that some evidence tells us more about the Jews of antiquity than does other evidence, should we not try to explain why this is so rather than reject such evidence beforehand? Why should language or onomastic practices not be good indicators of cultural change?

This brings us to our second point. Smallwood and Feldman have included some evidence at the expense of other evidence because they conceive of the relationship between Jew and non-Jews in rather rigid terms: Jews generally keep to themselves, and if they do not they will lose their Jewish identity and assimilate. Many scholars share this view. Solin called Jews who adopt the language of the country in which they are living "assimilationsbereit," while Sheppard concluded that Jews in Asia Minor who had undergone some degree of Hellenization were "lax."[3]

The fact that the term "isolation" is often used in conjunction with its opposite, namely "assimilation," makes it difficult to see the relationship of Jews and non-Jews in more organic terms. Yet why do we have to suppose on principle that fascination with another culture means that one wants to become identical to the representatives of that culture? Did not Rabban Gamaliel visit the (Roman) baths in Akko without being bothered in the least about the statue of a pagan goddess placed there? Why should we always conceive of the interaction between Jews and non-Jews in strictly religious terms, or, as Feldman has recently done, in terms of attack and counterattack and conversion of non-Jews to Judaism?[4]

In this book I have opted for a very different approach to determine if, and if so, how the Jews of ancient Rome were isolated. I have included discussions of archaeological evidence, inscriptions, and literary sources because I believe that all these different types of evidence inform in their own way about how Jews did and did not interact with their non-Jewish contemporaries. I do not believe that we have the means to decide beforehand that some types of material are *by definition* more representative or valid than other types of material. I have systematically compared these Jewish materials to non-Jewish

---

[3] Solin 1983, 777 and A. R. R. Sheppard, "Jews, Christians and Heretics in Acmonia and Eumenia," *Anatolian Studies* 29 (1979) 180.

[4] Feldman 1993 and see L. V. Rutgers, "Attitudes to Judaism in the Greco-Roman Period: Reflections on Feldman's *Jew and Gentile in the Ancient World*," forthcoming in *Jewish Quarterly Review*.

materials, because only in this way, I believe, can we determine what is specifically Jewish about these materials.

Although the types of evidence included in this study are very different in nature, it is all the more surprising to see that an analysis of all this evidence has led us, time and again, to reach the same conclusion: the use of artistic products, language, or literary forms that are generally Late Ancient goes hand in hand with a tendency to express an identity that is unmistakably Jewish.

In Chapter 2 we have seen, for example, that the Jews ordered their sarcophagi, gold glasses, and other artistic products from the same workshops as those that catered to pagans and Christians. We have also seen, however, that the use of artifacts which enjoyed popularity in Jewish and non-Jewish circles alike was combined, especially from the fourth century onwards, with a preference for iconographical themes that are decidedly Jewish. Similarly, we have seen that the construction of Jewish catacombs came of age in the same general period that Christians catacombs started to be constructed. We have also seen that the building history of both Jewish and early Christian catacombs developed along very similar lines. Yet, even though Jewish and early Christian catacombs look surprisingly alike, and even though they may have employed the same workmen, Jews and Christians were never buried together.

In Chapter 4 I have documented the extent to which Jewish onomastic practices changed after they had come under the influence of Roman name-giving practices. In this chapter we also saw, however, that Semitic names never disappeared entirely. I concluded that the importance of Semitic names did not lie in their actual number, but in the fact that they appear at all.

In Chapter 5 I then showed that in third- and fourth-century Rome, Jews employed exactly the same Greek and Latin in their inscriptions as did non-Jews. But I was also able to show that both languages, and Greek in particular, served to express ideas that can be found on Jewish inscriptions only.

In Chapter 6, finally, we saw that the *Collatio* and the *Letter of Annas to Seneca* are treatises that could appeal to Jews and non-Jews alike because of the language in which they were written and because of the problems they addressed. Yet, again, it could be shown that the underlying ideas expressed in these writings are typically Jewish. No pagan or Christian would have insisted on the importance of the Torah in such a way as did the author of the *Collatio*, just as the desirability of monotheism as expressed by the author of the *Letter of Annas to*

Seneca is most easily understood when it is seen against the larger background of Jewish religious reasoning.

Taking all this evidence into account, it becomes possible to draw a nuanced picture of how Roman Jews related to the world surrounding them. Instead of living in splendid isolation or longing to assimilate, the Roman Jews described in this book appear as actively and, above all, as self-consciously responding to developments in contemporary non-Jewish society. Interacting with non-Jews, Roman Jews did not give up their own identity. Rather, they freely borrowed elements from Roman culture, and in doing so they adapted such elements to their own needs. Thus they asserted and maintained their status as Jews. In this particular respect, Jews in third- and fourth-century Rome did not differ from Jews in Palestine, or, for that matter, from non-Jews.[5] As a result of recent studies we now know that Greek and Roman culture were not antithetical to local cultures. Rather, Greco-Roman civilization became an enabler which served to articulate or universalize non-Greek and non-Roman local traditions.[6]

In the previous chapters I have stressed several times that most of the materials discussed in this book were found in catacombs. In the private atmosphere of these catacombs, the need to represent oneself as socially and culturally respectable vis-à-vis Roman society as a whole was considerably less manifest than in tombs that were constructed above ground and that, as Trimalchio realized full well, were visible by all.[7] In the private, ill lit, and exclusively Jewish *ambiente* of the Jewish catacombs of Rome, external pressure played no role to speak of. Instead, here Jews represented themselves exactly as they saw themselves, namely as people who had access to the trappings of Roman culture, but who, at the same time, defined themselves in terms of involvement in Jewish daily life.[8]

---

[5] E. M. Meyers, "The Challenge of Hellenism for Early Judaism and Christianity," *Biblical Archaeologist* 55 (1992) 84-91.

[6] R. MacMullen, "Notes on Romanization," *Bulletin of the American Society of Papyrologists* 21 (1984) 161-77; G. Bowersock, *Hellenism in Late Antiquity* (Ann Arbor: University of Michigan Press, 1990) 5, 7-13, 67-68; M. Millet, *The Romanization of Britain. An Essay in Archaeological Interpretation* (Cambridge: Cambridge University Press, 1990) 1-2; S. Swain, "Hellenism in the East," *Journal of Roman Archaeology* 6 (1993) 461-66; P. W. M. Freeman, "Romanisation and Roman Material Culture," *Journal of Roman Archaeology* 6 (1993) 438-45.

[7] Petron., *Sat.* 71.

[8] Because the inscriptions from the Jewish catacombs of Rome represent only a minor fraction of the all the people who were buried there, one could argue that the "silent majority" did not share the views expressed in the epitaphs. Yet, they may very well have held the same views as the Jews whose inscriptions have survived.

It is all the more interesting to note that non-Jews may have viewed Jews in approximately the same way as the Jews defined themselves. This at least seems to be the implication of references to Jews contained in the *Codex Theodosianus*. According to the Theodosian Code, Jews held posts within the Roman administrative system until early in the fifth century C.E.[9] We do not know very much about these people. Apparently they were familiar enough with Roman administrative and legal practices to occupy such positions. At the same time, however, the people in question continued to be Jews and somehow they were recognizable as such. Until they banned Jews from the imperial service, Roman legislators saw no contradiction between a person's attachment to the Jewish community or the Jewish people and that person's role as a Roman official. Neither did the people who set apart seats for the Jews in the odeum in Aphrodisias, in the theatre of Miletos, or in the amphitheatre in Syracuse: again, the Jews we encounter here are people who fully participated in city-life, but who, at the same time, were consciously Jewish and who were also recognized by others as such.[10] Jews and non-Jews apparently agreed that there was nothing mutually exclusive about being Jewish on the one hand and being part of late Roman society on the other.

Having established that Jews and non-Jews interacted on different levels, it is much more difficult to measure the intensity, or even the nature, of the contacts between these two groups. Roman Jews may have employed Roman workshops, but would they also have accepted food, oil, or wine produced by non-Jews?[11] How far did socializing really go? How did the proselytes to Judaism we occasionally encounter on Jewish funerary inscriptions learn about Judaism? Conversely, did contacts with the larger outside world induce some Jews to change their attitude to the Jewish community or forsake the religion of their forefathers altogether? Were there any contacts between Jews and Jewish-Christians? Along similar lines, must the adoption of grave forms that are generally Late Ancient be taken as an indicator that Jewish burial customs and views about death and afterlife had

---

[9] E.g., *CTh* 16.8.16.

[10] For Aphrodisias, see Reynolds and Tannenbaum 1987, 132. For Syracuse, see Noy 1993, 200 (note that this inscription is incomplete). On Miletos, see H. Bellen, "*Synagoge toon Ioudaioon kai Theoseboon.* Die Aussage einer bosporanischen Freilassungsinschrift (CIRB 71) zum Problem der 'Gottesfürchtigen'," *Jahrbuch für Antike und Christentum* 8-9 (1965-66) 171-76.

[11] M. Dukan *et al.*, "Une inscription hébraïque sur amphore trouvée à Ravenne," *Revue des Études Juives* 143 (1984) 287-303 claims to have found in Ravenna an amphora that contained kosher wine, but fails to substantiate this claim.

changed too? Did Roman Jews hold notions about purity and defilement that corresponded in any way to the views on these matters expressed in contemporary rabbinic literature? What are the implications of the fact that Jews reused stones with both pagan and Jewish inscriptions for their own burials?[12]

It is very unfortunate that we cannot answer all these important questions. Sometimes we lack the evidence to answer them. At other times we do not have the means to resolve such issues. One of the reasons why interaction is such a difficult phenomenon to analyze is because it is a dynamic process that does not always fit into our analytical categories. Take, for example, the names borne by Jews. In Chapter 4 we saw that certain Greek names have Hebrew equivalents. It is conceivable that the Greek names in question were specifically chosen because they were direct translations of Hebrew names. Therefore, names that strike modern researchers as typically-Greek may thus have had very different connotations for the people who bore them: by preferring some Greek names over others, Jews who lived in a Latin speaking environment may have felt that they did not entirely give up more traditional onomastic practices. Although there is no evidence to prove this, it is in fact conceivable that the continuous use of certain Greek names by Jews will have lent a certain Jewish flavor to these names in the same way as in the nineteenth century the employment by Jews of names borrowed from the operas of Richard Wagner led certain people to believe, ironically, that these names were not Germanic at all, but rather typically Jewish.[13]

Along similar lines, words such as ὅσιος or ἀρχισυνάγωγος could be understood by everyone who spoke Greek. In Chapter 5 we saw, however, that in Jewish inscriptions from Rome these words assume an importance they never assumed in non-Jewish inscriptions.[14] Again, for a modern observer it is very difficult to determine when words or ideas lose their original Greek or Roman connotation and when they begin to assume a meaning that is specifically Jewish. That it is es-

---

[12] E.g. *CIJ* 33* (*recto*) and *CIJ* 302 (*verso*); *CIJ* 30* (*recto*) and *CIJ* 344 (*verso*); *CIJ* 9* (*recto*) and *CIJ* 148 (*verso*); *CIJ* 362 (*recto*) and 320 (*verso*) (both sides of this plaque may be Jewish); similarly *CIJ* 386 and 424. The formal characteristics of inscriptions such as *CIJ* 318, 412, 415, and 535 indicate that they were carved on reused stones. On the reuse of inscriptions, see also the Appendix, and see P. M. Nigdelis, "Synagoge(n) und Gemeinde der Juden in Thessaloniki: Fragen aufgrund einer neuen jüdischen Grabinschrift der Kaiserzeit," *Zeitschrift für Papyrologie und Epigraphik* 102 (1994) 298.

[13] M. van Amerongen, *De buikspreker van God. Richard Wagner* (Amsterdam: Arbeiderspers, 1983) 79.

[14] On *archisynagogoi*, see the remarks by Rajak and Noy 1993, 78 and 83-84.

sential to study in detail how words or ideas are transmitted from one culture to another may be evident; the fact that the word ὅσιος started to appear in Jewish epitaphs from Rome, and in these inscriptions only, may tell us more about how Jews had remained Jews than about how Jews had undergone the influence of Greek or Hellenistic culture.

Who were the Jews of Late Ancient Rome? The materials preserved in the Jewish catacombs of Rome suggest that the Jews buried there were neither assimilated nor isolated, but rather people who interacted freely with non-Jews. At the same time they were people who had a strong sense of identity. For all the presence of typically Roman elements in the Jewish epitaphs such as the onomasticon or references to age at death, the most explicit and therefore most important characteristic of these inscriptions is that Jews defined themselves in terms of what they did in the Jewish community.

The Jewish community consisted of various "synagogues." It is conceivable that differences existed between the different communities, but the surviving evidence does not permit us to reconstruct what kind of people belonged to individual communities or how such communities related to one another.[15] It is clear, however, that the evidence from the catacombs does not permit us to divide the Jews of ancient Rome into more Romanized and more traditional groups.

In the light of the evidence from the seemingly timeless world of the Jewish catacombs, or considering the good-humored nature of references to Jews in the *Scriptores Historiae Augustae*, it is easy to forget that the Jews of ancient Rome were continuously struggling to assert their Jewish identity. The Jews in third- and fourth-century Rome resembled their Renaissance descendants, in that, for both groups Jewish life was "not a carousel of servile imitation, but a perennial struggle for survival."[16] Long before Christianity started to manifest itself more visibly, Roman Jews had to come to terms with the question of what it meant to be a Jew in a non-Jewish society. In a world in which everyday contacts with non-Jews were inevitable and in which non-Jews were not infrequently amicably-disposed towards Jews, a minor-

---

[15] Imprecise is L. M. White, *Building God's House in the Roman World. Architectural Adaptation among Pagans, Jews, and Christians* (Baltimore and London: Johns Hopkins U.P., 1990) 61. R. MacMullen, "The Unromanized in Rome," in Cohen and Frerichs 1993, 54-56 maintains that in Rome Jewish synagogues were located on the "undesirable periphery," but this claim is only partially correct.

[16] See the important remarks of R. Bonfil, "The Historian's Perception of the Jews in the Italian Renaissance. Towards a Reappraisal," *Revue des Études Juives* 143 (1984) 59-82, esp. 79f. And see now R. Bonfil, *Gli Ebrei in Italia nell'epoca del Rinascimento* (Florence: Sansoni, 1991) 7-19, 89-109.

ity such as the Jews had to be particularly steadfast to avoid giving up their own distinctive identity.[17]

It is difficult to tell how much Roman Jews really felt "at home" in the Diaspora.[18] The finds from the Jewish catacombs present us with a misleadingly-static picture of exactly the period during which the Jews' position in society was slowly but unmistakably changing. Even though it was still illegal and even though "uninstitutionalized" violence was rather common in the later Empire as a whole, acts such the destruction of a synagogue in Rome in the late 380s must have profoundly shocked the Roman Jewish community. If, in the fourth century, Christianity might not have posed an immediate threat to the Jewish communities of antiquity, it certainly posed a new challenge. Jews realized that this was so. It was for precisely that reason that the author of the *Collatio* raised the arguments he or she raised in the way he or she did. In the late fourth century, Jews were still very much part of Late Ancient society. Yet although this process had not yet been completed, the late fourth and early fifth century was surely also a period during which new boundaries were being drawn. From a Christian perspective, Jews were slowly but steadily becoming "the quintessential other."[19]

It is difficult to tell how these changes in perception affected the way in which Jews viewed themselves. In Chapter 2 we saw that in the early fourth century, Jewish motifs become very pronounced on sarcophagi, wall paintings, and gold glasses. It remains difficult to determine, however, whether this change in iconographical preferences should be ascribed to external pressures alone or whether dynamics are at work here that are internal to the development of Jewish art in Late Antiquity. Because burial in the Jewish catacombs must have stopped sometime in the early fifth century, and because we have only such apologetic sources as the *Actus Sylvestri* we really do not fully know what happened in this period when Roman Jews were being increasingly marginalized. The fact that they stopped using catacombs suggests that the same developments that affected the Roman

---

[17] *Pace* Bickermann 1988, 251 and 256.

[18] This notion has been put forward by A. T. Kraabel, "The Roman Diaspora: Six Questionable Assumptions," *Journal of Jewish Studies* 33 (1982) 458.

[19] For the question of how Jews were perceived by others during an earlier period, see the important contributions by M. Goodman, "Nerva, the Fiscus Judaicus and Jewish Identity," *Journal of Roman Studies* 79 (1989) 40-44 and S. J. D. Cohen, "'Those Who Say They Are Jews and Are Not:' How Do You Know a Jew in Antiquity When You See One," in *id.* and Frerichs 1993, 1-46.

population at large also affected Roman Jews. It is very unfortunate that it is impossible to be more specific.

The Jewish catacombs provide us with a wealth of evidence. At the same time, however, the evidence is isolated in time and place. We can draw only a very patchy picture of what happened to Roman Jews who lived either before or after the period during which the Jewish catacombs of Rome were used. But as regards the Jews who were laid to rest in the catacombs, Abraham Berliner's phrase of a century ago still holds: "Sie wurden Juden, wurden aber auch Römer—römische Juden."[20]

---

[20] Berliner 1893, 93.

APPENDIX

DIS MANIBUS IN JEWISH INSCRIPTIONS FROM ROME

The inclusion of the pagan formula *DM* or *DMS* (*Dis Manibus Sacrum*, "to the gods of the underworld") in Jewish inscriptions has long baffled scholars. The excavator of the Jewish Monteverde catacomb, N. Müller, for example, discovered several inscriptions carrying this formula, but he did not know how to interpret it.[1] Several years later, J. B. Frey discovered many more inscriptions carrying the *DM* formula. He believed that most of them were not Jewish at all.[2] Frey also argued that the few Jewish inscriptions displaying the *DM* formula were likely to have originated in pagan workshops: he believed that the Jewish inscriptions in question were carved into stones that were "ready made" in that they already carried the *DM* formula.[3] Entering the Vigna Randanini catacomb in search of further Jewish inscriptions, Frey finally also observed that the *DM* formula never appeared in Jewish graffiti or painted inscriptions. For that reason he concluded, correctly, that the *DM* formula did not belong to the standard repertoire of epigraphic formulae used in Jewish funerary inscriptions from Rome.[4]

More recently, E. R. Goodenough proposed that Jews may have understood the *DM* formula as an abbreviation for *Deo Magno* or *Deo Magno Sacrum*. rather than for *Dis Manibus*.[5] Goodenough also observed that the appearance of *DM* in inscriptions does not provide sufficient grounds to consider such inscriptions *a priori* as pagan rather than as Jewish.[6] Still more recently, other scholars have tried to

---

[1] Müller 1912, 87.

[2] *CIJ* 6*, 7*,9*, 10*, 17*, 19*, 20*, 21*, 26*, 27*, 28*, 29*, 31*, 34*, 36*, 37*, 38*, 43*, 44*, 57*, 60*, 62*, and 63*.

[3] *CIJ* 287, 464, 524, and 531. *CIJ*, cxix. Other scholars have followed Frey, see Goodenough 1953-68, vol. 2, 139; van der Horst 1991, 43.

[4] J. B. Frey, "Inscriptions inédites des catacombes juives de Rome," *Rivista di Archeologia Cristiana* 5 (1928) 303.

[5] Goodenough 1953-68, 139, and, similarly, Le Bohec, 1981, 177.

[6] Goodenough 1953-68, 138. And cf. R. S. Kraemer, "On the Meaning of the Term 'Jew' in Greco-Roman Inscriptions," *Harvard Theological Review* 82 (1989) 41.

explain the appearance of the *DM* formula in Jewish inscriptions, but the question has not yet been settled definitively.[7]

Characteristic of most recent discussions of the *DM* formula in Jewish inscriptions is that scholars pay little or no attention to the precise archaeological context in which inscriptions carrying this formula have been found. It is true that excavation reports are usually not very helpful in this respect. Yet on one particular point, such reports provide important information. They indicate that in some cases inscriptions that carry the *DM* formula ended up in the Jewish catacombs only because they were reused there as filler to close up graves.

One such *DM* inscription was discovered *in situ* in 1906 by Müller, who noted that the inscription had been turned upside down so that the inscribed side was no longer visible at the time it served to seal off a Jewish grave.[8] By studying the way in which they had been deposited, Müller established that two other *DM* inscriptions he discovered in the Jewish Monteverde catacomb were nothing but reused materials.[9] Reuse is also certain in case of another inscription, which has been preserved in the collection of Jewish inscriptions now on display at the Musei Vaticani.[10] One side of this elegantly-carved inscription has nothing Jewish about it. It is written in Latin and carries, among other formulae, the abbreviation *DMS*. The other side of this marble plate carries a somewhat less elegantly-carved Jewish funerary inscription in Greek, but it lacks the *DM* formula. That the *DMS* inscription was reused by Jews rather than vice versa seems likely because the person who was responsible for the Greek (Jewish) side of the inscription took into account that one of the corners of the inscription had been broken off (perhaps this happened when the *DMS* inscription was removed for reuse as a Jewish inscription). In addition, fragments of stucco can still be seen on the *DMS* side of the inscription. This means that the *DMS* side of the inscription was turned towards the inside of the grave of the person commemorated in the Jewish-Greek, or visible, side of the inscription.

---

[7] E.g., Leon 1960, 345; Solin 1983, 657; Le Bohec, 1981 177; Kraemer (previous note) 41-42; *ead.*, "Jewish Tuna and Christian Fish: Identifying Religious Affiliation in Epigraphic Sources," *Harvard Theological Review* 84 (1991) 155-58; van der Horst 1991, 42-43.

[8] *CIJ* 34* and Müller and Bees 1919, 93 no. 104.

[9] *CIJ* 17* and 19*, see Müller 1886, 52-53. On the question of whether the hypogeum on the Via Appia Pignatelli is Jewish, see Vismara 1986.

[10] *CIJ* 36* (*recto*) and *CIJ* 148 (*verso*).

Inasmuch as we know of at least one other example where a *DM* inscription was reused by Jews by adding a Jewish inscription on the other side of the stone, reuse of *DM* inscriptions was not as unusual as one might perhaps expect.[11] In Chapter 2 we have seen that Roman Jews often relied on pagan workshops. It should therefore not come as a surprise to observe that they also reused pagan inscriptions.

When we now inspect the Jewish *DM* inscriptions collected by Frey, it may be observed that all these inscriptions lack information that would make an identification as Jewish inscriptions likely. Nor is it possible to ascertain the precise archaeological context of the inscriptions in question. *CIJ* 287, for example, was found outside the Jewish catacombs of Rome. It contains a dedication by Jews to someone who himself might not have been Jewish. *CIJ* 464 is a fragmentary inscription. Although it was found in the (Jewish) Monteverde catacomb, we simply cannot tell how it was found: was the piece found *in situ* and if so, was it reused?[12] To be sure, there is nothing Jewish about the inscription itself. Another inscription published by Frey, *CIJ* 524, concerns a *metuens*, not someone who was Jewish. Still another inscription in Frey's corpus, *CIJ* 531, is a pagan inscription that does not inform us about the use of *DM* among the Jews of Rome. Similarly unclear is the provenance of a *Dis Manibus* inscription discoverd by Fasola during his excavations in the upper Villa Torlonia catacomb.[13] Lacking any reference to Jews or Judaism, the inscription transcribed by Fasola certainly does not enable us to speak of "the Divine Spirits in a Jewish epitaph."[14] The inscription in question may simply be intrusive—which would not be surprising.[15]

Although there is no reason to suppose that the Jews of ancient Rome would, on principle, have refrained from using the *DM* formula, it is clear that few if any of them really liked this formula. It is certainly telling that *DM* does not appear in graffiti or *dipinti*—that is, in inscriptions that were incised or painted on a tomb by the Jews present directly after burial had taken place (the inscription was incised when the stucco was still wet). It is no less telling than none of the suppos-

---

[11] *CIJ* 9* (*recto*) and *CIJ* 148 (*verso*). For other examples of reused epitaphs, see Chapter 7.

[12] Paribeni in *Notizie degli Scavi di Antichità* 1919, 69 n. 22 is not helpful at all in this respect.

[13] Fasola 1976, 38.

[14] *Contra* Horsely 1981, 118.

[15] For the possibility of intrusive pieces in catacombs, see Chapter 2, section on sarcophagi.

edly Jewish *DM* inscriptions from Rome contains information that is typically Jewish: not a single *archisynagogos* or other community official is known to have received a *DM* inscription.[16]

It is exactly this lack of typically-Jewish elements that the Roman *DM* inscriptions have in common with supposedly Jewish *DM* inscriptions from other parts of the Roman world. Of all the *DM* inscriptions from North Africa collected by Le Bohec, there are only few if any that are arguably Jewish.[17] Similarly, a *DM* inscription from Pannonia may refer to someone who was a convert to Judaism rather than to a woman born Jewish.[18]

It is conceivable that in the future we will discover new materials which may show that Jews ordered *DM* inscriptions. The evidence presently available, however, indicates that the *DM* inscriptions enjoyed little popularity among the Jews of Rome, or, for that matter, among Jews elsewhere in the Roman world. Clearly, the *DM* inscriptions constitute too problematical a category to study how Jews adapted to or underwent influence from the non-Jewish world of Late Antiquity.

---

[16] As Solin 1983, 657 points out, the only *DM* inscription that possibly contains a Jewish name is *CIJ* 43*, but the evidence is not unequivocal: Ὠνείας ἀρχιερεὺς καὶ προφήτης.

[17] Possibly Jewish are Le Bohec 1981 nos. 12, 17, and 71. Not Jewish are Le Bohec 1981 nos. 10, 11, 46, 64, 77, and 81.

[18] Kraemer (above) 1989, 41 and 1991, 156 who follows S. Scheiber.

BIBLIOGRAPHY

Allenbach, J., et al., *Biblia patristica. Index des citations et allusions bibliques dans a littérature patristique.* 1. *Des origines à Clement d'Alexandrie et Tertullien* (Paris: CNRS, 1986).
―――, *Biblia patristica. Index des citations et allusions bibliques dans la littérature patristique.* 2. *Le troisième siècle (Origène excepté)* (Paris: CNRS, 1986).
―――, *Biblia patristica. Index des citations et allusions bibliques dans la littérature patristique.* 3. *Origène* (Paris: CNRS, 1980).
―――, *Biblia patristica. Index des citations et allusions bibliques dans la littérature patristique.* 4. *Eusèbe de Césarée, Cyrille de Jérusalem, Épiphanie de Salamine* (Paris: CNRS, 1987).
―――, *Biblia patristica. Index des citations et allusions bibliques dans la littérature patristique.* 5. *Basile de Césarée, Grégoire de Nazianze, Grégoire de Nysse, Amophiloque d'Iconium* (Paris: CNRS, 1991).
―――, *Biblia patristica. Supplément. Philon d'Alexandrie* (Paris: CNRS, 1982).
Alt, A., *Die griechischen Inschriften der Palaestina Tertia westlich der 'Araba* (Berlin and Leipzig, 1921).
Angelis d'Ossat, G. de, *La geologia delle catacombe romane* (Vatican City: Pontificio Istituto di Archeologia Cristiana, 1943).
Applebaum, S., *Jews and Greeks in Ancient Cyrene* (Leiden: Brill, 1979).
Aringhi, P., *Roma subterranea novissima* (Rome, 1651).
Avigad, N., *Beth She`arim. Report on the Excavations During 1953-1958.* Volume III. *Catacombs 12-23* (Jerusalem: Israel Exploration Society, 1976).

Basnage, J., *Histoire de l'église depuis Jésus-Christ jusqu'a présent* (Rotterdam: Leers, 1699).
Becker, E., *Malta Sotterranea. Studien zur altchristlichen und jüdischen Sepulkralkunst* (Strassburg, 1913).
Bees, N., *Die Inschriften der jüdischen Katakombe am Monteverde zu Rom* (Leipzig: Harrassowitz, 1919).
Berliner, A., *Geschichte der Juden in Rom* (Frankfurt on the Main: Kaufmann, 1893).
Beyer, H. W. and H. Lietzmann, *Die jüdische Katakombe der Villa Torlonia in Rom* (Berlin: de Gruyter, 1930).
Bickerman, E. J., *The Jews in the Greek Age* (Cambridge Mass. and London: Harvard U.P., 1988).
Bischoff, B., *Anecdota Novissima. Texte des vierten bis sechzehnten Jahrhunderts* (Stuttgart: Anton Hiersemann Verlag, 1984).
Blumenkranz, B., *Die Judenpredigt Augustins. Ein Beitrag zur Geschichte der jüdisch-christlichen Beziehungen in den ersten Jahrhunderten* (Basel: Helbing and Lichtenhahn, 1946).
Bohec, Y. Le, "Inscriptions Juives et Judaïsantes de l'Afrique romaine," *Antiquités Africaines* 17 (1981) 165-207.
―――, "Juifs et Judaïsants dans l'Afrique romaine. Remarques onomastiques," *Antiquités Africaines* 17 (1981) 209-29.
Bokser, B. M. "Todos and Rabbinic Authority in Rome," in J. Neusner et al. (eds.), *New Perspectives on Ancient Judaism* 1. *Religion, Literature, and Society in Ancient Israel, Formative Christianity and Judaism* (New York and London: Lanham, 1987) 117-30.
Bonfil, R., "The Historian's Perception of the Jews in the Italian Renaissance. Towards a Reappraisal," *Revue des Études Juives* 143 (1984) 59-82.
Bowersock, G., *Hellenism in Late Antiquity* (Ann Arbor: University of Michigan Press, 1990).
Boyaval, B., "Remarques sur les indications d'ages de l'épigraphie funéraire grecque," *Zeitschrift für Papyrologie und Epigraphik* 21 (1976) 217–43.
Brandenburg, H., "Überlegungen zu Ursprung und Entstehung der Katakomben Roms," *Jahrbuch für Antike und Christentum Erg. Band* 11 (1984) 11-49.
Brooks, R., *The Spirit of the Ten Commandments. Shattering the Myth of Rabbinic Legalism* (San Francisco: Harper & Row, 1990).

Brooten, B., *Women Leaders in the Ancient Synagogue* (Chico: Scholars Press, 1982).
Buhagiar, M., *Late Roman and Byzantine Catacombs and Related Burial Places in the Maltese slands* (1986) = BAR International Series 302.
Burn, A. R., "Hic breve vivitur. A Study of the Expectation of Life in the Roman Empire," *Past and Present* 4 (1953) 2–31.

Cantera, F., and J. M. Millás, *Las inscripciones hebraicas de España* (Madrid: C. Bermejo, 1956).
Chilton, B., "The Epitaph of Himerus from the Jewish Catacomb of the Via Appia," *Jewish Quarterly Review* 79 (1989).
Clauss, M., "Probleme der Lebensalterstatistiken aufgrund römischer Grabinschriften," *Chiron* 3 (1973) 395–417.
Coale, A. J., and P. Demeny, *Regional Model Life Tables and Stable Populations* (New York: Academic Press, 1983).
Cochrane, E., *Historians and Historiography in the Italian Renaissance* (Chigago and London: University of Chigago Press, 1981).
Cohen, S. J. D. and E. S. Frerichs (eds.), *Diasporas in Antiquity* (Atlanta: Scholars Press, 1993).
Colafemmina, C., "Archeologia ed epigrafia ebraica nell'Italia meridionale," in *Italia Judaica. Atti del I convegno internazionale. Bari 18-22.V.1981* (Rome: Ministero per i beni culturali e ambientali, 1983) 199-210.
Cotton, H., "The Guardianship of Jesus son of Babatha: Roman and Local Law in the Province of Arabia," *Journal of Roman Studies* 83 (1993) 94-108.
Cracco Ruggini, L., "Ebrei e Romani a confronto nell'Italia tardoantica," in *Italia Judaica. Atti del I convegno internazionale, Bari 18-22.V.1981* (Rome: Ministero per i beni culturali e ambientali, 1983) 38-65.
_____, "Intolerance: Equal and Less Equal in the Roman World," *Classical Philology* 82 (1987) 187-205
_____, "La lettera di Anna a Seneca nella Roma pagana e cristiana del IV secolo," *Augustinianum* 28 (1988) 301-25.

Deckers, J. G., et al., *Die Katakombe "Santi Marcellino e Pietro." Repertorium der Malereien* (Vatican City: Pontificio Istituto di Archeologia Cristiana, 1987).
_____, "Wie genau ist eine Katakombe zu datieren?" in *Memoriam Sanctorum Venerantes = FS V. Saxer* (Vatican City: Pontificio Istituto di Archeologia Cristiana, 1992) 217-38.
Deichmann, F. W., *Einführung in die christliche Archäologie* (Darmstadt: Wissenschaftliche Buchgesellschaft, 1983).
Duncan-Jones, R., "Age–Rounding, Illiteracy and Social Differentiation in the Roman Empire," *Chiron* 7 (1977) 333–53.
_____, "Age–Rounding in Greco–Roman Egypt," *Zeitschrift für Papyrologie und Epigraphik* 33 (1979) 169–77.
_____, *Structure and Scale in the Roman Economy* (Cambridge: Cambridge U. P., 1990).
Duval, N. (ed.), *L'onomastique latine = Colloques internationales du CNRS no. 564* (Paris: CNRS, 1977).

Eichner, K., "Die Produktionsmethoden der stadtrömischen Sarkophagfabrik in der Blütezeit unter Konstantin," *Jahrbuch für Antike und Christentum* 24 (1981) 85-113.
Engemann, J., "Bemerkungen zu römischen Gläsern mit Goldfoliendekor," *Jahrbuch für Antike und Christentum* 11-12 (1968-69) 7-25.
_____, "Altes und Neues zu Beispielen heidnischer und christlicher Katakombenbildern im spätantiken Rom," *Jahrbuch für Antike und Christentum* 26 (1983) 128-51.
_____, "Christianization of Late Antique Art," *The 17th International Byzantine Congress. Major Papers*, 3-8.VIII.1986 (New Rochelle: Caratzac, 1986) 83-105.
Ery, K. K., "Investigations on the Demographic Source Value of Tombstones Originating from the Roman Period," *Alba Regia* 10 (1969) 51–67.

Fasola, U. M., "Le due catacombe ebraiche di Villa Torlonia," *Rivista di Archeologia Cristiana* 52 (1976) 7-62.
Feldman, L. H., "Hengel's *Judaism and Hellenism* in Retrospect," *Journal of Biblical Literature* 96 (1977) 371-82.
_____, *Jew and Gentile in the Ancient World. Attitudes and Interactions from Alexander to Justinian* (Princeton: Princeton U.P., 1993).
Ferretto, G., *Note storico-bibliografiche di archeologia cristiana* (Vatican City: Tipografia Poliglotta Vaticana, 1942).
Ferrua, A., "Sulla tomba dei Cristiani e su quella degli Ebrei," *La Civiltà Cattolica* 87:4 (1936) 298-311.
Frey, J. B., "Nouvelles inscriptions inédites de la catacombe juive de la Via Appia," *Rivista di Archeologia Cristiana* 10 (1933) 27-50 and 386.
Frier, B., "Roman Life Expectancy: Ulpian's Evidence," *Harvard Studies in Classical Philology* 86 (1982) 213–51.
_____, "Roman Life Expectancy: The Pannonian Evidence," *Phoenix* 37 (1983) 328–344.
Funkenstein, A., *Perceptions of Jewish History* (Berkeley: University of California Press, 1993).
Fuks, G., "Where Have All the Freedman Gone? On an Anomaly in the Jewish Grave-Inscriptions from Rome," *Journal of Jewish Studies* 36 (1985) 25-32.

Garrucci, R., *Cimitero degli antichi Ebrei scoperto recentemente in Vigna Randanini* (Roma, 1862).
_____, *Dissertazioni archeologiche di vario argomento* (Roma, 1864-65).
Giardina, A. (ed.), *Società romana e impero tardoantico* (Bari: Laterza, 1986).
_____, (ed.), *Società romana e impero tardoantico. II. Le merci. Gli insediamenti* (Bari: Laterza, 1986).
Goodenough, E. R., *Jewish Symbols in the Graeco-Roman Period* (New York: Pantheon Books, 1953-68).
Goodman, M., "Nerva and the *Fiscus Judaicus*," *Journal of Roman Studies* 79 (1989) 40-44.
Graetz, H., *Die Konstruktion der jüdischen Geschichte* (Berlin: Schocken, 1936 [1846]).
Guyon, J., *Le cimetière aux deux lauriers. Recherches sur les catacombes romaines* (Rome: Pontificio Istituto di Archeologia Cristiana and École Française de Rome, 1987).

Hachlili, R., *Ancient Jewish Art and Archaeology in the Land of Israel* (Leiden: Brill, 1988).
Hälvä-Nyberg, U., *Die Kontraktionen auf den lateinischen Inschriften Roms und Afrikas bis zum 8.Jh.n.Chr.* (Helsinki: Suomalainen-Tiedeakatema, 1988).
Hengel, M., *Judaism and Hellenism. Studies in Their Encounter in Palestine During the Early Hellenistic Period* (Philadelphia: Fortress Press, 1974).
Henten, J. W. van, and P. W. van der Horst (eds.), *Studies in Early Jewish Epigraphy* (Leiden: Brill, 1994).
Herzog, E., "Le catacombe degli Ebrei in Vigna Randanini," *Bulletino dell'Instituto di Corrispondenza Archeologica* 1861, 91-104.
Hesberg, H. von, and P. Zanker (eds.) *Römische Gräberstraßen. Selbstdarstellung-Status-Standard* (Munich: Bayerische Akademie der Wissenschaften, 1987).
Hinard, F. (ed.), *La mort, les morts et l'au-delà dans le monde romain. Actes du colloque de Caen 20–22 nov. 1985* (Caen, 1987).
Hoffmann, C., *Juden und Judentum im Werk deutscher Althistoriker des 19. und 20. Jahrhunderts* (Leiden: Brill, 1988).
Hoheisel, K., *Das antike Judentum in christlicher Sicht* (Wiesbaden: Harassowitz, 1978).
Hopkins, K., "On the Probable Age Structure of the Roman Population," *Population Studies* 20 (1966-67) 245–64.
Horsley, G. H. R., *New Documents Illustrating Early Christianity* 1 (North Ryde: Macquarie University, 1981).
Horst, P. W. van der, *Ancient Jewish Epitaphs. An Introductory Survey of a Millennium of Jewish Funerary Epigraphy (300 B.C.E.-700 C.E.)* (Kampen: Kok Pharos, 1991).

Huttunen, P., *The Social Strata in the Imperial City of Rome. A Quantative Study of the Social Representation in the Epitaphs Published in the Corpus Inscriptionum Latinarum, Volumen IV* (Oulu: University of Oulu, 1974).
Hyamson, M., *Mosaicarum et Romanarum Legum Collatio* (London: Oxford U. P., 1913).

Jones, A. H. M., *The Later Roman Empire, 284-602: A Social, Economic and Administrative Survey* (Baltimore: Hopkins U. P., 1964).
Juster, J., *Les Juifs dans l'Empire Romain* (Paris: Geuthner, 1914).

Kajanto, I., *Onomastic Studies in the Early Christian Inscriptions of Rome and Carthage*. Acta Instituti Romani Finlandiae II: 1 (Helsinki: Tilgmann, 1963).
———, *A Study of the Greek Epitaphs from Rome*. Acta Instituti Romani Finlandiae II: 3 (Helsinki: Tilgmann, 1963).
———, *The Latin Cognomina* = *Commentationes Humanarum Litterarum* 36.2 (1965).
———, *Supernomina. A Study in Latin Epigraphy* = *Commentationes Humanarum Litterarum* 40:1 (1966).
———, *On the Problem of the Average Duration of Life in the Roman Empire* (Helsinki: Suomalainen Tiedeakatemia, 1968).
Kanzler, R., "Scoperta di una nuova regione del cimitero giudaico della Via Portuense," *Nuovo Bulletino di Archeologia Cristiana* 21 (1915) 152-57.
Keil, J., and A. Wilhelm, *Monumenta Asiae Minoris Antiquae*. III. *Denkmäler aus dem Rauhen Kilikien* (Manchester: Manchester U. P., 1931).
Keyfitz, N., *Population. Facts and Methods of Demography* (San Francisco: Freeman, 1971).
Klein, S., *Tod und Begräbnis in Palästina zur Zeit der Tannaiten* (Berlin: Itzkowski, 1908).
———, *Sefer ha–Yishuv* (Jerusalem: I. Ben–Zvi, 1977).
Kloner, A., *The Necropolis of Jerusalem in the Second Temple Period* (Unpublished Ph.D. dissertation: Hebrew University of Jerusalem, 1980) (Hebrew).
Koch, G., and H. Sichtermann, *Römische Sarkophage* (Munich: Beck, 1982).
Konikoff, A., *Sarcophagi from the Jewish Catacombs of Ancient Rome. A Catalogue Raisonné* (Stuttgart: Steiner, 1986).
Kraabel, A. T., "The Roman Diaspora: Six Questionable Assumptions," *Journal of Jewish Studies* 33 (1982) 445-64.
Kraemer, R. S., "Non–Literary Evidence for Jewish Women in Rome and Egypt," *Helios* 13 (1986) 85–101.
———, *Her Share of the Blessings. Women's Religions Among Pagans, Jews, and Christians in the Greco-Roman World* (New York and Oxford: Oxford U. P., 1992).

Langmuir, G. I., *History, Religion and Antisemitism* (Berkeley: University of California Press, 1990).
———, *Toward a Definition of Antisemitism* (Berkeley: University of California Press, 1990).
Leon, H. J., "The Jewish Catacombs and Inscriptions of Rome: an Account of Their Discovery and Subsequent History," *Hebrew Union College Annual* 5 (1928) 299-314.
———, "The Jews of Venusia," *Jewish Quarterly Review* 44 (1953-54) 267-84.
———, *The Jews of Ancient Rome* (Philadelphia: The Jewish Publication Society of America, 1960).
Levine, L. I., *The Rabbinic Class of Roman Palestine in Late Antiquity* (Jerusalem: Y.I. Ben-Zvi, 1989).
Liebeschütz, H., *Das Judentum im deutschen Geschichtsbild von Hegel bis Max Weber* (Tübingen: Mohr, 1967).
Liebs, D., *Die Jurisprudenz im spätantiken Italien (260-640 n. Chr.)* (Berlin: Dunker and Humbolt, 1987).
Linder, A., *The Jews in Roman Imperial Legislation* (Detroit and Jerusalem: Wayne State U. P. and the Israel Academy of Sciences and Humanities, 1987).
Lüderitz, G., *Corpus jüdischer Zeugnisse aus der Cyrenaika* (Wiesbaden: Dr. Ludwig Reichert, 1983).

Maioli, M. G., "Caratteristiche e problematiche delle necropoli di epoca tarda a Ravenna e in Romagna," *XXXV Corso di Cultura sull'Arte Ravennate e Bizantina* (Ravenna 1988) 315-356.
Mann, V. B. (ed.), *Garden and Ghettos. The Art of Jewish Life in Italy* (Berkeley: University of California Press, 1989).
Marucchi, O., "Di un nuovo cimitero giudaico sulla Via Labicana," *Dissertazioni della Pontificia Accademia Romana di Archeologia* 2 (1884) 499-532.
―――, *Breve guida del cimitero giudaico di Vigna Randanini* (Roma, 1884).
―――, *Di un nuovo cimitero giudaico sulla Via Labicana* (Rome, 1887).
―――, *Le catacombe romane* (Roma: Libreria dello Stato, 1933).
Mazar, B., *Beth She`arim. Report on the Excavations During 1936-1940*. Vol. I. Catacombs 1-4 (Jerusalem: Massada Press, 1973).
Mazzoleni, D., "Les sépultures souterraines des Juifs d'Italie," *Les Dossiers de l'Archéologie* 19 (1976) 82-98.
Meyers, E. M., *Jewish Ossuaries. Reburial and Rebirth* (Rome: Biblical Institute Press, 1971).
―――, "The Challenge of Hellenism for Early Judaism and Christianity," *Biblical Archaeologist* 55 (1992) 84-91.
Millar, F., "The Jews of the Graeco-Roman Diaspora Between Paganism and Christianity, AD 312-438," in J. Lieu *et al.* (eds.), *The Jews Among Pagans and Christians in the Roman Empire* (London and New York: Routledge, 1992) 97-123.
―――, *The Roman Near East 31 BC-AD 337* (Cambridge and London: Harvard U.P., 1993).
Mittman, S., *Beiträge zur Siedlungs -und Territorialgeschichte des nördlichen Ostjordanlandes* (Wiesbaden: Harrassowitz, 1970).
Momigliano, A., "Ancient History and the Antiquarian," *Journal of the Warburg and Courtauld Institutes* 13 (1950) 285-315.
―――, *Alien Wisdom. The Limits of Hellenization* (Cambridge: Cambridge U.P., 1971).
―――, "The New Letter by 'Anna' to 'Seneca'," *Athenaeum* 63 (1985) 217-19.
Moretti, L., "Iscrizioni greco-giudaiche di Roma," *Rivista di Archeologia Cristiana* 50 (1974) 213-19.
Morris, I., *Death Ritual and Social Structure in Classical Antiquity* (Cambridge: Cambridge U.P., 1992).
Mulder, M. J., and H. Sysling (eds.), *Mikra. Text, Translation, Reading and Interpretation of the Hebrew Bible in Ancient Judaism and Early Christianity* (Assen and Philadelphia: Van Gorcum-Fortress, 1988).
Müller, N., "Le catacombe degli Ebrei presso la Via Appia Pignatelli," *Römische Mitteilungen* 1 (1886) 49-56.
―――, *Die jüdische Katakombe am Monteverde zu Rom* (Leipzig: Fock, 1912).
―――, "Cimitero degli antichi Ebrei posto nella Via Portuense," *Dissertazioni della Pontificia Accademia Romana di Archeologia* 12 (1915) 205-318.
―――, and N. Bees, *Die Inschriften der jüdischen Katakombe am Monteverde zu Rom* (Leipzig: Fock, 1919).

Nahon, G., *Inscriptions hebraïques et juives de France médiévale* (Paris: Corlet 1986).
Naveh, J., "Another Jewish Aramaic Tombstone from Zoar," *Hebrew Union College Annual* 56 (1985) 103–16.
―――, "The Fifth Jewish Aramaic Tombstone from Zoar," *Liber Annuus* 37 (1987) 369–71.
Nock, A. D., "Cremation and Burial in the Roman Empire," *Harvard Theological Review* 25 (1932) 321-59.
Nordberg, H., *Sylloge Inscriptionum Christianarum Veterum Musei Vaticani = Acta Instituti Romani Finlandiae* 1.2 (1963).
Noy, D., *Jewish Inscriptions of Western Europe*. Vol. 1. *Italy (excluding the City of Rome), Spain and Gaul* (Cambridge: Cambridge U.P., 1993).
Overman, J. A. and R. S. MacLennan (eds.), *Diaspora Jews and Judaism. Essays in Honor of, and in Dialogue with A. Thomas Kraabel* (Atlanta: Scholars Press, 1992).

Paribeni, R., "Iscrizioni del cimitero giudaico di Monteverde," *Notizie degli Scavi di Antichità* 46 (1919) 60-70.
_____, "Catacomba giudaica sulla Via Nomentana," *Notizie degli Scavi di Antichità* 46 (1920) 143-55.
Pfanner, M., "Über das Herstellen von Porträts. Ein Beitrag zu Rationalisierungsmaßnahmen und Produktionsmechanismen von Massenware im späten Hellenismus und in der römischen Kaiserzeit," *Jahrbuch des Deutschen Archäologischen Instituts* 104 (1989) 157-257.
Pietri, C., *Roma Christiana. Recherches sur l'église de Rome, son organisation, sa politique, son idéologie de Miltiade à Sixte III (311-440)* (Rome: Ecole Française, 1976).

Rabello, A. M., "Alcune note sulla 'Collatio Legum Mosaicarum et Romanarum' e sul suo luogo d'origine," in *Scritti sull'ebraismo in memoria di Guido Bedarida* (Florence, 1966) 177-86.
_____, "Sul l'ebraicità dell'autore della 'Collatio Legum Mosaicarum et Romanarum'," *La Rassegna Mensile di Israel* 33 (1967) 339-49.
_____, "Sul Decalogo 'Cristianizzato' e l'autore della Collatio Legum Mosaicarum et Romanarum," *La Rassegna Mensile di Israel* 55 (1989) 133-35.
Rahmani, L. Y., "Ancient Jerusalem's Funmerary Customs and Tombs," *Biblical Archaeologist* 44-45 (1981-82);
Reeg, G., *Die Geschichte der zehn Märtyrer* (Tübingen: Mohr, 1985).
Reekmans, L., *Die Situation der Katakombenforschung in Rom* (Opladen: Westdeutscher Verlag, 1979).
_____, "Spätrömische Hypogea," in O. Feld and U. Peschlow (eds.), *Studien zur Spätantiken und Byzantinischen Kunst FS Deichmann* 2 (Bonn: Habelt, 1986) 11-37.
Romanelli, P., "Una piccola catacomba giudaica di Tripoli," *Quaderni di Archeologia della Libia* 9 (1977) 111-18.
Rossi, G. B. de, "Scoperta d'un cimitero giudaico presso l'Appia," *Bulletino di Archeologia Cristiana* 5 (1867) 16.
_____, *La Roma sotterranea cristiana* (Rome: Salvucci, 1864).
Ruderman, D. B., *Essential Papers on Jewish Culture in Renaissance and Baroque Italy* (New York and London: New York U. P., 1992).
Rutgers, L. V., "Ein in situ erhaltenes Sarkophagfragment in der jüdischen Katakombe an der Via Appia," *Jewish Art* 14(1988) 16-27.
_____, "Überlegungen zu den jüdischen Katakomben Roms," *Jahrbuch für Antike und Christentum* 33 (1990) 140-57.
_____, "Archaeological Evidence for the Interaction of Jews and non-Jews in Late Antiquity," *American Journal of Archaeology* 96 (1992) 101-18.

Salomies, O., *Die römischen Vornamen. Studien zur römischen Namengebung* (Helsinki: Societas Scientiarum Fennica, 1987).
Sanders, E. P., *Judaism. Practice and Belief. 63 B.C.E.-66 C.E.* (London and Philadelphia: SMC and Trinity Press International, 1992).
Schäfer, P., *Studien zur Geschichte und Theologie des rabbinischen Judentums* (Leiden: Brill, 1978).
Scheiber, A., *Jewish Inscriptions in Hungary from the Third Century to 1686* (Budapest and Leiden: Akadémiai Kiadó Budapest and Brill, 1983).
Schmidt, T. M., "Ein jüdisches Goldglas in der frühchristlich-byzantinischen Sammlung," *Forschungen und Berichte* 20-21 (1980) 273-80.
Schneider-Graziosi, G., "La nuova Sala Giudaica nel Museo Cristiano Lateranense," *Nuovo Bulletino di Archeologia Cristiana* 21 (1915) 13-56.
Schrage, E. J. H., "La date de la 'Collatio Legum Mosaicarum et Romanarum' étudié d'après les citations bibliques," in J. A. Ankum *et al.* (eds.), *Mélanges Felix Wubbe* (Fribourg: U.P. Fribourg, 1993) 401-17
Schreckenberg, H., *Die christlichen Aversus-Judaeos-Texte und ihr literarisches Umfeld (1.-11.Jh.)* (Frankfurt on the Main-Bern: Lang, 1982).
Schulz, F., "Die Anordnung nach Massen als Kompositionsprinzip etc." *Atti del congresso Internazionale di diritto romano (Bologna e Roma 17-27.IV.1933)* (Pavia: Successori, 1935) vol. 2, 11-25.

―――, "Die biblischen Texte in der Collatio Legum Mosaicarum et Romanarum," *Studia et Documenta Historiae et Iuris* 2 (1936) 20-43.
―――, "The Manuscripts of the *Collatio Legum Mosaicarum et Romanarum*," in M. David et al. (eds.), *Symbolae ad jus et historiam antiquitatis pertinenetes, Julio Christiano van Oven dedicatae (Symbolae van Oven)* (Leiden: Brill, 1946) 313-32.
―――, *History of Roman Legal Science* (Oxford: Clarendon, 1946).
Schürer, E., *The History of the Jewish People in the Age of Jesus Christ (175 B.C.-A.D. 135)* (Edinburgh: Clark, 1976-1986).
Schwabe, M., and B. Lifshitz, *Beth She`arim* Volume II. *The Greek Inscriptions* (New Brunswick: Rutgers U. P., 1974).
Serrao, E., "Nuove iscrizioni da un sepolcreto giudaico di Napoli," *Puteoli* 12/13 (1988-89) 103-17.
Smallwood, E. M., *The Jews under Roman Rule from Pompey to Diocletian* (Leiden: Brill, 1976).
Smith, J. Z., "Fences and Neighbors: Some Contours of Early Judaism," in W. S. Green (ed.), *Approaches to Ancient Judaism* (Chico: Scholars Press, 1980) 1-25.
Smits, N., *Mosaicarum et Romanarum Legum Collatio* (Haarlem: Tjeenk Willink, 1934).
Solin, H., *Beiträge zur Kenntnis der griechischen Personennamen in Rom* (Helsinki: Societas Scientiarum Fennica, 1971).
―――, "Juden und Syrer im westlichen Teil der römischen Welt. Eine ethnisch-demographische Studie mit besonderer Berücksichtigung der sprachlichen Zustände," *Aufstieg und Niedergang der römischen Welt* II, 29.2 (1983) 587-789.
Spigno, L., "Considerazioni sul manoscritto Vallicelliano G. 31 e la Roma sotterranea di Antonio Bosio," *Rivista di Archeologia Cristiana* 51 (1975) 281-311.
―――, "Della Roma sotteranea del Bosio e della su biografia," *Rivista di Archeologia Cristiana* 52 (1976) 277-301.
Stefano Manzella, I. di, "L. Maecius archon, centurio alti oridinis. Nota critica su *CIL*, VI, 39084 = *CII*, I, 470," *Zeitschrift für Papyrologie und Epigraphik* 77 (1989) 103-112
Steinby, M., "Ziegelstempel von Rom und Umgebung," *RE Supplementband* XV (1978) 1489-1531.
Stemberger, G., *Die römische Herrschaft im Urteil der Juden* (Darmstadt: Wissenschaftliche Buchgesellschaft, 1983).
Stern, M., *Greek and Latin Authors on Jews and Judaism* (Jerusalem: Israel Academy of Sciences and Humanities, 1974-84).
Strack, H. L., and G. Stemberger, *Introduction to the Talmud and Midrash* (Edinburgh: T&T Clark, 1991).

Testini, P., *Archeologia cristiana* (Bari: Laterza, 1980 [1958]).
Toaff, A., "Matia' ben Cheresh e la sua accademia rabbinica di Roma," *Annuario di Studi Ebraici* 2 (1964) 69-80.
Toynbee, J. M. C. *Death and Burial in the Roman World* (Ithaca: Cornell U. P., 1971).

Urbach, E. E., "The Rabbinical Laws of Idolatry in the Second and Third Centuries in the Light of Archaeological and Historical Facts," *Israel Exploration Journal* 9 (1959) 149-65 and 229-45.
―――, *The Sages. Their Concepts and Beliefs* (Cambridge Mass. and London: Harvard University Press, 1975).
―――, "The Role of the Ten Commandments in Jewish Worship," in B. Segal and G. Levi (eds.), *The Ten Commandments in History and Tradition* (Jerusalem: Magnes Press 1990) 161-189.

Visconti, C. L., "Scavi di Vigna Randanini," *Bulletino dell'Instituto di Correspondenza Archeologica* 1861, 16-22.
Vismarra, C., "I cimiteri ebraici di Roma," in Giardina 1986, vol. 2, 351-89.
―――, "Orientali a Roma. Nota sull'origine geografica degli Ebrei nelle testimonianze di età imperiale," *Dialoghi di Archeologia* 5 (1987) 119-21.
Vogelstein, H. and P. Rieger, *Geschichte der Juden in Rom* (Berlin: Mayer and Müller, 1896).

Volterra, E., "Collatio Legum Mosaicarum et Romanarum," *Atti della Reale Accademia Nazionale dei Lincei. Memorie* 6.3.1 (1930).

Wegner, J. R., *Chattel or Person? The Status of Women in the Mishnah* (New York: Oxford U. P., 1988).
Williams, M. H., "The Organization of Jewish Burials in Ancient Rome in the Light of Evidence from Palestine and the Diaspora," *Zeitschrift für Papyrologie und Epigraphik* 101 (1994) 165-82.
Wischmeyer, W., "Die Epistula Anne ad Senecam. Eine jüdische Missionsschrift des lateinischen Bereichs," in J. van Amersfoort and J. van Oort (eds.), *Juden und Christen in der Antike* (Kampen: Kok, 1991) 72-93.

Yerushalmi, Y. H., *Zakhor. Jewish History and Jewish Memory* (Seattle and London: University of Washington Press, 1982).

INDEX

Abraham in early Christian literature 220
Actus Sylvestri 211-13
Ambrosiaster 212, 229-31
amulets 88-9
antiquarians 21, 27
Apollonia 102
Aphrodisias 157, 205
Aringhi, P. 16-8, 32
ascia 89
assimilation 260-68
Augustine 224-5, 231-2

Babatha's archive 249
Baronio, C. 6-7, 10-11
Basnage, J. 18-21
Benjamin of Tudela 3-5
Bernal, M. xv-xvi
Beth She'arim 89, 143-7
Biblia Patristica 234
Blumenkranz, B. 117-8
Bohec, Y. Le 87
Boldetti, M. 22-3
Bollandists 15
Bosio, A. 8-13, 20, 38
brickstamps 98
Brooks, R. 244
burial (see also catacombs)
   enchytrismos 60-1
   in catacombs 10-11, 17, 34-41, 51-2, 92-5
   columbaria 51, 97
   cremation 97
   inhumation 51, 97
   of Jews in Diaspora 65-6
   kokhim 61-6, 93
   secondary burial 61

catacombs, Christian
   S. Callisto 58, 59
   Domitilla 51, 55
   Pretestato 78
   Priscilla 57
   Ss. Marcellino e Pietro 59, 82
   S. Ermete ai Parioli 82
   S. Sebastiano 51, 55
   Vigna Sanchez 7
catacombs, Jewish
   archaeological excavation of 30-42

dating of xvii-xviii, 92-3, 96-8
discovery of 9, 32-3
formal appearance of 53-7
intrusive materials 77-9
Monteverde 2-3, 9, 19, 31, 57, 73, 145-7, 176-8, 188
soil types 57-8
tomb types 58-65
Vigna Cimarra 56
Vigna Randanini 53, 61, 145-6, 176-8, 188
Villa Labicana 56
Villa Torlonia 55, 74, 145-6, 149-50, 176-81
Catania 204
cave of Machpelah 10, 31
Celarevo 66
church of St. Basilio 2
Codex Gregorianus 215
Codex Hermogenianus 215
Collatio 210-53
   date 215-7, 250
collegia 200
common Judaism 208
common workshop-identity 67-73, 76, 89-95
Counterreformation 5-14

Doclea 65

Edfu 152
Egra 66
Engemann, J. 83
Enlightenment 25

Fasola, U. M. 43
Fides Isaacis ex Iudaeis 212
Fragmenta Vaticana 215
Fuks, G. 166-9

Garrucci, R. 35-7, 39
Gibbon, E. 28
gold glasses 81-5
   with Temple of Jerusalem 82
   with Torah shrine 83
   pie zeses 83
Goodenough, E. R. 33, 44
Grand Tour 25-6
Graetz, H. 45

Gütschow, M. 78

Herzog, E. 34-5
Heyne, C. G. 29
Hincmar, bishop of Reims 239
historiography, ecclesiastical 6
hypothetical life table 129-31

inscriptions, Jewish (see also names, and women)
    age-rounding 119-24
    contractions 187-8
    Dis Manibus Sacrum 269-72
    epithets 191-9
    Greek, post-Classical 185, 189
    Hebrew 202
    Jewish community officials 150, 199
    Latin, vulgar 186, 188-9
    linguistic aspects 184-91
    references to age at death 100-38
    sex ratios 132-4
    survival rate 118-9
inscriptions (Christian) 199-200
    Greek and Latin in 183-4
interaction (see also isolation, and Romanization 170-5, 260-8
isolation (see also interaction, and Romanization) 45-9, 260-8

Jerome 231
Jerusalem
    Tomb of the Prophets 64

Korykos 66

lamps 85-8, 91
language (see also inscriptions) 176-201
Leon, H. J. xviii, 44, 48, 64, 139-43, 176-8
Letter of Annas to Seneca 253-6, 257-8
Liebs, D. 216, 217, 238
literacy 123, 189

Magdeburg Centuries 6
Malta 65
Mar Abha 237-8
Marangoni, G. 23
Marchi, G. 30-2
Mattiah ben Heresh 203-4
Maurists 15
Medusa, representation of 88
menorah, representation of 93-4, 182

Michaelis, J. D. 29
minim 245
Mithras, representations of 70
model life table 124-8
Momigliano, A. 45, 48
Morin, G. 212
mortality 111
Moses in early Christian literature 221-3
Müller, N. 2-3, 33, 39

names 139-75
    agnomina 163, 167
    duo and tria nomina 158-63
    etymology of 171-3
    filiation 164
    of Jews in Aphrodisias 157
    of Jews in Beth She`arim 153-4
    of Jews in Edfu 151-3
    of Jews in Rome 143-51
    of Jews in Venosa 156-7
    Semitic names 163-4
    of slaves 166-9
    supernomina 167
    theophoric names 162-3
    transliterations of Hebrew names 143-4
    women's names 165-6
Naples 65
Neri, F. 8
Nock, A. D. 96-7
Noto Antica 65

Palmyreans in Rome 70, 105
Paul, commentaries on 226-33
Pentateuch
    in early Christian literature 219-33
    in Josephus 242-3
    in Philo 240-42
    in rabbinic literature 243-6
Porta Portese 2
Porto Torres 65

Rabat 65
Rabbinic Judaism 203-9
Romanization (see also interaction, and isolation) 64, 139-43, 176-79
Rome
    Jews in medieval 1-5
    Jews in Renaissance 12, 13-4
Rossi, G. B. de 34, 37-8

Said, E. xv-xvi
Samaritans 251-2
Sanders, E. P. 208

INDEX

sarcophagi   69, 77-81
   identification of Jewish   77-9
   lead   90
   lenos   75
   season sarcophagi   79, 93-4
   strigiles   80
   terracotta   59
   Urania sarcophagus   59, 80
Sardinia   65
Schultze, V.   38-9
Shema   244
social history   121, 131, 165-9, 189-91
Solin, H.   47
Story of the Ten Martyrs   3-5
strenae   85
Styger, P.   40-1
Sulcis   65
Syrian-Roman Law Book   236-7

Taranto   102
Tell el-Yehoudieh   109, 112, 114, 116, 119, 120, 131
Tertullian   220
Teucheira   66, 109, 112, 114, 116, 119, 120, 131
Theudas, see Todos
Todos   204-5, 207

Tortosa   102
Tripoli   66

Urbach, E.   244

Venosa   155-7
Verus Israel   17, 24
Victorinus   228-9
Volterra, E.   210, 214, 218

wall painting   73-7
   incrustation   74
   Cubiculum II   75
   Painted Rooms I and II   44, 54-5, 73
   Painted Room III   73-5
   Painted Room IV   59, 74, 80
Winckelmann, J. J.   27-8
Wissenschaft des Judentums   31, 43
women, Jewish   117-8, 122, 131-36, 165-6

Zoar   102

SCHOLARS' LIST

Through its Scholars' List Brill aims to make available to a wider public a selection of its most successful hardcover titles in a paperback edition.

Titles now available are:

AMITAI-PREISS, R. & D.O. MORGAN, *The Mongol Empire and its Legacy.* 2000. ISBN 90 04 11946 9, price USD 29.90

COHEN. B., *Not the Classical Ideal.* Athens and the Construction of the Other in Greek Art. 2000. ISBN 90 04 11712 1, price USD 39.90

GRIGGS, C.W., *Early Egyptian Christianity* from its Origins to 451 CE. 2000. ISBN 90 04 11926 4, price USD 29.90

HORSFALL, N., *A Companion to the Study of Virgil.* 2000. ISBN 90 04 11870 5, price USD 27.90

JAYYUSI, S.K., *The Legacy of Muslim Spain.* 2000. ISBN 90 04 11945 0, price USD 54.90

RUTGERS, L.V., *The Jews in Late Ancient Rome.* Evidence of Cultural Interaction in the Roman Diaspora. 2000. ISBN 90 04 11928 0, price USD 29.90

TER HAAR, B.J., *The Ritual and Mythology of the Chinese Triads.* Creating an Identity. 2000. ISBN 90 04 11944 2, price USD 39.90

THOMPSON, T.L., *Early History of the Israelite People* from the Written & Archaeological Sources. 2000. ISBN 90 04 11943 4, price USD 39.90

WOOD, S.E., *Imperial Women.* A Study in Public Images, 40 BC – AD 68 2000. ISBN 90 04 11950 7, price USD 34.90

YARBRO COLLINS, A., *Cosmology & Eschatology in Jewish & Christian Apocalypticism.* 2000. ISBN 90 04 11927 2, price USD 29.90